The Psychology of Expertise

Robert R. Hoffman
Editor

The Psychology of Expertise
Cognitive Research and Empirical AI

With 31 Figures

Springer-Verlag
New York Berlin Heidelberg London Paris
Tokyo Hong Kong Barcelona Budapest

Robert R. Hoffman, Ph.D.
Department of Psychology
Adelphi University
Chapman Hall-Room 2
Garden City, NY 11530
USA

Library of Congress Cataloging-in-Publication Data
The psychology of expertise: cognitive research and empirical AI/
edited by Robert R. Hoffman.
p. cm.
Based on papers presented at a conference on expert systems and the psychology of expertise, held at Adelphi University on May 5, 1989.
Includes bibliographical references and indexes.
ISBN 0-387-97686-8
1. Expertise—Congresses. 2. Expertise—Research—Methodology—Congresses. 3. Knowledge acquisition (Expert systems)—Congresses. 4. Expert systems (Computer science)—Congresses. 5. Artificial intelligence—Congresses. 6. Cognitive science—Congresses.
I. Hoffman, Robert R.
BF378, E94P77 1992
153—dc20 91-33158

Printed on acid-free paper.

Production managed by Hal Henglein; manufacturing supervised by Jacqui Ashri.
Typeset by Best-set Typesetter Ltd., Hong Kong.
Printed and bound by Edwards Brothers, Inc., Ann Arbor, MI.
Printed in the United States of America.

9 8 7 6 5 4 3 2 1

ISBN 0-387-97686-8 Springer-Verlag New York Berlin Heidelberg
ISBN 3-540-97686-8 Springer-Verlag Berlin Heidelberg New York

This volume is dedicated to Olin Mintzer, expert aerial photo interpreter at the "Engineer Topographic Laboratories" of the US Army Corps of Engineers, Ft. Belvoir, Virginia. The opportunity to work with Olin was both a privilege and an honor. From the perspective of the researchers at the ETL, my job was to "model Olin's brain." From my perspective, I was a fish well out of water. Olin not only gave me oxygen, he taught me how to fly. For performing this transmutation, and for all the other life-giving transformations he has wrought on students and colleagues, I hereby promote him to the rank of Wizard.

Preface

Each year since 1982 Adelphi University has sponsored a conference on applied experimental psychology. The first six conferences were broad in scope in that each included a number of topical sessions. Around 1986, however, we started to think about holding a series of specialized or topical conferences, with some connection to applied experimental psychology (to be sure) but also with a workshop atmosphere in that groups of researchers would be collected so that they could hash out important issues. One such conference was on the perception of illusory contours (Petry & Meyer, 1987). Another marked the centennial of the psychology of learning (Gorfein & Hoffman, 1987). Another focused on behavior and social attitude change (Curtis & Stricker, 1990). Yet another focused on research on semantic ambiguity (Gorfein, 1989).

For 1988, we decided to focus on expert knowledge and the application of experimental psychology to expert system development. It had become clear that there was a need to get certain people together. Some of these "certain" people were computer scientists and some were psychologists, but all shared an empirical approach to the problem that artificial intelligence (AI) has encountered. Many had conceived of conducting empirical (if not experimental) analyses of knowledge elicitation issues, and most had gotten their hands dirty in knowledge elicitation.

The conference titled "Expert Systems and The Psychology of Expertise" was held at Adelphi on May 5, 1989. There were 14 participants, including Keynote Speaker Robert Sternberg. Each participant gave a 30-minute talk, followed by discussions. The conference was small (about 40 attendees) and had a workshop atmosphere. Actually it was rather intense. It was certainly jam-packed with discussions.

Many of the debates were stimulated and nurtured by Stephen Regoczei of Trent University. He came to the conference feeling hesitant to speak up, being one of a few fish out of water. But his questions were perhaps more novel and challenging than those that might be proffered by one of the "experts" on expertise.

I should point out here that the focus on experimental psychology and empirical approaches did not exclude the work by computer scientists on automated knowledge acquisition (cf. Gaines & Boose, 1988). Such work was described at the conference and is represented in this volume. However, as such conferences always go, I was not able to gather all of the "certain people" I had initially set out to collect. I will not name the research psychologists and computer scientists who were contacted, clearly expressed an interest, but were unable to attend—mostly because of their being overloaded already, the bane of us all. I want to take this opportunity to thank them for encouraging me as I developed the conference plan. They may all rest assured that their work and ideas appear in this volume in spirit, and that their contributions form a valuable component of the psychology of expertise.

A number of the conference participants were unable, for a variety of reasons, to contribute to this volume. Bill and Bev Thompson of the Knowledge Garden, A. Michael Burton of the University of Nottingham, and Jeff Bradshaw of Boeing Corporation all made valuable contributions to the conference, and I am confident that the spirit of their work is preserved in this volume.

Speaking on behalf of everyone who was involved with the conference, I would like to thank the corporations that employ many of the participants and attendees, for allowing their people to participate and for supporting them in terms of travel and lodging expenses: Bell Communications Research, Boeing Corporation, IBM, Klein Associates, the Knowledge Garden, Universal Technology Corporation, and the U.S. Air Force.

Expenses for some participants were covered in part by their host academic institutions and in part by the conference budget of the dean of the College of Arts and Sciences of Adelphi University. Our special thanks therefore go to Adelphi, and Dean Sean Cashman. Thanks also to Marianne Walters, now of Felician College, who, as the dean's assistant, helped with the details of budgeting and reimbursements.

I want to thank Eve Bhumitra and Jean Beck, both graduate students in experimental psychology at Adelphi, for helping with the details of conference management (transportation, visual aid setup, etc.). They are each hereby promoted to the rank of Invaluable-to-the-Hectic.

A very special thanks goes to David Gorfein, chairman of the Department of Psychology at Adelphi. As chairman, he recently endured the labors of Hercules. He nurtured the idea for the conference from its inception through my subsequent devastation of our department's copying budget. The final efforts involved in preparing this volume were made especially efficient through his support.

In retrospect, I'm both pleased and shocked by this volume. My thanks go out to the contributors for their prompt preparation of manuscripts and for enduring the multiple rounds of my infamous marginal "green scratchings." Thanks also to the editorial staff at Springer-Verlag. A final word of thanks goes to Pat Carey, Eleanor Shaw, and Angela Mavaro of the Department of Psychology of Adelphi, who typed the bibliographies that appear at the end of this volume, and to Robin Akerstrom of Marine Safety, International, who assisted in the preparation of the subject and author indexes.

A classic dig on psychologists is, "Give a psychologist a hammer and he'll think everything's a nail." This is a reference to the experimentalist's inclination to beat problems to death with fancy factorial statistical analyses. The analogous tendency in computer science is for researchers to assume that the only way to solve problems is to build more programs. These biases can be self-serving, of course. Psychologists generate more experiments; computer scientists generate more compilers. But when the disciplines dovetail, as they do in "cognitive science," the research can seem especially self-serving. The clearest example of this is the bounty (to put it *mildly*) of reports on problem solving by computer programmers. (Will we see computer models of the knowledge of expert computer programmers?) I do not really mean to denigrate that particular research topic; it is important. But surely there are other things in the world to study. It is clear that all of us, psychologist and computer scientist, need to expand our conceptual horizons . . . and our capabilities. It also seems clear that, for topics in expertise, it helps to work together. It is hoped that this volume contributes to the spirit of interdisciplinary research, both basic and applied.

Expert systems technology may or may not evolve significantly beyond rule-based operations and their brittleness problems (Keyes, 1989). But imagine an empirically based "knowledge elicitation methods palette" that enables one to specify the methods needed to capture the knowledge and skills of experts in a given domain. Imagine the benefits to society of an advanced technology for representing and disseminating the knowledge and skills of the best corporate managers, the most seasoned pilots, or the most renowned medical diagnosticians! It is hoped that this volume contributes in some way to that enterprise.

This volume should be of interest to cognitive psychologists and computer scientists, and to those who are out in the trenches developing expert systems. It should be of interest to anyone who is pondering the nature of expertise and the question of how it can be studied scientifically.

Long Island, NY — Robert R. Hoffman

References

Curtis, R. C., & Stricker, G. (Eds.). (1990). *How people change inside and outside therapy*. New York: Plenum.

Gaines, B. R., & Boose, J. H. (Eds.). (1988). *Knowledge acquisition for knowledge-based systems*. New York: Academic Press.

Gorfein, D. S. (Ed.). (1989). *Resolving semantic ambiguity*. New York: Springer-Verlag.

Gorfein, D. S., & Hoffman, R. R. (Eds.). (1987). *Memory and learning; The Ebbinghaus centennial conference*. Hillsdale, NJ: Erlbaum.

Keyes, J. (1989, November). Why expert systems fail. *AI Expert*, pp. 50–53.

Petry, S., & Meyer, G. E. (Eds.). (1987). *The perception of illusory contours*. New York: Springer-Verlag.

Contents

Contributors

Jack R. Adams-Webber
Department of Psychology
Brock University
St. Catherines, Ontario, L2S 3A1 Canada

Mark R. Adler
Digital Equipment Corporation
111 Locke Drive
Marlborough, MA 01752 USA

Francis S. Bellezza
Department of Psychology
Ohio University
Athens, OH 45701-2979 USA

Norman R. Brown
Department of Psychology
Carnegie-Mellon University
Pittsburgh, PA 15213 USA

Robert L. Campbell
Department of Psychology
Clemson University
Clemson, SC 92634-1511 USA

Stephen J. Ceci
Department of Human Development and Family Studies
College of Human Ecology
Cornell University
Ithaca, NY 14853-4401 USA

Nancy J. Cooke
Department of Psychology
Rice University
Houston, TX 77001 USA

Mary P. Czerwinski
Compaq Computer Corporation
Houston, TX 77269-2000 USA

Lia A. DiBello
Laboratory for Cognitive Studies of Work
City University of New York
New York, NY 10036-8099 USA

Mícheál Foley
Computer Services
University College Galway
Galway, Ireland

Kenneth M. Ford
Institute for Human and Machine Cognition
Division of Computer Science
University of West Florida
Pensacola, FL 32514-5750 USA

Peter A. Frensch
Department of Psychology
University of Missouri
Columbia, MO 65211 USA

Sallie E. Gordon
Department of Psychology
University of Idaho
Moscow, ID 83843 USA

Alan S. Gunderson
Digital Equipment Corporation
111 Locke Drive
Marlborough, MA 01752 USA

Anna Hart
Faculty of Science
Lancashire Polytechnic
Preston, Lancashire, PR1 2TQ
England

Graeme Hirst
University of Toronto
Toronto, Ontario, M5S 1A4
Canada

Robert R. Hoffman
Department of Psychology
Adelphi University
Garden City, NY 11530 USA

Gary A. Klein
Klein Associates, Inc.
582 E. Dayton-Yellow Springs
Road
Fairborn, OH 45324-3987 USA

Robert Mack
User Interface Institute
IBM Watson Research Center
Yorktown Heights, NY 10598 USA

Karen L. McGraw
Cognitive Technologies
2315-B Forest Drive #9
Annapolis, MD 21401 USA

David S. Prerau
GTE Laboratories, Inc.
40 Sylvan Road
Waltham, MA 02254 USA

Stephen B. Regoczei
Computer Studies
Trent University
Peterborough, Ontario, K9J 7B8
Canada

Jill Burdett Robinson
Customer Support Centre
IBM United Kingdom
Laboratories, Ltd.
Hursley, United Kingdom

Ana Ruiz
Cornell University
Ithaca, NY 14853 USA

Robert M. Schumacher
User Technology Services, Inc.
1010 E. Coolidge Avenue
Wheaton, IL 60187 USA

John F. Sowa
IBM Systems Research Institute
500 Columbus Avenue
Thornwood, NY 10594 USA

Robert J. Sternberg
Department of Psychology
Yale University
New Haven, CT 06520 USA

Part I
Introduction

1
Doing Psychology in an AI Context: A Personal Perspective and Introduction to This Volume

Robert R. Hoffman

The reason that this book exists is because trees are hot in winter infrared photography. Over my Christmas holiday in 1979 I visited a close friend, physicist Walter Carnahan of Indiana State University. I know him to be a specialist in optics and particle physics, which is why I was perplexed when I found him pondering a computer graphic display which merely showed a crazy patchwork of colors. The display seemed to have nothing to do with particle physics, so I figured he had a new software toy or something. But no, this was Science.

"This is a thermogram," he said, "taken over Terre Haute, Indiana, from about a thousand feet. In winter. It's part of an energy conservation project."

Well, I knew what a thermogram was, and although the false-color coding scheme was new to me, I tried my hand at interpretation. "Is this green blob a tree?" I asked.

"No, it's a house. Actually, one with pretty good insulation."

"Oh," was the only word I could muster. But pointing to another region I asked, "So is this yellow blob an uninsulated house?"

"No, that's a tree. Trees are about the hottest things around in winter photography."

The color code, displayed off to one side of the image, seemed to be something like a rainbow, with warmer temperatures being coded as white to yellow to red, cooler temperatures as greens, and cold as blue to black. But it wasn't a perfect rainbow. The hottest white appeared slightly pinkish. The yellows were just awful, and interspersed among the darker blues was a sort of olive-gray color.

"There are a number of things called 'interpretation anomalies' in thermography," Walter explained. "If a tall building has lots of windows that leak heat, then looking down on it, all the whites and yellows make it look as if the sides are ballooning out at you." Later at his home, his family and I had a chance to play with a portable infrared camera, one that enabled us to fiddle with the color coding. (I use the word *play* advisedly. Not only are such cameras expensive, but they have to be primed with liquid nitrogen.) We performed all manner of experiments, but just simply watching a person through the camera was like nothing I'd ever seen before. The closest I can come to describing the dynamic panoply is that it's like looking at a world of ghosts painted in a Peter Max psychedelic of flowing colors. Ghostlike residues of human activity appear everywhere, such as handprints left glowing on doors. Especially odd were the infrared reflections. For example, a candle could be seen reflected, not in a mirror or window, but in an ordinary wall! Infrared affords a very unusual way of looking at the world.

Fortunately, the experimental psychologist in me is not concerned about making a fool of himself by saying something stupid in the presence of a person who knows tensor calculus. So I asked, "In your aerial thermograms, who decided how to code the different temperature ranges in the different colors?"

Walter had to think about that for a moment . . . and he then said, "I don't know. Probably some engineer up in Michigan where they have the plane and the IR camera and the computers that process the data."

At that moment I had an intuition that the field called *remote sensing* might be fertile ground for human factors research. That possibility intrigued me because remote sensing displays are often beautiful, and I love beautiful things. Furthermore, I'd long wanted to somehow mix my work as experimental psychologist with my love for things that are related to space travel—I'm a science fiction nut.

The first step was to learn about remote sensing. The parent discipline of remote sensing is called aerial photo interpretation, in which aerial black-and-white photographs are interpreted in an elaborate process called terrain analysis. In 1983 I applied for a summer grant to learn from the masters of this craft, the experts at the U.S. Army Engineer Topographic Laboratories (ETL) at Fort Belvoir, Virginia. Their response to my application was basically: "Glad you applied. It turns out, we need an experimental psychologist to help us figure out ways of extracting the knowledge of our expert image interpreters . . . because we want to build expert systems."

I thought to myself, "Huh?" At that time the hype about expert systems was just building up steam, and I knew next to nothing about them. Well, at the ETL I managed to learn about both remote sensing and about expert systems. Eventually, my intuitions about applying experimental psychology to remote sensing bore some fruit (Hoffman, 1990, 1991; Hoffman & Conway, 1989). With regard to expert systems, I discovered that no one was doing anything about the "knowledge acquisition bottleneck" problem.

Expert system developers had learned that building a prototype expert system (with, say, 100 or so rules and perhaps some dozens of core concepts), and getting the prototype to actually do something, can take anywhere from weeks to perhaps a few months. But specifying the knowledge that is to be captured in the system's knowledge base and inference engine can take week after week of exhausting interviews. This knowledge elicitation effort must be conducted *before* any system is built, and therefore it takes the system developers away from what they really like to do—program computers.

At the time, summer of 1983, there was a large literature of case study reports (conference papers, journal articles, edited volumes) on various people's experiences while developing expert systems. A number of review articles had also appeared. Looking across the reports, many (most?) developers of expert systems had taken it for granted that knowledge is to be elicited through unstructured interviews (see Cullen & Bryman's 1988 survey). Many system developers apparently did not even conceive of the idea that an interview can be "structured" beforehand, or that adding structure could make the interviewing more efficient overall. In retrospect, it is not surprising that knowledge acquisition had become a bottleneck.

A second general feature of the literature is that it was (and still is) chock-full of inconsistent advice about knowledge elicitation. Some researchers recommend interviews with small groups; others say that such interviews can be disastrous—because disagreements with experts can cause serious problems for the system developer. Some recommend using "test cases" to probe for knowledge; others recommend the use of questionnaire ratings scales. Some say that experts are often pompous and defensive; some say they are usually not. Some say that experts are often unable to verbalize their knowledge; some say this is rarely a significant problem. And so on, each researcher generalizing from, and swearing by, his or her particular experience, with no hard data as back up.

I read that some computer scientists even recommended creating "automated knowledge acquisition" or interviewing tools. At the time, this struck me as being somewhat self-serving, if not mere overkill. These people were saying, basically, "We can't just jump in and start programming the expert system that we started out to build, so let's take one step back and build yet *another* program, one that can do the interview for us!" I felt that there must be simple, flexible solutions—*human* ones. Experimental psychologists have been studying learning for over 100 years. They are old hands at bringing people into the lab; teaching them some stuff (like lists of words or sentences); and then testing their memorial, perceptual, and cognitive processes. "Surely," I thought, "one could tinker with an expert's usual task, turning it into what psychologists call a 'transfer design.' With that, one could pull out an expert's knowledge, and in the long run do so more easily than by stepping back and writing a whole additional program just to conduct interviews!"

Perhaps. But at the time, *all* of the pronouncements in the literature on expert systems had to be taken on faith. This included the views of philosophers, both pro-AI and anti-AI, who seem rarely to get their hands dirty in either programming or empirical research. Most unfortunate of all was the fact that people in the AI community, both academics and nonacademics, rarely regarded the issues they encountered as problems that begged for an *empirical* solution (Mitchell & Welty, 1988).

The most frustrating manifestation of the nonempirical outlook was that everyone *talked* about the knowledge elicitation bottleneck, but nobody ever *did* anything about it. To my amazement, no one reported any informative details about the methods that they actually used in their knowledge elicitation sessions. Indeed, something like a Methods section never appears in reports on expert system development projects—unless method is equated with system architecture. Typically, the first paragraph of a report on an expert system development project mentions that knowledge elicitation was a problem, but then the second paragraph launches into an omnivorous discussion of the system architecture and its remarkably clever qualities and memory search algorithms.

Enter the experimental psychologist. Experimental psychologists love to get their hands dirty in the nits and picks of behavioral data. Such dirt is shunned by the hands of programmers. Not only that, experimental psychologists seem to think *differently* from programmers. For example, the idea of various kinds of "structured interview" comes so naturally to experimentalists that it is perhaps one of the most reinvented of wheels. There is a copious social science literature on interviewing techniques, the classic work being decades old. In general, the idea of holding off on programming or modeling in order to design and conduct the necessary background psychological research—this is something that comes naturally to the experimental psychologist who finds himself or herself in a computational context.

These words come easily now, but at the time I was conducting my original research this was all a bit scary. I'd been spiritually prepared to dive into applied contexts by one of my mentors, James Jenkins, when I was a postdoc at the Center for Research on Human Learning at the University of Minnesota. But to suddenly *be* a fish out of water. . . .

The library available to me at the ETL was full of information about remote sensing, of course. But, as one might also expect, it was a bit slim in its psychology selection. So I had to begin at the beginning and trust some of my intuitions. Fortunately, not all of the wheels I ended up with were reinventions.

First, I generated a functional (i.e., cognitively based) scheme for classifying various alternative knowledge elicitation methods, by incorporating ideas from the psychologist's learning and problem-solving laboratory as well as methods that had been used by developers of expert systems. I also generated some measures that would permit comparisons of knowledge elicitation methods. (To do so, I had to rethink the usual definition of data "validity," and frankly, that wasn't terribly easy.) The next step was to apply these preliminary ideas in a number of studies of experts—the ETL expert interpreters of remote sensing images and U.S. Air Force experts at airlift planning and scheduling. This work was reported in *The AI Magazine* in the summer of 1987.

It was about then that I learned of research being conducted at the University of Nottingham by Michael Burton and his colleagues

(Burton, Shadbolt, Hedgecock, & Rugg, 1987). Burton et al.'s work not only knocked my socks off, it also reknitted them. For one thing, I learned that I was no longer a lone voice in the wilderness; they, too, were empirically comparing alternative knowledge elicitation methods. Although Burton et al. were not as concerned as I was with figuring out ways to calculate the relative efficiency of various knowledge elicitation methods, their work was ambitiously experimental in that a number of variables were systematically manipulated. For example, they looked at the effect of personality variables (such as introversion-extraversion) on performance at various knowledge elicitation tasks. Like me, they looked at interviews and various kinds of "contrived" tasks, and with regard to the major conclusions, we were both pretty much on the same track: Methods other than unstructured interviews should be used in knowledge elicitation. Special contrived tasks, which somehow alter the expert's habitual procedures, can be surprisingly efficient at eliciting refined knowledge. On a broad scale, we both knew that progress *could* be made on addressing the knowledge elicitation bottleneck problem.

At that point, early 1988, it became clear that there was a need for a general overview of the relevant psychological literature on expertise. Hypotheses about expertise were being discussed in diverse literatures (e.g., judgment and decision-making research, expert systems development research, psychological research on learning and memory, etc.), with no attempt being made to cross disciplines in order to integrate and compare hypotheses. As it turns out, the relevant literature is huge (see Hoffman & Deffenbacher, in press; Hoffman, Shanteau, Burton, & Shadbolt, 1991).

The psychological study of expertise dates at least as far back as the World-War-I-era research on highly skilled machine operators, railway motormen, airplane pilots, and other professionals. Methods commonly used today in human factors studies of task performance (including human-computer interaction) can be dated at least as far back as the late 1800s, when Hugo Münsterberg studied the efficiency of movements in the use of a pocket watch and Frederick W. Taylor researched the problem of designing efficient shovels. The "think-aloud" method that is now commonly used in the study of the problem-solving process (and the "protocol analysis" method that is used to analyze the data) can be traced back to educational research on the acquisition and development of knowledge by Edouard Claparède (1917) and subsequent research by Karl Duncker in the 1930s (Duncker, 1945). In fact, it was pioneer psychologists such as Duncker and Alfred Binet who deserve credit for generating nearly all of the core concepts of modern theories of reasoning (e.g., concepts that are now called means-end analysis, reasoning by analogy, goal decomposition, concept-driven versus data-driven processing, and a host of other core ideas). (For the historical details, see Hoffman & Deffenbacher, in press; Hoffman et al., 1991.)

A considerable body of modern research represents the efforts of experimental psychologists to understand, in some detail, the knowledge, reasoning, and skills of experts in such domains as medicine, physics, and computer programming. Indeed, Kenneth Hammond (e.g., Hammond, 1966) and James Shanteau (e.g., Shanteau & Phelps, 1977) had been studying the cognition of experts long before the advent of expert systems. (For a review of the research see Chi, Glaser, & Farr, 1988; or Hoffman et al., 1991.)

In their expert system projects, David Prerau and his colleagues at GTE Laboratories, Inc., and Allison Kidd and her colleagues (she's now at Hewlett-Packard), had paid great attention to the fine details of knowledge elicitation—everything from dealing with management to the problems encountered in scheduling meetings with experts (Kidd, 1987; Kidd & Cooper, 1985; Kidd & Welbank, 1984; Prerau, 1985, 1989). Anna Hart had published one of the few books on "how to do" expert systems in which attention was truly devoted to how to do knowledge elicitation (Hart, 1986). Karen McGraw, whose research focused on the problem of multiple experts, was working on a similar book on knowledge acquisition (McGraw & Harbison-Briggs, 1989).

A number of experimental psychologists had been developing an interest in expert systems from a human factors perspective, such as Dianne Berry (Berry & Broadbent, 1986); Nancy Cooke (Cooke & McDonald, 1986); Sallie Gordon (1988); Gary Klein (1987); and of course Michael Burton, who had by then successfully prodded me into revamping the knowledge elicitation methods classification scheme (see Shadbolt, Hoffman, Burton, & Klein, 1991; Hoffman, 1989).

The Conference

Given this surge of interest, the timing seemed right for a conference, and a meeting titled "Expert Systems and The Psychology of Expertise" was held at Adelphi University in May of 1989. The major goals of the conference were to discuss alternative knowledge elicitation methods and to discuss *frankly* the assumptions, merits, and weaknesses of each method. Participants were challenged to reach some consensus about the relative efficiency of various knowledge elicitation methods. They were challenged to generate ideas about how to match knowledge elicitation methods to particular needs or particular domain types. Finally, they were encouraged to say outrageous things and present their newest ideas or research.

It was my hope that we could put the so-called knowledge acquisition bottleneck problem behind us and announce what I felt was the *real* bottleneck, the "training bottleneck." This bottleneck challenges educators, psychologists, and AI researchers alike. In the early days of expert systems, some said the purpose of an expert system is to replace the expert. But expertise is rare because it is an achievement. We need more experts, not fewer. A theme or underlying spirit of the conference may be summarized as follows: *Independent of the hype about expert systems, the elicitation, analysis, preservation, and dissemination of expert knowledge is a worthwhile enterprise.*

As should happen at a good conference, most of us learned that our a priori imaginings and overreactions were in some ways correct and in some ways incorrect. Knowledge elicitation is still a tough problem. We lack detailed empirical knowledge about expert reasoning across diverse domains. Indeed, we are still not exactly sure what distinguishes expertise from other manifestations of cognition (Klein & Hoffman, in press). Therefore, prescriptions or proscriptions about "how to do" knowledge elicitation remain elusive in their all-important details.

Looking back on my conference notes, I see a number of interesting specific ideas about knowledge elicitation that stemmed from the conference. (Not to mention the serious grandaddy question, What do we really mean by *knowledge*?) With regard to the specific question, What types of knowledge elicitation methods work best? we will need a classification of domain types as well as a classification of elicitation methods. Furthermore, both classifications will have to be based on the cognitive functions required of the expert (i.e., such variables as complexity, constructability, and uncertainty will be involved in the analysis of domains). An implication of this is now called the *differential access hypothesis*—that different knowledge elicitation methods will tap into different kinds of knowledge (e.g., declarative, procedural, metaknowledge, etc.). This hypothesis beckons us to further experimentation (Shadbolt et al., 1991).

Training is indeed recognized as a critical problem, one that is closely related to the work on expert systems (i.e., the need for expert systems to "explain" their reasoning). But those who are studying knowledge elicitation from an empirical standpoint have only just begun to approach training issues and address themselves to the large body of relevant work on instructional design. Those who concern themselves with instructional design, on the other hand, have recognized for some time that the design of learning materials presupposes that the materials were derived from the knowledge of someone who knows, that is, an expert (e.g., Glaser, 1976).

The Topics and Goals of This Volume

I did not push the participants to prepare chapters in advance of the conference, because I wanted them to come undistracted. From past experience I had learned that there can be great rewards in simply getting people together to hash out interesting ideas and to work on tough problems. Indeed, I had no plan to do a book. But a number of people (both attendees and participants) asserted that a book would be a good idea.

It was true that no one had really pulled together all of the research on knowledge elicitation. The extant volumes on expertise and expert systems were disappointing in that they were too AI-oriented in their focus on expert system architectures and the problems of knowledge representation. Edited volumes on psychological theory and research focused mostly on problem solving by physicists and computer programmers. Such research had already become well-known through journal articles. Both types of edited volumes neglected the gory details of knowledge elicitation.

I wanted this volume (like the conference) to represent a number of facets of the area—computer scientists' work with experts, psychologists' research, cognitive theory, alternative viewpoints about methodology, etc. I believe that the volume does that quite successfully. I also wanted the volume to include new material, and not reprints of older material.

The volume contains chapters written by people who were unable to attend the conference, such as David Prerau and Nancy Cooke, but few of the chapters actually represent the conference presentations verbatim. Basically, the conference allowed everyone to lay their cards on the table and try to focus on interesting or important issues. With the conference as a background context, the participants then sallied forth with the task of writing a chapter on topics of their choice—things they really cared about.

The Organization of This Volume

In Chapter 2, Stephen Regoczei and Graeme Hirst accomplish the tough introductory work, by asking what we mean by *knowledge*. They consider the alternative answers from both a psychological and a computational standpoint. Chapter 2 both introduces the specific topics of the volume and presents some of the challenging ideas that were debated at the conference. Their task in preparing chapter 2 (and Regoczei's task in preparing the final overview chapter) was truly Herculean. Despite my encouragement to "keep it simple," complexities and subtleties of the issues kept reappearing. To help make sense of things, Regoczei and Hirst illuminate the various metaphors that underlie conceptions of what "knowledge" is.

Part II includes chapters by Cooke and by Schumacher and Czerwinski, which describe psychological theories of cognition and summarize some of the psychological research that is pertinent to expertise. This includes the research on "mental models" and much of the recent research on expertise in problem solving. Sowa's chapter discusses the problem of representing knowledge from a computational standpoint, which is very sensitive to the psychological and functional aspects of knowledge representations and the language we use to express the knowledge.

Part III focuses on knowledge elicitation methods. Gordon considers the tough problem of eliciting "tacit" or "unconscious" knowledge. Ford and Adams-Webber discuss the theory behind the rating grid knowledge elicitation method (i.e., "repertory grids"), a method that is particularly well suited to implementation in automated knowledge acquisition. Prerau, Adler, and Gunderson discuss the knowledge elicitation methods they used in developing expert systems for GT&E. McGraw's chapter treats the gory details of how to document and manage the knowledge acquisition process. Such details as those discussed in the chapters by Prerau et al. and McGraw should be useful to the system developers who are out there in the trenches. Finally, the chapter by Klein surveys various knowledge elicitation methods, focusing on their strengths and weaknesses. Klein discusses his own success with the "Critical Decision Method" and reminds us of the perspective beyond expert systems: the general utility of preserving and disseminating the knowledge of experts.

Part IV consists of three chapters that illustrate how psychological research methods can be used in the study of expertise. The first two chapters demonstrate that it can be easier to "bring expertise into the laboratory" than one might suppose. Sternberg and Frensch describe their studies of expert bridge players and examine expertise in terms of its strengths and weaknesses relative to the functionality of cognitive processes. Francis Bellezza describes his research on the memory capacity of experts, in which he likens expert knowledge to a "mnemonic" memory technique. The third chapter in this section, by Ceci and Ruiz, represents an extension of their seminal study of expert horserace handicappers. Their initial work focused on the relation of expertise to general intelligence. Here, they report a study in which they investigated the transfer of expertise from one domain (handicapping) into another (stock market analysis). Their concern is with "what it takes" in terms of intelligence to be an expert and to transfer knowledge and experience from one domain to another.

Part V focuses on the development of expertise, which is usually conceived of in terms of the simple difference between novices and experts. Foley and Hart summarize what is known about expert-novice differences and then focus on the implications of these differences for knowledge elicitation. They also present a new study of their own, on expert-novice differences in program debugging. Thus, like other chapters in the volume, Foley and Hart's chapter illustrates the application of psychological experimental methods in the study of expert knowledge. So, too, does the second chapter in Part V, that by Mack and Robinson. Their focus is on what computer users need to know in order to accomplish their tasks. In their empirical case study, Mack and Robinson employed the "question-asking" method, which complements the "think-aloud" method that is traditionally used in the study of problem solving. They also consider some broader implications for knowledge elicitation and human-computer interaction. In the third chapter in Part V, Campbell, Brown, and DiBello present a case for a developmental approach to expertise in programming. They begin true to the empirical spirit, with the results from some structured interviews with experts and analyses of audiotaped think-aloud diaries. The "constructivist developmental" theory of expertise they propose goes considerably beyond the simple novice-expert distinction.

In Part VI, Stephen Regoczei faces three tough tasks: He must be Surveyor, Critic, and Clarion. As Surveyor, he pulls out some of the key points in each chapter. As Critic, he points to the outstanding problems, both theoretical and methodological, with the extant research on expertise. As Clarion, he points to the lingering issues and practical problems still begging for empirical solutions.

The appendixes consist of bibliographies that list recent publications pertinent to expertise and expert systems. Separate bibliographies cover the following: psychological reviews and theories, experimental and empirical investigations of expertise, knowledge elicitation methodology, automated knowledge acquisition, expertise in programming, AI theory and philosophy of expert systems, applications of expert systems, and the programming and verification of expert systems. Within the scope of the sources, including interlibrary loan resources, that were available to me, I tried to make the bibliographies as complete and up-to-date as possible. Special attention was paid to *published* works: articles in journals, single-author books, and edited volumes. I avoided citations of conference presentations and proceedings, for two related reasons. First, copies of the proceedings of various esoteric technical groups and organizations are not routinely available, even in the best of libraries. Second, and much to my chagrin, precise details about papers and presentations are often insufficiently specified in the References sections of the *other* publications in which they are cited (e.g., page numbers are not provided, place of publication is not provided, etc.). The size of the bibliographies in this volume could easily have been doubled, were various (published and unpublished) conference presentations included. I preferred to focus on references for which complete information was available (i.e., published works), even though that might entail leaving out a few important works, or leaving out reports on work that is in progress.

References

Berry, D. C., & Broadbent, D. E. (1986). Expert systems and the man-machine interface. *Expert Systems*, *3*, 228–230.

Burton, A. M., Shadbolt, N. R., Hedgecock, A. P., & Rugg, G. (1987, December). "A formal evaluation of knowledge elicitation techniques for expert systems." Paper presented at the Expert Systems '87 Conference, Brighton, England.

Chi, M. T. H., Glaser, R., & Farr, M. J. (Eds.). (1988). *The nature of expertise*. Hillsdale, NJ: Erlbaum.

Claparède, E. (1917). La psychologie de l'intelligence. *Scientia*, *22*, 215–251.

Cooke, N. M., & McDonald, J. E. (1986). A formal methodology for acquiring and representing expert knowledge. *Proceedings of the IEEE*, *74*, 533–550.

Cullen, J., & Bryman, A. (1988). The knowledge acquisition bottleneck: Time for a reassessment? *Expert Systems*, *5*, 216–255.

Duncker, K. (1945). On problem solving. *Psychological Monographs*, *58* (Whole No. 270), 1–113. (L. S. Lees, Trans.; original work published in German in 1935).

Glaser, R. (1976). Cognitive psychology and instructional design. In D. Klahr (Ed.), *Cognition and instruction* (pp. 303–315). Hillsdale, NJ: Erlbaum.

Gordon, S. E. (1988, January). The human factor in expert systems. *AI Expert*, pp. 55–59.

Hammond, K. R. (1966). Clinical inference in nursing: A psychologist's viewpoint. *Nursing Research*, *15*, 27–38.

Hart, A. (1986). *Knowledge acquisition for expert systems* (1st ed.) New York: McGraw-Hill.

Hoffman, R. R. (1987, Summer). The problem of extracting the knowledge of experts from the perspective of experimental psychology. *The AI Magazine*, *8*, 53–64.

Hoffman, R. R. (1989, April). A survey of methods for eliciting the knowledge of experts. In C. R. Westphal & K. M. McGraw (Eds.), *Special Issue on Knowledge Acquisition*, *SIGART Newsletter* (No. 108, pp. 19–27). Special Interest Group on Artificial Intelligence, Association for Computing Machinery, New York, NY.

Hoffman, R. R. (1990). Remote perceiving: A step toward a unified science of remote sensing. *GEOCARTO International*, *5*, 3–14.

Hoffman, R. R. (1991). Human factors psychology in the support of forecasting: The design of advanced meteorological workstations. *Weather & Forecasting*, *6*, 98–110.

Hoffman, R. R., & Conway, J. A. (1989). Psychological factors in remote sensing: A review of some recent research. *GEOCARTO International*, *4*, 3–21.

Hoffman, R. R., & Deffenbacher, K. C. (in press). A brief survey of the history of applied cognitive psychology. *Applied Cognitive Psychology*.

Hoffman, R. R., Shanteau, J., Burton, A. M., & Shadbolt, N. R. (1991). *The cognition of experts*. Unpublished manuscript, Department of Psychology, Adelphi University, Garden City, NY.

Kidd, A. L. (Ed.). (1987). *Knowledge acquisition for expert systems: A practical handbook*. New York: Plenum.

Kidd, A. L., & Cooper, M. B. (1985). Man-machine interface issues in the construction and use of an expert system. *International Journal of Man-Machine Studies*, *22*, 91–102.

Kidd, A. L., & Welbank, M. (1984). Knowledge acquisition. In J. Fox (Ed.), *Expert systems: Infotech state of the art report*. Oxford: Pergamon Infotech.

Klein, G. A. (1987). Applications of analogical reasoning. *Metaphor and Symbolic Activity*, *2*, 201–218.

Klein, G. A., & Hoffman, R. R. (in press). Seeing the invisible: Perceptual-cognitive aspects of expertise. In M. Rabinowitz (Ed.), *Applied Cognition*. Hillsdale, NJ: Erlbaum.

McGraw, K. L., & Harbison-Briggs, K. (1989). *Knowledge acquisition: Principles and guidelines*. Englewood Cliffs, NJ: Prentice-Hall.

Mitchell, J., & Welty, C. (1988). Experimentation in computer science: An empirical view. *Interna-*

tional Journal of Man-Machine Studies, *29*, 613–624.

Prerau, D. S. (1985, Summer). Selection of an appropriate domain for an expert system. *The AI Magazine*, *6*, 26–30.

Prerau, D. S. (1989). *Developing and managing expert systems: Proven techniques for business and industry*. Reading, MA: Addison-Wesley.

Shadbolt, N. R., Hoffman, R. R., Burton, A. M., & Klein, G. A. (1991). *Eliciting knowledge from experts: A methodological analysis*. Unpublished manuscript, Department of Psychology, University of Nottingham, England.

Shanteau, J., & Phelps, R. H. (1977). Judgment and swine: Approaches in applied judgment analysis. In M. F. Kaplan & S. Schwartz (Eds.), *Human judgment and decision processes in applied setting* (pp. 255–272). New York: Academic Press.

2
Knowledge and Knowledge Acquisition in the Computational Context

Stephen B. Regoczei and Graeme Hirst

Introduction

The enterprise of artificial intelligence (AI) has given rise to a new class of software systems. These software systems, commonly called expert systems, or knowledge-based systems, are distinguished in that they contain, and can apply, knowledge or some particular skill or expertise in the execution of a task. These systems embody, in some form, humanlike expertise. The construction of such software therefore requires that we somehow get hold of the knowledge and transfer it into the computer, representing it in a form usable by the machine. This total process has come to be called knowledge acquisition (KA). The necessity for knowledge representation (KR)—the describing or writing down of the knowledge in machine-usable form—underlies and shapes the whole KA process and the development of expert system software.

Concern with knowledge is nothing new, but some genuinely new issues have been introduced by the construction of expert systems. The processes of KA and KR are envisaged as the means through which software is endowed with expertise-producing knowledge. This vision, however, is problematic. The connection between knowledge and expertise itself is not clearly understood, though the phrases *knowledge-based system* and *expert system* tend to be used interchangeably, as if all expertise were knowledgelike. This haziness about basics also leads to the unrealistic expectation that the acquisition of knowledge in machine-usable form will convey powers of expert performance upon computer software. These assumptions are questionable. For a deeper understanding, we must clarify the concepts KA and KR, and the concept of knowledge itself, as they are used in the computer context. That is the first goal of this chapter. The second goal is to explicate the issues involved in KA and show how they are amenable to research by experimental or cognitive psychologists.

The chapter will be organized as follows. In the second section we will set the stage for cross-disciplinary discussion by sketching the history of AI and KA. In the third section, we try to answer the question, What is knowledge? by examining the various approaches that people have taken in trying to grasp the nature of knowledge. In the fourth section, we discuss the KA problem. In particular, we present a model of the KA process to reconcile and pull together the various approaches to KA that are found in the literature. This basic model of KA will be used in the commentaries chapter (chapter 17) to compare the contributions to this volume. In the present introductory chapter, we outline some crucial current issues, especially those that could be fruitfully addressed by experimental psychologists, and as a conclusion we try to point to some future directions for research.

Two Camps

Discussions of expert systems and the psychology of expertise invariably address one of two audiences: One group is mainly interested in human psychology; the other is interested in machines and applications. What seems obvious to one group may provoke outraged opposition, dismay, or disbelief in the other camp, and such exclamations as, "Nobody could seriously believe such things!" Hence, this chapter, as well as much of this volume, is an exercise in cross-cultural mediation.

To start on this work of communicating across the gap, it is helpful to look at some of the basic assumptions of the two camps. Although people in psychology and AI are both building models, an important distinction in attitudes toward model building has to be drawn. At times, psychologists, while perceiving themselves to be scientists, will say, "AI is not a science." And this may be true, but it is not a stigma to be ashamed of. Born in the world of software, AI is not sciencelike. It is something else—and realizing its non-natural-science nature and its artificial orientation could, we believe, actually make it *more* acceptable to the empirical psychologist.

As McDermott (1981), himself one of the eminent practitioners of AI, has remarked:

> As a field, artificial intelligence has always been on the border of crackpottery. Many critics . . . have urged that we are over the border. . . . Unfortunately, the necessity for speculation has combined with the culture of the hacker in computer science . . . to cripple our self-discipline. (p. 143)

But, continues McDermott, "In a young field, self-discipline is not necessarily a virtue" (p. 143).

McDermott (1981) is not unduly worried about respectability, although he would like to correct a few faults. McDermott's attitude is widely shared in AI. The fact is that AI saves itself from disgrace by having a strong engineering orientation—although *gadgeteering* or *bricolage* may be more apt terms to use. People in AI design and build things, like engineers. This realization necessitates our invoking the distinction between science and engineering, and their differing world views: two very different sets of attitudes, both crucially affected by the conceptual frameworks that are brought to the task. The scientist asks, "What can I observe? What can I find out?" The engineer asks, "What can I build? How well will it work?" Building models is engineering-style work, and this is the basic attitude of the computer software developer.

Having described these two camps, we should strike a hopeful note. These two camps can and do communicate: In fact, we have to qualify our description by saying that often each will use the techniques of the other. For example, physicists sometimes think like empirical scientists and sometimes like applied mathematicians. Sometimes they run experiments to "find out" what there is, and sometimes they turn creative and "manufacture" new elementary particles. The balance between gadgeteering and empirical field work is the key to future success. This is where the applied psychologist can team up with the AI practitioner.

AI is more like a science of the artificial (Simon, 1969) than are the natural or social sciences. This fact influenced its historical development. We proceed to look at this history.

Gadgeteering (versus field work) put its stamp on the development of the field of AI. For example, one goal of AI is to emulate, or build models of, cognition. (AI's having been called *cognitive simulation* in the early days testifies to this basic objective.) Thus, one might have expected the field to start with careful empirical investigations of cognition; follow this primary research with the development of conceptual structures, theories, and models; and conclude with an implemented version of the software that would enable a computer to simulate some aspects of cognition. But the field didn't develop this way. It started from an environment permeated with mathematics, physics, and engineering. This heritage slanted things in a peculiar way. Its discourse became saturated with words and phrases such as *intuitively obvious*, *self-evident*, and *axiomatic*. Empirical research, or careful examination of the world, is not part of the research practice, or extends only so far as to make sure that the artifacts function, the machines work, and the bridges don't fall down. Social interaction, such as interviewing people as they engage in the activities that were to be simulated, would have been considered pointless. In such an environment, task analysis and armchair speculation dominated.

The results are well known. Chess playing, theorem proving, solving mathematical puzzles and similar "toy problems" (Gardner, 1985) were held up as the true benchmarks of intelligence. Because "raw IQ" was to be independent of experience and training, there was no need for any content knowledge in the early systems. Basic number-crunching power and excellence in reasoning, searching, and problem solving are what produced outstanding performance, or so it was believed.

The years went by. As the pioneers seasoned and the researchers themselves became more experienced, the realization dawned that experience acquired over years of practice may have something to do with skillful performance (Gardner, 1985). It was realized, slowly at first but with greater conviction later, that explicit knowledge was required. Ironically, this still did not focus attention on KA. Rather, knowledge *representation* became the salient issue. Graduate students were encouraged to look at knowledge representation techniques such as semantic networks or first-order logic. Where the knowledge came from was not an issue: Knowledge representation was central, and KA was as yet a nontopic.[1] Knowledge representation was therefore indistinguishable from so-called knowledge engineering work. Little thought was given to effective knowledge elicitation techniques or to the psychology of the expert. Nor was the psychology of the KA analyst considered. In fact, the psychology of the analyst is still a neglected area (see below). Knowledge is considered to be objective, and the analyst is expected to be an impartial, detached observer devoid of any psychology.

Knowledge-based systems and rule bases were designed and built at the cost of great effort. Looking back at the experience with such systems as DENDRAL, MYCIN, and DIP-ADVISOR (Hayes-Roth, Waterman, & Lenat, 1983), it came to be that the acquisition of knowledge for the systems was a significant part of the development effort (Hoffman, 1987). From the point of view of KA, early efforts in expert systems development were misleading and set the wrong precedent. For example, if a doctor himself is developing the medical expert system, such as MYCIN, then there seems to be no need for KA—why should he acquire knowledge when he already has it? All he needs is someone to help with the programming: the so-called knowledge engineer. Elicitation also seems to be easy: The doctor will simply tell—or so it was thought—everything he knows to the "knowledge engineer."

This particularized view of the KA-KR process held back further developments for several years and obscured the true nature of the difficulties encountered in developing expert systems. The need for heuristics to make brutal search techniques tractable forced the issue. Eventually, attention was focused on knowledge acquisition.

In effect, the field backtracked to visit the stages of development that were skipped in the early days. If we extrapolate this retracing, we may predict that the next step will be to explore the usable portions of cognitive psychology for insight and inspiration regarding both the process of knowledge acquisition and the architecture of cognition, in order to suggest design ideas for further artificial intelligence work. However, before we can take up this topic, we must first examine more closely just what we mean by knowledge and knowledge acquisition.

What Is Knowledge?

One way of answering the question "What is knowledge?" is to say that knowledge is a concept consisting of a cluster of associated metaphors. A basic goal of this chapter is to highlight the nature of the metaphors that underlie the discourse about knowledge and KA.

The most fundamental metaphor is that knowledge is a substance: It is a metaphoric substance that people possess and that enables them to perform at expert levels. Knowledge is like a distilled essence that somehow transcends the more dilute and less valuable "data" or "information," and mysteriously manages to account for expertise. This substance, like a liquid, can be poured from one mind-container into another (Reddy, 1979). KA is thus taking the substance knowledge from the head of an informant and transferring it into the head of the analyst, or knowledge engineer, who then suitably reformulates it and pours it into the computer.

Of course, this is a very simplistic metaphor. It suggests, for example, that when the analyst gets knowledge from the informant, the informant has less of it left. This "fluid" model also breaks down when we consider knowledge as composed of chunks and discrete pieces that are perhaps embedded in a more porridgelike matrix.

In another metaphor, knowledge is an organic being: We may say that a person's knowledge is "growing," and we may think of new knowledge being created by social interaction, by science, by creative thinkers, and even by everyday living. This metaphor can easily be misleading by suggesting that knowledge takes on a life of its own.[2]

In spite of their simplicity, metaphors like these should be taken very seriously because they express the manner in which people approach the topics of knowledge and KA. The metaphors specify the "default" frameworks that people would normally be expected to use when thinking or talking about these topics. But although such metaphors are helpful in many cases, when pushed too far they can produce problems, incongruities, and breakdowns. Illuminating and overcoming some of these breakdowns is a major task of this chapter.

General Theories of Knowledge

Philosophers have worried for centuries about the nature of knowledge. A commonly cited conclusion is that knowledge is "justified true belief." The problem for KA is that this notion is not readily operationalizable. Beliefs are proposition-like, and they are not as robust as we would like them to be. In trying to elicit beliefs, we obtain statements or verbal expressions of the propositions that constitute these beliefs. We were after beliefs, but all we got were *words*. The assumption that statements and beliefs can somehow be put in a one-to-one correspondence may not be valid in practice. In fact, the situation is much worse: The informant will invariably make contradictory, inconsistent, and incoherent statements. Are we to conclude from the contradictory statements that the expert's knowledge is also inconsistent? The mismatch between what is uttered as a statement and what may be believed as a proposition is one difficulty. The second major difficulty arises from the qualifiers of *true* and *justified*. It seems that the onus is on the knowledge acquisition analyst to check for justification and for truth, and presumably to reject those statements of the informant that do not pass the test. No one has yet been able to suggest a workable technique that mere mortal analysts and informants can use in the field.

It may be suggested that a way of getting around the problem is to develop "belief acquisition" to parallel, and perhaps to serve as a prior stage to, "bona fide" knowledge acquisition, at least bona fide under this notion of knowledge.

A more operationalizable definition of knowledge is Newell's (1982) identification of knowledge with goals, aims, objectives, plans, and purposeful behavior. We consider this a philosophical theory, although Newell would probably be insulted by the suggestion. But what are the other knowledges to be called if only goals and plans qualify as "real" knowledge?

An AI-inspired view of knowledge sees it as a form of writing or some writinglike representation inscribed in the mind of the cogniting agent, be it human or machine. To impart knowledge to the machine, we have to decipher the writing within the mind of the informant and translate it or transform it in such a way that we can represent and use it in a machine. After it is in the machine, knowledge is construed as a representation within a symbolic processing context, that is, within a physical symbol system (Newell & Simon, 1972). The knowledge is construed as the representation of real-world knowledge, so this approach has its roots in naive realism and Tarski-like semantics.

Specific Kinds of Knowledge

We can perhaps further clarify the nature of knowledge by drawing a few distinctions along some specific dimensions.

Personal, Public, and Objective Knowledge

Our first distinction is between personal knowledge; public knowledge; and objective, eternal knowledge. This can be best illustrated by using examples from mathematics. A natural number n, such as 1,234,567,891,234, 567, might or might not be a prime number. If a person knows it to be a prime, that is personal knowledge. If at a given time, a culture's mathematical knowledge is elaborate enough to have rules and techniques for factorizing such numbers or to be able to test them for divisibility using simple rules, then for this culture it may be public knowledge that the number is or is not a prime, although perhaps only a handful of individuals may actually know it. Some people would argue that n's being prime or not prime, and its primeness one way or the other, is objective and eternal knowledge (Frege, 1950). Such knowledge is also often referred to as *the truth*, or *real knowledge*. Objective knowledge is believed to be independent of individuals or cultural groups, indeed independent of any knower, and hence independent of psychological, sociological, historical, or anthropological considerations. (The concept of "objective" knowledge is closely linked to intuitions about the factual nature of physical objects. Applying it to platonic ideas reflects a commitment to a kind of realism.)

Now let us consider not mathematics, but a less esoteric field such as medicine. The private knowledge of a physician consists of his or her knowledge of medical practice. This professional knowledge is partly a reflection of public knowledge in which the physician shares by virtue of his or her formal education and his or her own stock of experiential knowledge. Prerau (this volume) draws a similar distinction.

Thus, public knowledge can be construed as that which is well known, or as a union of the "essential" components of individuals' personal knowledges. Hence, public knowledge is a social construct and a type of consensual knowledge. It can be looked upon as being encoded in public documents such as textbooks and encyclopedias. Since World War II, Western society's emphasis on the "hard" sciences and mathematics has led to an almost exclusive focus on objective knowledge. In such a climate, KA did not naturally arise; one does not find a solution to a problem in operations research by going around and asking people. As we saw above, the need for KA was recognized only very late in the history of AI.

Book Learning and Hard Experience

We said above that the possession of knowledge somehow accounts for expertise, even if expertise is partially attributional (Sternberg & Frensch, this volume). But there is more to this story. It is commonly recognized that there is a distinction between "book learning," on the one hand, and knowledge that has been acquired through direct experience, on the other. This experiential knowledge has been "field tested," that is, improved, selected, and tested in action. This distinction is supported not only by an underlying folk model of values but also by experiment and empirical observation (e.g., Prerau, this volume). Book learning is considered to be not as reliable and not as strong as knowledge gathered through direction action. This differential valuation is seldom emphasized but shows up in indirect ways, such as when surprise is expressed at realizing—in spite of contrary expectations—the power of book learning in actual action (Prerau, this volume).

Knowing What and Knowing How

As our discussion so far has implied, there are numerous distinctions of various types of knowledges. Unfortunately, AI currently seems to have few formal or even informal characterizations of knowledge types (Kidd, 1987, p. 3). This is a serious gap, one that we hope will be filled in the near future (Chandrasekaran, 1988). Nevertheless, there is a distinction that is rather popular, and that is the distinction between task knowledge and domain knowledge, or "how-to" knowledge versus "what is" knowledge (Buchanan & Smith, 1989). Intuitively, we distinguish between a description of what is around as, on the one hand, and a list of directives as to what actions to take, on the other. This distinction is prominently enshrined in computer science in the distinction between algorithms and data structures. At times, this distinction is referred to as *procedural* versus *declarative*. (This is unfortunate, as will be argued in chapter 17. It confounds the underlying concepts: Task knowledge, for example, can be represented either declaratively or procedurally.)

Other Ways of Knowing

It can be argued that there is no such thing as knowledge in itself, only knowledge *of* something. If this position is taken, then the word *knowledge* is short for *world knowledge*, that is, knowledge that is true of the world. It is sometimes contrasted with heuristic knowledge, which consists of rules of thumb that are likely to give good results but do not guarantee the reading of "truth."

Knowledges may be distinguished according to their functional use (e.g., strategic knowledge, planning knowledge, etc.). For this classification the question is, How are they used? not, What are they?

In the Western tradition, we tend to concentrate on knowledge that is scientific, analytic, logical, or rational. Other ways of knowing—such as the holistic, the aesthetic, or the ethical—are liable to be looked upon as of merely marginal concern. Yet aesthetic judgment is crucial in mathematical work and pivotal in design work. Sometimes a design is adopted or a plan is accepted because it "looks beautiful" or it "feels right." These judgments, seemingly based on "gut feeling," may be crucial to the practice of some types of expertise. Considering its importance, little systematic work on aesthetic judgment has been conducted—certainly little within the computational context.

There is a kind of knowledge, however, that seems to occupy an especially important position in our everyday activity: This is background, or commonsense, knowledge. This strange kind of contextual knowledge is something that all of us are claimed to possess, and yet we find it difficult to capture. Often this kind of knowledge is looked upon as a magical ingredient in AI and is used to explain inadequacies: If a particular piece of software cannot perform a task, it is because it does not "have" commonsense knowledge. This tacit, contextual knowledge is considered to be required in temporal and spatial reasoning, and also in bodies of expertise called naive, as in "naive physics" (Hayes, 1984).[3]

Finally, we should address the elusive boundaries between data or information and knowledge. Factual knowledge, such as the assertion that the temperature of an object is 18.3 °C, comes closer to data or information than knowledge. We can separate out what could be called *ontological knowledge*, such as that an object exists, that this object has a temperature, and that the temperature can be measured. But the fact that the temperature happens to be 18.3 °C as opposed to, say, 20.5 °C is something that is more datalike. Capturing this kind of information is more like data capture than knowledge acquisition. Hence, we can distinguish between ontological knowledge (i.e., the kind of things that may exist in an object domain) and the factual information, or inventory-like knowledge, that specifies what actually exists in a given situation at a given moment.

Computational Representations of Knowledge

Whatever knowledge is, if a computer program is to make use of it, the knowledge must be somehow present in the machine. In AI, one speaks of knowledge *representations*, or *formalisms*, or *encoding*, which reflects the frequent computer-science concern of carefully distinguishing content from form. In this section, we will look at some of these representation methods. In all methods, an important consideration for AI is that representations permit automatic theorem-proving techniques to derive new inferences and conclusions from the knowledge.

Rules

One simple, representation-oriented theory considers knowledge to be a collection of facts about some domain, together with some rules—also called heuristics—about drawing appropriate conclusions (Feigenbaum, 1983). To the outsider it may seem strange that a collection of rules can be legitimately labeled as knowledge. But a conditional statement, such as an if-then structure, does seem to encompass knowledge of a sort. The rule "if (X), then (Y)" tells us that if X is true, then we know that Y is true.

Needless to say, psychologists—and even experts!—often find this somewhat simplistic, if not insulting. A veteran expert may be somewhat taken aback by the suggestion that a young "knowledge engineer' is going to "write down" in the form of simple rules the fruits of his hard-earned expertise gained over many years. Nevertheless, the AI community favors the simple approach mainly because it is implementable; in other words, with this approach it becomes possible to develop actual working software. We should sympathize with this position. Implementability is a very severe constraint on flights of fancy, or even on realistic attempts to include enough cognitive phenomena to make the software approximate a real thinking agent. From the perspective of psychological verisimilitude, calling rulelike structures *knowledge* may still seem strange, but at least the drive for simplicity becomes understandable.

Historically, knowledge-as-rules was one of the early paradigms of knowledge-based systems: It was hypothesized that we do indeed perform knowledge-based reasoning by applying if-then rules. Note that if-then structures were familiar in programming even before

```
FRAME:         cat
A-KIND-OF:     animal
PHYSIOLOGY:    mammal
FUR:           yes
CAN-BE-PET:    yes
SIZE:          typically 5 kg.
```

FIGURE 2.1. A simple "frame."

knowledge-based systems were conceived. But in general programming, a conditional statement is construed as a control structure with an active role, not a passive component that encapsulates knowledge. With expert systems, the rule base of if-then conditional statements on the one hand, and the inference engine on the other, were separated. The rules, together with the facts, form the knowledge base—this is construed as a passive module. The active component is the inference engine, which "chains through" the rule base. The inference engine, in direct contrast to the knowledge base, is construed as being devoid of "cognitive" content.

Frames

A frame is a data structure that represents a kind of entity or a specific instance of a kind; hence frames can be considered as surrogates for concepts. A frame includes the name of the entity and a number of slots for attributes, each of which may have a value. In addition, special information about the entity, required for its use, may be associated with the frame. Figure 2.1 shows a (simplistic) frame for cats.

Frames are connected in taxonomic hierarchies, with the understanding that a frame inherits all of the attributes of frames above it in the hierarchy. Thus, it is not necessary to specify in the frame for cats any of the information already given in the frames for animals or mammals. But inherited information can be overridden; for example, mammals might be generally specified to be quadrupeds, the exceptions being noted in the few appropriate frames.

A notion closely related to frames is the *semantic network* (although *taxonomic network* might be a better term), which is essentially a notational variant of a frame hierarchy. Semantic networks in turn are closely related to the various memory structures studied in cognitive psychology. Thus, we see frames as an attempt to capture a psychologically authentic representation of knowledge.

Frames are also the natural representations for the schemata of the psychologist, except that schemata may be quite general and can contain more than textual material. Should frames include pictures or other nontextual components? These extensions will surely be added as soon as increased technological capability enables us to process them in satisfactory ways. The generalization of hypertext ideas to encompass hypermedia is certainly one such direction. In the meantime, generalizations of frames such as event schemata—better known as scripts—have been successfully used to simulate goal-directed or expectation-driven behavior (Schank & Abelson, 1977).

Knowledge as Propositions

Knowledge may be written as a set of formulas or propositions in a language, such as first-order logic, amenable to automatic theorem-proving techniques. This, in effect, is a version of logic programming. This approach prefers an assertional, nonprocedural style over the procedural or functional programming paradigm. The programming language PROLOG was developed to suit this approach. Although usable for general programming, it is often used in AI to build a knowledge base of assertions or postulates from which further knowledge can then be derived.

Repertory Grids

Repertory grids (Kelly, 1955) are fairly well known in psychology, but are just becoming well known in AI. They are based on the theory of personal constructs, as described by Ford and Adams-Webber (this volume). As a way of representing knowledge, grids are not quite in the mainstream of AI. Nevertheless, upon comparing them to the above "bona fide" knowledge representation techniques, one finds grids to be neither more nor less natural. The repertory grid was turned into a knowledge acquisition tool largely by the work of Boose, Gaines, and Shaw (Shaw & Gaines, 1987). Although it seems far away from what one might intuitively think of as knowledge, the repertory grid serves a useful purpose in KA by bringing to the surface knowledge that the informant might never have thought he or she had, or was not sure of, or had never tried to express. Thus, repertory grids can be thought of as sometimes accessing tacit knowledge or, to put it slightly differently, helping the informant in verbalizing and conceptualizing the subconceptual.

The repertory grid notion of knowledge is the only one that is explicitly oriented toward KA. As a knowledge representation technique, it can be computationally handled, either directly or by transforming the knowledge into rules, frames, or propositional formulas.

What Is KA?

In the computational context, KA is considered to be a particular, clearly specifiable process. It is characterized by some type of interview between an informant and an analyst. The informant is the expert, and the analyst is sometimes referred to as a KA analyst or a knowledge engineer. There may be numerous variations on this informant-analyst interviewing interaction, but in all cases the knowledge being collected is to be embedded—eventually—in a knowledge-based system. This distinguishes KA from superficially similar activities such as systems analysis or the establishment of requirements specifications for a new software system. The goal of KA is very specific: The knowledge is to be acquired for the purpose of "enriching" the software with domain and task knowledge. This knowledge must be recorded or represented in an eventually implementable form using a knowledge representation technique processable by computer. The main features of KA in the computational context follow from this goal and this requirement.

Although there are other kinds of activities in which knowledge is acquired (e.g., studying a book, reading a map), it makes sense in the context of this volume to restrict the meaning of the term *knowledge acquisition* (*KA*) to that described above. Considerable confusion would be generated if we applied the term indiscriminantly to activities as diverse as spying, journalistic interviewing, police interrogation, or even to cramming for an exam the night before. Unfortunately, the phrase *knowledge acquisition* is already beginning to be overused. A recently published work on textbook writing and reading boasts the title "KA from text and pictures" rather than a more modest but also more honest version: "Understanding and learning from text and pictures in textbooks." One may not be able to stop the proliferation, but it may be helpful to keep in mind a parochial meaning of the phrase.

Metaphors of, or for, KA

The metaphor clusters associated with KA follow closely the metaphors for knowledge that we described above. To start off, let us look at Kidd's (1987) description of KA as ascribed to Feigenbaum and McCorduck (1983):

> A popular view of KA [sic] has been to consider experts' heads as being filled with bits of knowledge and the problem of KA [sic] as being one of "mining those jewels of knowledge out of their heads one by one" (Feigenbaum & McCorduck, 1983). The underlying assumption is that some magical one-to-one correspondence exists between the verbal comments made by an expert during an elicitation session and real items of knowledge that are buried within his head. . . . Once these comments are transcribed onto a piece of paper, they are considered to be nuggets of the truth about that domain. (p. 2)

We now know this to be overly simplistic, but these metaphors work well for a broad range of cases and situations and for a broad range of people with different cognitive styles. The metaphors need close reexamination only if they cause a breakdown and do not function as expected (Winograd & Flores, 1986). We saw above some of the questions we can ask about the knowledge component of the metaphors: Does knowledge come in pieces? Do these pieces have an internal structure? Are the pieces alike, or are they of different sorts? If not, is knowledge more like porridge, with perhaps a few lumps in it? The analysis of the metaphors may reveal many of the hidden assumptions, but it scarcely touches the main problem.

The problem is that the necessity of expressing knowledge by using language introduces a whole range of possible breakdowns. Even if knowledge itself does not come in pieces, language does. What the expert says—words, phrases, expressions, and sentences—are all piece like and discrete, and inevitably there is a mismatch between the possibly continuous, aggregate, or conglomerate-like substance on the one hand, and the hard-edged, bony concepts and linguistic expressions on the other (Sowa, 1984). Nor is there a clear separation between the two. For a good analogy, one can think about porridge with some lumps in it, or concrete aggregate before it crystalizes, or the conglomerate of the geologist, or nuts in a chocolate bar. The one-to-one correspondence between what is being said and what is being thought is a piece of wishful thinking—a romantic illusion that has great simplifying value and great attraction but that will not stand up to actual practice.

A Basic Model for KA

What actually happens during KA? Before we offer a model of the KA process, it is probably helpful to start with a simple overview. KA can be divided roughly into three phases: elicitation, explication, and formalization (Regoczei & Hirst, 1989). The analyst elicits verbal expressions of the expert informant's knowledge and helps to refine and organize this knowledge in cooperation with the expert informant. The basic assumption is—and this needs validation—that the expert's expressed knowledge accounts for his or her expertise. The result is a formalized partial model of the informant's knowledge. This knowledge is "written down"; that is, it is represented or described in a machine-usable form. The

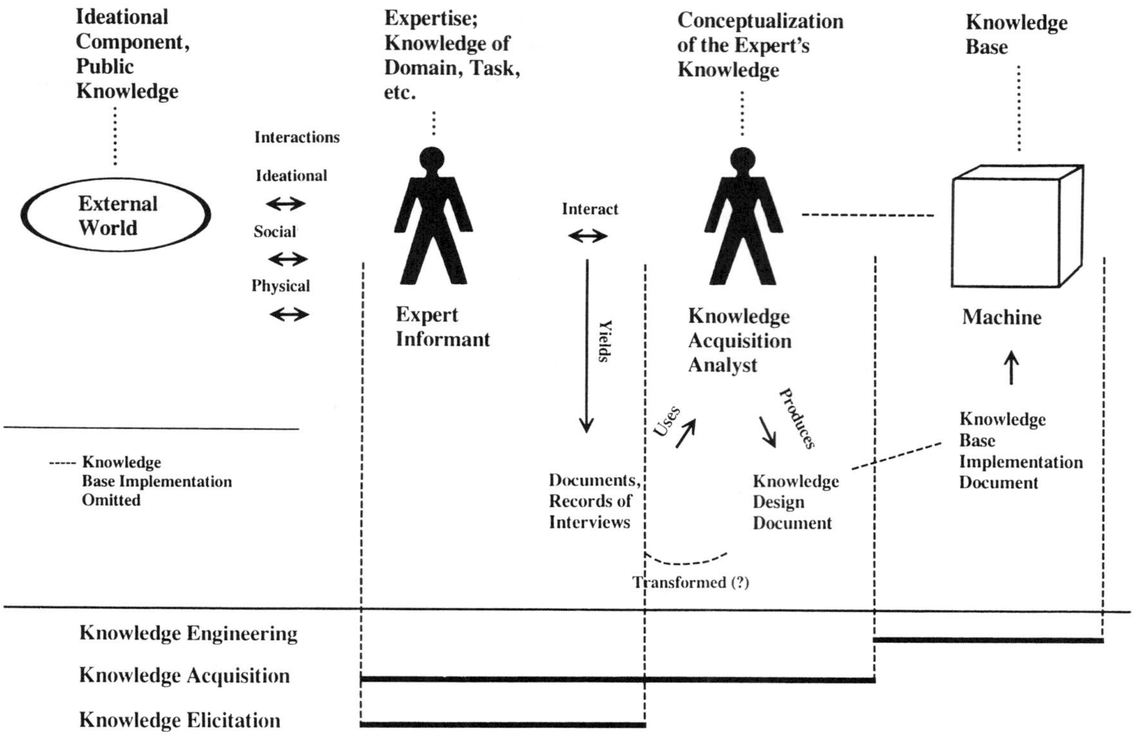

FIGURE 2.2. Expert system implementation.

phrase *expertise modeling* (Feigenbaum, 1983) aptly summarizes the process.

KA is only one part of system building, and so for the sake of clarity we must first locate the KA process within the larger context of the full implementation of expert systems. This is shown in Figure 2.2. The elicitation of knowledge from the expert and the recording of this knowledge in what we are calling the knowledge design document are only the "front end" to some difficult systems work, usually conducted by highly specialized programmers, to implement on the computer the knowledge base within the expert system. Thus, we distinguish the earlier KA phase (i.e., the obtaining of the knowledge) from the knowledge base implementation phase, during which the knowledge is transformed, usually by a software engineer, into computer-usable form. (In this chapter, we do not concern ourselves with implementation issues.)

The KA process is summarized in Figure 2.3. The expert can be considered to interact with the external world at three levels: physical, social, and ideational. It is the excellence of this interaction, as judged by others in the social context (Sternberg & Frensch, this volume), that gets the person designated an expert. The knowledge he or she is thought to possess is to be acquired and recorded: first in a publicly examinable form and later to be recast in a machine-usable form. The expert would typically express his knowledge in words, but he may also use diagrams, formulas, tables, and other symbolic material. On the basis of natural language input, plus the other supporting material obtained by the analyst (including numerical data if repertory grids are used), the analyst is to conceptualize and reconceptualize the knowledge of the informant. During this conceptualization stage, the analyst adds additional structure to what he or she construes to be the expert's knowledge, that is, what he or she attributes to the expert. The analyst records his or her conceptualization of the knowledge in what we are calling the knowl-

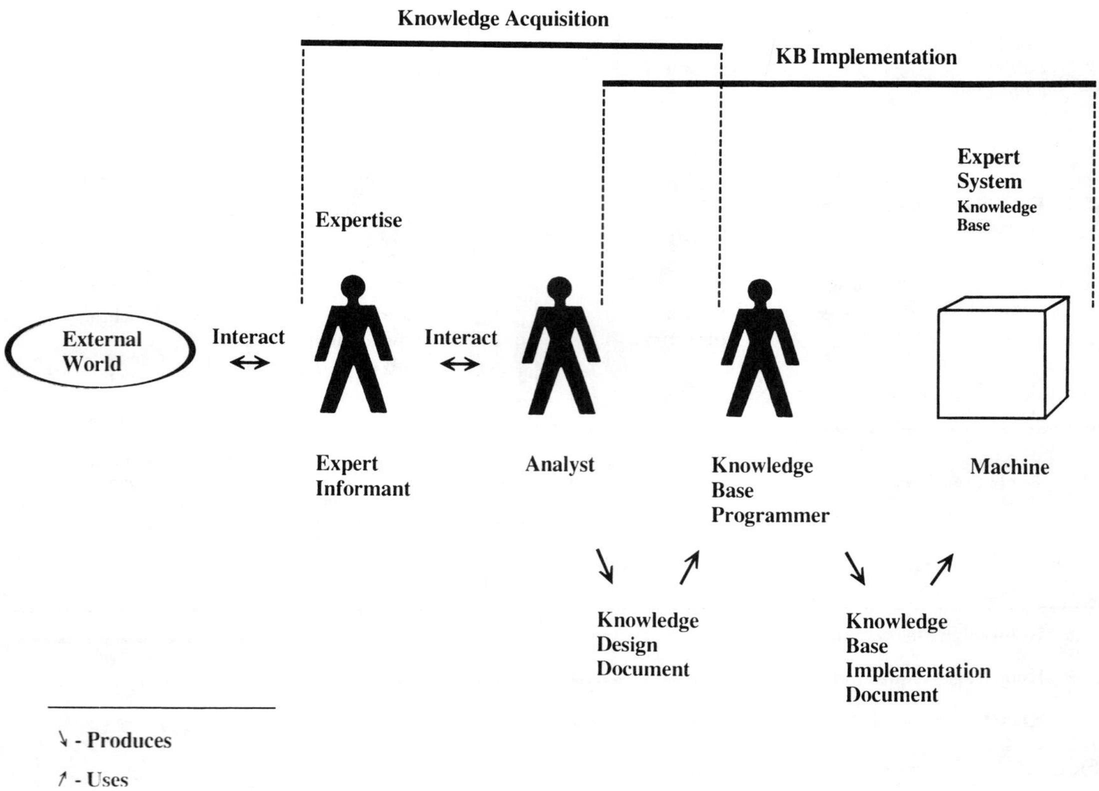

FIGURE 2.3. The knowledge acquisition process.

edge design document (Fig. 2.2). This document contains a formalized partial model of the expert's knowledge as conceptualized by the analyst. It is this document that is the final output of the KA process.

Although the knowledge design document would contain descriptions of the overall structure of the expert's knowledge as a kind of a skeletal framework, as well as some of the knowledge itself, it typically would not contain all of the detailed knowledge that eventually would be needed for a fully functioning expert system. Detailed knowledge, as well as revisions and updates, can be captured during later phases of the development cycle.

The most common misconception about KA is that the analyst copies something that is already there. We should note that the analyst does *not* "acquire" something that is already "there" in the expert's head, ready to be recorded. Rather, a new entity is *created*: A body of knowledge is constructed cooperatively by the expert and the analyst, with each contributing his or her share. The knowledge design document should capture the essential elements of the informant's knowledge and have sufficient structure—provided largely by the analyst—to make it usable in the next stage to implement the knowledge base. The caveat is that the process, although it creates something new, should not introduce serious distortions into the expert's knowledge.

In the commentaries chapter (chapter 17), we will return to the misconception that KA is discovery, as opposed to creation. It is misleading to say that the analyst acquires some ready-made substance called knowledge that is residing in the head of the expert informant. It has to be emphasized, contrary to what one reads in the literature, that typically there are no rules or frames in the expert's head. The contents of the knowledge design document and the knowledge base implementation document have to be constructed by the analyst out of the material provided by the expert.

The conceptualization, by the analyst, of the expert's knowledge is the critical phase of the KA process. To carry out this task, the analyst interacts with the informant. They talk to each other. The transcripts of these interactions are to be preserved as a permanent record of the interviews, and together with the supporting documents they form the raw material out of which the formal model/representation of the knowledge design document is to be fashioned. During their interactions, analyst and informant begin to think along the same lines. They begin to use the same vocabulary and concepts. In mental-model terms (Schumacher & Czerwinski, this volume), they begin to harmonize their mental models (Regoczei & Plantinga, 1987). The analyst is forming mental models of the expert's mental models, and vice versa. Through the exchange of natural language utterances—all recorded!—they verify whether their mental models match in essential respects.

To summarize the terminology laid out in Figure 2.3, elicitation is the establishing of a permanent record of the informant-analyst interview. The initial elicitation attempts are followed by explications, elaboration, and formation of conceptual models on the part of the analyst. The process culminates in the formalization of a partial model of the expert's knowledge and recording of this knowledge in the knowledge design document. This constitutes the KA phase of the project.

It is part of the implementation stage to reformulate the knowledge design document so that the knowledge is expressed in ways that are compatible with the particular software running on the host machine. The result of this work is the knowledge base implementation document. This reformulation is a task for the implementation phase of the project. (It should not be allowed to interfere with, and possibly distort, the knowledge elicitation process). Here lies a useful role for the term *knowledge engineering*: to refer to the later stages of the expert system implementation cycle, including the reformulation of the contents of the knowledge design document such that it becomes suitable for machine use.

Knowledge will be represented differently in the three key documents. The permanent record of the interviews, the knowledge design document, and the knowledge base implementation document (Fig. 2.3) will all use different knowledge representation techniques. The permanent record will use a natural language, such as English, and some tables and diagrams. The knowledge design document will use more formal notations, such as logic or a formal diagramming technique, such as conceptual graphs (Sowa, 1984). The actual formulation of the knowledge in the knowledge base implementation document will be in the form required by the expert system shell or other software running on the computer.

The main issue in KA, as we understand it in this chapter, is the externalization of knowledge: This means the "writing down" of the knowledge in a form that can eventually be turned into machine-usable representations.

The Participation of Psychologists

Psychologists, we believe, have an important role to play in expert system development. Because of the "cognitive turn" in psychology, the old impediments to working with knowledge and expertise have been removed. As Jonathan Miller (1983), himself one of the great KA practitioners (in contexts broader than the computational), has remarked:

> In its understandable effort to be regarded as one of the natural sciences, psychology paid the unnecessarily high price of setting aside any consideration of consciousness and purpose, in the belief that such concepts would plunge the subject back into the swamp of metaphysical idealism. Research was designed on "positivistic" lines, so that the emphasis inevitably fell on measurable stimuli and observable behavior. It soon became apparent that such a program could not be sustained, and that psychology would begin to stagnate if research failed to take account of the inner state of the living organism. . . . [As] Bruner points out . . . , ". . . all was not well in this positivistic heaven." (p. 32)

Fortunately, taking into account the "inner state of the living organism" no longer threatens anyone with the abyss of metaphysics. In fact, inner states can be formalized within automata theory—one of the sciences of the artificial—and this is perhaps as concrete and as "unmetaphysical" as one can get. Attitudes in psychology have certainly changed over the years, and the chapters in this volume indicate how much the range of interest has been broadened.

We can now even look back upon a "tradition" of KA within psychology. The elicitation phase of the KA process is specifically within the domain of expertise of the psychologist, and we can anticipate further work in the area. Psychologists will not be implementing expert systems. Rather, they will work on the knowledge-elicitation phase of the project as members of a multidisciplinary team. In the short run, we expect that knowledge elicitation and the validation of the acquired knowledge is where empirical, experimental, applied psychology is going to be most helpful. In the long run, we can anticipate that ideas and methods from psychology will have a profound effect on the changing architecture of future expert systems (see, e.g., Sternberg & Frensch, this volume). Additionally, we can anticipate that insight from psychological work that probes the complex nature of expertise and expertise transfer will suggest new uses for the acquired expertise, such as the preserving of expertise through a general "knowledge medium," as suggested by Klein (this volume).

It is quite likely that psychologists will play an active role in the knowledge-elicitation phase of the knowledge acquisition process, and that their insight will help make expert systems more usable for actual users (Hart & Foley, this volume). The division of labor and the use of development teams combined of several diverse professional skill sets seem like inevitable developments as expert systems become larger and more sophisticated. One can foresee an exciting era of cooperation between AI and psychology based on the necessity of delivering empirically sound, workable systems that are usable by real people who are ecologically situated in actual contexts.

Notes

1. Further reasons for the extensive knowledge representation literature, as compared to the more modest work in knowledge acquisition, are given by Regoczei and Plantinga (1987).
2. A more detailed description of various theories and metaphors of knowledge can be found in Regoczei and Hirst, 1989, and Hoffman, Cochran, and Nead, 1990.
3. The appelation *naive* is intended to indicate that such knowledge is wrong. Conversely, "commonsense" knowledge is often assumed to be basically correct although fallible.

References

Buchanan, B. G., & Smith, R. G. (1989). Fundamentals of expert systems. In A. Barr, P. R. Cohen, & E. A. Feigenbaum (Eds.), *The handbook of artificial intelligence* (pp. 149–192). Reading, MA: Addison-Wesley.

Chandrasekaran, B. (1988). Generic tasks as building blocks for knowledge-based systems. *The Knowledge Engineering Review*, *3*, 183–210.

Feigenbaum, E. A. (1983). Knowledge engineering: The applied side. In J. E. Hayes & D. Michie (Eds.), *Intelligent systems: The unprecedented opportunity* (pp. 37–55). New York: Halstead Press.

Feigenbaum, E. A., & McCorduck, P. (1983). *The fifth generation*. Reading, MA: Addison-Wesley.

Frege, G. (1950). *The foundations of arithmetic*. (J. L. Austin, Trans.). New York: The Philosophical Library.

Gardner, H. (1985). *The mind's new science: A history of the cognitive revolution*. New York: Basic Books.

Hayes, P. (1984). The second naive physics manifesto. In J. Hobbs (Ed.), *Formal theories of the commonsense world*. Norwood, NJ: Ablex.

Hayes-Roth, F., Waterman, D., & Lenat, D. (1983). *Building expert systems*. Reading, MA: Addison-Wesley.

Hoffman, R. R. (1987, Summer). The problem of extracting the knowledge of experts from the perspective of experimental psychology. *The AI Magazine*, pp. 53–66.

Hoffman, R. R., Cochran, E. L., & Nead, J. (1990). Cognitive metaphors in experimental psychology.

In D. Leary (Ed.), *Metaphors in the history of psychology* (pp. 173–229). Cambridge: Cambridge University Press.

Kelly, G. A. (1955). *The psychology of personal constructs*. New York: Norton.

Kidd, A. L. (1987). *Knowledge acquisition for expert systems*. New York: Plenum.

McDermott, J. (1981, Summer). Rl: The formative years. *The AI Magazine*, pp. 21–28.

Miller, J. (1983). *States of mind*. Toronto: Methuen.

Newell, A. (1982). The knowledge level. *Artificial Intelligence*, *18*, 87–127.

Newell, A., & Simon, H. A. (1972). *Human problem solving*. Englewood Cliffs, NJ: Prentice-Hall.

Reddy, M. J. (1979). The conduit metaphor: A case of frame conflict in our language about language. In A. Ortony (Ed.), *Metaphor and thought* (pp. 284–324). Cambridge: Cambridge University Press.

Regoczei, S., & Hirst, G. (1989). *SORTAL: A sortal analysis assistant for knowledge acquisition*. Technical Report, Department of Computer Science, University of Toronto, Toronto, Ontario, Canada.

Regoczei, S., & Plantinga, E. (1987). Creating the domain of discourse: Ontology and inventory. *International Journal of Man-Machine Studies*, *27*, 235–250.

Schank, R., & Abelson, R. (1977). *Scripts, plans, goals, and understanding*. Hillsdale, NJ: Erlbaum.

Shaw, M. L. G., & Gaines, B. R. (1987). An interactive knowledge-elicitation technique using personal construct technology. In A. L. Kidd (Ed.), *Knowledge acquisition for expert systems* (pp. 109–136). New York: Plenum.

Simon, H. A. (1969). *The sciences of the artificial*. Cambridge, MA: MIT Press.

Sowa, J. (1984). *Conceptual structures: Information processing in mind and machine*. Reading, MA: Addison-Wesley.

Winograd, T., & Flores, F. (1986). *Understanding computers and cognition*. Norwood, NJ: Ablex.

Part II
Cognitive Theory and Expertise

3 Modeling Human Expertise in Expert Systems

Nancy J. Cooke

Introduction

The goal of cognitive psychologists interested in human expertise is to identify the cognitive structures and processes that are responsible for skilled performance within a domain. It is assumed that these structures and processes maintain some degree of generality regardless of the specific content of knowledge (i.e., the specific facts and rules making up the knowledge base). On the other hand, the goal of those interested in expert systems is often quite different. For knowledge engineers, the development of the expert system is the goal, and the explanation of expertise is typically secondary at most. Also, because knowledge, in the form of specific facts and rules, is assumed to be the power behind expert systems (Minsky & Papert, 1974), knowledge engineers tend to be less concerned with general characteristics of expert knowledge than they are with the specific content of that knowledge. As a consequence, knowledge acquisition, the process of transferring knowledge from a source of expertise (either human or textual) to the expert system, is of paramount importance to the development of expert systems, but unfortunately it is also a major bottleneck in expert system design. Although the bulk of the research that has been done in cognitive psychology on expertise does not directly address the elicitation of specific facts and rules from human experts, methods that have been used in cognitive psychology to study memory organization and expertise can be applied to the knowledge elicitation problem. Furthermore, there are numerous arguments for considering the knowledge engineering implications of cognitive research related to the general structures and processes that underlie expertise.

The purpose of this chapter is to bridge the gap between cognitive psychology and knowledge engineering by pointing to some cognitive methods that have direct application to knowledge engineering, as well as to some research that is relevant to expert system design. The following section addresses methodologies used in cognitive psychology that can also be used to elicit expert knowledge. The remainder of the chapter reviews research on expertise. This work is organized according to the relevant cognitive faculties of pattern recognition, memory and knowledge organization, problem solving, and learning. The aim is to familiarize the reader with research in this area and to suggest some possible implications of this work. Whereas new knowledge acquisition methodologies are eagerly accepted by knowledge engineers, the implications of cognitive research are sometimes less apparent and are considered as a last resort, if at all. There are, however, several arguments for the application of cognitive research to knowledge engineering.

First, although the development of a successful expert system does not require the modeling of human expertise—because there is no other form of expertise—it is difficult to build such systems that are not inspired by the characteristics of human expertise. Such inspirations are generally based on the designer's intuitions rather than on data. Because our intuitions are often inaccurate (Nisbett & Wilson, 1977), empirical data should at least be considered in making these decisions.

Second, cognitive research on expertise has much to say about the representation of expert knowledge. Although the same knowledge structures and processes may occur across different knowledge contents, the content seems quite dependent on the nature of the structure that embodies it and the processes that operate on it. In other words, the content depends on the knowledge representation. The same concept may take on a completely different meaning when represented in a different form. Therefore, the expert's representation of knowledge should be considered when selecting an expert system architecture and coding the knowledge. In addition, human representation of knowledge needs to be considered if the machine is to communicate with a human.

Finally, the limitations, suboptimal behaviors, and biases that have been studied extensively in cognitive psychology can reveal important mechanisms of expertise, just as studies of the limits of short-term memory reveal chunking mechanisms (Miller, 1956) and studies of decision-making biases reveal heuristics (Kahneman, Slovic, & Tversky, 1982). Thus, strengths can be discovered by attending to the limits of "the system." In addition, human limitations need to be considered for an effective allocation of tasks to human and machine. The limitations covered in this chapter are those that are particularly relevant to expert performance and thus exclude more general limitations such as limits on attentional resources, limitations on memory capacity, and physical limits imposed by the sensory and motor systems.

In summary, methods from cognitive psychology can conceivably be used to help overcome the knowledge acquisition bottleneck, and because many center on memory organization, they can also be used to elicit the representation of expert knowledge, as well as content. In addition, research on expertise has numerous implications for knowledge engineering.

Psychological Methods for Knowledge Elicitation

The process of eliciting and representing expert knowledge typically involves a knowledge engineer and a domain expert. The knowledge engineer works with the expert to identify the facts and rules pertinent to the domain. The knowledge engineer may interview the expert and ask for this information directly, or may observe the expert as she solves a problem or performs a task in her domain. Often the expert will be asked to "think aloud," to verbalize her thought processes while solving the problem. The resulting protocol can be analyzed later for specific facts, rules, and heuristics, as well as any recurring themes, ambiguities, or inconsistencies. Although many automated tools have been recently developed to aid in the knowledge acquisition process (Boose & Gaines, 1988), most facilitate the representation or the coding of information that has already been elicited from an expert. The elicitation of expert knowledge is associated with numerous difficulties, including the problem of eliciting knowledge that appears to be intuitive or implicit.

People cannot always give complete or accurate reports on their mental processes (Nisbett & Wilson, 1977), although some situations are less prone to these limitations than others (Ericsson & Simon, 1984). It can be especially hard for the expert to convey some types of knowledge, such as procedural knowledge. To illustrate this difficulty, consider trying to list the steps required to tie a shoe. For this reason, much expertise is learned by watching and doing instead of through explicit instruction. Minsky (1981) stresses that "self-awareness is a complex, but carefully constructed illusion" (p. 99) and that "only in the exception, not the rule, can one really speak of what one knows" (p. 100).

Psychologists have long been aware of problems related to verbal reports and introspection on one's own mental processes (Claparède,

TABLE 3.1. Symmetrical distance matrix.[1]

	Bird	Robin	Eagle	Penguin	Sparrow
Bird	0	1	2	4	2
Robin	1	0	3	5	2
Eagle	2	3	0	4	3
Penguin	4	5	4	0	5
Sparrow	2	2	3	5	0

[1] 1 = related/close, 5 = unrelated/distant.

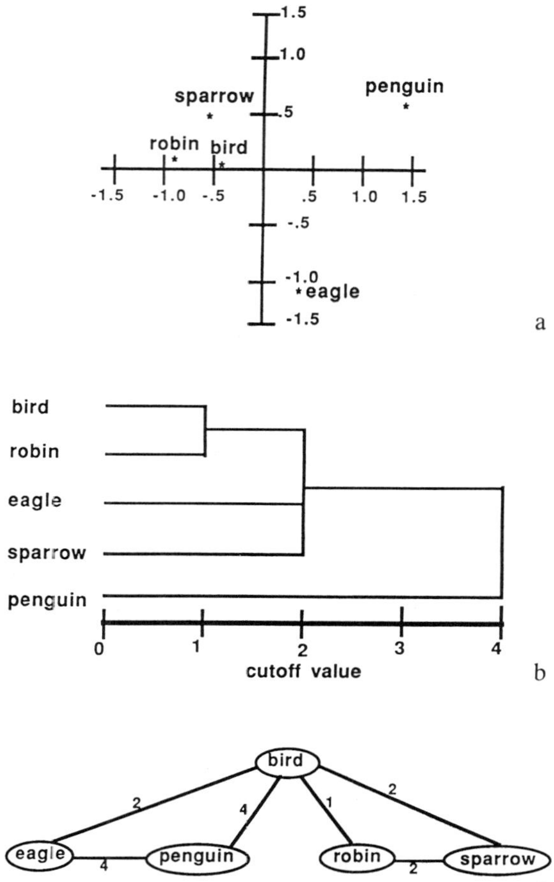

FIGURE 3.1. Two-dimensional scaling (a), hierarchical cluster analysis (b), and Pathfinder network analysis (c) of data in Table 3.1.

1934; Duncker, 1945). As a consequence, researchers interested in cognitive structures and processes have developed a number of "indirect" measures to get at this information. For instance, the difference in time that it takes to respond in two different tasks is assumed to reflect differences in the complexity of the mental processes associated with those tasks (Taylor, 1976). In addition, the pattern of errors that people make on various tasks provides information about their mental processes (Ashcraft, 1989).

Psychological scaling techniques that have been used to reveal the structure of knowledge provide an alternative to the traditional knowledge elicitation techniques (Butler & Corter, 1986; Cooke & McDonald, 1986, 1987; Gammack & Young, 1985). In general, scaling techniques require relatedness estimates for all pairs of a set of items. For the purposes of illustration a set of pairwise proximities for five "bird" concepts are provided in Table 3.1. Small proximities indicate that the pair is related, whereas large indicate that the pair is unrelated. These estimates can be derived in a number of ways, including similarity ratings, co-occurrences in a sorting task, confusion probabilities, or co-occurrence in an event record.

The estimates are then submitted to any one of several scaling procedures, which in turn generate a structural representation of the estimates. Multidimensional scaling (MDS) algorithms (Kruskal & Wish, 1978; Shepard, 1962a, 1962b) generate an n-dimensional spatial layout in which dimensions correspond to features common to the whole set of items. A two-dimensional Kruskal MDS solution of the proximities in Table 3.1 is presented in Figure 3.1a. Inspection of the solution suggests that the horizontal dimension represents size and that the vertical dimension, although much more difficult to interpret, appears to represent ferocity. Cluster analysis routines (Johnson, 1967) produce a hierarchical structure of the same items (see Figure 3.1b), whereas network scaling techniques such as Pathfinder (Schvaneveldt, 1990; Schvaneveldt & Durso, 1981; Schvaneveldt, Durso, & Dearholt, 1985) generate a graph structure that is not restricted to hierarchical relations (see Figure 3.1c). In this type of structure, concepts are represented as nodes, and links between nodes indicate relations between concepts.

Thus, scaling techniques provide an indirect means of identifying an expert's representation of knowledge. Cooke, Durso, and Schvaneveldt (1986) and Cooke (1990a) have shown that proximities derived from both MDS and Pathfinder are predictive of performance in recall and categorization tasks and that Pathfinder is predictive independent of the ratings used to derive it. Thus, this evidence supports the psychological validity of the structures. Generally, these techniques reveal the structure of knowledge; however, with certain types of relatedness estimates (e.g., co-occurrence of events in an event record), they can reveal information about process. For instance, McDonald and Schvaneveldt (1988) obtained records of on-line command usage for users of the UNIX operating system. They derived proximities for pairs of commands from the frequency of co-occurrence of the commands. These data were submitted to various scaling procedures. Because Pathfinder is able to handle asymmetries, it provides a particularly useful representation of procedural knowledge in the form of common command sequences used to perform tasks.

Because of the focus on representation in cognitive psychology, techniques such as Pathfinder and MDS are geared toward the elicitation of representation or structure rather than the specific content of that representation. For example, the techniques do not address how the set of items to be scaled is obtained in the first place or how the semantic labels of network links or MDS dimensions are identified.

Cooke (1989) discusses and compares methods that can be used to elicit items or concepts in a domain. These include having the expert list concepts or steps, transcribing concepts from a knowledge engineering interview, or asking the expert to list chapter headings and subheadings in a hypothetical book on the domain. Cooke (1989) found that the techniques differed in terms of the number and type of ideas elicited by each. More specifically, the interview and chapter-listing tasks both generated a relatively large number of ideas in comparison to the other two tasks, but these differences diminished when redundant and irrelevant ideas were not included. Most of the concepts generated across all four techniques could be classified as either concepts, general rules, or procedures. There were very few facts, conditional rules, or explanations generated. However, the type of knowledge generated was dependent on the elicitation task used. For instance, the chapter-listing task generated mostly concepts, whereas the step-listing task generated mostly general rules and procedures. Cooke (1989) concluded that factors such as the importance of covering the domain, the type of knowledge desired, and the man-hours available should be considered in the selection of one or more of these techniques.

The steps that should be taken between scaling representation and expert system design are also unclear. Cooke (1990b) outlined a methodology by which Pathfinder links can be labeled with semantic relations after the network is generated. Basically, after the Pathfinder network is generated, pairs of concepts that are directly linked in the network are isolated from the structure and given to the expert to sort (in this study each pair was presented on an index card). The expert was told to think of how the pair was related and to sort the pairs into piles based on this relation. Thus, at the end of the sorting task, linked pairs were in groups that according to the expert shared the same relation. These data were then combined with those of other experts, and cluster analysis was performed on the frequencies with which a pair of links occurred in the same pile. The relation underlying a cluster of related links was then identified by another group of experts. In this way, a taxonomy of semantic relations was created for a specific network. More work along these lines is needed in order to adapt the scaling methodology to the knowledge engineering problem. However, the methodology offers a reasonable alternative to the direct methods of thinking aloud and interviews. In addition to methods, cognitive psychology has a great deal to offer knowledge engineering in terms of research on expertise.

Research on Expertise

Although research on general skills (e.g., problem solving and decision making) has a long history in cognitive psychology, the study of expertise within a very specific problem domain (e.g., chess, medical diagnosis) is a relatively new research area. The influential work on chess expertise by Chase and Simon was published in 1973. During the years since, there has been a flurry of work in the area. A large portion of this work is descriptive; that is, expert behavior is described and often compared to novice behavior. Glaser and Chi (1988, pp. xvii–xx) summarize characteristics of expertise that are robust and generalizable across a variety of domains:

1. Experts excel mainly in their own domains.
2. Experts perceive large meaningful patterns in their domain.
3. Experts are fast; they are faster than novices at performing the skills of their domain, and they quickly solve problems with little error.
4. Experts have superior short-term and long-term memory.
5. Experts see and represent a problem in their domain at a deeper (more principled) level than novices; novices tend to represent a problem at a surperficial level.
6. Experts spend a great deal of time analyzing a problem qualitatively.
7. Experts have strong self-monitoring skills.

These characteristics are based on numerous experiments in which expert and novice differences are contrasted in a variety of domains including chess (Chase & Simon, 1973a; de Groot, 1966); computer programming (Adelson, 1981, 1984; Cooke & Schvaneveldt, 1988; McKeithen, Reitman, Rueter, & Hirtle, 1981; Shneiderman, 1976); bridge (Engle & Bukstel, 1978); physics (Chi, Feltovich, & Glaser, 1981; Larkin, 1981); radiology (Lesgold, Rubinson, Feltovich, Glaser, Klopfer, & Wang, 1988; Myles-Worsley, Johnston, & Simons, 1988); judicial decision making (Lawrence, 1988); and electronics (Egan & Schwartz, 1979). Some of the characteristics, such as the fourth one, are generalizations that require qualification. It is not the case that experts have better memories in general than novices, but that they have better memories for domain-related information. Despite the progress that has been made in research on this topic, much remains to be done.

In general, we seem to know quite a bit about characteristics of expertise, but unfortunately, we know very little about the cognitive structures and processes underlying these characteristics (Glaser & Chi, 1988). Part of this limitation is probably due to the fact that the available research relies almost entirely on cross-sectional designs and employs a rather limited representation of tasks and measures. Thus, research results tend to center on discrete expert-novice differences on specific tasks rather than continuous changes in cognition as a skill is acquired.

Another weakness of much of the research concerns the definition of *expertise*. The term *expert* has been used to refer to a variety of types of knowledge and behavior. An expert is typically viewed as a specialist in a particular domain (e.g., chess, medical diagnosis, physics). However, if experience is the only criterion, then expertise could also be defined from a developmental viewpoint as an extreme along a continuum that ranges from infancy to adulthood. Similarly, expertise could be defined from a skills viewpoint, by which practiced individuals in visual search experiments would be considered experts (e.g., Shiffrin & Schneider, 1977). It seems appropriate, however, to confine expertise to a domain that has a broader scope than the visual search task, yet a scope that is more restricted than general knowledge.

Even if a definition of expertise is agreed upon, the identification of experts within a domain can be problematic. Typically, researchers have used experience (e.g., number of classes, number of years on the job) as a measure of expertise. Unfortunately, it is not always the case that experience is perfectly correlated with expert performance. For instance, there are chess players who maintain the same rating for years. Also, there are many cases in which children achieve high levels of skill within a domain with very little practice. Posner (1988) points out that there are other factors, such as motivation and individual capabilities, that are relevant to expert performance.

Despite these limitations, much progress has been made in this area within the last 15 years. In the following sections of this chapter, research on expertise is categorized according to the psychological faculty that is implicated (i.e., pattern recognition, memory, problem solving, decision making, or learning). However, this is by no means the only possible classification of this research.

Pattern Recognition

Pattern recognition processes involve the identification of objects, scenes, words, sounds, and speech through the rapid encoding of external environmental input into an internal code that is associated with the stimulus. In general, pattern recognition processes are needed to recognize and categorize stimuli in the environment (Best, 1989). These processes enable humans to immediately recognize a chair, the word *cat* written in several handwriting styles, a kitchen, and speech in a noisy background. A combination of bottom-up or data-driven processes and top-down or conceptually driven processes are normally thought to account for the successful recognition of environmental input (Matlin, 1989). Bottom-up processes start pattern recognition analysis with the information in the stimulus and continue analysis until the stimulus matches a concept. For example, when a perceiver is trying to read a scribbled word, bottom-up processes would operate first at a feature level. Once the features had been identified, they would be combined in order to identify individual letters and eventually the word. Top-down processes inititiate from the concept generated by the perceiver's expectations and knowledge and continue until a matching stimulus is found. In the case of the scrawled word, the perceiver may know that the sentence that contains the word is about pets, and from this information and knowledge about the *pet* concept the perceiver can generate some expectations of what the word should be (e.g., goldfish, dog, cat) and proceed to check the stimulus for these possibilities. A combination of bottom-up and top-down processes allows the perceiver to make the most use of the environmental and conceptual information.

Of course, all humans can be considered skilled in pattern recognition, for they begin to acquire this skill from day one and practice it throughout their lives. Similarly, experts in a particular domain have accumulated much experience with stimuli in that domain and as a result are skilled at recognizing very domain-specific patterns.

Perceptual Chunking in Chess

De Groot (1966) and Chase and Simon (1973a) conducted some of the first psychological studies on expertise. They investigated expert-novice differences in the domain of chess by having chess players reconstruct the positions of pieces on a chess board, after a brief presentation (5 seconds), and found that experts could recall many more pieces than novices could. Chase and Simon compared reconstruction of a meaningfully arranged board to a randomly arranged board and found no differences in the recall accuracy of experts and novices. They concluded that although experts do not have a greater memory capacity than novices, they are superior at reconstructing the board because they perceive the pieces in terms of larger groups or "chunks," in comparison to the weaker players, who perceive the pieces in terms of fewer and smaller chunks. Thus, related pieces are quickly recognized as a cluster by the experts and stored in short-term memory as a single entity (i.e., the chunk label). Because short-term memory is limited to about seven items (Miller, 1956), this chunking strategy greatly increases the number of pieces that can be recalled.

According to Chase and Simon, the ability to quickly encode a complex configuration into familiar chunks is due to chess-specific patterns stored in long-term memory. Thus, whereas novices must deduce relationships and structure through conscious reasoning processes, experts are able to perform the same task more quickly by using unconscious pattern-recognition processes. This difference is analogous to the difference between a beginning reader who carefully attends to each word of a sentence and a skilled reader who processes larger units with very little effort. Chase and Simon (1973b) relate chunking to chess skill by assuming that chunks are associated with particular moves. According to this view, perception is the key to chess still.

This recognition-association theory of chess skill is not universally accepted. Charness (1976) and Frey and Adesman (1976) questioned the assumption that chunk labels are stored exclusively in short-term memory. They replicated the recall results of Chase and Simon (1973a, 1973b) but found no decrement in recall performance in conditions in which rehearsal was prevented (by either instructions or an interpolated task) after presentation of the board and before subsequent recall. If this information was held only in short-term memory it should have been disrupted by the interpolated tasks, just as memory for consonant trigrams suffers a deficit when rehearsal is prohibited. The authors concluded that the information necessary for the recall of the chess positions must be stored in long-term memory.

Other researchers have also questioned the significance of perceptual chunking to expert performance. Holding and Reynolds (1982) found that the quality of chess moves was unrelated to recall performance. In another study, Holding and Pfau (1985) found differences in move choice and evaluation without corresponding recall differences, suggesting a dissociation between the cognitive processes underlying recall for chess positions and those underlying move selection. Therefore, it is possible that skilled pattern recognition is only a prerequisite of expertise and does not discriminate between experts at various levels of skill.

Despite the fact that the recognition-association theory is controversial, the results of Chase and Simon (1973a) have been replicated in a number of other domains. Reitman (1976) investigated expert-novice differences in the game of go and found that experts were better able to reconstruct a meaningful board than novices and that differences disappeared when the board was randomly arranged. Engle and Bukstel (1978) also investigated differences among bridge players of varying skill levels. The authors point out that bridge skill is interesting, for it is more closely associated with recall skills such as card counting than is chess or go. Results again replicated previous findings that experts benefit from meaningful structure but are no different from novices with unstructured card arrangements. Similar results have also been obtained in the domain of music (Sloboda, 1976). These results all indicate that experts do not simply have a larger memory capacity, but must have cognitive structures that are different from novice structures, enabling them to organize and chunk information in a more efficient fashion.

It should be noted that there have been several failures to replicate this seemingly robust finding. Norman, Brooks, and Allen (1989) claim that the recall results have not been replicated in the domain of medicine, and Vessey (1988) failed to find a relation between recall of a computer program and expertise as measured by program debugging time.

In general, it seems clear that skilled pattern recognition is an important component of expertise, particularly in domains such as chess and electronics that rely heavily on perception. In these domains, experts recall meaningful domain-related material much better than novices do, suggesting that the material is perceived by experts in an efficient manner. On the other hand, it seems premature to completely attribute differences in skill (even in the domain of chess) to perceptual chunking.

Conceptual Chunking

Some researchers have argued that chunking is not strictly a perceptual process, but rather includes conceptual components (Cooke, Atlas, Lane, & Berger, 1991). Egan and Schwartz (1979) replicated the traditional expert-novice recall results with electronics technicians and the reconstruction of symbolic circuit drawings. Results indicated that experts attempted to recall random drawings (i.e., a drawing with random placement of circuit symbols) in terms of units that were functionally related. In addition, the experts were faster than the novices on between-chunk transitions and often characterized the entire display in a matter of seconds. These results are difficult for the perceptual chunking hypothesis to explain. Instead, Egan and Schwartz proposed a conceptual chunking hypothesis that would, according to them, overcome many problems of perceptual chunking. The authors suggested that expert skill is more conceptual than perceptual chunking assumes. Rather than perceiving the stimulus in terms of very specific groups of related elements, Egan and Schwartz assumed that experts can identify a concept that is more abstract and characterizes an entire display. This higher level concept could then be used to relate perceptual chunks.

In other studies, Goldin (1978a) and Lane and Robertson (1979) found that recall of chess pieces with a semantic orienting task (e.g., choosing a move) was superior to that orienting task that focused on more perceptual aspects of the board (e.g., counting pieces). These results suggest that the recall effect is not purely perceptual, but must also involve some higher level cognitive structure. Goldin (1978b) proposed that this structure could take the form of a prototype, an average of several related instances. In support of this proposal she found that experts recognized typical positions better than atypical positions.

In another study Cooke, Atlas, Lane, and Berger (1991) presented meaningful board configurations to experienced chess players. The presentations were either preceded or followed by a brief verbal description of the board. Performance in the description-before condition was superior to that in the description-after condition, which suggests that the high-level information must have facilitated the pattern recognition processes by providing experts with expectations. Thus, high-level knowledge, beyond that of a chunk of related pieces, seems to play a role in pattern recognition of experts. This view of conceptual chunking places more emphasis on top-down processes than does perceptual chunking. It is likely that top-down, conceptual chunking processes work in conjunction with bottom-up, perceptual chunking processes.

Conclusions and Implications

One way that experts differ from novices is in their ability to quickly recognize meaningful patterns within their domain of expertise. This ability has been attributed to domain-specific knowledge (i.e., chunks, high-level knowledge, or both) that facilitates recognition, organization, and classification of typical situations and thereby imposes meaning on the environment. This knowledge generates expectations, and consequently time is saved because verification of expectancies can take the place of the deliberate reasoning used by novices.

Implications for the design of expert systems are relatively straightforward. The modeling of the pattern-recognition processes used by human experts could serve to quickly identify the problem structure or to decompose the problem into familiar chunks, thus reducing the solution space. For example, the expert system DENDRAL analyzes chemical structures based on a chemical formula and a mass spectrogram. The mass spectrogram is pictorial in nature and is interpreted by a series of if-then rules relating to the heights of its peaks (Barr & Feigenbaum, 1982). However, it could also be initially analyzed in a more global fashion, just as the expert chess player perceives the chess board. This type of analysis could be implemented using pattern recognition techniques from artificial intelligence. Specific data-driven input could activate relevant patterns, and these could guide further analysis. The results of such a process may lead to a quick initial classification of the mass spectrogram. This type of early classification could focus search by initially restricting it to expected structures.

Simon and Barenfeld (1969) have implemented a model of chess based on the perceptual processes of chess players. Their system, PERCEIVER, was able to simulate the sequence of eye movements of a chess master in a preliminary analysis of the problem structure. PERCEIVER employs largely bottom-up processing, in which attention is directed toward attack and defense relations, and in this way models acquisition of the initial data needed to drive pattern selection processes. Simon and Gilmartin (1973) combined the ideas behind PERCEIVER with EPAM, a network model of learning, in a system called Memory-Aided Pattern Perceiver (MAPP). It uses a pattern-matching process in which salient pieces are identified and matched to stored patterns in memory. Based on this simulation, it was estimated that chess masters have about 50,000 patterns stored in memory. The MAPP simulation did not perform as well as experts in a typical recall task, but the overall pattern of pieces recalled and not recalled compared fairly well with the pattern of expert recall. Based on the research reviewed in this section, it is possible that improvements in these models would be seen with the incorporation of top-down processing.

Holding (1985) discusses several chess programs (e.g., PARADISE) that also use pattern recognition processes as well as knowledge about plans. Unfortunately, these programs cannot compete with chess programs that exploit the advantages of the computer by using brute-force, forward-search strategies. However, this difference is not discouraging given that we do not have a complete understanding of human expertise. Perhaps additional research in this area will offer further clues that can be used to improve chess-playing programs that operate more along the lines of humans.

Meanwhile, it seems that expert systems could benefit to some extent from the information that is already known. It is important for knowledge engineers to keep in mind the fact that human experts can rely on skilled perceptual processes and often do to a great extent. It is not at all easy for experts to introspect about pattern recognition processes, making this type of information particularly inaccessible to the knowledge engineer. In most cases (e.g., medical diagnosis), the perceptual aspects of problem solving are left to the human expert. However, connectionist modeling techniques may offer a possible means of representing expert pattern recognition (Rumelhart & McClelland, 1986). One component of expertise that is more accessible than pattern recognition is knowledge organization.

Memory and Knowledge Organization

Cognitive psychologists assume that information in memory is organized in such a way as to facilitate retrieval and storage operations (but see Landauer, 1975). For instance, in theories assuming a network representation of knowledge (e.g., Collins & Loftus, 1975), related concepts (e.g., canary, bird) are "closer" in memory than unrelated concepts (e.g., canary, truck). That is, memory is organized according to meaning, although there may be other organizing principles as well. Indeed, semantic relatedness has been found to affect processes involving access to memory such as word recognition and sentence comprehension (Meyer & Schvaneveldt, 1971). The purpose of this section is to specify the role of knowledge organization in expert problem solving. Does memory organization of domain-related information change as a function of expertise? If so, what are the specific differences between novice and expert memory organizations? The research presented in this section also emphasizes the importance of considering the representation of expert knowledge, in addition to the content of that knowledge.

Elaboration and Integration

Several models of memory organization propose that declarative knowledge is stored in a graph structure with concepts represented as nodes and relations represented as links between nodes. These links are often labeled with the specific type of relation (e.g., property, superordinate). According to Anderson's (1983) ACT* model, activation spreads from one node to other related nodes via the nearby links. Thus, redundant connections between concepts improve the likelihood that this particular information will be retrieved. Because an expert is able to see more relations among concepts than a novice, concepts in an expert's network should be richer in interconnections than those in a novice's network. These rich interconnections allow experts to retrieve domain-related information very quickly.

On the other hand, there are problems inherent in the preceding account of expert retrieval. When several propositions or facts are learned about one concept, they become linked to that concept. The more propositions that are learned, the greater the "fan," or the number of connections to a single concept. (The network structure that represents these connections resembles a fan.) According to Anderson's (1983) model, a large fan results in dissipation of total activation in the network and thus slow retrieval. Indeed, in experimental conditions in which more facts are learned about a concept, retrieval of a fact takes longer than in conditions with smaller fans (Reder & Anderson, 1980). Results such as these imply, perhaps counterintuitively, that the more that is known about a concept, the longer it will take to retrieve information about that concept. In other words, expertise should lead to slower retrieval of domain-related information.

Hayes-Roth (1977) and Reder and Anderson (1980) point out that the fan effect implies that the date of one's own birthday should take longer to retrieve than the date of Lincoln's birthday, because one surely has stored more propositions about oneself than about Lincoln. Likewise, retrieving information about well-known people should take longer than retrieving information about little-known people. These predictions are clearly counterintuitive. Hayes-Roth suggests that interference effects due to fanning may be reduced by overlearning and strengthening of associations. The strengthening of associations leads to "unitization," or integration of the associated material, which reduces the interference normally caused by several unintegrated items.

This paradox was also investigated by Smith, Adams, and Schorr (1978). These authors found that interference caused by the fan effect was reduced when the to-be-learned material could be thematically integrated. For instance, the predicates of the propositions *Dan entered the restaurant* and *Dan was seated by the waitress* can be meaningfully integrated. Such predicate relations exemplify a script, a stereotyped situation in which events are temporally ordered, resulting in specific expectations. The results of Smith, et al. were taken as evidence that scripted material is organized in memory in an integrated fashion, which thus reduces the fan effect.

Anderson (1981) investigated the effects of prior knowledge on retrieval of information and found, in accord with the fan effect, that when subjects were asked to retrieve facts about fictitious people (e.g., "Carol Norman frequents auctions"), fact retrieval times were slower for those who were presented with prior knowledge about the individual than for those who were not. In other words, Anderson (1981) found the fan effect. On the other hand, individuals retrieved facts about well-known people (e.g., "Elizabeth Taylor is an actress") faster than they retrieved facts about fictitious people in the prior knowledge condition. For the well-known condition, it was expected that retrieval times would be slower than in the fictitious prior knowledge condition, because there should be more preexperimental knowledge stored about well-known people than was given in the brief experimental description about fictitious people. Anderson suggested that some hierarchical fan arrangements make search more efficient in the former case or that perhaps the well-learned facts have a stronger level of activation.

Thus, the meaningful integration of material can overcome retrieval interference caused by the fan effect. However, this same integration can result in another form of interference. Maxwell (1983) points out that integrating materials meaningfully in a script can reduce retrieval interference, but when materials are scripted, unlearned materials that are related to the script interfere in comparison to materials unrelated to the script. This interference can be explained by an elaboration process in which the script-based inferences are associated with learned materials. The result is that unlearned script-related materials take longer to verify (whether or not they had been previously presented) in a recognition task than unrelated items. Maxwell found that scripted materials both reduce interference due to integration and increase elaboration interference.

In summary, some form of integration of information seems necessary to reduce the fan effect. Hayes-Roth (1977) proposes that integration occurs by the strengthening of associated connections with overlearning, whereas Smith et al. (1978) and Maxwell (1983) suggest that integration occurs with thematically related or scripted materials. Anderson (1981) explains fan effect attenuation by either an increase in spreading activation (similar to the strengthening proposed by Hayes-Roth) or a hierarchical organization. Clearly, experts overcome fan effects, and it is possible that this is accomplished by a combination of strengthening of associations and organization of material in memory. In fact, novices do not demonstrate fan effect attenuation, as do experts, which suggests that for novices information is not well integrated (Maxwell, 1983). In the next section, research directed at uncovering some specific memory organization schemes of experts is reviewed.

Organization and Skilled Memory

In most cases, integration or organization is not consciously mediated, but develops naturally over the course of knowledge acquisition. Could conscious strategies be employed to organize information and improve memory? Several investigators have addressed this issue. For instance, Chase and Ericsson (1981) investigated the characteristics of skilled memory of a single individual, "SF." SF was able to drastically improve his memory for strings of digits in 2 years, increasing his digit span from 7 to 80 digits. Verbal protocols from this task illustrate some important characteristics of expertise and memory. SF was able to extend his recall due to the use of a mnemonic scheme (see Bellezza, this volume). For instance, chunks of four number were encoded into running times that were meaningful to SF. In order to keep the order of the chunks straight, SF organized the running times into a hierarchical retrieval structure. Thus, the intermediate knowledge states (running times and similar mnemonic codes) were stored in this structure and could be accessed in order.

In another study, Ericsson and Polson (1988) investigated the skilled memory of a waiter who was able to correctly recall 20 dinner orders from individuals at different tables. They found that the waiter used encoding strategies and retrieval structures similar to those used by SF. Interestingly, they also found that the waiter's skilled memory transferred to orders presented in a different sequence as well as to completely different material (e.g., animals, time, flowers, metals). They pointed out that this result is difficult to explain by memory for purely domain-specific patterns suggested by Chase and Simon (1973a, 1973b). Instead the participants in these studies seem to have a domain-independent structure and set of procedures for encoding and retrieving the material. Thus, these results suggest not only that memory organization can be learned but also that memory characteristics of expertise can be domain-independent.

Categorization

Chunking seems to play a major role in expertise, and research on the categories of experts and novices reveals the organizing schemes underlying their chunks. Categories are important to knowledge representations because they enable generalizations to be made about category members and because they facilitate search through a knowledge base by grouping facts and rules that are likely to be used together. Adelson (1981) found that expert and novice programmers categorize lines of a program in different ways. Novices form categories based on superficial features such as syntax, but experts' categories are based on program function or more semantic aspects.

Similarly, Chi, Feltovich, and Glaser (1981) found that experts in physics categorize problems according to physics principles, but novices according to surface features of the problem.

Not only do experts have different categories than novices, but also experts' categories are less distinct than those of novices. Murphy and Wright (1984) investigated differences in the representation of categories for individuals who differed in levels of expertise in clinical psychology. The clinicians were required to list attributes of three categories of childhood disorders. Findings from this study indicated that experts agreed with each other more on category attributes and listed more features for each category than novices did. However, the experts' attributes often overlapped across two or three categories. Murphy and Wright proposed that the experts may have a more realistic view of the domain than novices do. In cases in which category boundaries are fuzzy in reality, novices are generally trained on prototypical cases. Experience would lead to an increased number of exceptions to the stereotype and thus to a broadened view of categories.

In general, studies have shown that experts and novices tend to organize knowledge differently. Like SF and the waiter discussed above, experts can increase the efficiency of their memories by using a particular organizing scheme. In order to illustrate how knowledge organization relates to specific domains of expertise, some studies on experts and novices in the domain of computer programming will be discussed in the following section.

Memory Organization of Computer Programmers

McKeithen, Reitman, Rueter, and Hirtle (1981) had computer programmers recall ALGOL-reserved words in a multitrial cued-recall task and derived tree structures from the recall protocols that illustrated the organization of the reserved words. They found that expert-novice differences were not evidenced in structural aspects of organization, such as depth of nesting or amount of subjective organization, but rather appeared in the content of the organization. Experts organized the lists of words according to their meaning in an ALGOL program. Novices organized the lists according to mnemonics or surface features. For example, the words IF, THEN, and ELSE were grouped in an expert's representation and IF, IS, OF, and OR in a beginner's representation. Also, experts agreed more among themselves on the organization than did novices.

In a similar study, Adelson (1981) had expert and novice programmers memorize lines of programming code. She found that in comparison to novices experts recalled more, "had" large chunks, "had" greater subjective organization (consistency in recall over trials for a single individual), agreed more with each other on recall order, and chunked according to programs or routines. Novices, however, clustered programming lines according to syntax or surface features. That is, the novices' cluster diagram contained groups of RETURN, IF, and ASSIGNMENT statements instead of groups of functionally related statements.

Cooke (1983) investigated expert-novice computer programmer differences for a set of abstract programming concepts (e.g., SUBROUTINE, CHARACTER, SORT). Individuals were classified as expert, intermediate, novice, or naive programmers based on their programming experience. Proximity ratings for concept pairs were collected and used to derive Pathfinder network and multidimensional scaling representations. (Both of these scaling techniques were discussed previously in the section on psychological techniques for knowledge elicitation.) Experts tended to organize concepts along a structure-function dimension, whereas naive programmers based their organizations on familiarity. For instance, concepts such as SEARCH and REPETITION, which have meanings outside of the programming domain, were located closer in their organizations than concepts such as PARAMETER and ALGORITHM. Interestingly, intragroup agreement was high for both experts and naive groups, but low for intermediate and novice groups. The latter finding suggests that naive programmers, as well as experts, were using some shared, well-defined organization for the basis of their ratings. Perhaps the naive organization is based on shared misconceptions about programming. The lack of agreement evidenced in intermediate and novice groups indicates that no common basis for organizing these concepts in memory exists for these groups.

Based on the above results, it appears that expert and novice programmers differ greatly in their organization of programming knowledge. Adelson (1984) further investigated this difference. She hypothesized that experts represent programs in an abstract form in which emphasis is placed on what the program does, whereas novices represent programs in a more concrete form and focus on the methods by which operations are accomplished. In order to test this hypothesis, Adelson (1984) manipulated the form of an experimental representation (concrete or abstract) that accompanied a program, as well as the type of question (concrete or abstract) to be answered about the program. For instance, in the first experiment the representations consisted of flowcharts that either described the output of the program (abstract) or described how the program functioned (concrete). Likewise, the questions concerned what the program did (abstract) or how the program functioned (concrete). Inappropriate mental sets (e.g., abstract representation and a concrete question) and appropriate mental sets (e.g., abstract representation and an abstract question) were presented to both expert and novice programmers. Adelson found that for inappropriate conditions, experts made more errors than novices on concrete questions, but novices made more errors on abstract questions. For appropriate conditions, this interaction of question type with expertise was not present, which suggests that both groups were able to use the appropriate information that was given to answer the question. Adelson concluded that experts represent programs abstractly, allowing for substitution of a variety of algorithms for the same operations. The finding that experts represent material at an abstract operational level as opposed to a level of concrete procedures may also explain why experts have difficulty describing procedural information.

The results presented here indicate that organization of memory differs for experts and novices, and that this difference is manifested in memory organization principles. Based on recall protocols, proximity ratings, and comprehension tests, expert programmers seem to organize domain-related information at a deeper, more abstract level than novices do. That is, experts organize around programming semantic features instead of surface or syntactic features. Also, experts demonstrate more intragroup agreement, which suggests that this particular type of organization is not idiosyncratic.

Conclusions and Implications

An efficient and task-appropriate knowledge organization appears to underlie many expert processes such as chunking, elaboration, and preliminary classification of a situation or problem. Likewise, artificial expertise may depend on an efficient representation of knowledge.

A major hurdle in the design of more complex expert systems is the organization and retrieval of knowledge from large knowledge bases (Barr & Feigenbaum, 1982). If all of the future uses of a piece of knowledge were known at the time of its storage, facts and rules could be encoded according to particular contexts. However, it is difficult to anticipate all of the future uses of knowledge. People share this problem with computers. Studies have shown that the context in which information is originally encoded is important for the future retrieval of that information (Tulving & Thomson, 1973).

Based in part on the research cited in this chapter, it is apparent that humans rely on a semantic organization of memory in which concepts that are meaningfully related are close in memory. Currently, knowledge is often represented in expert systems in a treelike structure, for example, the MYCIN (Shortliffe, 1976) AND/OR goal tree. The significance of studies on human expertise would be to elucidate the organizational schemes used by humans. For instance, expert associations in computer programming are semantic or functional in flavor. This might be a reasonable way to organize information in a programming expert system. In MYCIN, disease hypotheses could be organized in the tree based on expert physicians' judgments of relatedness. Representation of knowledge could also take into account categorical information such as category attributes or dimensions.

The use of meaning to organize a knowledge base is not a new idea. Barr and Feigenbaum (1982) discuss the use of contexts as a way to organize knowledge bases. In this type of structure, the data base is divided into several different contexts, each representing a different situation in the world. The contexts are organized in a tree structure, and thus contexts can be changed as needed. The idea of contexts is similar to the notion of categories or scripts.

Organization by meaning or context is one way to model human memory in an expert system, making retrieval from large knowledge bases more efficient. In addition, the research reviewed in the section on elaboration and integration emphasizes the importance of other features of memory that are especially important for expertise. Experts are able to handle large amounts of knowledge and are able to overcome some of the problems associated with prior knowledge, such as retrieval interference, by integrating or chunking information in a meaningful way. The idea of contexts is related to the chunking idea, but perhaps it would be useful to chunk within contexts. Strengthening of associations could also be implemented by assigning weights to particular paths of a tree that are closely associated or frequently used together. In general, the strength of the psychological research on memory organization and expertise is its ability to point out the specific organizing principles or expert knowledge. This information is at least equal in importance to knowledge content.

Memory organization underlies all aspects of expertise, including expert problem solving, which is discussed in the next section. Problem solving can be thought of as a search through a problem-state space in order to get from the initial problem state to some goal state (Newell & Simon, 1972). Much of the research on problem solving has focused either on the strategies for searching the problem space or on *problem representation*—the way the problem is represented as a problem space. The research reviewed in this section on expert memory organization is very relevant to issues of expert problem representation. In general, the problem representation serves as a bridge between memory organization and problem solving. Thus, in the next section some research on memory organization will be discussed in order to illustrate its relevance to problem solving.

Problem Solving

The research on computer programming is illustrative of some of the empirical findings relevant to memory and expertise. Similarly, there have been a number of studies on expert-novice differences in physics that address problem solving and expertise. The findings from these studies will be discussed first, followed by some research that addresses specific issues relevant to problem representation and strategies.

Physics Expertise

Chi and her colleagues conducted a series of studies on physics expertise (Chi, Feltovich, & Glaser, 1981; Chi, Glaser, & Rees, 1981). One of the tasks they used was a sorting task in which individuals who had had different degrees of physics training were required to sort physics problems into meaningful categories. Physics experts sorted the problems according to physics principles that were relevant to the problem solution, whereas novices sorted according to key words in the problem, objects mentioned, or other surface features. These results support the claim of differences in memory organization discussed in the previous section. In another experiment, physics experts and novices were asked to elaborate on each of their category labels. Experts' elaborations were more detailed than those of novices and consisted of more procedural or condition-action rules. Furthermore, in another experiment, in which individuals summarized a book chapter (while having access to the book), novices demonstrated a lack of declarative knowledge (i.e., the laws of physics) in comparison to experts. Thus, not only do experts and novices organize their knowledge differently, but experts also have access to *more* knowledge, both declarative and procedural.

Chi and her colleagues performed another experiment aimed at identifying the cues or features that individuals initially extracted from the problem statement. Physics experts and novices described their basic approach to categorizing the problem. Interestingly, the verbal protocols indicated that individuals in both groups attended to identical cues. However, novices used the features to directly classify problems, whereas experts used the features to cue other relevant knowledge that was then used in categorization. The protocols also revealed that novices have difficulty formulating physics problems at a level that is more abstract than specific equations and less abstract than vague procedural statements.

The expert-novice differences discussed to this point are largely differences in amount and organization of physics knowledge. How do these differences relate to problem-solving expertise? The concept of problem representation serves as a bridge between knowledge organization and problem solving. A problem representation is a cognitive structure that, if appropriate, will lead to successful and efficient problem solving. According to Chi, Feltovich, and Glaser (1981), it is derived from the problem itself and from domain-specific knowledge. First, an initial qualitative analysis of the problem allows it to be categorized, activating problem schemata associated with that category. Problem schemata are structures within the knowledge base that embody the problem category and knowledge associated with that catagory. In other words, schemata are ways of organizing the knowledge base by subdividing it according to problem category, and in this sense they are similar to conceptual chunks, which were discussed earlier. After a schema is verified as the appropriate one, it is then used to build the problem representation for that problem. Thus, the problem representation is an instantiation of a problem schema. Chi, Feltovich, and Glaser concluded that the experts' rich and sophisticated categories must be indicative of rich problem schemata and, consequently, a useful problem representation.

Chi and her colleagues also found expert-novice differences not only in the way that problems were represented but also in the search strategies used to navigate through the problem space. Think-aloud protocols taken while individuals solved physics problems indicated that novices used a means-ends strategy by which they searched for principles or equations that would directly satisfy the unknowns. If values were unknown, then finding these values would be made the subgoal. On ther other hand, experts generally worked forward, not backward, generating known values from the given variables until the solution was reached. However, as will be discussed later, research on strategies in other problem domains has indicated that forward search is not always associated with expert problem solving (Simon & Reed, 1976).

Larkin (1981) simulated these strategic differences. Her novice simulation, "Barely ABLE," uses a general means-ends strategy and selects principles from the knowledge base that will satisfy the unknowns. The expert simulation, "ABLE," learns by working problems. Gradually, it builds up knowledge about each physics principle that facilitates the recognition of specific situations that require the use of that principle. Thus, the context of the problem is directly related to certain principles. These models were able to account for experts' and novices' selection order of principles and the time needed to solve the problem.

The research reviewed in this section highlights some major differences between experts and novices in physics. Knowledge-based differences exist in that experts possess more knowledge than novices and organize it differently. It is hypothesized that these knowledge-based differences underlie differences in the specific representation of a physics problem, as well as the approach to problem solving. Although all individuals see the same problem features as relevant, experts tend to take problem representation one step beyond the literal problem and relate these cues to physics principles. In addition to knowledge-based differences, strategic differences also exist in the direction in which the solution space of a problem is searched. In the next section some specific findings on problem representation and problem solving strategies are reviewed.

Problem Representation and Strategies

In their work on problem perception, Schoenfeld and Herrmann (1982) overcame a methodological problem common to expert-novice studies by testing the same students at different points in their development (i.e., they used a longitudinal approach instead of the typical cross-sectional approach). Students sorted math problems before and after taking a class. In one condition, students were taught math-problem-solving techniques in the class, and in another (the control condition) they were taught computer programming techniques. Findings based on this longitudinal design replicate previous physics findings discussed above: The experimental group, which had taken the problem-solving class, demonstrated superior performance in comparison to the control group. In addition, the experimental group sorted math problems based on the deep structure of the problem (math principles), and the control group sorted problems based on surface features, such as terms and objects mentioned in the problem. Sorting results were also compared to problem classification by expert mathematicians, and it was found that students in the experimental condition more closely approximated the sorting strategy of experts than did students in the control group. These results support the view that problem representation changes as a function of expertise. Furthermore, previous differences in problem representation cannot solely be attributed to other factors that vary with expertise, such as age or aptitude.

As previously mentioned, early research on problem solving centered on the strategies by which a problem solver searched through a problem-state space. However, increasing interest in expert problem solving in complex domains led to the realization that there was more to problem solving than sophisticated search strategies (Glaser & Chi, 1988). Consequently, the emphasis on general strategies was replaced by a focus on domain-specific knowledge. It seems that the shift in emphasis caused many to lose sight of the fact that strategic differences in experts and novices do exist. In fact, although early studies of de Groot (1966) in the domain of chess indicated that there were no search differences between experts and novices, Charness (1981) later reported a study in which depth of search in chess was greater for experts than novices.

Simon and Reed (1976) have studied the issue of strategies and shifts in strategies with additional knowledge or experience. They investigated and simulated strategies that humans used to solve the "missionaries and cannibals" problem. In this problem, the goal is to get a group of people across a river in a boat of limited capacity without letting the number of cannibals exceed the number of missionaries. Like Schoenfeld and Herrmann (1982), Simon and Reed (1976) investigated strategy changes with expertise longitudinally. Specifically, they looked at how strategies were affected by a hint in the form of a subgoal (the number of people on each side of the river at an intermediate solution stage) and by successful completion of one trial. Both of these factors caused shifts in strategies from a simple balancing strategy in which the number of missionaries and cannibals on both sides of the river is balanced, to a means-ends strategy. The simulation was also able to successfully predict the steps used to solve the problem. These results suggest that strategies differ not only between individuals but also within the same individual. Problem-solving strategies are likely to become more efficient with experience.

In addition to search strategies, high-level strategies such as those involved in planning have also been investigated. Much of this work has been done in the domain of computer programming. Soloway, Ehrlich, Bonar, and Greenspan (1982) and Soloway, Adelson, and Ehrlich (1988) present empirical evidence that indicates that the identification of an appropriate plan is crucial to program comprehension and composition. Experts, for instance, can use their knowledge of plans to "fill in the blank" of a computer program, whereas novices cannot. However, if the program is not planlike, then the experts cannot make use of their plan knowledge, and performance decreases to the level of novice programmers. In addition, errors that the experts make in these cases tend to be consistent with a plan. Thus, strategic differences occur not only in the general method and direction of search through the problem space but also in high-level, domain-specific plans for achieving the goal.

More often than not, in the design of an expert system a general strategy such as backward chaining is used to search the data base. However, experts may differ in the strategies they employ, depending on the problem situation. In solving physics problem, experts search in a forward direction, but in solving the missionaries and cannibals problem, experience leads to a backward searching strategy. Strategies are dependent on the problem as well as expertise. Therefore, care should be taken in expert system design to reveal strategies used by experts in that domain.

Conclusions and Implications

In general, differences in expert and novice problem solving can be attributed to (a) the representation of the problem and (b) strategies or heuristics used to solve the problem. Although search strategies and heuristics have played a role in expert systems, more importance should be placed on subtle strategy shifts discussed by Simon and Reed (1976). For instance, MYCIN (Shortliffe, 1976) relies exclusively on backward chaining, but there may be some situations in which the data are more reliable or available, making forward chaining more appropriate. More important, the strategies used by experienced physicians should be considered in this decision.

The initial perception and classification of problems appear to play a major role in human problem solving. A problem's representation could be matched to previous problem representations or even to abstract schemata. Solutions could then be generated and tested based on the closest matches. Research on expertise sheds some light on the problem representations of experts and thus provides useful clues for the design of expert systems. Again, the importance of problem representation to expert problem solving suggests that knowledge engineers should strive to elicit not only the content but also the structure or organization of an expert's knowledge. The psychological scaling techniques discussed earlier provide a means of eliciting this type of information.

Problem solving is a high-level cognitive process that involves a number of subprocesses such as representation, planning, and search. In addition, the problem solver is often required to select from a number of alternatives and in doing so must make judgments that involve factors such as likelihood and frequency. There is a large literature on this specific component of problem solving (Arkes & Hammond, 1986; Hogarth, 1987; Kahneman, Slovic, & Tversky, 1982; Nisbett & Ross, 1980). The following section is devoted to a review of the findings from this literature on decision making that are most relevant to expertise.

Decision Making

Research on pattern recognition, memory, and problem solving has centered on the cognitive structures and processes that underlie behavior in these areas. Thus, expert-novice differences are described in terms of differences in structures and processes. Interestingly, much of the research on decision making has a different flavor. For the most part, this research has focused on the decision-making errors that people make, along with possible explanations for their deviations from optimal responses.

It is possible to evaluate decision-making performance, because we have formalisms such as Bayes' theorem that we can use as a standard with which to compare human performance. Thus, there is a focus on decision-making performance that often supersedes concerns of structures and processes that underlie this performance. Additionally, as Johnson (1988) points out, the research on expertise generally assumes superior performance on the part of the expert. As a consequence, there are relatively few studies that investigate the decision-making performance of experts (but see Phelps & Shanteau, 1978). In the few studies that have examined this issue, it has been found that although experts may surpass novices in their decision-making performance, they are by no means optimal decision makers in the sense that their predictions do not approach the performance of a linear regression model (e.g., Johnson, 1988).

In this section, the work on decision-making heuristics and biases will be reviewed, along with hypothesized explanations for decision-making limitations. Insofar as these heuristics and biases are common to all individuals, regardless of expertise, they are applicable to the design of expert systems. Whenever possible, results from studies on expertise and decision making will also be discussed.

Heuristics and Biases

It has been widely observed that humans, even experts in a field, often fail to make optimal decisions. *Optimal* in this sense refers to decisions based on statistical probabilities and formalisms such as Bayes' theorem. It is assumed (possible incorrectly) that the formalism that is selected as a standard leads to optimal outcomes. Instead of using these optimal strategies, decision makers tend to rely on a variety of heuristics or rules of thumb that result many times in less-than-optimal decisions.

The sense in which heuristics are discussed in the decision-making literature is slightly different from the sense in which they are referred to in artificial intelligence and cognitive science. To the latter disciplines, heuristics are rules that can be applied to an artificial system to prune search or evaluate alternatives. In this sense, heuristics are helpful, and sometimes necessary, to avoid combinatorial explosions in a search space. On the other hand, heuristics in the recent human decision-making literature are often associated with biases, and these biases are emphasized in the psychological literature. Generally, heuristics lead to satisfactory solutions, but in some instances they do not (Simon, 1981). In these circumstances, the heuristic would be manifested as a bias.

It is important to understand heuristics and biases in order to overcome human limitations and to possibly exploit expert system resources in these areas. Of course, the mistakes that humans make when using these heuristics should be avoided in artificial expertise, but in many cases these strategies may aid in constraining search, in a preliminary evaluation of alternatives, or in a sophisticated guessing strategy when no other information is available. In the next section, two common heuristics, representativeness and availability, are discussed; this discussion is then followed by some possible explanations for these findings, including the possibility that they are the result of artificial tasks.

Representativeness and Availability

When individuals are asked to judge the likelihood that an object belongs to a certain class, they often use the representativeness heuristic (Kahneman & Tversky, 1973). Instead of basing predictions on prior probabilities, sample size, and validity of information, people frequently rely on the similarity of the object to a class or, in other words, how representative the object is of the class.

Kahneman and Tversky (1973) investigated the predictions made by people under varying sets of circumstances. When people judged the likelihood that an individual belonged to a certain college, based on a written description of the person, Kahneman and Tversky (1973) found that the judged similarity of an individual to a group correlated with judgments ($r = .97$) much more than the base rate correlated with judgments ($r = -.65$). For example, when given a description of an individual and asked to judge the likelihood that the person is an engineering or humanities major in college, people will consistently rely on the description and ignore the fact that there are more humanities majors than engineering majors. If the description fits the stereotype of an engineering student, people will judge the individual to be an engineer regardless of the known validity of the description and the base rate. Experts are not immune to this bias. Performance data and verbal reports collected while experts made decisions about applicants for medical internships or predicted stock prices have suggested that experts ignore base rate information and focus on case-specific data instead (Johnson, 1988).

In addition to base rate, the representativeness heuristic is also relied on at the expense of several other relevant variables. For instance, people ignore sample size when making a prediction using this heuristic (Tversky & Kahneman, 1982). They instead have the tendency to place too much faith on small sample sizes. Furthermore, people do not always regress their predictions toward the mean as uncertainty increases (Tversky & Kahneman, 1982). In short, the representativeness heuristic emphasizes similarity, which often results in the neglect of other relevant information.

The availability heuristic is used to make judgments about frequency or the probability that an event will occur. Tversky and Kahneman (1982) point out that when people are asked whether there are more words that start with the letter *r* or in which the third letter is an *r*, they typically, yet erroneously, choose the first option. Tversky and Kahneman (1982) relate this to the ease of searching memory for words that start with *r*. Generally, the availability heuristic is the tendency to rely on instances that are easily retrieved from memory.

Hypothetically, biases result from applying the availability heurisic when the most frequent event is not the most salient, familiar, or easy to imagine. A bias called illusory correlation can also be explained by inappropriate application of the availability heuristic. People tend to overestimate the frequency of co-occurrence of two events (i.e., illusory correlation) in situations in which the association between two events is very strong or salient (Tversky & Kahneman, 1982). Thus, in this case availability of the association is mistaken for correlation.

Why do people retain these decision-making biases? Couldn't they refine their heuristics to overcome some of these limitations? Several explanations have been offered for the existence of decision-making biases.

Explaining Suboptimality

First, it should be noted that lab experiments, the source of most of these findings, place people in artificial situations in which errors are likely to occur (Ebbesen & Konečni, 1980). Often, the information is limited to a vague description, and artificial constraints are put on the problem that reduce the ability of the human to use different and possibly correlated information to make a decision. Thus, the extent of the biases may be exaggerated relative to the successful use of these heuristics in the real world. Nevertheless, it is important to recognize the limits of the human information-processing system in order to understand the mechanisms behind its strengths.

Another explanation for suboptimality concerns the absence of feedback or knowledge of results. In many real-world situations in which a prediction or judgment is made, the correctness of the judgment is never discerned. For instance, in selecting candidates for a job, one might select certain candidates based on the prediction that they will perform well in the job. How can one decide whether the prediction is accurate? First of all, the group of individuals not selected will never have the chance to perform, so there is no comparison group of candidates who were not selected. Of course, the percentage of successful candidates selected could be compared to the percentage of unsuccessful candidates selected, but success could only be defined in terms of the population selected, not the total population. In addition, a "Pygmalion" effect, or self-fulfilling prophecy may occur, in which a person placed in a situation in which he is expected to succeed will succeed. Thus, the selection process may actually cause the success. Problems such as these obscure knowledge of results, making it difficult to discern correct predictions.

Einhorn (1982) discusses other underlying causes of the use of suboptimal heuristics. Basically he feels that learning is usually based on experience and induction, whereas rules such as Bayes' theorem are not easily induced from experience, but need to be learned deductively. Because this information does not directly follow from experience, it is more difficult of grasp.

In addition, there is evidence that information in memory is organized by content, not by general task structure or problem type. Rumelhart and Norman (1981) support this view in a discussion of the Wason "four card" problem. The problem consists of four cards with letters on one side and numbers on the other. In addition, a rule (e.g., if a card has a vowel on one side, it has an even number on the other) accompanies the cards, which are exposed on one side only. The task is to decide which of four cards would need to be turned over to test the validity of this rule. People frequently turn over the wrong cards (e.g., they turn over the card with the consonant on the face, demonstrating the reasoning fallacy "denial of the antecedent"). However, when this problem is stated in terms of a required signature on the back of an invoice totaling over $30, performance greatly improves. Presumably because memory is organized by content of the problem (surface features) and not problem structure, people have a hard time abstracting the task structure that is related to optimal rules. Thus, even if they learn Bayes' theorem, they may not notice its applicability to specific situations.

Kahneman and Tversky (1982) summarize the effect of context on decision making by saying, "It appears that the actual reasoning process is schema-bound or content-bound so that different operations or inferential rules are available in different contexts" (p. 499). On the other hand, the abstraction of task structure seems to be an area in which experts excel (e.g., Chi, Feltovich, & Glaser, 1981). Thus, experience in a particular domain seems to result in a transition from a content-dependent problem representation to a content-independent or generic representation. Differences in problem representations would predict that within the domain of expertise, experts should not show decision-making biases to the same extent as novices. Any existing expert biases might be attributed to some remaining context-dependent features of the problem representation.

Conclusions and Implications

Research on decision making has uncovered some heuristics and pitfalls of human decision making. As mentiond before, these heuristics are often successful in the real world. Often, the events that come to mind most readily (i.e., the availability heuristic), *are* the most frequent events. At the least, these heuristics enable individuals (including experts) to make sophisticated guesses without having to engage in complex calculations.

Perhaps it would be useful to extend these heuristics to the expert system environment for use in situations in which evidence is incomplete or a quick solution is required. For example, the representativeness heuristic could be used to make a preliminary hypothesis based on little evidence. A hypothesized disease could be generated based on the similarity of the patient's symptoms to symptoms associated with a particular disease. This type of heuristic could also be used to implement data-driven activation of schemata. It may be useful to quickly classify a situation without a full-scale analysis of it in order to benefit from top-down processes. The important difference between the artificial and human expert in this situation would be that the artificial expert would not stop when a similar match was found among the schemata. Instead, it would verify and test the hypothesis using more information as needed. The representativeness heuristic would merely be an initial stab at the problem.

Finally, humans do possess some decision-making skills that have not been successfully reproduced by current artificial intelligence technologies. Dawes (1982), in a discussion of the superior performance of linear models over expert judges, points out that the human expert is not good at integrating large amounts of information but is good at selecting important predictor variables and coding these in some meaningful way. Johnson (1988) similarly reports data that indicate that experts make better predictions than novices because they focus on important cues. Even so, experts were surpassed in predictive ability by a linear regression model.

Integration of information does not seem to be the only weakness inherent in expert predictions. Often, the cues that they select are case-specific, at the expense of more mundane but highly predictive cues such as base rate information. For example, Lawrence (1988) investigated judicial decision-making expertise and found that experts' sentences were more individualized that the novices' sentences, which suggests that the experts attended to exceptions and atypical information. Johnson (1988) argues that the linear model is superior to experts not only because of its superior integration capability but also because it includes the mundane information that the expert ignores. On the other hand, linear models ignore the rare but often important cues to which the experts attend.

These results have implications for the allocation of duties between expert system and human expert. Experts are poor at integrating information, and thus the expert system should be responsible for this task. Furthermore, it would be unwise to attempt to elicit this type of information from a human expert. On the other hand, experts are generally good at identifying the important variables that are to be integrated. These variables should be identified in the knowledge elicitation phase of expert system development. Particular emphasis should be placed on very rare but important variables, for it is precisely this information that differentiates expert decision making from a linear model.

The elicitation of cues that affect predictions is not always straightforward. The knowledge engineer typically asks the human expert to identify the important pieces of information that affect the prediction in question. At this point, the knowledge engineer should keep in mind that humans have the tendency to ignore mundane information, and consequently the engineer should prompt the expert for these variables. One approach to this problem involves an iterative process of linear modeling of the expert's decisions. A situation could be presented to an expert, who is asked to make a prediction about the outcome and to identify the factors that were considered in making this prediction. Then, a linear model based on these factors could be used to predict the outcome of future situations. Each time, the expert could also be asked to make a prediction, and with each iteration new factors could be identified and the model refined. Finally, because the expert is so good at identifying unusual cues that may be overlooked during system design, the expert should always be allowed to interact with the system in order to augment the decision-making process by adding "special-case" information.

The topics that have been reviewed thus far (i.e., pattern recognition, memory, problem solving, and decision making) involve cognitive processes that differ depending on expertise within a domain. Given that domain experts are different from novices in many ways, to what can we attribute these differences? Individual differences such as motivation and aptitude may account for some of the difference, but a large contributor to expertise is simply a lot of practice or experience within the domain. In the next section, the process of acquiring experience (i.e., learning) is reviewed, with an emphasis on domain expertise.

Learning

Learning enables an individual to make the transition from novice to expert. Thus, expert learning processes are not normally compared to novice learning processes in the way that expert and novice problem-solving strategis are compared. Learning is usually thought of in more general terms, as a process that is invariant among individuals who differ in degree of experience. However, some have proposed that learning mechanisms and strategies evolve as one "learns to learn" (Langley & Simon, 1981).

Others have demonstrated that changes in the acquisition of domain-related knowledge occur with expertise (Chiesi, Spilich, & Voss, 1979). In a series of experiments, Chiesi et al. presented baseball passages to individuals who possessed either high or low knowledge of baseball High-knowledge individuals demonstrated superior recall and recognition of the baseball passages in comparison to the low-knowledge individuals. This superior performance was explained by the mapping of new information onto an existing knowledge structure consisting of goals, states, and actions. Thus, the expert-novice memory differences discussed previously affect learning by enabling experts to assimilate new knowledge into an existing, domain-appropriate structure (see also Bellezza, this volume).

Although experts may be more efficient at acquiring new information than novices are, other studies have shown that learning or experience is not always associated with improved performance across all tasks. Recently, Lesgold et al. (1988) investigated X-ray diagnosis by individuals who varied in level of expertise and found that performance (interpretation of X-rays) was not monotonically related to experience. Unlike expert radiologists, radiologists at intermediate levels of experience (i.e., third- and fourth-year residents) were less likely than novices (first- and second-year residents) to correctly interpret the X-rays. Not only does this result emphasize the importance of evaluating performance when identifying a person as an expert, but it also suggests that learning (as measured by performance on specific tasks) may be more complex than continuous refinement of structures and processes. Lesgold et al. hypothesize that early learning is perceptual and is qualitatively different from later cognitive learning, although the late stages are dependent on the products of earlier perceptual learning. That is, experts invoke perceptual schemata to interpret an X-ray, but unlike intermediates, experts test this selection rigorously. Intermediates, on the other hand, sometimes reshape their perceptions to fit the schema, whereas novices just rely on their perceptions. Thus, the decline in performance at intermediate levels of expertise is explained by the shift from perceptual learning to cognitive learning. More research needs to be done in order to identify and explain changes in learning processes that occur with experience in a domain.

The remainder of this discussion of learning focuses on various explanations of the learning process. In these theoretical accounts, the learning mechanisms themselves do not differ as a function of the amount of prior knowledge. The first section addresses the issue of the creation of new knowledge representations through analogical reasoning, and the next two sections deal with restructuring existing knowledge.

Learning by Analogy

A number of investigators have suggested that analogies play a major role in learning (Gentner, 1983; Gentner & Gentner, 1983; Gick & Holyoak, 1980; Gick & Holyoak, 1983; Rumelhart & Norman, 1981; Vosniadou & Ortony, 1989). According to Rumelhart and Norman, learning by analogy enables new knowledge representations or schemata to be created from existing ones. The authors view schemata as procedures that detect features in the environment that are relevant to the concept that the schema represents. Initially, similarity between an unknown situation and a known situation (i.e., recognition) causes activation of an existing schema. A new schema is then modeled based on the existing schema. Rumelhart and Norman view additional learning as a "fine tuning" of the schema in which differences between it and the parent schema from which it was derived are filtered out. They stress that this type of learning is rapid but vulnerable to errors. For example, misconceptions in learning a text-editing system are often based on incorrect extensions of a particular analogy, such as, "The computer is like a typewriter" (Halasz & Moran, 1981).

Learning by analogy could be a useful addition to automated knowledge acquisition in expert systems (Carbonell, 1986; Klein,

1987). In situations in which a new type of problem or new concept is to be learned, an analogy based on established knowledge in the knowledge base may guide learning in a more efficient way. For instance, expert systems in medicine, after acquiring a knowledge base that covers several diseases, could use analogical reasoning to learn additional diseases. Either the appropriate parent disease schemata could be pointed out by a knowledge engineer, or preliminary learning of basic facts could trigger activation of the most similar disease schema. As soon as an existing schema has been selected, experience in diagnosing could lead to a "fine tuning" of a new schema based on misconceptions and correct inferences made using the parent schema. Thus, this type of learning mechanism enables the expert system to generalize across various problem domains. After a knowledge base on infectious blood diseases is established, the same knowledge could be extended through analogy to liver or heart disease. Of course, the more similar the new domain is to the old, the less fine tuning will be required of the new schemata. Developments in case-based reasoning, in which specific cases are learned and are referred to as new cases are presented, make use of some of these ideas (Klein, 1987).

Whereas learning by analogy results in the creation of new knowledge structures from old ones, other theories of learning focus on the restructuring of existing knowledge. In the next two sections, two different views of this restructuring process are reviewed.

Changes in the Unit of Cognition With Learning

Modification of an existing memory structure can occur through reorganization of knowledge representations or through the addition or deletion of concepts in the structure. Knowledge assembly theory (Hayes-Roth, 1977) assumes that knowledge is represented in a network with nodes and links, and that knowledge is activated by a spreading activation process similar to that proposed and empirically supported by Collins and Loftus (1975). In addition, it is assumed that there are qualitative changes in structural and procedural aspects of knowledge with experience. According to this theory, knowledge is initially represented in low-level memory structures that are activated in an all-or-none fashion. Learning causes these basic representations to become strengthened and associated with other related representations. Gradually these associations are strengthened to the point of "unitization," which causes the entire configuration of representations to be activated in an all-or-none fashion. Therefore, learning acts to change the cognitive unit and in general causes related concepts to be integrated.

Hayes-Roth (1977) relates the knowledge assembly theory to many empirical findings, including the chess (Chase & Simon, 1973a) and go (Reitman, 1976) results discussed previously. In terms of Hayes-Roth's theory, chess and go players initially represent knowledge as separate but associated low-level units (e.g., a single chess piece, explicit definitions, and simple rules). As the players gain experience, associations among related pieces are strengthened and eventually encoded in an all-or-none fashion. Thus, the knowledge assembly theory provides an explanation for the chunking differences found among experts and novices.

The knowledge assembly theory is also suggestive of a method for implementing learning in automated knowledge acquisition systems. The actual representation of knowledge could change as the system learns. As relations are strengthened between concepts, the concepts could be unitized and, as a result, accessed together in the future. Thus, with increasing experience the system could have access to clusters of related concepts. In addition, Hayes-Roth's theory has implications for knowledge elicitation. Specifically, it suggests that a fact or rule elicited from a particular knowledge assembly should cue other facts and rules that are also in that assembly. On the other hand, if integration of the cognitive unit happens at the expense of the individual facts and rules that make up the unit, then it should be difficult to elicit the knowledge that comprises the knowledge assembly. The theory of skill acquisition described in the next section makes the latter prediction.

Learning and Knowledge Compilation

Anderson (1983) divides learning processes into three stages. The first stage is the interpretive stage, in which factual information or declarative knowledge related to the domain is learned. In order to be used, however, this knowledge must first be interpreted by general purpose production rules. Productions take the form of if-then rules in which the *if* part (i.e., the condition) must be satisfied in order for the *then* part (i.e., the action) to occur. These particular productions are domain-independent and apply general strategies such as goal decomposition and backward chaining to the declarative knowledge. For example, in solving geometry proofs, an interpretive production might be, "If the goal is to do a list of problems, then set as a subgoal to do the first problem in the list."

Anderson refers to the second major stage of learning as *knowledge compilation* and thus welcomes the analogy to computer compilation. This stage involves the grouping of individual domain-specific productions. According to Anderson (1983), the use of productions, and consequently this second stage, is significant in the development of expertise. Knowledge compilation involves the two subprocesses of *composition* and *proceduralization*. Composition combines a group of related single productions into one large production. This macroproduction can be thought of as a chunk of productions, and so the process of composition is similar to Hayes-Roth's (1977) unitization of knowledge representations. The second subprocess, proceduralization, involves the direct embedding of domain-specific factual knowledge into a production rule. The proceduralization process is the point at which Anderson's and Hayes-Roth's theories seem to diverge. Whereas unitization and compilation processes simply group related knowledge, Anderson's proceduralization process actually changes the original form of the knowledge. Thus, the knowledge may be (but doesn't have to be) inaccessible in its original form (e.g., we may be able to dial a familiar phone number but unable to verbally report its declarative form). The whole compilation process results in faster performance, decreased verbal rehearsal, and smoother performance due to decreased step-by-step processing. Domain-specific facts are proceduralized and integrated into macroproductions.

Finally, Anderson's (1983) third stage involves a fine tuning of the task-specific procedure. A generalization process increases the breadth of the production rules, whereas discrimination narrows the scope to which they apply. Strengthening results in increased activation of successful production rules and weakened activation of unsuccessful ones.

Conclusions and Implications

Several implications can be drawn based on the above studies. The concept of knowledge representation underlies all of the above research, but the research differs in the process that is assumed to act on this representation. In learning by analogy, processes of schema activation and fine tuning occur to acquire new knowledge; in Hayes-Roth's (1977) unitization process, strengthening and association integrate information in the representation; in Anderson's (1983) compilation process, composition of rules and proceduralization operate on knowledge to integrate it and alter its declarative form. Just as learning theories assume different processes to account for the novice-to-expert transition, a variety of expert system technologies approach knowledge acquisition differently. Knowledge acquisition in expert system development can be achieved through (a) the elicitation of knowledge from a human expert or other information source, (b) the refining of the knowledge base through human interaction with the system via explanation facilities, and (c) the implementation of machine learning techniques. Implications of learning research for each of these approaches is discussed in this section.

Anderson's (1983) theory of skill acquisition provides a possible explanation for some of the difficulty that experts have in verbally expressing their knowledge. That is, knowledge in the original declarative form is proceduralized and grouped with other relevant knowledge. Whereas this process leads to the speed and efficiency (i.e., automaticity) characteristic of skilled behavior, it is likely to be at the expense of the original knowledge. In other words, after a body of knowledge is compiled or automatic, it may be difficult to decompose into its original constituents. For instance, after we learn how to ride a bike or play tennis, it is difficult to demonstrate the skill in a step-by-step fashion. Indeed, in order to do this, most people run through the compiled procedure, attending to specific components in an attempt to isolate them. Similarly, the elicitation of knowledge from experts may be difficult because so much of the knowledge is in a compiled form.

Although difficult, the elicitation of compiled procedural knowledge is not impossible (but see Gordon, this volume). One approach to the problem is to rely less on experts' verbal reports of behavior and more on observation of the actual behavior. Procedural scaling representations can be obtained from data in the form of co-occurrence frequencies of actions. Still, in some cases, the behavior may be so automatic that isolating events becomes difficult. In this situation it may be beneficial to observe experts perform the task as they train novices. It is also possible that novices, because they are still in the declarative stage of learning, would be able to verbalize some of this knowledge.

Another approach to knowledge acquisition is the continual modification of the expert system through interaction with a human expert. The idea of iteration, or a continual refining of the knowledge base, is also common to the various theories of learning discussed above. Explanation facilities play an important role in the interactive transfer of expertise from a human expert to an expert system. Such facilities describe the expert system's chain of reasoning to the human expert. This information is critical to the expert for the locating bugs in the system and for modifying incorrect or inconsistent rules or facts.

According to Hayes-Roth, Waterman, and Lenat (1983), a problem with current explanation facilities is the lack of abstracted knowledge or basic principles. In many systems, explanation involves simply listing the rules used to get to a solution. However, an enumeration of production rules is often cryptic and more complex than needed for an adequate explanation. Often an abstract account of the rules or principles provides a better explanation. This situation is analogous to the teacher-student situation in which the teacher abstracts the concept and explains it at a more general level because a detailed explanation would only serve to confuse the student. The lack of basic principles and abstracted knowledge is partially due to the fact that expert system development either ignores human expert processes or is initiated at the expert level without consideration of the learning process.

Cross (1983) added a body of commonsense knowledge to an existing mathematical expert system in an attempt to improve its explanations for math solutions that typically involved complex computations. The commonsense heuristics enabled the system to provide qualitative reasoning based on the more complex knowledge. Cross notes that this type of qualitative reasoning, as opposed to enumeration of equations and rules, was similar to the way humans learn. One problem inherent in this approach is the identification of the appropriate level of knowledge. It is also likely that one level of qualitative reasoning would not be sufficient for users at all stages. On the other hand, good teachers adjust their lectures according to the particular audience, so it seems reasonable to design expert systems with similar flexibility.

Much of the research on learning is also directly relevant to the machine learning approach to knowledge acquisition. These techniques provide an alternative to the elicitation of knowledge directly from a human expert. Instead, machine learning techniques are employed that enable the system to learn new rules or facts based on the facts and rules in the current knowledge base and some general learning algorithms (Michalski, Carbonell, & Mitchell, 1986). In a similar vein, some have argued that the intuition of experts is best modeled using connectionist techniques (Dreyfus & Dreyfus, 1987). In this situation, as soon as inputs and outputs to the system are specified, the machine can "learn" the input-output relationships by manipulating connection weights in a complex pattern of activation. In the discussion of pattern recognition, these connectionist techniques were mentioned as a possible approach to modeling this component of expertise.

Unfortunately, the knowledge elicited from the human expert may look nothing like the knowledge generated by machine learning or connectionist techniques. In fact, it is difficult to abstract any rules, concepts, or features from a connectionist model. This mismatch between human and machine representation may lead to communication problems between the human and the expert system interface along the lines of the explanation problems discussed above (Michie, 1982). One solution to this mismatch would be through the development of machine learning algorithms that more closely model human learning processes. Of course, more research is needed in order to discover the mechanisms that underlie human learning. An alternative would be the development of better interfaces between the human and systems built via the machine learning approach. Thus, instead of modeling human learning in order to generate a knowledge base that can be understood by the human expert, human explanation would be modeled and would act as a translator between the cryptic knowledge base and the human expert. In fact, it is likely that there is not a perfect correspondence between the tacit knowledge that results from human learning and the explicit explanations given for decisions based on this knowledge.

In summary, by gaining a better understanding of human learning, the knowledge acquisition process can be facilitated in a number of ways. Research and theories of learning suggest possible reasons for some knowledge elicitation difficulties, as well as some ideas for overcoming these difficulties. However, research on human learning seems most applicable to a very different approach to knowledge acquisition. Instead of eliciting knowledge from a human, the machine can acquire the knowledge on its own or with the aid of an expert. This approach may be successful in situations in which human knowledge is most difficult to elicit, although the success is at the expense of a knowledge base that can be easily understood by human experts.

Conclusions

Throughout this chapter the implications of human expertise for expert system design and knowledge elicitation have been summarized, with an emphasis on attending to both strengths and limitations of expertise. The assumption underlying this emphasis is that a descriptive approach to knowledge engineering is necessary. This is not to say that optimal strategies should be avoided in expert systems but instead that there are many advantages to modeling human expertise. Also, it may be necessary to simulate human cognition in an expert system if the system is ever to get some tasks done at all, let alone done well.

Elaborative interference and decision-making biases were two of the limitations discussed. Elaboration occurs when inferences are made about information that is learned, and thus related new facts get confused with old facts. This is a limitation when a task involves distinguishing new facts from old facts. However, the processes of elaboration and drawing inferences are also important features of human intelligence (Anderson, 1985). This is one example of how the study of limitations can reveal mechanisms underlying general intelligence.

The study of biases in decision making is another example of how powerful heuristics can be revealed by analysis of limitations. Although one would not want to model biases in an expert system, the heuristics could be useful when the appropriate information is unavailable or when speed is of critical importance. Perhaps it is impossible to model heuristics without also modeling the associated biases. At the least, users of an expert system should be made aware of the biases that occur as a result of using these rules of thumb. The decision to incorporate a heuristic should be based on the costs and benefits associated with it.

Not only is an understanding of cognitive limitations important for revealing cognitive strengths, but it is also necessary in order to improve artificial systems and overcome these weaknesses. The purpose of expert systems is to improve upon human decision making and problem solving, especially in areas in which humans make mistakes. Thus, expert systems should eventually be designed to produce optimal solutions whenever possible. However, in order to overcome human weaknesses, it is necessary to discover these weaknesses and understand their causes. In other words, to write a better computer program, it is necessary to locate the bugs in the old one and to discover the causes of the bugs.

Furthermore, an understanding of the representation of the facts and rules used by an expert is required in order to communicate with an expert on his or her terms. For example, the exhaustive structure generator in DENDRAL and the backward chaining of MYCIN do not necessarily mimic the processes used by chemists and diagnosticians in solving problems. Thus, when an explanation of this process is required, it is not very meaningful to the domain expert (Aikins, 1983; Barr & Feigenbaum, 1982; Ellis, 1989). This problem has become even more obvious in the design of intelligent computer-aided instruction (ICAI) systems. These systems structure teaching based on the student's misconceptions and strengths, and they embody expert knowledge about the domain, knowledge of teaching strategies, and knowledge about the student and his or her knowledge (Sleeman & Brown, 1982). Teaching, like explaining, needs to be done at the level that may differ substantially from the computational level.

Whereas there are numerous methods in cognitive psychology applicable to knowledge engineering, the research presents a less-than-complete picture of expertise. However, the work done thus far identifies some expert-novice differences and can account for some aspects of skilled performance, as well as some limitations and biases that are associated with this skilled performance. These findings have numerous implications for the design of expert systems. Further research should more fully address the cognitive structures and processes underlying expert-novice differences. In addition, longitudinal studies of expertise would greatly add to an understanding of the developmental functions of components of expertise. Finally, although the study of expertise requires the selection of a particular task domain, it would be worthwhile to integrate findings across domains in order to generate a comprehensive theory of expertise. The area of knowledge elicitation and expert systems provides an excellent test bed for cognitive theories and methods. Thus, an increasing understanding of human expertise should suggest new approaches to expert system design and knowledge elicitation. The implementations of these ideas will in turn suggest strengths and limitations of the research and in doing so will advance research and theory of human expertise.

Knowledge engineers can also benefit from a better understanding of human expertise and the methods of cognitive psychology. Cognitive psychologists study the implicit structures and processes of human cognition and in doing so have developed a number of techniques that have direct applications as knowledge elicitation techniques. Asking the expert "how he does it" questions provides only one, often impoverished, source of information. Other, more indirect methods (e.g., psychological scaling of ratings, reaction time measures, observations of task performance) may be preferable for eliciting tacit knowledge. In general, knowledge elicitation should not be viewed as skilled interviewing but rather as problem solving that involves making inferences about and integrating information from a variety of techniques.

In summary, research in cognitive psychology on human expertise points out (a) the strengths and limitations of expert performance that are relevant to the development of artificial expertise; (b) expert-novice differences in knowledge representation and search strategies that have implications for the architecture of expert systems; (c) issues relevant to the expert system-user interface, such as explanation and automated knowledge acquisition; and (d) methods applicable to knowledge elicitation. Increased communication between the areas of knowledge engineering and cognitive psychology would not only advance expert systems technology but would also provide a test bed for cognitive theories of expertise.

References

Adelson, B. (1981). Problem solving and the development of abstract categories in programming languages. *Memory & Cognition*, *9*, 422–433.

Adelson, B. (1984). When novices surpass experts: The difficulty of a task may increase with expertise. *Journal of Experimental Psychology: Learning, Memory, and Cognition*, *10*, 483–495.

Aikins, J. (1983). Prototypical knowledge for expert systems. *Artificial Intelligence*, *20*, 163–210.

Anderson, J. R. (1981). Effects of prior knowledge on memory for new information. *Memory & Cognition*, *9*, 237–246.

Anderson, J. R. (1983). *The architecture of cognition*. Cambridge, MA: Harvard University Press.

Anderson, J. R. (1985). *Cognitive psychology and its implications*. New York: W. H. Freeman.

Arkes, H. R., & Hammond, K. R. (Eds.). (1986). *Judgment and decision making: An interdisciplinary reader*. Cambridge: Cambridge University Press.

Ashcraft, M. H. (1989). *Human memory and cognition*. Glenview, IL: Scott, Foresman.

Barr, A., & Feigenbaum, E. A. (Eds.). (1982). *The handbook of artificial intelligence* (Vol. 2). Los Altos, CA: William Kaufmann.

Best, J. (1989). *Cognitive psychology*. St Paul, MN: West Publishing Co.

Boose, J. H., & Gaines, B. R. (1988). *Knowledge acquisition tools for expert systems* (Vol. 2). Cambridge, MA: Academic Press.

Butler, K. A., & Corter, J. E. (1986). The use of psychometric tools for knowledge acquisition: A case study. In W. Gale (Ed.), *Artificial intelligence and statistics* (pp. 295–319). Reading , MA: Addison-Wesley.

Carbonell, J. G. (1986). Derivational analogy: A theory of reconstructive problem solving and expertise acquisition. In R. S. Michalski, J. G. Carbonell, & T. M. Mitchell (Eds.), *Machine learning: An artificial intelligence approach* (pp. 371–392). Los Altos, CA: Morgan Kaufmann.

Charness, N. (1976). Memory for chess positions: Resistance to interference. *Journal of Experimental Psychology: Human Learning and Memory*, *2*, 641–653.

Charness, N. (1981). Search in chess: Age and skill differences. *Journal of Experimental Psychology: Human Perception and Performance*, *7*, 467–476.

Chase, W. G., & Ericsson, K. A. (1981). Skilled memory. In J. R. Anderson (Ed.), *Cognitive skills and their acquisition* (pp. 141–189). Hillsdale, NJ: Erlbaum.

Chase, W. G., & Simon, H. A. (1973a). Perception in chess. *Cognitive Psychology*, *5*, 55–81.

Chase, W. G., & Simon, H. A. (1973b). The mind's eye in chess. In W. G. Chase (Ed.), *Visual information processing*. New York: Academic Press.

Chi, M. T. H., Feltovich, P. J., & Glaser, R. (1981). Categorization and representation of physics problems by experts and novices. *Cognitive Science*, *5*, 121–152.

Chi, M. T. H., Glaser, R., & Rees, E. (1981). *Expertise in problem solving*. Technical report, Learning Research and Development Center, University of Pittsburgh.

Chiesi, H. L., Spilich, G. J., & Voss, J. F. (1979). Acquisition of domain-related information in relation to high and low domain information. *Journal of Verbal Learning and Verbal Behavior*, *18*, 257–273.

Claparède, E. (1934). Genèse de l'hypothèses. *Archives de Psychologie*, *24*, 1–155.

Collins, A. M., & Loftus, E. F. (1975). A spreading activation theory of semantic processing. *Psychological Review*, *82*, 407–428.

Cooke, N. J. (1989). The elicitation of domain-related ideas: Stage one of the knowledge acquisition process. In C. Ellis (Ed.), *Expertise and explanation: The knowledge-language interface*. (pp. 58–75). Chichester, England: Ellis Horwood.

Cooke, N. J. (1990a). Empirically defined semantic relatedness and category judgment time. In R. Schvaneveldt (Ed.), *Pathfinder associative networks: Studies in knowledge organization*. (pp. 101–110). Norwood, NJ: Ablex.

Cooke, N. J. (1990b). Using Pathfinder as a knowledge elicitation tool: Link interpretation. In R. Schvaneveldt (Ed.), *Pathfinder associative networks: Studies in knowledge organization*. (pp. 227–239). Norwood, NJ: Ablex.

Cooke, N. J., Atlas, R. S., Lane, D. H., & Berger, R. C. (1991). "The role of high-level knowledge in memory for chess positions." Unpublished manuscript, Rice University, Houston, TX.

Cooke, N. J., & Schvaneveldt, R. W. (1988). Effects of computer programming experience on network representation of abstract programming concepts. *International Journal of Man-Machine Studies*, *29*, 407–427.

Cooke, N. M. (1983). "Memory structures of expert and novice computer programmers: Recall order vs. similarity ratings." Unpublished master's thesis, Department of Psychology, New Mexico State University, Las Cruces, NM.

Cooke, N. M., Durso, F. T., & Schvaneveldt, R. W. (1986). Recall and measures of memory organization. *Journal of Experimental Psychology: Learning, Memory, and Cognition*, *12*, 538–549.

Cooke, N. M., & McDonald, J. E. (1986). A formal methodology for acquiring and representing expert knowledge. *Proceedings of the IEEE: Special Issue on Knowledge Representation*, *74*, 1422–1430.

Cooke, N. M., & McDonald, J. E. (1987). The application of psychological scaling techniques to knowledge elicitation for knowledge-based systems. *International Journal of Man-Machine Studies*, *26*, 533–550.

Cross, S. E. (1983). A qualitative reasoning approach to mathematical and heuristic knowledge integration. *Proceedings of CHI '83 Conference on human factors in computing systems* (pp. 186–189). New York: Association for Computing Machinery.

Dawes, R. M. (1982). The robust beauty of improper linear models in decision making. In D. Kahneman, P. Slovic, & A. Tversky (Eds.), *Judgment under uncertainty: Heuristics and biases* (pp. 391–407). New York: Cambridge University Press.

de Groot, A. D. (1966). Perception and memory versus thought: Some old ideas and recent findings. In B. Kleinmuntz (Ed.), *Problem solving: Research, method, and theory* (pp. 19–50). New York: Wiley.

Dreyfus, S. E., & Dreyfus, H. L. (1987). Towards reconciliation of phenomenology and AI. In D. Partridge & Y Wilks (Eds.), *A source book on the foundations of AI*. London: Cambridge University Press.

Duncker, K. (1945). On problem solving. *Psychological monographs*, *58* (5, Whole No. 270).

Ebbeson, E. B., & Konečni, V. J. (1980). On the external validity of decision-making research: What do we know about decisions in the real world? In T. S. Wallsten (Ed.), *Cognitive processes in choice and decision behavior* (pp. 21–45). Hillsdale, NJ: Erlbaum.

Egan, D. E., & Schwartz, B. J. (1979). Chunking in recall of symbolic drawings. *Memory & Cognition*, *7*, 149–158.

Einhorn, H. J. (1982). Learning from experience and suboptimal rules in decision making. In D. Kahneman, P. Slovic, & A. Tversky (Eds.), *Judgment under uncertainty: Heuristics and biases* (pp. 268–283). New York: Cambridge University Press.

Ellis, C. (1989). Explanation in intelligent systems. In C. Ellis (Ed.), *Expert knowledge and explanation: The knowledge-language interface* (pp. 108–126). Chichester, England: Ellis Horwood.

Engle, R. W., & Bukstel, L. (1978). Memory processes among bridge players of differing expertise. *American Journal of Psychology*, *91*, 673–689.

Ericsson, K. A., & Polson, P. G. (1988). A cognitive analysis of exceptional memory for restaurant orders. In M. T. H. Chi, R. Glaser, & M. J. Farr (Eds.), *The nature of expertise* (pp. 23–70). Hillsdale, NJ: Erlbaum.

Ericsson, K. A., & Simon, H. A. (1984). *Protocol analysis: Verbal reports as data*. Cambridge, MA: MIT Press.

Frey, P. W., & Adesman, P. (1976). Recall memory for visually presented chess positions. *Memory & Cognition*, *4*, 541–547.

Gammack, J. G., & Young, R. M. (1985). Psychological techniques for eliciting expert knowledge. In M. A. Bramer (Ed.), *Research and development in expert systems* (pp. 105–112). London: Cambridge University Press.

Gentner, D. (1983). Structure mapping: A theoretical framework for analogy. *Cognitive Science*, *7*(2), 155–170.

Gentner, D., & Gentner, D. R. (1983). Flowing waters or teeming crowds: Mental models of electricity. In D. Gentner & A. L. Stevens (Eds.), *Mental models* (pp. 99–129). Hillsdale, NJ: Erlbaum.

Gick, M. L., & Holyoak, K. J. (1980). Analogical problem solving. *Cognitive Psychology*, *12*, 306–355.

Gick, M. L., & Holyoak, K. J. (1983). Schema induction and analogical transfer. *Cognitive Psychology*, *15*, 1–38.

Glaser, R., & Chi, M. T. H. (1988). Overview. In M. T. H. Chi, R. Glaser, & M. J. Farr (Eds.), *The nature of expertise* (pp. xv–xxviii). Hillsdale, NJ: Erlbaum.

Goldin, S. E. (1978a). Effects of orienting tasks on recognition of chess positions. *American Journal of Psychology*, *91*, 659–671.

Goldin, S. E. (1978b). Memory for the ordinary: Typicality effects in chess memory. *Journal of Experimental Psychology: Human Learning and Memory*, *4*, 605–616.

Halasz, F., & Moran, T. P. (1981). Analogy considered harmful. *Proceedings of the Human Factors in Computer Systems Conference*, Gaithersburg, MD.

Hayes-Roth, B. (1977). Evolution of cognitive structures and processes. *Psychological Review*, *84*, 260–278.

Hayes-Roth, F., Waterman, D. A., & Lenat, D. B. (Eds.). (1983). *Building expert systems*. Reading, MA: Addison-Wesley.

Hogarth, R. (1987). *Judgement and choice*. New York: John Wiley & Sons.

Holding, D. H. (1985). *The psychology of chess skill*. Hillsdale, NJ: Erlbaum.

Holding, D. H., & Pfau, H. D. (1985). Thinking ahead in chess. *American Journal of Psychology*, *98*, 271–282.

Holding, D. H., & Reynolds, R. I. (1982). Recall or evaluation of chess positions as determinants of chess skill. *Memory & Cognition*, *10*, 237–242.

Johnson, E. J. (1988). Expertise and decision under uncertainty: Performance and process. In M. T. H. Chi, R. Glaser, & M. J. Farr (Eds.), *The nature of expertise* (pp. 209–228). Hillsdale, NJ: Erlbaum.

Johnson, S. C. (1967). Hierarchical clustering schemes. *Psychometrika*, *32*, 241–254.

Kahneman, D., Slovic, P., & Tversky, A. (Eds.). (1982). *Judgment under uncertainty: Heuristics and biases*. New York: Cambridge University Press.

Kahneman, D., & Tversky, A. (1973). On the psychology of prediction. *Psychological Review*, *80*, 237–251.

Kahneman, D., & Tversky, A. (1982). On the study of statistical intuitions. In D. Kahneman, P. Slovic, & A. Tversky (Eds.), *Judgment under uncertainty: Heuristics and biases* (pp. 493–508). New York: Cambridge University Press.

Klein, G. A. (1987). Applications of analogical reasoning. *Metaphor and symbolic activity*, *2*, 201–218.

Kruskal, J. B., & Wish, M. (1978). *Multidimensional scaling*. London: Sage Publications.

Landauer, T. K. (1975). Memory without organization: Properties of a model with random storage and undirected retrieval. *Cognitive Psychology*, *7*, 495–531.

Lane, D. H., & Robertson, L. (1979). The generality of the levels of processing hypothesis: An application to memory for chess positions. *Memory & Cognition*, *7*, 253–256.

Langley, P., & Simon, H. A. (1981). The central role of learning in cognition. In J. R. Anderson (Ed.), *Cognitive skills and their acquisition* (pp. 361–380). Hillsdale, NJ: Erlbaum.

Larkin, J. H. (1981). Enriching formal knowledge: A model for learning to solve textbook physics problems. In J. R. Anderson (Ed.), *Cognitive skills and their acquisition* (pp. 311–334). Hillsdale, NJ: Erlbaum.

Lawrence, J. A. (1988). Expertise on the bench: Modeling magistrates' judicial decision-making. In M. T. H. Chi, R. Glaser, & M. J. Farr (Eds.), *The nature of expertise* (pp. 229–260). Hillsdale, NJ: Erlbaum.

Lesgold, A., Rubinson, H., Feltovich, P., Glaser, R., Klopfer, D., & Wang, Y. (1988). Expertise in a complex skill: Diagnosing X-ray pictures. In M. T. H. Chi, R. Glaser, & M. J. Farr (Eds.), *The nature of expertise* (pp. 311–342). Hillsdale, NJ: Erlbaum.

Matlin, M. W. (1989). *Cognition*. New York: Holt, Rinehart, and Winston.

Maxwell, K. J. (1983). *A scripts analysis of fact retrieval from memory*. Unpublished doctoral dissertation, Department of Psychology, New Mexico State University, Las Cruces, NM.

McDonald, J. E., & Schvaneveldt, R. W. (1988). The application of user knowledge to interface design. In R. Guindon (Ed.), *Cognitive science and its applications for human-computer interaction* (pp. 289–338). Hillsdale, NJ: Erlbaum.

McKeithen, K. B., Reitman, J. S., Rueter, H. H., & Hirtle, S. C. (1981). Knowledge organization and skill differences in computer programmers. *Cognitive Psychology*, *13*, 307–325.

Meyer, D. E., & Schvaneveldt, R. W. (1971). Facilitation in recognizing pairs of words: Evidence of a dependence between retrieval operations. *Journal of Experimental Psychology*, *90*, 227–234.

Michalski, R. S., Carbonell, J. G., & Mitchell, T. M. (Eds.). (1986). *Machine learning: An artificial intelligence approach*. Los Altos, CA:

Morgan Kaufmann.

Michie, D. (1982). The state of the art in machine learning. In D. Michie (Ed.), Introductory readings in expert systems (pp. 208–229). London: Gordon and Breach.

Miller, G. A. (1956). The magical number seven, plus or minus two: Some limits on our capacity for processing information. *Psychological Review, 63*, 81–97.

Minsky, M. (1981). K-Lines: A theory of memory. In D. A. Norman (Ed.), *Perspectives on cognitive science*. Norwood, NJ: Ablex.

Minsky, M., & Papert, S. (1974). *Artificial intelligence*. Condensed lectures, Oregon State System of Higher Education, Eugene, OR.

Murphy, G. L., & Wright, J. C. (1984). Changes in conceptual structure with expertise: Differences between real-world experts and novices. *Journal of Experimental Psychology: Learning, Memory, and Cognition, 10*, 144–155.

Myles-Worsley, M., Johnston, W., & Simons, M. (1988). The influence of expertise on X-ray image processing. *Journal of Experimental Psychology: Learning, Memory, and Cognition, 4*, 553–557.

Newell, A., & Simon, H. A. (1972). *Human problem solving*. Englewood Cliffs, NJ: Prentice-Hall.

Nisbett, R. E., & Ross, L. (1980). *Human inference: Strategies and shortcomings of social judgment*. Englewood Cliffs, NJ: Prentice-Hall.

Nisbett, R. E., & Wilson, T. D. (1977). Telling more than we can know: Verbal reports on mental processes. *Psychological Review, 8*, 231–259.

Norman, G. R., Brooks, L. R., & Allen, S. W. (1989). Recall by expert medical practitioners and novices as a record of processing attention. *Journal of Experimental Psychology: Learning, Memory, and Cognition, 15*, 1166–1174.

Phelps, R. H., & Shanteau, J. (1978). Livestock judges: How much information can an expert use? *Organizational Behavior and Human Performance, 21*, 209–219.

Posner, M. I. (1988). Introduction: What is it to be an expert? In M. T. H. Chi, R. Glaser, & M. J. Farr (Eds.), *The nature of expertise* (pp. xxix–xxxvi). Hillsdale, NJ: Erlbaum.

Reder, L. M., & Anderson, J. R. (1980). A partial resolution of the paradox of interference: The role of integrating knowledge. *Cognitive Psychology, 12*, 447–472.

Reitman, J. S. (1976). Skilled perception in Go: Deducing memory structures from interresponse times. *Cognitive Psychology, 8*, 336–356.

Rumelhart, D. E., & McClelland, J. L. (1986). *Parallel distributed processing: Explorations in the microstructure of cognition*. Cambridge, MA: MIT Press.

Rumelhart, D. E., & Norman, D. A. (1981). Analogical processes in learning. In J. R. Anderson (Ed.), *Cognitive skills and their acquisition* (pp. 335–359). Hillsdale, NJ: Erlbaum.

Schoenfeld, A., & Herrmann, D. (1982). Problem perception and knowledge structures in expert and novice mathematical problem solvers. *Journal of Experimental Psycology: Learning, Memory and Cognition, 8*, 584–494.

Schvaneveldt, R. W. (Ed.). (1990). *Pathfinder associative networks: Studies in knowledge organization*. Norwood, NJ: Ablex.

Schvaneveldt, R. W., & Durso, F. T. (1981, November). *Generalized semantic networks*. Paper presented at the meeting of the Psychonomic Society, Philadelphia, PA.

Schvaneveldt, R. W., Durso, F. T., & Dearholt, D. W. (1985). *Pathfinder: Scaling with network structures*. Memorandum in Computer and Cognitive Science (no. MCCS–85–9). Computing Research Laboratory, New Mexico State University, Las Cruces, NM.

Shepard, R. N. (1962a). Analysis of proximities: Multidimensional scaling with an unknown distance function. I. *Psychometrika, 27*, 125–140.

Shepard, R. N. (1962b). Analysis of proximities: Multidimensional scaling with an unknown distance function. II. *Psychometrika, 27*, 219–246.

Shiffrin, R. M., & Schneider, W. (1977). Controlled and automatic human information processing: II. Perceptual learning, automatic attending, and a general theory. *Psychological Review, 84*, 127–190.

Shneiderman, B. (1976). Exploratory experiments in programmer behavior. *International Journal of Computer and Information Sciences, 5*, 123–143.

Shortliffe, E. H. (1976). *Computer-based medical consultations: MYCIN*. New York: American Elsevier.

Simon, H. A. (1981). Cognitive science: The newest science of the artificial. In D. A. Norman (Ed.), *Perspectives on cognitive science* (pp. 13–25). Norwood, NJ: Ablex.

Simon, H. A., & Barenfeld, M. (1969). Information processing analysis of perceptual processing in problem solving. *Psychological Review, 76*, 473–483.

Simon, H. A., & Gilmartin, K. (1973). A simulation of memory for chess positions. *Cognitive Psychology, 5*, 29–46.

Simon, H. A., & Reed, S. K. (1976). Modeling strategy shifts in a problem solving task. *Cognitive Psychology*, *8*, 86–97.

Sleeman, D., & Brown, J. S. (Eds.). (1982). *Intelligent tutoring systems*. New York: Academic Press.

Sloboda, J. A. (1976). Visual perception of musical notation: Registering pitch symbols in memory. *Quarterly Journal of Experimental Psychology*, *8*, 1–16.

Smith, E. E., Adams, N., & Schorr, D. (1978). Fact retrieval and the paradox of interference. *Cognitive Psychology*, *10*, 438–464.

Soloway, E., Adelson, B., & Ehrlich, K. (1988). Knowledge and processes in the comprehension of computer programs. In M. T. H. Chi, R. Glaser, & M. J. Farr (Eds.), *The nature of expertise* (pp. 129–152). Hillsdale, NJ: Erlbaum.

Soloway, E., Ehrlich, K., Bonar, J., & Greenspan, J. (1982). Tapping into TACIT programming knowledge. *Proceedings of CHI '82 Conference on Human Factors in Computing Systems* (pp. 52–57). New York: Association for Computing Machinery.

Taylor, D. A. (1976). Stage analysis of reaction time. *Psychological Bulletin*, *83*, 161–191.

Tulving, E., & Thomson, D. M. (1973). Encoding specificity and retrieval processes in episodic memory. *Psychological Review*, *80*, 352–373.

Tversky, A., & Kahneman, D. (1982). Judgement under uncertainty: Heuristics and biases. In D. Kahneman, P. Slovic, & A. Tversky (Eds.) *Judgment under uncertainty: Heuristics and biases* (pp. 3–20). New York: Cambridge University Press.

Vessey, I. (1988). Expert-novice organization: An empirical investigation using computer program recall. *Behaviour and Information Technology*, *7*, 153–171.

Vosniadou, S., & Ortony, A. (Eds.). (1989). *Similarity and analogical reasoning*. Cambridge: Cambridge University Press.

4
Mental Models and the Acquisition of Expert Knowledge

Robert M. Schumacher and Mary P. Czerwinski

There is a well-known fable about six learned blind men and an elephant. In this fable, each blind man encounters the elephant from a different perspective and develops his own opinion as to the nature of this beast: One grabs his ear and thinks it is a fan, another feels his tail and believes it to be a rope, and so on. The fable concludes with each blind man steadfastly holding his own narrow opinion. Among the many metaphorical elephants in cognitive science, one that stands out is the concept *mental model*. In this paper we examine this elephant by exploring the multiplicity of claims made about mental models. We do not pretend to be any less blind than other researchers who have tackled these problems, nor do we claim to have the only answers to many of the questions we raise. What we hope is that by taking a fresh look at mental models research we can provide some unification of concepts and point to promising methodologies and applications. We hope that by examining the issues surrounding mental models we raise the awareness of those who would try to use mental models in construction of knowledge-based systems (see Davis, 1982).

We have organized this chapter in the following way. First we discuss definitions of mental models, followed by a detailed discussion of several properties of mental models. Next we consider four sets of research that, we believe, provide interesting evidence for mental models. Last, we present a theoretical account of how mental models are acquired.

Mental Models: Concepts and Definitions

The concept of a mental model is one that many cognitive scientists and human factors specialists use in explaining certain behaviors, but hardly one that everyone would accept. As a working definition we will define a mental model as a collection of knowledge about a physical device, system, or process. Some efforts have even been made to develop a logical proof of the existence of mental models; for instance, Conant and Ashby (1970) argue that the brain must "model" the environment in order to survive. The two main sources of research on mental models come from cognitive science and human factors psychology. In cognitive science, researchers have used the mental model construct to explain such things as how people learn to use in-fix notation calculators (e.g., Halasz & Moran, 1983; Young, 1983); operate simulated devices (e.g., Kieras & Bovair, 1984; Schumacher & Gentner, 1988); and understand physical processes (e.g., McCloskey, 1983). In human factors, mental models have been used to describe how people learn computer systems (e.g., Carroll & Olson, 1988; Rumelhart & Norman, 1981); comprehend programming concepts (e.g., Cooke & Schvaneveldt, 1988; Mayer & Bayman, 1981); interact with everyday devices (Norman, 1988); and perform complex process control tasks (e.g., Moray, 1986, 1988; Veldehuysen & Stassen, 1977; Wickens & Kessel, 1979, 1980).

If nothing else, mental models provide an organizing principle of knowledge (e.g., Bower, Clark, Lesgold, & Winzenz, 1969); that is, they are a framework for thinking about a device. To illustrate, in the typical study of mental models, one group of participants is given procedural instructions on how to control a device (the procedural group), and another group is given "how-it-works" knowledge (the model group). The groups are then given the same task to perform, and performance data (e.g., time to complete a task) are recorded. Usually, participants in the model group show improved performance relative to those in the procedural group (e.g., Halasz & Moran, 1983; Kieras & Bovair, 1984; Mayer, 1976; Schumacher & Gentner, 1988). Halasz and Moran (1983), for instance, showed that students who learned the conceptual model of a pushdown stack were better able to solve complex problems on an in-fix notation calculator than students who simply learned procedures for solving the problems.[1] The interpretation is that the model group possessed a better mental model of the device than the procedural group, which enabled the model group to better perform the tasks.

Despite the appeal of the construct of a mental model, it can often become a crutch that provides little explanatory or predictive power in user-centered system design. For instance, as Rouse and Morris (1986) point out, if a mental model is made synonymous with the general knowledge needed to carry out the task, then it is of little scientific value. Further compounding this problem is the myriad of ways in which researchers have defined the construct. While we believe that the concept of a mental model can be useful, it must be defined clearly and uniformly. Thus, we take seriously the notion that defining the concept is a crucial aspect of cognitive theory. As soon as a definition is arrived at, identifying paradigms for direct applications of mental models to knowledge engineering and user-centered design problems becomes easier.

Classes of Definitions of Mental Models

Definitions of mental models fall into at least three classes: descriptions of mental models as collections of knowledge structures, descriptions of mental models as metaphors and analogies, and process descriptions of how users interact with complex systems. After discussing these definitions of mental models, we will provide a definition that encompasses the various classes of descriptions.

Mental Models as Collections of Knowledge

Some descriptions of mental models tend to be quite broad. For instance, Veldehuyzen and Stassen (1977) have defined a mental model to include knowledge about the system to be learned; knowledge about the sorts of disturbances likely to be encountered when interacting with the system; and knowledge about the criteria, strategies, and so forth associated with the task. In other words, the human operator combines a large number of knowledge structures concerning the system, the task, and his or her role within that task, to carry out the duties required. This is the mental model. Similarly, Norman (1983, 1986) describes a mental model in terms of knowledge structures but claims they are messy, incomplete, and indistinct. Wickens (1984) describes an "internal" model as a hypothetical construct which accounts for an operator's behavior during sampling, scanning, and planning while operating a system. These definitions consider the mental model as an overarching collection of knowledge.

Other definitions hold that mental models guide human-machine interaction through an understanding of how a device or system works in terms of its structure and function (e.g., de Kleer & Brown, 1983; Halasz & Moran, 1983; Kieras & Bovair, 1984; Moray, 1988; Schumacher & Gentner, 1988; Williams, Hollan, & Stevens, 1983).[2] What needs to be made explicit is exactly how this understanding is achieved (or seems to be lacking), and to what level. In this way mental models should also be viewed in relation to other knowledge structures. This means that mental models are sometimes considered a subset of schemata (e.g., Rumelhart & Ortony, 1977). According to this view, mental models are sometimes considered a subset of schemata in that mental models refer specifically to physical devices, systems, and processes as opposed to non-physical-entity-oriented knowledge structures. The interesting properties of mental models derive not because they are a unique knowledge structure, but because they bring together knowledge in unique and interesting ways.

Mental Models as Analogies and Metaphors

The role and value of analogies as mental models is interesting. Analogies as mental models are used in two subtly different ways (e.g., Collins & Gentner, 1986; Gentner & Gentner, 1983; Holland, Holyoak, Nisbett, & Thagard, 1986). The most common usage is to consider that the mental model of a device has certain similarities to the mental model of an analogous device (e.g., a computer text-editor is like a typewriter).[3] Mental models that are a function of analogies arise from the mapping of the objects (e.g., the keyboard) and relations (e.g., moving text) involved between a known system and a novel system (e.g. Collins & Gentner, 1986; Gentner & Gentner, 1983), as well as the user's goals. The analogies-as-mental-models approach can serve to show how mental models are acquired and how the use of analogies can benefit a learner's first interactions with a system. There is much evidence (and intuition) that learners of computer systems derive most of their early knowledge from learning by analogy and example. Carroll and Thomas (1982) have argued that in learning, users adapt new knowledge structures through the metaphorical extension of old knowledge structures. Rumelhart and Norman (1981) have also claimed that learning a new system is easiest if the user draws an analogy between the new system and one that is already familiar to the user. There is also evidence that experts also draw upon examples and case studies (e.g., Hinsley, Hayes, & Simon, 1977). Clement (1988) reported that the protocols of advanced graduate students and professors in physics revealed heavy reliance on analogies during problem solving. He linked this analogical problem solving to the progressive construction of a mental model.

Some regard the use of analogy in this way as being potentially dangerous. Halasz and Moran (1982) argue that although analogies may help a user to learn and understand a new system, they are not always the best way to teach new users. They suggest that analogies may hinder a novice user from developing a good understanding of the system. Halasz and Moran recommend that the appropriate tool for imparting a mental model of a new system is to provide an abstract conceptual model, not an analogy. The conceptual model could be shaped to contain specific information about the current system, which would help the user with his or her required interactions without the baggage that analogies can bring. We have no argument with the idea that providing a good conceptual model can be a fruitful technique for imparting a mental model, but we believe that the indictment of analogy as potentially dangerous reflects a lack of appreciation for the mechanisms of analogical learning. Analogy is a powerful and ubiquitous tool, and it is the responsibility of the designer and instructor to pick and use metaphors carefully in order to aid in system understanding.

The second way in which analogies and mental models have been linked together in cognitive theory is when the mental model is considered an analogy between the domain it represents and a person's cognitive representation of the domain (e.g., the reference to morphisms by Holland et al., 1986). The implication is that every mental representation of the outside world must be an analogy to the outside world. In reality, there may be no difference between a mental model that is considered an analogy to the world it represents and a mental model regarded as a process description or seen as a collection of knowledge. It may simply be that some use the term *analogy* (perhaps unfortunately) as a terminological convenience. (See Hoffman, Cochran, & Nead, 1989; and Hoffman & Nead, 1983; for an interesting philosophical treatment of these issues.)

Brown and de Kleer (1981; de Kleer & Brown, 1983) define mental models as topologies of device models, which we take to mean a mapping between the user's mental model and the physical presentation of the real system. This characterization is interesting because it is really only the observable parts of the system (i.e., the interface and documentation) that the user has available that allow inferences about how the system works, especially if no how-it-works information is observable. Brown and de Kleer believe strongly that one distinguishing property of mental models is the ability to simulate the device in the mind's eye (see also Collins & Gentner, 1986; Williams et al., 1983). Wickens and Kessel's (1980) definition is consistent with this definition of a mental model, although Wickens and Kessel argue that the experimental evidence suggests that the user's mental model often represents an idealized abstraction of the real system. Using the mental model as an analogy of the system, the user employs various heuristics (by borrowing from familiar domains or instances) rather than representing the true system and its operations.

Note that the descriptions and definitions of mental models vary according to whether they refer to interacting knowledge structures or whether they derive from analogies between systems. These two characterizations of models can be seen as complementary, and they form the basis for our thinking on acquisition of mental models. For example, analogies can be viewed as a bridge between a user's knowledge of a familiar system and unfamiliar systems. After the user gains experience with a system, however, he or she may rely less on example and analogy. Specific knowledge about that system may be pooled and organized into better formulated knowledge structures.

Mental Models as Process Descriptions

Some of the work in mental models is involved with rigorous "modeling" of a user's mental model of a system, or at least providing a process description of a mental model (e.g., Jagacinski & Miller, 1978; see Wickens, 1984). According to Moray (1986, 1988; Moray & Reeves, 1987; see also Kieras & Polson, 1985; Engelbeck, 1989), a mental model is a representation of a set of semi-independent subsystems into which the total (physical) system can be decomposed. Moray suggests that a theory of design for intelligent displays and decision aids can be developed through the use of a "lattice" theory of mental models. If the mental model is regarded as a lattice, and the role of the components of the system is viewed as a way to provide paths in the lattice that will be otherwise inaccessible to the operator, then theory can guide the design of better systems. Because we discuss Moray's research in much more depth later on in this paper, we defer further discussion until then. There have been other classificatory efforts relating to mental models. In the next section we talk about two other approaches.

Other Characterizations of Mental Models

Carroll and Olson (1988) provide a breakdown of mental models into surrogate, metaphor, glass-box, and network models. A surrogate model (first described by Young, 1983) is one that completely matches the target system's behavior and does not assume that the way in which output is produced in the surrogate is the same as it is produced in the target system. The same kind of assumptions are made in modeling operator performance in optimal control theory (see Wickens, 1984) and in some knowledge-based systems where reasoning does not putatively follow the veridical processes of an expert in solving a problem. Metaphor models are like those described in the analogy section above. Glass-box models mimic the system exactly, but unlike surrogate models they provide some semantic interpretation for the internal components. Network models try to incorporate what the user knows about the system (typically expressed as production rules) with a representation of the system itself. This is represented in the generalized transition network approach of Kieras and Polson (1985).

Rouse and Morris (1986) found in their review of the literature that taxonomic efforts to characterize mental models tend to produce attribute-oriented approaches to particular tasks. Thus, they found in trying to develop a taxonomy that the literature reflected the idiosyncratic nature of the domains studied and the background of the researchers. Nevertheless, Rouse and Morris identified two prominent dimensions that described the differences in the research efforts. The first is the nature of the model manipulation, and the second is the level of behavioral discretion given to the research participant. By the nature of the model manipulation they mean whether or not the subject is actually cognizant of the model he or she is using. For instance, in Clement's (1988) experiments with physics students, he gets reports of the participants visualizing and mentally manipulating objects. In other tasks, such as process control, the operator may have no ability to introspect on component pieces of the task. In terms of the level of behavioral discretion, the participant may either be a monitor and have no control over the task domain (e.g., observing the levels of system parameters), or the participant may have full control over the task domain (e.g., driving a car). Rouse and Morris believe these dimensions also highlight methodological differences in the study of mental models. Inferential methods provide good descriptions of models when the participant has little behavioral discretion in the task, whereas verbalization methods are better for tasks where there is an explicit manipulation (i.e., much behavioral discretion).

Now that we have presented some definitions and, briefly, some classifications of mental models, in the next section we offer a definition that we think links together the many threads we have discussed.

Defining a Mental Model: Constraints and Attributes

We define a mental model as a cognitive representation of a complex, physical, dynamic device, system, or process that allows an operator to understand and explain system components and their interactions, and to predict system outcomes from system input. The many facets of this definition need to be spelled out more precisely. In the following paragraphs we justify the three key parts of the definition: physicality, complexity, and dynamism. It is only fair to let the reader know that we do not refer directly to these constraints within the context of the model that we present later; however, they do constitute the foundation of our thinking, and as such we feel it is important that they be presented.

For the first constraint, we concern ourselves strictly with cognitive representations that deal with devices, systems, machines, and so forth—in short, physical entities and processes. We do not deny that one could have a mental model of marriage or a mental model of the social environment in the workplace, but we simply choose to consider only those domains whose interactions are physical in nature. Granted, this is an arbitrary restriction, but it is one that is directed toward the way that most mental models researchers have carved up the world. This constraint is also motivated by how one defines an expert. We would argue that experts in computer systems are easier to define than experts in marriage.

Second, the physical entities should be constrained further to be complex. Saying that one has a mental model of a fishhook seems to miss the point. Complexity is a necessity.

Third, the entities must be not only complex and physical but also dynamic. Consider the girders of a modern office tower made of thousands of pieces of iron welded together. The structure is complex and physical, but it is not dynamic. A dynamic system has parameters with input values, intermediate states, and output values, which can be mapped into a psychological understanding of system function.

Of course, these constraints—physicality, complexity, and dynamism—represent continua. The physicality of the system does not constrain it to wires, diodes, and transistors however. For instance, the constraints would not restrict us from discussing a doctor's mental model of blood pressure. Nor do the constraints set restrictions on the level of description or granularity (barring violation of the simplicity constraint). Physicists have mental models, some based on widely accepted theories that rest on known physical laws, of both atomic and galactic systems. Neither do these constraints restrict one to having a model of something that does not exist. One might very well have a mental model of a computer system architecture that is not implemented yet satisfies all the constraints.

We believe that mental models have had ascribed to them many attributes aside from these constraints. In the next two sections we lay out some of these attributes.

Stable Versus Derived Models

Are models already present in memory or constructed at the moment they are needed? A model that is enduring, or built up over time, we call a *stable model*, and one that is seemingly constructed on the fly we call a *derived model*. For example, Bainbridge (1981) studied expert operators in the steel-making process. The experts' models were acquired over much experience with a stable system. In some sense, these models seem to be prestored in memory. Other models have the flavor of being computed or constructed from reasoning about a process or system (see Clement, 1988); much of the work on naive physics has this flavor (e.g., McCloskey, 1983). For example, one question we often ask people is, What causes a sonic boom? (A typical answer is, The sound waves can't get out of the way fast enough.) Unless the person queried is a physicist, it is unlikely that a stable model is retrieved for just this case. This distinction between stable and derived models is not often made, but it is not hard to think of examples of where there might be a difference between the two. A stable model might support everyday interaction with a system, whereas a derived model could be used to explain behavior that is out of the common stream of experience with a particular device or system, such as a fault condition. Any theory of mental models must recognize and account for stable and derived models. Another attribute that must be considered is multiple models.

Multiple Models

Many researchers have suggested that people can possess more than one model of a system (e.g., Norman, 1983; Williams et al., 1983). Collins and Gentner (1986) interpreted research participants' statements as indicating that their models of the evaporation process consisted of analogies from different domains, such as a rocket ship leaving the earth or a room crowded full of people. There are at least three kinds of multiple models a person might have of a domain: complementary, parallel, or abstracted.

First, if a domain, like a computer system, is sufficiently complex, then people may develop many local models that do not overlap the same part of the system to any great degree. These multiple models are called *complementary models*. For instance, borrowing from Gentner and Gentner (1983), students could be told that the hydraulic analogy for how water moves through pipes is better for reasoning about batteries in electrical circuits than the moving-crowds analogy. They could also be told that the moving-crowds analogy is better for reasoning about resistors than the hydraulic analogy. Thus, two separable parts of the same system—batteries and resistors—are best covered by two different analogical models. Gentner and Gentner showed that using a hydraulic analogy allowed subjects to do better on questions relating to batteries than on questions relating to resistors, and using a moving-crowds analogy allowed (different) subjects to do better on resistors than batteries. Conceivably, if a student knows these facts, then invoking the optimal analogy when reasoning about batteries or resistors will lead to the best results.

Second, a person might have two analogous models of the same part of the system; these kinds of multiple models are *parallel* (Schumacher, 1989). One of Kempton's (1986) participants compared his home thermostat to a gas pedal, an electric mixer, and a water faucet (these specific comparisons have problems, as we will see). Thus, parallel models are those where a user has several different models, all at roughly the same level of abstraction to help the user understand the same (sub)system. It would not be unusual for these models to be inconsistent with each other. Collins and Gentner's (1986) participants reflected this kind of inconsistency in explaining evaporation: The same participant invoked different models to explain the same process.

The third type of multiple model, *abstracted models*, is a variant of parallel models. One can speak of multiple models that cover the same system or device but are at different levels of description (e.g., Rasmussen, 1986). Rasmussen argues that one can talk of the same device at several levels of abstraction: physical form, physical functions, generalized functions, abstract function, and functional purpose. He writes,

> A change of level of abstraction involves a shift in concepts and structure for representation as well as a change in the information suitable to characterize the state of the function or operation at various levels of abstraction. Thus, an observer asks different questions of the environment depending on the nature of the currently active internal representation. (p. 121)

While it seems plausible and even desirable (experimentally) to assume that a user may have only one mental model of a device, it should be clear that people might have several models of a system. Furthermore, one might speculate that novices rely more on multiple models of a particular device than experts, because uncertainty with system capabilities might lead to trying out several approaches to learning the device. This, in turn, leads to the user's spontaneously recognizing surface and deep similarities with other systems and devices.

Attempts to specify "the one" mental model a person might have of a device seem relatively naive. The best we can do is approximate at some level how users think about the devices with which they interact. In the following section, we discuss some recent efforts to understand mental models, paying particular attention to the methods used to elicit knowledge.

Evidence for Mental Models

In addition to the many definitions of mental models, there are a variety of methods for generating data about mental models. Many of these methods complement those for eliciting information for knowledge-based systems. If expert reasoning involves the generation and use of mental models, then we need to describe, elicit, and represent the expert's mental models in order to build an expert system. In knowledge engineering, Hoffman (1987) lists five typical ways in which information is collected for knowledge-based systems: structured interviews, familiar-tasks method, limited-information tasks, constrained-processing tasks, and the tough-cases method. In structured interviews experts are probed for their verbalizable knowledge about the task; information is also gathered from texts, manuals, and so forth. This is one of the most common methods. In the familiar-tasks method experts provide analysis (verbally) about tasks that they usually perform and the methods they use to make decisions. In limited-information tasks the expert is given only a subset of the usual information needed to make a decision, and the expert's reasoning processes are examined for useful insights. In constrained-processing tasks experts are asked to work under time pressure or some other (artificial) constraint to gain insight into the reasoning processes. In the tough-cases method the expert is given a difficult problem to solve so that the knowledge engineer may observe extraordinary processing. The first two methods require a great deal of expert time and knowledge engineer time; however, they can yield useful data. The last three approaches can make the expert feel uncomfortable and may seem artificial, but they too provide insight into the reasoning processes needed for the task.

Hoffman (1987) also provides criteria by which elicitation representation techniques should be evaluated and compared. These criteria are the simplicity of the task, the simplicity of the materials needed for the task, the duration of the task, the flexibility of the task in terms of materials and experts, the similarity of the task to the expert's real task, whether the data provide evidence for reasoning strategies, the ease with which the data can be placed into machine-ready form, and the efficiency of the task. Based on these criteria, the above methods of eliciting expert knowledge representations are far from ideal. Elicited knowledge often results in a patchwork of facts and rules instead of an integrated network of knowledge (Davis, 1982).

Mental models research methods can be useful in knowledge engineering. Mental models have been explored through a number of methods for collecting various kinds of data: verbal protocol (e.g., Bainbridge, 1979); various performance measures (e.g., Kieras & Bovair, 1984); and similarity judgments (e.g., Schvaneveldt et al., 1985). On the whole, these efforts have been directed toward validating the concept of a model or testing the properties of a model. But one could ask whether some paradigms in mental models can complement or be adapted to knowledge-based systems. In pursuit of this question, we examine some evidence for the existence of mental models, highlighting methods for probing mental models. For each of the four empirical approaches to mental models, we describe the domain and the method, and then evaluate its results in terms of both mental models per se and the relationship of the work to knowledge elicitation of mental models.

Observation and Protocol: Kempton

A set of observational experiments by Kempton (1986) illustrates the fact that people's behavior is systematically related to the kinds of examples they draw upon—that is, people use analogy-based problem solving. In his studies, Kempton observed how people controlled their home thermostat. Some of the people operated the device correctly. Their mental model of the device fit the way in which the thermostat was intended to be operated, that is, as a closed-loop feedback device. If the room temperature was 68 °F and the desired temperature was 75 °F, the users would set their thermostat to 75 °F. Others possessed what Kempton called an accelerator model of the thermostat. With this model, if the temperature of the room is 68 °F and the user wanted the room to be 75 °F, the user would set the thermostat to 80 °F, thinking that the room would heat up faster. In addition to data on actual settings of the thermostats, he recorded cases where people verbally compared the thermostat to various familiar devices: an electric mixer, an automobile, and a water faucet. The users are borrowing the model of how they operate these other devices to explain, albeit incorrectly, how to control their home heating system.

Kempton's simple example of mental models of home heat control illustrates two points. First, naive models are routinely drawn from familiar devices, especially if the device is particularly opaque. These familiar devices are often superficially similar (i.e., share many obvious features) to the domain of current interest, but not always. At one extreme, the conceptual model of a standard push-button telephone is the same regardless of which of the millions of telephones one is discussing. The fact that one has never used telephone *X* does not make it difficult to use telephone *X*, because one can draw on the mental model of other telephones one has used. At the other extreme, quite similar conceptual models may share few superficial commonalities. Although a dentist's drill and a windmill share few superficial commonalities, the underlying conceptual model for them can be quite similar. Kempton argues that because our world is filled with more accelerator devices than feedback devices, people will more often have a mental model of a thermostat consistent with accelerator devices than with other feedback devices. He also showed the power of using observation and protocol analysis as data-gathering methods. However, it can be hard to get a detailed picture of a user's mental model of a complex system from protocol analysis and observation alone. (See Bainbridge, 1979, 1986, for a discussion of these issues.)

In evaluating protocol techniques for knowledge engineering, observation of experts solving typical tasks (under a variety of conditions) combined with protocol analysis is a popular method of gathering information about expert knowledge (Hoffman, 1987). The strength of this technique as a means of eliciting expertise is that the expert is in fact performing typical tasks. However, there are two serious drawbacks to observation alone or observation augmented with verbal protocols. First, the problem of not being able to elicit useful verbalization is cited by many (e.g., Nisbett & Wilson, 1977; but see Ericsson & Simon, 1984). Also, the tedium of transcribing and coding the protocols makes it a less-than-easy means of understanding the expert's model of a system. The second, and more serious, problem is that observation alone, unless it is done over a protracted period of time, will not yield the full range of behaviors that an expert performs. Simply asking experts to "think out loud" may not be sufficient either, because it is unclear as to how closely what experts say they do matches what they actually do.

Performance Tasks: Wickens and Kessel

In a series of experiments, Wickens and Kessel (1979, 1980, 1981; Kessel & Wickens, 1982) contrasted the kinds of internal models (as they termed them) developed by monitors of systems versus controllers of systems. In Wickens and Kessel's tasks, participants controlled the system dynamics of a two-dimensional pursuit display (e.g., an air traffic control screen) while monitoring for system faults. Other participants simply monitored the same system for faults. Wickens and Kessel found that these two sets of participants differed in their ability to detect faults: System controllers detected faults better than system monitors. Wickens and Kessel asserted that system controllers deduce malfunctions by detecting subtle changes in system performance through cues that are not available to people trained only to be monitors. One concrete manifestation of this is that experienced nuclear power plant operators can detect certain vibrations in the floor of the control room that indicate trouble, whereas an inexperienced operator is unable to use this cue (Curry, 1981).

One can assume from Wickens and Kessel's studies that the kinds of models people possess are dependent on the way in which those models are acquired. In practical terms, this means that people who are monitoring systems (e.g., supervisors) should spend sufficient time in controlling a system before moving to monitoring that system. Control knowledge appears to degrade relative to the time out of the loop (Wickens, 1984). It is also interesting to note in their studies that the transfer of control skills was asymmetric (Kessel & Wickens, 1982). Skill at controlling the system transferred well to monitoring tasks, but skill at monitoring did not transfer well to control tasks. The reader might wish to consider that one of the ramifications of our technological age is to automate control of systems and place people right into monitoring tasks in such systems as nuclear power plants. Based on the aforementioned results, this could have grave consequences in fault conditions.

The implications of Wickens and Kessel's findings for knowledge engineering and expertise are obvious. Wickens and Kessel show the importance of using performance data for describing expert knowledge. Showing that there is a different kind of expertise developed on the same system with people doing relatively similar tasks is interesting and important. However knowledge of a system is extracted, many types of experts should be consulted lest elements of the task go unincluded. Matching the expert to the kind of task the system is expected to perform is critical; otherwise, the data on which cues are used, how the task is actually performed, and so forth may be flawed.

Theory and Performance Tasks: Moray

In an example of more theoretically driven research, Moray (1988) suggested that a formal theory of mental models can be established using Ashby's (1956) general theory of systems. Because Moray's (1988) is one of the few analytic theories of mental models, we will spend some time describing his theory and data. Moray (1988) believes that a mental model can be regarded as a homomorph of a real system. Because homomorphs are logical decompositions of a system into its subsystems, Moray postulates that users may represent complex systems in just this way. The claim is that users reduce cognitive load by grouping system variables into subsystems if those variables are observed to covary. The longer variables are observed to covary, the stronger the linking between the variables within the homomorph. Also, two variables residing within different subsystems cannot be considered as covarying. This is due to the fact that the user need only consider the system with respect to the lower order groupings of variables, and not in terms of its many isolated (yet interacting) variables. According to Moray's (1988) theory, then, a mental model is the decomposition of a system into its "natural" subsystems. Moreover, Moray (1988) argues that with practice these subsystems become so strongly bound together that it becomes difficult for an experienced user to think of new subsystems or a rearrangement of existing subsystems. Moray (1988) believes these theoretical properties could help explain "cognitive lockup" during fault conditions, because operators often have difficulty considering certain systematic relationships under these situations.

Homomorph theory suggests a vehicle for trapping a user's mental model of a complex system. In one task (Moray & Reeves, 1987), operators controlled a display of eight bar graphs; the goal was to prevent each of the continuously varying graphs from exceeding its bounds. Some of the bar graphs were correlated with each other, providing the operator with observable "subsystems." (The operator was informed prior to participating that certain bar graphs would fall into subgroups and that the discovery of these groupings would make the task easier). Operators could move bars around and change their color to facilitate controlling the device. If the operator believed she or he had the correct decomposition, then she or he would query the computer for correctness until all of the correct groupings were discovered.

After the operator had learned the correct groupings, three different kinds of faults were introduced (i.e., regroupings of the bars): (a) One variable, represented as a bar graph, from one group could become "isolated"—that is, it would no longer be correlated with any of the other groups; (b) two subsystems that were uncorrelated became linked into a bigger subsystem; and (c) one variable from one group became correlated with another group.

Moray (Moray & Reeves, 1987) predicted that the existence of a well-established homomorph would prevent the participants from noticing changes in the covariations. This is exactly what he found. In analyzing participants' response times to notice the regroupings, as well as the response times to discover the original groupings, three main results were obtained. First, there were large individual differences for the rate at which the original bar graph groupings were learned. Second, reassigning bars to groups after an incorrect grouping took longer to learn than learning the original groupings. Last, of the changes instantiated after the original grouping was found, noticing that a bar from one group had become correlated with another group took the longest time, and noticing that two previously uncorrelated groups were now linked took the second longest amount of time. Interestingly, all changes took longer to notice than it took to discover the initial homomorph groupings.

The implication of Moray's (1984, 1988) work for knowledge engineers is that knowledge relies on a theoretical framework. If the knowledge engineers understand the framework and the mapping of the expert's knowledge into the framework, then predicting how the expert operator will behave should be a product of the theory. More work needs to be done in terms of whether or not the assumptions behind homomorph theory are supported (e.g., the hierarchical nature of systems) and how widely applicable they might be to other systems and processes. Nevertheless, the paradigm used by Moray (1984, 1988) is a rich one and a useful tool for exploring mental models. A knowledge engineer might use this theory by understanding the kinds of systems and subsystems that are built into a domain. By applying Moray's (1984,1988) techniques (specifically, the lattice decomposition) a conceptual graph of an operator's understanding of the system can be constructed with some confidence.

Judgment Tasks: Schvaneveldt et al.

Schvaneveldt et al. (1985) take a different empirical and theoretical approach to the study of mental models. They focus on the assumption that the organizational properties of concepts in memory have critical impact on performance. The domain in which they worked was air combat maneuvering. Their goal was to understand and measure the conceptual structures of both expert and novice fighter pilots. Flight concepts for two different maneuvers were identified, and novices and experts made similarity judgments on all pairs of concepts for both maneuvers. The resulting data were submitted to multidimensional scaling and network analysis routines, yielding relational and organizational information about participants' impressions of the concepts. The authors showed that given this multivariate output they could reliably discriminate among groups of pilots, predict group membership of pilots, and detect subtle differences in novice and expert pilots.

Schvaneveldt et al. (1985) showed that these techniques could be used for extracting expertise for knowledge-based systems. Their approach has several advantages over other methodologies at eliciting and representing semantic knowledge (Barnett, 1989). First, multivariate techniques are objective and quantitative. A second advantage is their ease of application and interpretation. The expert is asked to help select key concepts, rate those concepts, and interpret the output. The amount of expert time is minimal. Thus, multivariate techniques are not as costly as qualitative methods for either the expert or the interviewer. Also the rating task (a) does not require the expert to state a rule or heuristic, and (b) is less cumbersome to analyze and interpret than protocol analysis. The last advantage is that composite solutions from a group of experts can be made; moreover, expert matrices can be correlated and compared (in some cases) to view consistency among experts. Group solutions can be used to generate predictions and can be tested against individual solutions. As Schvaneveldt et al. have shown, the group solution can also provide information as to how nonexpert performance differs from expert performance. One of the problems with interview techniques is that people tend to talk very differently, and at different levels, about what they know. With similarity judgments, the data can be combined for a group solution or examined individually to provide different opinions.

The main disadvantage of multivariate techniques is concept selection. If the concepts and assumptions are not chosen carefully and correctly, the representation might not reflect any knowledge at all. Furthermore, if pairwise similarity ratings are used, the number of concepts is constrained to be quite small (less than 30). (The number of pairwise comparisons is $N^2 - N$, if one is assuming asymmetry, and $(N^2 - N)/2$, if not.) There are other techniques for acquiring similarity ratings from large sets of concepts (up to several hundred concepts), but these are not considered as reliable as those obtained by similarity ratings. Most domains probably have a large number of interrelated concepts that are candidates for this technique, and selecting the correct subset of concepts can be difficult. A second disadvantage is that it is unclear how to combine several related networks of concepts. For instance, if a financial expert were asked to rate the similarities of macrofinancial concepts (e.g., to what extent does the bond market influence the stock market?), then how would the answer be integrated with a network that only deals with stock market concepts? Or to an individual stock? This seems to be a fruitful area of research. Third, the mapping of predicates to links in the network is nontrivial. Just because a link between the bond market and the price of currency may exist, knowing what predicate to use (or even believing in that link) is not entirely obvious. Although we can assert that a link between concepts exists, it needs to be given a meaningful name from a predetermined list of predicates. Where one might hope that the predicate-to-link mapping would be one-to-one, it might not be; thus, there could be significant problems in moving from the derived psychological representations to the computer representation. Last, scaling tasks are artificial and unfamiliar. They force experts into conforming a subset of their knowledge to the requirements of the task. When the knowledge engineer is building a knowledge base using scaling techniques, these deficiencies will require that the data derived from judgments be supplemented by other standard knowledge elicitation tasks.

In sum, the techniques described by Schvaneveldt et al. (1985) are enticing. When key assumptions and concepts of domains can be identified, the relationships between concepts can be identified through a series of similarity comparisons. Although there are some legitimate concerns here about the assumptions of scaling representations of similarity (see Tversky, 1977), it seems that uncovering the structure of what people know about a device using scaling is precisely the point of studying mental models. For instance, Schvaneveldt et al. were able to differentiate novices from experts solely on the basis of similarity judgments and to reliably identify those novices who were subjectively felt by the experts to be the most promising young pilots. It should be noted that the data obtained by these techniques are not always as clean as that described by Schvaneveldt et al., but others have found reasonable results by applying this technique (e.g., Hanisch, Kramer, Hulin, & Schumacher, 1988).

In closing this section, two points come out. First, there are many ways to probe people's mental models of systems. Second, and more important, the best routes in the study of mental models are those for which there are theoretical bases that make clear predictions. To this end, Schvaneveldt et al.'s (1985) method describes what people know if the concepts can be identified, but not how they learned what they know. Moray's (1986,1988) theory makes predictions concerning expert operator performance but is not developed enough to account for learning. What is lacking is a theoretical account of how people acquire mental models. If we understand how people acquire mental models, then we should have a better understanding of how to exploit mental models for knowledge engineering and user-centered design.

Toward a Theory of the Acquisition of Mental Models

We will next present a preliminary qualitative theory of mental model acquisition with an eye toward an analytical account. The unique aspect of our approach to mental models is that we consider the process of using a mental model to be one that reflects cyclic changes in memory retrieval. Thus, our approach to this theoretical effort is inspired by memory theories found in the psychological literature. As with most memory theories, the key element to the theory is the nature of the representation. The plan for the rest of this section is to present three stages that we believe describe how a user acquires a mental model of a device. We see this sequence as an evolution from where the learner principally relies on the superficial features of a domain but later relies on the structural or causal features of the domain with increasing expertise. Following that, we move into the most important part of this paper, where we present our approach to an analytical account of mental models. We end up with a brief discussion of how understanding a user's model of a system can help in improving interactions with that system.

Three distinct stages of mental model acquisition are postulated (Schumacher, 1987; see also Forbus & Gentner, 1986): pretheoretic, experiential, and expert. In forming our thoughts we discovered that a number of memory and categorization models from cognitive psychology could be seen as providing close descriptions of how we envisioned each of the three stages, although no single model encompassed the characteristics of all three stages.

1. In the *pretheoretic* stage, the understanding of system performance centers on retrieval of similar instances in memory—specifically, the superficial similarities of the current system that are shared with (relevant to user's goals) previously learned systems. (By *superficial similarities* we mean those similarities that are irrelevant or secondary to the user's tasks, e.g., font size or color; *structural similarities* are those that directly relevant to the user's task.) These shared superficial features support interaction with the current system to the extent that the superficial features are correlated with the system's structural features. When shared superficial and structural features are nil or in conflict, performance is reduced (e.g., Gentner & Toupin, 1986; Medin & Ortony, 1989; Ross, 1984; Schumacher & Gentner, 1988). The pretheoretic stage is seen as primarily a stage where mental models are a collection of experiences that are retrieved from memory based on similarities to prior events. The categorization theories of Medin and Schaffer (1978) and Nosofsky (1985) describe how instances are organized in memory in terms of their similarity in psychological space. We would like to borrow from these views in describing how humans might store similar instance experiences in memory. Suppose someone compares an experience with a new system to other prior system experiences (including the exact same system) in memory on the basis of salient features. Suppose further that if two systems are similar across a variety of features, events would then be stored in a like manner, mainly on the basis of superficial similarities. The implication is that both the retrieval of old knowledge and the storage of new information would be dependent on similarities between old and new systems in memory. Especially in initial interactions with a system, the user may rely heavily on superficial similarities (because they are presumably more salient) to guide retrieval and storage of new system information. We assume that similarity between attributes of systems is measured psychologically, much as described by Nosofsky.

2. In the *experiential* stage, some understanding of causal relationships emerges even if this understanding is not supported by superficial features. For example, Schumacher and Gentner (1988) had participants control a simulated steamship. With small amounts of practice, subjects discovered a direct relationship between the "speed" gauge and the "temperature" gauge of the system: When speed increased, so did temperature. Late in the experiential stage, the user may develop expertise with these specific parts of the system, but this knowledge may not transfer easily to other, similar systems (Schumacher & Gentner, 1988). Schumacher and Gentner found that when subjects transferred from the training device to a device that did not share superficial similarities but was functionally isomorphic with the old system, transfer was reduced relative to the case where superficial similarities were shared. The implication is that the mental model of the system must be ingrained before transfer can be effective. The experiential stage is where we assume that abstraction begins to take place. Hintzman's (1986) MINERVA model of memory accounts for this notion of abstraction over time. Although MINERVA represents an exemplar theory of memory retrieval, it paradoxically proved to be able to show how higher order abstractions might be formed and stored. Hintzman's is not the only theory that argues that abstraction arises from recognition of common features across instances (see Anderson, 1987, for good discussion of this), but it is one that is formulated well for our purposes. Much of the heart of our model was derived from ideas about how MINERVA works.

3. The *expert* stage is reached when the user makes abstractions across various system representations. In this stage, the user easily recognizes systemic patterns of behavior and retrieves old system knowledge. Knowledge can therefore be easily transferred across instantiations of the system even with systems that are superficially dissimilar.

The expert stage is described by abstraction and facile recognition of patterns of system behavior (Anderson, 1987; Dreyfus & Dreyfus,

1986). This final stage is well characterized by Moray's (1984, 1988) lattice theory of mental models described earlier. The main components of Moray's theory that apply to this stage of mental model acquisition are that the experts' models rely less on irrelevant detail than novices' models, and expert models are assumed to be high-level abstractions of system knowledge.

An Analytical Account

As with all analytical theories, being able to simulate the cognitive process described above could prove helpful not only in honing in more closely on the nature of mental model acquisition but also in making predicitions concerning novice versus expert interactions with new and old systems. Therefore, prompted by the possible utility of an analytical account in helping us specify mental model acquisition, we next describe our first attempt toward this goal.

We assume that each encounter with a system has associated with it a vector of features stored in memory. While we do not suppose to understand precisely what a "feature" is, let us consider, for expository purposes, things like "font size," "copy," "save," "diskette," and "remove" to serve as features. Features might be relations between objects, descriptions of objects, or objects themselves. Even user goals in a given context, as well as the context, could be considered features within this framework. The size and boundary conditions of the feature vectors we will also not speculate on, except to say that we will assume there are limits to the number of features in an episode: For computational simplicity, each episode vector has a fixed length. Because the feature vector is a part of the user's model of the system, the user includes in each vector both relational and object-based features. We assume that there are parameters that weight the user's attention toward either the relational or superficial features of an episode. Retrieval processes are assumed to correspond those described by Raajmakers & Shiffrin (1981), where a stochastic sampling of similar items stored in memory takes place. The final assumption is that once a feature vector (or number of feature vectors) has been retrieved given a probe vector, the probe and feature vector(s) "merge," forming an abstraction vector. After this abstraction process, all vectors—those retrieved from memory, the probe vector, and the abstraction vector—are then stored in long-term memory.

During any task, we see the user as probing memory with feature sets (vectors) from the current task. The probes are then compared to other vectors in memory. Novices' vectors have attentional weights that emphasize the object-based features, whereas the experts' weights emphasize relational features. Given that a probe vector of features enters memory, feature vectors from previously encountered systems (or the same system on a prior occasion, most likely) are retrieved based on their proximity in psychological space across all features, modified by their weights. This process is modeled using a variation of Luce's choice rule (see Engelbeck, 1989; Raajmakers & Shiffrin, 1981). The user is assumed to stochastically sample from the retrieved system vectors. Each time the user successfully probes memory with a vector, or retrieves a particular system vector, the attentional weights of both those vectors shift in favor of the relational features. The commonalities and differences between the two vectors are retained (i.e., stored in memory), resulting in an abstraction across exemplars. According to the theory, a user will more often retrieve abstracted system information from memory with experience. With increasing interaction with a system, a user would more often retrieve previously stored vectors that share relational features with the current system. Thus, the movement from a system novice to a system expert is the shifting of attentional weights from surface to relational features associated with a system and the subsequent retrieval of similar system information from memory. Within certain boundaries, then, an expert should be able to retrieve more abstracted, relational system information from memory than a novice during interaction with a new system.

An example might help to illustrate how this account might work during a person's encounter with a new system. Let us suppose that a UNIX®[3] expert is suddenly placed in a situation where he or she must learn to use a Macintosh computer for a period of time. Let us further assume that the user would like to learn the procedure most analogous to navigating along a hierarchical path in UNIX® on the Macintosh®.[4] The UNIX® user might be deeply embedded within a Macintosh® application and pull open an "OPEN" file scrollbox. Seeing that the file being searched for is not listed within the scrollbox, the user clicks on the folder icon above the list. Two other folders are pulled down from that folder. At this point, the user uses this set of system features to probe memory about how to use this system information: folders, files, lists, mouse, scrolling, selecting, and so forth. Probing memory with values for these system features results in some feature vectors relating to other Macintosh encounters, because the user probed memory with surface features used heavily in the Macintosh interface. If the user had probed memory with the knowledge that the pull-down list contained a hierarchical organization showing the "path" that the user could follow (i.e., relational features), then abstracted information from both UNIX® and Macintosh® system use could have been retrieved from memory.

This prospectus is obviously incomplete. For instance, one problem is that with the vector notation used to describe this theory, there is no way to represent relational structure, which we view as critical. This is one problem that lattice theory handles easily, for expert knowledge (Moray, 1988). One possibility for extending our model to handle relational structure is to move toward a graph-type representational structure so that the relations between vectors can be handled within the framework. Another problem is what to do with system features in the probe vector that do not match any similar system information retrieved from memory. It is well known that the dissimilarities encountered between objects serve an important role during similarity judgments (e.g. Tversky, 1977; Tversky & Gati, 1982). We are currently working on possible solutions to this problem so that dissimilar system features might serve an "orienting" function. System dissimilarities may provide the grounds from which valuable hypothesis-testing is generated.

Nevertheless, a number of interesting theoretical tests of the theory seem possible and are currently being formulated. For instance, the three stages of mental model acquisition are being investigated in the laboratory. The theory would predict that if users are relying on the retrieval of old system knowledge to help them to learn about a new system, with varying levels of expertise different sorts of old system knowledge will be retrieved from memory. A cued-recall test format could be used to determine the system knowledge available during various stages of mental model acquisition. Also, the model strongly predicts that abstraction of system information should take place as expertise with the system emerges. This can be tested using recall and recognition techniques. For example, in a recognition test, new system information that is "prototypical" of previously encountered systems should seem familiar to expert users. These predictions, among others, are currently under study.

We believe that analytic simulations of users' interactions with systems can and should be used to aid in the iterative design and testing of user interfaces. Data collected (using traditional recognition and recall techniques, among others) while users learn and transfer system information can be iteratively used to build detailed descriptions of user mental models. The data collected can be used to strengthen the theoretical assumptions about users' mental models. In turn, simulating user-system interaction can then lead to quick predictions about how users will be able to learn and transfer information between systems, as well as what surface and relational information is most important.

Using Mental Models in User-Centered Design

The last area that we address is one of mapping what we can find out about mental models into what we can do to improve performance.

Clearly, there are benefits for knowing what someone will do with certain knowledge. For instance, if one knows that a process control operator's model of a system is different (in some ways) from a troubleshooter's model of a system, training regimens can be designed to accommodate and support their respective tasks. Knowledge about information deemed important by users can also be directed toward effective training sessions and user guides.

If our theoretical assumptions are supported, early user-system interaction should be characterized by the user's relying on superficial features of the system. Perhaps the user will spend most of his or her time attempting to recognize and operate these features of the new system. With some experience, however (and this can be simulated), the user should no longer be concentrating on these surface features. In fact, the superficial features become irrelevant during interaction with the system. The interface should provide more than one interaction "mode" so that important relational features are now easily accessible and clearly displayed. The number of adapting modes is an empirical question and could also be simulated using the theory. Finally, knowing the kinds of features important in user models of a system could help the user interface designer in providing useful associations and analogies to the user. As Halasz and Moran (1982) argue, providing the user with a conceptual model can be productive for learning. In order to determine which associations and analogies might be deemed most useful, theoretical simulations could be run and then matched with usability data later.

Knowing the kinds of models people have or will generate gives system designers and human factors practitioners the leverage to design better systems for those models. Knowing the level of expertise a particular user has will be important as more intelligent user interfaces are designed, for example, for expert systems. A quantitative theory of mental model acquisition and maintenance, as discussed above, could prove beneficial in designing flexible interface communication that adapts to the user as his or her interactions with the system give way to different levels of system understanding.

Notes

1. It is usually argued at this point that the procedural group might form a model too. Yes, that is undoubtedly the case, given the ongoing discussion. However, the hypothesis of the experiment is that the model group will have a *better* model than the procedural group.
2. One of the best recent books for presentation of how-it-works knowledge for a wide variety of machines is by Macauley (1988).
3. The typewriter/text-editor analogy is not a particularly good or powerful one by most accounts (e.g., Gentner, 1983, 1989). The reason is that the base concept (or the well-known concept, the typewriter) is not very rich relative to the target term. Thus, few sophisticated predictions can be made by borrowing this mental model. Furthermore, the fact that this association is quite often made simply reflects the fact that domains that share many features are likely to be those that are compared, even (and perhaps especially) those that share many superficial features (e.g., Gentner & Landers, 1985).
4. UNIX is a registered trademark of AT&T Information Systems.
5. Macintosh is a registered trademark of Apple Computer, Inc.

References

Anderson, J. R. (1987). Skill acquisition: Compilation of weak-method problem solutions. *Psychological Review*, *94*, 192–210.

Ashby, W. R. (1956). *Introduction to cybernetics*. London: Chapman and Hall.

Bainbridge, L. (1979). Verbal reports as evidence of the process operator's knowledge. *International Journal of Man-Machine Studies*, *11*, 411–436.

Bainbridge, L. (1981). Mathematical equations or processing routines. In J. Rasmussen & W. B. Rouse (Eds.), *Human detection and diagnosis of system failures* (pp. 259–286). New York: Plenum.

Bainbridge, L. (1986). Asking questions and accessing knowledge. *Future Computing Systems*, *1*, 143–149.

Barnett, B. (1989). *Information processing components and structural knowledge representations in pilots' judgments*. Unpublished doctoral dissertation, University of Illinois at Urbana-Champaign, Champaign, IL.

Bower, G. H., Clark, M. C., Lesgold, A. M., & Winzenz, D. (1969). Hierarchical retrieval schemes in recall of categorized word lists. *Journal of Verbal Learning and Verbal Behavior*, *8*, 323–343.

Brown, J. S., & de Kleer, J. (1981). Mental models of physical mechanisms and their acquisition. In J. R. Anderson (Ed.), *Cognitive skills and their acquisition* (pp. 285–308). Hillsdale, NJ: Erlbaum.

Carroll, J. M., & Olson, J. R. (1988). Mental models in human-computer interaction. In M. Helander (Ed.), *Handbook of human-computer interaction* (pp. 45–66). New York: North-Holland.

Carroll, J. M., & Thomas, J. C. (1982). Metaphor and the cognitive representation of computing systems. *IEEE Transactions on Systems, Man, and Cybernetics*, *12*, 107–116.

Clement, J. (1988). Observed methods for generating analogies in scientific problem solving. *Cognitive Science*, *12*, 563–586.

Collins, A., & Gentner, D. (1986). How people construct mental models. In D. Holland & N. Quinn (Eds.), *Cultural models in language and thought* (pp. 243–265). Cambridge: Cambridge University Press.

Conant, R. C., & Ashby, W. R. (1970). Every good regulator of a system must be a model of that system. *International Journal of Systems Science*, *1*, 89–97.

Cooke, N. M., & Schvaneveldt, R. (1988). Effects of computer programming experience on network representations of abstract programming concepts. *International Journal of Man-Machine Studies*, *29*, 407–427.

Curry, R. E. (1981). A model of human fault detection for computer dynamic processes. In J. Rasmussen & W. B. Rouse (Eds.), *Human detection and diagnosis of system failures* (pp. 171–183). New York: Plenum.

Davis, R. E. (1982). *Expert systems: Where are we? And where do we go from here*? (AI Memo #665) Cambridge, MA: MIT AI Lab.

de Kleer, J., & Brown, J. S. (1983). Assumptions and ambiguities in mechanistic mental models. In D. Gentner & A. Stevens (Eds.), *Mental models* (pp. 155–190). Hillsdale, NJ: Erlbaum.

Dreyfus, H. L., & Dreyfus, S. E. (1986). *Mind over machine*. New York: Macmillan.

Engelbeck, G. (1989). *The acquisition and retention of a cognitive skill*. Unpublished doctoral dissertation, University of Colorado at Boulder, Boulder, CO.

Ericsson, K. A., & Simon, H. A. (1984). *Protocol analysis: Verbal reports as data*. Cambridge, MA: MIT Press.

Forbus, K., & Gentner, D. (1986). Learning physical domains: Towards a theoretical framework. In R. M. Michalski, J. Carbonell, & T. Mitchell (Eds.), *Machine learning: An artificial intelligence approach* (Vol. 3, pp. 311–348). Los Altos, CA: Morgan Kaufmann.

Gentner, D. (1983). Structure-mapping: A theoretical framework for analogy. *Cognitive Science*, *10*, 157–170.

Gentner, D. (1989). The mechanisms of analogical learning. In S. Vosniadou & A. Ortony (Eds.) *Similarity and analogical reasoning* (pp. 199–241). London: Cambridge University Press.

Gentner, D., & Gentner, D. R. (1983). Flowing waters or teeming crowds: Mental models of electricity. In D. Gentner & A. Stevens (Eds.), *Mental models* (pp. 99–129). Hillsdale, NJ: Erlbaum.

Gentner, D., & Landers, R. (1985, November). Analogical reminding: A good match is hard to find. *Proceedings of the International Conference on Systems, Man and Cybernetics* (pp. 607–613).

Gentner, D., & Toupin, C. (1986). Systematicity and surface similarity in the development of analogy. *Cognitive Science*, *10*, 277–300.

Halasz, F., & Moran, T. P. (1982). Analogy considered harmful. *Human Factors in Computing Systems Proceedings*. Washington, DC: National Bureau of Standards.

Halasz, F., & Moran, T. P. (1983). Mental models and problem solving in using a calculator. *Proceedings of the CHI '83 Conference on Human Factors in Computer Systems* (pp. 212–216).

Hanisch, K., Kramer, A. F., Hulin, C., & Schumacher, R. (1988). Novice-expert differences in the cognitive representation of system features: Mental models and verbalizable knowledge. *Proceedings of the 32nd Annual Meeting of the Human Factors Society* (pp. 219–223).

Hinsley, D. A., Hayes, J. R., & Simon, H. A. (1977). From words to equations: Meaning and representation in algebra and word problems. In M. A. Just & P. A. Carpenter (Eds.), *Cognitive processes in comprehension* (pp. 62–68). Hillsdale, NJ: Erlbaum.

Hintzman, D. (1986). "Schema abstraction" in a multiple-trace memory model. *Psychological Review*, *93*, 411–428.

Hoffman, R. R. (1987). The problem of extracting the knowledge of experts from the perspective of experimental psychology. *AI Magazine*, *8*, 53–67.

Hoffman, R. R., Cochran, E. L., & Nead, J. M. (1989). Cognitive metaphors in experimental psychology. In D. Leary (Ed.), *Metaphors in the history of psychology*. Cambridge: Cambridge University Press.

Hoffman, R. R., & Nead, J. M. (1983). General contextualism, ecological science, and cognitive research. *The Journal of Mind and Behavior*, *4*, 507–561.

Holland, J., Holyoak, K. J., Nisbett, R. E., & Thagard, P. (1986). *Induction: Processes of inference learning, and discovery*. Cambridge, MA: MIT Press.

Jagacinski, R. J., & Miller, R. A. (1978). Describing the human operator's internal model of a dynamic system. *Human Factors*, *20*, 425–433.

Kempton, W. (1986). Two theories of home heat control. *Cognitive Science*, *10*, 75–90.

Kessel, C. J., & Wickens, C. D. (1982). The transfer of failure-detection skills between monitoring and controlling dynamic systems. *Human Factors*, *24*, 49–60.

Kieras, D. E., & Bovair, S. (1984). The role of a mental model in learning to operate a device. *Cognitive Science*, *8*, 255–274.

Kieras, D. E., & Polson, P. G. (1985). An approach to the formal analysis of user complexity. *International Journal of Man-Machine Studies*, *22*, 365–394.

Macauley, D. (1988). *The way things work*. Boston, MA: Houghton Mifflin.

Mayer, R. E. (1976). Some conditions of meaningful learning for computer programming: Advance organizers and subject control of frame order. *Journal of Educational Psychology*, *67*, 725–734.

Mayer, R. E., & Bayman, P. (1981). Psychology of calculator languages: A framework for describing differences in users' knowledge. *Communications of the ACM*, *24*, 511–520.

McCloskey, M. (1983). Naive theories of motion. In D. Gentner & A. Stevens (Eds.), *Mental models* (pp. 299–324). Hillsdale, NJ: Erlbaum.

Medin, D., & Ortony, A. (1989). Psychological essentialism. In S. Vosniadou & A. Ortony (Eds.), *Similarity and analogical reasoning* (pp. 179–196). London: Cambridge University Press.

Medin, D., & Schaffer, M. (1978). A context theory of classification learning. *Psychological Review*, *85*, 207–238.

Moray, N. (1986). Intelligent decision aids, mental models and the theory of machines. In E. Hollnagel, G. Mancini, & D. D. Woods (Eds.), *Intelligent decision support in process environments* (pp. 273–291). New York: Springer-Verlag.

Moray, N. (1988). *A lattice theory of mental models of complex systems* (Report No. EPRL–88–08). Engineering Psychology Research Laboratory, University of Illinois.

Moray, N., & Reeves, T. (1987). Hunting the homomorph: A theory of mental models and a method by which they may be identified. *Proceedings of the International Conference on Systems, Man, and Cybernetics* (pp. 594–597).

Nisbett, R. E., & Wilson, T. D. (1977). Telling more than we can know: Verbal reports on mental processes. *Psychological review*, *84*, 231–259.

Norman, D. A. (1983). Some observations on mental models. In D. Gentner & A. Stevens (Eds.), *Mental models* (pp. 7–14). Hillsdale, NJ: Erlbaum.

Norman, D. A. (1986). Cognitive engineering. In D. Norman & S. Draper (Eds.), *User-centered system design: New perspectives on human-computer interaction* (pp. 31–61). Hillsdale, NJ: Erlbaum.

Norman, D. A. (1988). *The psychology of everyday things*. New York: Basic Books.

Nosofsky, R. (1985). Choice, similarity, and the context theory of classification. *Journal of Experimental Psychology: Learning, Memory, and Cognition*, *10*, 104–114.

Raajmakers, J. G. W., & Shiffrin, R. M. (1981). Search of Associative Memory. *Psychological Review*, *88*, 93–134.

Rasmussen, J. (1986). *Human information processing in man-machine systems*. New York: North-Holland.

Ross, B. (1984). Remindings and their effects in learning a cognitive skill. *Cognitive Psychology*, *16*, 371–416.

Rouse, W. B., & Morris, N. M. (1986). On looking into the black box: Prospects and limits in the search for mental models. *Psychological Bulletin*, *100*, 349–363.

Rumelhart, D. E., & Norman, D. A. (1981). Analogical processes in learning. In J. R. Anderson (Ed.), *Cognitive skills and their acquisition* (pp. 335–359). Hillsdale, NJ: Erlbaum.

Rumelhart, D. E., & Ortony, A. (1977). The representation of knowledge in memory. In R. C. Anderson, R. J. Spiro, & W. E. Montague (Eds.), *Schooling and the acquisition of knowledge* (pp. 99–135). Hillsdale, NJ: Erlbaum.

Schumacher, R. M. (1987). Acquisition of mental models. In J. Flach (Ed.), *Proceedings of the Fourth Annual Mid-Central Human Factors/ Ergonomics Conference* (pp. 142–148). New York: Spring-Verlag.

Schumacher, R. M. (1989). *Factors affecting access to analogical similarity*. Unpublished doctoral dissertation, University of Illinois at Urbana-Champaign, Champaign, IL.

Schumacher, R. M., & Gentner, D. (1988). Transfer of training as processing routines. *IEEE Transactions on Systems, Man, and Cybernetics*, *18*, 592–600.

Schvaneveldt, R., Durso, F., Goldsmith, T., Breen, T., Cooke, N., Tucker, R., & DeMaio, J. (1985). Measuring the structure of expertise. *International Journal of Man-Machine Studies*, *23*, 699–728.

Tversky, A. (1977). Features of similarity. *Psychological Review*, *84*, 327–352.

Tversky, A., & Gati, I. (1982). Separability, similarity and the triangle inequality. *Psychological Review*, *89*, 123–154.

Veldehuyzen, W., & Stassen, H. G. (1977). The internal model concept: An application to modeling human control of large ships. *Human Factors*, *19*, 367–380.

Wickens, C. D. (1984). *Engineering psychology and human performance*. Columbus, OH: Charles Merrill.

Wickens, C. D., & Kessel, C. (1979). The effects of participatory mode and task workload on the detection of dynamic system failures. *IEEE Transactions on Systems, Man and Cybernetics*, *13*, 24–31.

Wickens, C. D., & Kessel, C. (1980). Processing resource demands of failure detection in dynamic systems. *Journal of Experimental Psychology: Human Perception and Performance*, *6*, 564–577.

Wickens, C. D., & Kessel, C. (1981). Failure detection in dynamic systems. In J. Rasmussen & W. B. Rouse (Eds.), *Human detection and diagnosis of system failures* (pp. 155–169). New York: Plenum.

Williams, M. D., Hollan, S. D., & Stevens, A. (1983). Human reasoning about a simple physical system. In D. Gentner & A. Stevens (Eds.), *Mental models* (pp. 131–154). Hillsdale, NJ: Erlbaum.

Young, R. M. (1983). Surrogates and mappings: Two kinds of conceptual models for interactive devices. In D. Gentner & A. Stevens (Eds.), *Mental models* (pp. 35–52). Hillsdale, NJ: Erlbaum.

5
Conceptual Analysis as a Basis for Knowledge Acquisition

John F. Sowa

In high school algebra, "word problems" are notoriously difficult. Students who can solve complex equations have trouble analyzing an English sentence to determine the significant variables and relationships. Professional programmers face the same difficulty. The greatest source of errors is the mapping from informal specifications to a formal language. For expert systems, the problem is even worse. Many of them deal with subjects that have never been formalized, such as medical diagnosis, oil well exploration, or automobile registration. In analyzing those subjects, knowledge engineers have no formal theories to guide them. As an aid to formalization, conceptual analysis provides general techniques for analyzing knowledge on any subject. This chapter presents conceptual analysis as a method of analyzing informal knowledge expressed in natural language as a preliminary stage to encoding it in a knowledge representation language. For the examples in this chapter, conceptual graphs are used as the primary knowledge representation language, but the techniques could be applied to any other artificial intelligence (AI) language.

Analyzing Knowledge Expressed in Language

Knowledge acquisition is the process of eliciting, analyzing, and formalizing the interconnected patterns of thought underlying some subject matter. In elicitation, the knowledge engineer must get the expert to articulate tacit knowledge in natural language. In formalization, the knowledge engineer must encode the knowledge elicited from the expert in the rules and facts of some AI language. Between those two stages lies conceptual analysis: the task of analyzing the concepts expressed in natural language and making their implicit relationships explicit. Conceptual analysis must be closely integrated with the other two stages, because the knowledge engineer must do some analysis in deciding what questions to ask the expert and in formalizing the answers in an AI language. Despite the central position of conceptual analysis, most discussions of knowledge acquisition tend to ignore it or illustrate it with a few ad hoc examples. This chapter concentrates on conceptual analysis as an important subject in its own right. An understanding of the techniques can heighten knowledge engineers' sensitivity to the nuances of language, give them guidelines for organizing knowledge, and help them avoid common errors.

A deep understanding of logic and language is a prerequisite for conceptual analysis. Logic is essential because every knowledge representation language is a thinly disguised version of logic; language is essential because the primary means of communication between the expert and the knowledge engineer is natural language. Besides logic and language, some philosophical sophistication is also necessary. The word *ontology*, which usually appears only in abstruse philosophical tomes on the nature of being, is now widely used in theoretical discussions of knowledge acquisition. For the Cyc project, one of the most ambitious knowledge acquisition projects ever undertaken (Lenat & Guha, 1990), the project director has even advertised for *ontological engineers*. With prerequisites like logic, language, and philosophy, conceptual analysis requires a considerable amount of training. Yet some of the techniques have already been implemented in semiautomated tools for knowledge acquisition (Sowa, 1989). With a proper set of tools, the computer itself can become an important aid for helping knowledge engineers learn and do conceptual analysis. But before any technique can be automated, the tool developers themselves must know how to use it with pencil and paper.

Conceptual analysis determines the general principles that define an expert system. In performing inferences, an expert system combines those general principles with specific facts about particular instances. As an example, consider the next three sentences:

Wallets are used to carry money. (1)
Charlie lost his wallet on the train. (2)
Charlie lost all the money in his wallet. (3)

Sentence (1) is a general principle that might be found in a dictionary; it could be encoded in a rule that would apply to any instance of a wallet. Sentence (2) is a specific fact about a particular instance; it might be stored in a database, but it is not a general rule. Sentence (3) is an inference that an expert system might derive by combining the general principle (1) with the specific fact (2).

The general principles constitute the *semantic memory* of an expert system—its basic knowledge about the world. It defines the *ontology*, the catalog of all the object types that the system can reason about. It also includes all the rules that it reasons with. Specific facts constitute *episodic memory* about situations at particular times and places. Episodic memory makes up the operational knowledge that is stored in a data base, acquired from a user during a dialog, or derived by inference in answer to a question. Semantic memory includes five kinds of information:

1. *Taxonomy*: a hierarchy of types and subtypes of concepts and relations.
2. *Definitions*: necessary and sufficient conditions that define new types of concepts and relations in terms of lower level or more primitive types.
3. *Constraints*: general principles that must be true of the instances of those concepts, including their internal structure and their external relationships.
4. *Schemata*: associated relationships that are expected for various concept types. For each type, a schema states the defaults and expectations rather than the definitions and constraints.
5. *Behavior*: rules that govern the actions by and upon each type of object and the interactions of collections of objects.

As an example, a definition of *automobile* would state the necessary and sufficient conditions that define the term: "a four-wheeled, self-propelled vehicle designed for transporting passengers." In that definition, *vehicle* is the genus or supertype, and the modifiers state the conditions that distinguish automobiles from other kinds of vehicles. Besides the defining constraints, there may be legal and practical constraints, for example, on size (big enough to carry a human being but small enough to drive on typical roads) or on the engine (powerful enough to attain highway speeds but with a limited amount of polluting fumes). Definitions and constraints must be true, but for different reasons: If the definition of *automobile* were false for some object, then that object would not be called an automobile; if the other constraints were false, some law would be violated, or the thing could not be used for its normal purpose. Besides the requirements stated in the definitions and constraints, schemata state defaults and background information: A typical automobile costs $12,000 when new; most families in the United States own one or two; they may be status symbols; teenagers covet them; they are essential for people living in Los Angeles, but not in New York City.

Although conceptual analysis is essential for expert system design, it is not a new technique.

The anthropologist Clifford Geertz (1983) observed that similar methods of analysis are necessary for interviewing a native in a foreign culture or an expert in an unfamiliar discipline:

> The problem of how a Copernican understands a Ptolemaian, a fifth republic Frenchman an *ancien regime* one, or a poet a painter is seen to be on all fours with the problem of how a Christian understands a Muslim, a European an Asian, an anthropologist an aborigine, or vice versa. We are all natives now, and everybody else not immediately one of us is an exotic. What looked once to be a matter of finding out whether savages could distinguish fact from fancy now looks to be a matter of finding out how others, across the sea or down the corridor, organize their significative world (pp. 151).

In the computer field, conceptual analysis is also called *conceptual modeling* and *enterprise analysis*. But whatever it is called, the problems and techniques are essentially the same.

In knowledge engineering, as in anthropology, the most fundamental assumptions are usually left unsaid. When asked about them, the expert tends to say, "Oh, but that's obvious! Of course! You mean you didn't know that!?!" The knowledge engineer may have to spend weeks asking probing questions to dig out the underlying assumptions. Some experts may be highly articulate, but others may have a limited ability to express the distinctions verbally. Following is an excerpt from a conversation I had with an expert on malfunctions of a particular device:

> Knowledge engineer: I'm not sure what you want the expert system to do—determine whether a malfunction has occurred, determine what caused it, or determine what action to take in order to correct it.
> Expert: What's the difference?

The expert was highly skilled in fixing the device, but he had never thought of his actions as separate stages or analyzed them verbally. But the knowledge engineer must do that analysis in order to design an expert system that performs a similar function. If the analysis is incomplete or inaccurate, the resulting system may contain arbitrary restrictions, inconsistent data, or limitations that make future extensions impossible.

This chapter will present techniques of conceptual analysis for finding and representing semantic distinctions. The following section introduces two graphical representations: concept maps (Novak & Gowin, 1984) as an informal notation for quickly capturing relationships; and conceptual graphs (Sowa, 1984) as a system of logic for analyzing distinctions in detail and representing them precisely. The third section introduces co-occurrence patterns as empirical tests for distinguishing various conceptual features and relations. The remaining sections of the chapter apply conceptual analysis to some language features that are often represented inadequately and inconsistently in databases and knowledge bases: names, types, colors, measures, and roles. Although the analysis requires some linguistic sophistication, it could be supported by semiautomated aids that would help knowledge engineers learn the methods and apply them consistently.

Representing Concepts and Relations

The analysis of language begins with words. The key words in an interview or specification document express the basic concepts of an application. They map to predicates in logic, domains in a database, slots in frames, attributes in production rules, or classes in Smalltalk. But words and the concepts associated with them are not isolated symbols; they are interconnected in complex patterns of thought:

Religion: sin, kosher, taboo, karma.
Automobiles: carburetor, ignition, brake drum.
Finance: tax shelter, depreciation, puts and calls.

None of these words can be understood in isolation. Even a dictionary is of little help to a person who has no knowledge of related concepts. The entry for *sin*, for example, might define it as a transgression against God. But that introduces the concepts of transgression and God. A transgression is a violation of a

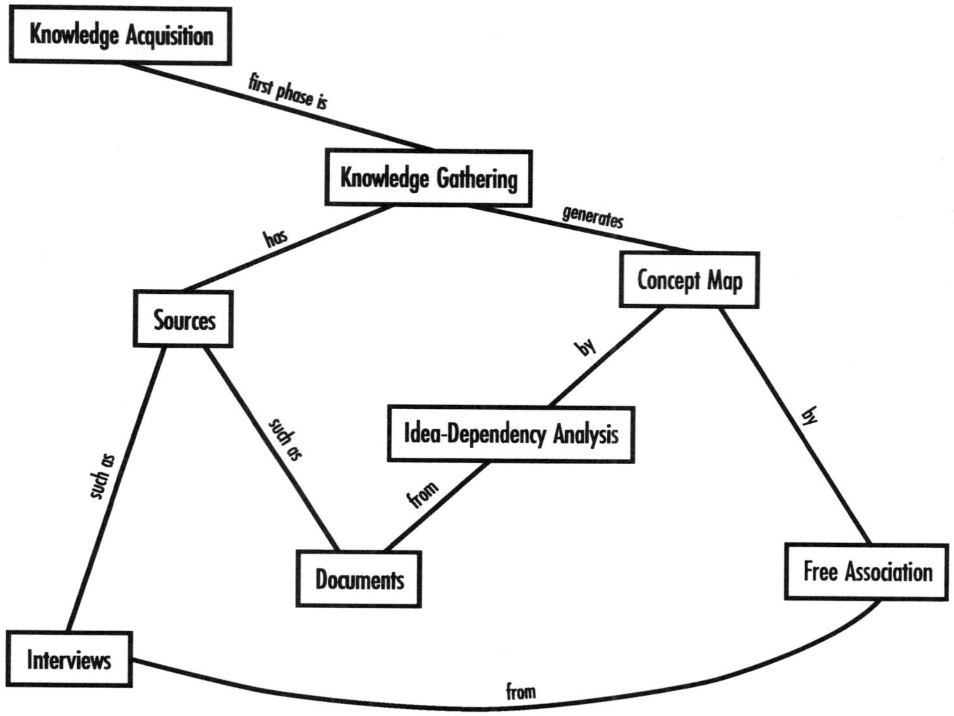

FIGURE 5.1. Concept map for knowledge gathering.

law, but that raises questions about how God gives laws and how they differ from human laws or laws of physics. A few more steps lead to the concepts of heaven and hell and eventually all of theology. Understanding automobile ignition requires knowledge about the electrical system, the fuel system, and even the operation of the engine itself. Puts and calls cannot be understood without detailed knowledge about the stock market and how it operates. In every field of human endeavor, from cooking and fashion to topology and quantum mechanics, the basic concepts can only be understood in relation to other concepts in tightly organized structures of thought. Knowledge acquisition may begin with words, but it must also find the connections that link those words in larger structures.

In eliciting knowledge from experts or gathering it from other sources, informal techniques of brainstorming and free association help the knowledge engineer find related concepts. Drawing graphs or networks of concepts is a good way to record the associations. Unlike traditional outlines, graphs allow a change of focus at any time: Any node in a graph can become a new center for growing a new pattern of relations. The ability of graphs to grow freely corresponds to the way associations can branch off in any direction. Buzan (1974) advocated such networks as an aid for taking notes, solving problems, drafting speeches, and even organizing meetings. The educational psychologists Novak and Gowin (1983) developed similar networks, which they called *concept maps*. Figure 5.1 shows a concept map for "knowledge gathering" and related ideas. The following are a few informal guidelines for drawing concept maps:

As you think of a concept, write it down and draw a box around it.

Keep the concept descriptors short—usually one or two words.

Draw lines linking the new concept to other concepts that are already on the map.

Write a word or phrase on each link to show how the concepts are related.

Novak and Gowin have successfully taught concept mapping to children as early as first grade. It helps to draw their attention to key concepts and relationships. Teachers have used it to assess the concepts a child already knows and the points where the child is still confused.

Concept maps can serve as a common notation for the expert and the knowledge engineer. Their informality allows ideas to be written down as soon as anyone thinks of them, and their flexibility allows them to accommodate the special symbols and equations used in various disciplines. Yet simplicity, which is their major strength, is also their major weakness: The lack of a formal structure makes them unable to express all of logic; the absence of detailed rules forces every user to develop special conventions that are incompatible with everybody else's conventions; and their highly abbreviated style makes them incapable of capturing all of the details expressed in natural language. In short, concept maps are a good tool for knowledge elicitation, but for the analysis and formalization stages a more precise formal language is necessary.

Concept maps can be made into a formal language by imposing constraints on how they are written and used. For database design, entity-relationship (E-R) diagrams are stylized concept maps that express constraints on relations. In AI, semantic networks are more general than E-R diagrams and express a wider range of relationships. Conceptual graphs are a version of semantic networks designed as a complete system of logic: They have a direct mapping to and from English; they can be translated to and from other AI formalisms; and they can support automated knowledge acquisition tools. In this chapter, they will be used as the notation for capturing and formalizing the results of conceptual analysis. As an example of the mapping from language to conceptual graphs, consider the following sentence, which might appear in a specification document:

Every employee is hired by some manager on some date.

The analysis starts with the *content words* in the sentence: *employee*, *hire*, *manager*, and *date*. In any database or knowledge base, these words would map to some significant feature. In conceptual graphs, they map to *type labels* for the associated concepts. The labels are written as upper-case character strings: EMPLOYEE, HIRE, MANAGER, and DATE. Those labels must then be organized in a type hierarchy: MANAGER is a subtype of EMPLOYEE, which is a subtype of PERSON; HIRE is a subtype of ACT; and DATE is a subtype of TIME. Subtypes may be shown either in a tree diagram or with the < symbol:

MANAGER < EMPLOYEE < PERSON
HIRE < ACT
DATE < TIME

The other words in the sentence—*every*, *is*, *by*, *some*, and *on*—are usually called *function words*. Unlike the content words that map to concepts, function words map to conceptual relations or to quantifiers inside a concept node. The words *every* and *some* represent quantifiers; the auxiliary verb *is* and the ending *-ed* mark the passive voice, which indicates that [HIRE] is linked to [EMPLOYEE] by the PTNT (patient) relation; and *by* indicates that [HIRE] is linked to [MANAGER] by the AGNT (agent) relation. The preposition *on* maps to different relations in different contexts; in this example, the type DATE, which is a subtype of TIME, indicates that *on* represents the PTIM (point in time) relation. Figure 5.2 shows a conceptual graph for that sentence.

In the concept [EMPLOYEE: $\forall$], the type label is followed by a colon and a *referent field*, which contains the quantifier $\forall$ for *every*. The word *some* represents an existential quantifier. The ordinary existential $\exists$ is the default; the

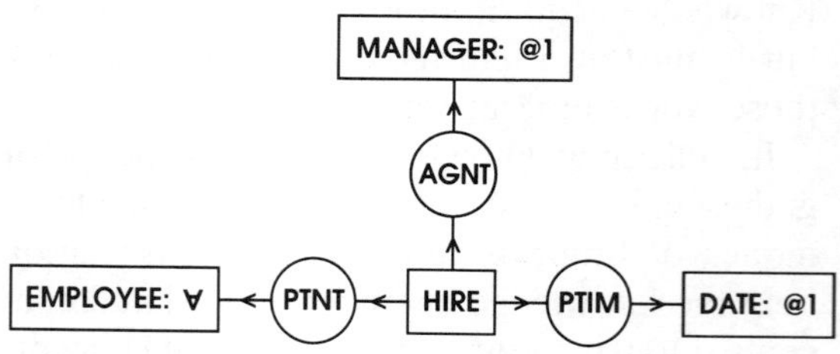

FIGURE 5.2. A conceptual graph.

phrase *some manager* would be represented by the concept [MANAGER], with nothing in the referent field. That concept corresponds to the formula $(\exists x)$manager(x), which means that there exists one or more individuals x of type manager. If no further information were available, the default would be assumed for both *some manager* and *some date*. But whenever the word *some* is used, the knowledge engineer should ask questions to determine what kind of existential quantifier is intended:

Is each employee hired by exactly one manager?
Is it possible for an employee to be hired by several managers?

In predicate calculus, the unique existential quantifier ∃! is used to show exactly one; in conceptual graphs, the symbol @1 is used in the referent field: [MANAGER: @1].

To map conceptual graphs to the predicate calculus, the *formula operator* ϕ is used. In the mapping, ϕ assigns a constant or a quantified variable to each concept node. The symbol ∀ in the referent field maps to the quantifier ∀; the symbol @1 maps to the quantifier ∃!; and the concept [HIRE] with nothing in the referent field maps to the default ∃. Type labels map to one-place predicates, and conceptual relations map to predicates with as many arguments as there are arcs attached to the circles. With these conventions, ϕ maps Figure 5.2 to the following formula:

$(\forall x)(\exists! y)(\exists! z)(\exists w)(\text{employee}(x) \supset (\text{manager}(y) \land \text{date}(z) \land \text{hire}(w) \land \text{agnt}(w, x) \land \text{ptnt}(w, y) \land \text{ptim}(w, z)$.

To save space, there is also a linear notation for conceptual graphs where the boxes are written as square brackets and the circles are written as rounded parentheses. Since the concept [HIRE] in Figure 5.2 has three arcs attached to it, the graph cannot be drawn on a straight line. Instead, [HIRE] is followed by a hyphen to show that the relations attached to it are continued on subsequent lines:

```
[HIRE]-
  (AGNT) → [MANAGER: @1]
  (PTNT) → [EMPLOYEE: ∀]
  (PTIM) → [DATE: @1].
```

In this notation, a conceptual graph looks very much like a frame: The relations correspond to attributes, the referent fields are slots for holding values, and the type labels are constraints on the possible values that can fill those slots. A particular fact—such as *John Smith was hired by Mary Brown on July 14, 1988*—would be represented with the names of the individuals in the referent fields:

```
[HIRE]-
  (AGNT) → [MANAGER: Mary Brown]
  (PTNT) → [EMPLOYEE: John Smith]
  (PINT) → [DATE: July 14, 1988].
```

Although this graph looks like a frame, conceptual graphs are more general than frames, because they can represent all of the quantifiers and operators of logic. They can also be used to define new types of concepts and relations. In a definition, one or more concepts in the graph are marked as *formal parameters*. The following is a definition of the new relation type DOH, which represents date of hire:

```
relation DOH(x,y) is
  [HIRE]-
    (AGNT) → [MANAGER]
    (PTNT) → [EMPLOYEE: *x]
    (PTIM) → [DATE: *y].
```

In this definition, DOH is the label of a two-place conceptual relation, where the variables x and y mark the two parameters. The first parameter is the concept [EMPLOYEE], and the second is [DATE]; the asterisks on $*x$ and $*y$ in the concept nodes distinguish them from names. With this definition, DOH could then be used to relate an employee to a date:

[EMPLOYEE: John Smith] → (DOH) → [DATE: July 14, 1988].

The arrow pointing toward the relation marks the first parameter, and the arrow pointing away marks the second. Note that the definition of DOH also included two other concepts: [HIRE] and [MANAGER] as well as the relations (AGNT), (PTNT), and (PTIM). Those concepts and relations can be recovered by expanding DOH. Definitions are a mechanism for hiding information that is not immediately relevant; when needed, the hidden information can always be recovered.

Co-occurrence Patterns

Words with certain semantic features tend to occur in one kind of expression, but not another. Linguists take advantage of this tendency by looking for *co-occurrence patterns*—phrase or sentence patterns that systematically occur with words that have a certain feature. The co-occurrence patterns serve as an empirical test for the presence of that feature. This technique, which is a standard tool in linguistics, can also be used in knowledge engineering. As an example, consider the following pattern:

a book, three books
an elephant, three elephants
an idea, three ideas

Those phrases sound normal, but the following seem odd or incorrect:

a water, three waters
a butter, three butters
a happiness, three happinesses

A phrase like *three waters* may be used, but only to indicate three kinds, not three instances: *You have a choice of three waters: Perrier, Deer Park, or New York City tap*. This co-occurrence pattern highlights a distinction between two kinds of nouns: *book*, *elephant*, and *idea* are *count nouns* that occur with the article *a* in the singular and can be used normally in the plural; but *water*, *butter*, and *happiness* are *mass nouns* that do not occur with *a* in the singular and have a different meaning in the plural. The distinction corresponds to a semantic difference that is important for knowledge representation: Count nouns refer to things that can be counted, but mass nouns refer to substances that can only be measured. Although the distinction is usually quite sharp, some nouns like *cake* may be used as both count nouns and mass nouns: As a count noun, *a cake* refers to an object, but as a mass noun, *cake* refers to the substance of which a cake is made. This is not an exception to the rule, but an example of a word with two different meanings.

The presence or absence of an article in English signifies an important semantic distinction, but many languages, such as Chinese and Russian, have no articles. Each language has its own means of showing semantic distinctions, and different languages emphasize different ones. Although Russian does not mark definite and indefinite noun phrases as clearly as English, it is much more specific about marking perfective versus imperfective verbs. Another feature of English is the distinction between the simple present tense *walks* and the progressive form *is walking*. That distinction is the basis for useful co-occurrence tests that have no direct counterpart in languages, such as French or German, that lack a progressive form. In every language, there are co-occurrence patterns that can be used to test for semantic features, but the patterns in one language do not always have a direct translation into other languages.

In English, verbs have three syntactic relations that are signaled by position: subject, object, and indirect object. Yet those syntactic relations express different semantic or conceptual relations that could be distinguished by co-occurrence patterns. The verbs *eat* and *like*, for example, both occur in the following pattern:

I *ate* the doughnut.
I *liked* the doughnut.

This pattern distinguishes transitive verbs from intransitive verbs like *sleep* or *walk*. Yet there are still finer distinctions to be drawn. The following patterns sound natural or normal with the verb *eat*, but they sound odd with the verb *like*:

I am *eating* the doughnut.
I am *liking* the doughnut.
What I did was *eat* the doughnut.
What I did was *like* the doughnut.

These co-occurrence patterns distinguish actions from states: verbs in the *eat*-class express actions, while verbs in the *like*-class express states. Other verbs that can occur in the same patterns as *eat* include *mash*, *bake*, *dunk*, *discard*, and *explode*. Some verbs that occur in the same patterns as *like* include *own*, *admire*, and *know*.

In conceptual graphs, the distinction would be marked by different conceptual relations instead of the single syntactic relation of subject: Actions have agents, represented by the AGNT relation; states, however, are linked to their subjects by EXPR (experiencer) if they are mental states or by STAT (state) if they do not depend on any mental experience:

[PERSON: #I] ← (AGNT) ← [EAT] → (PTNT) → [DOUGHNUT: #].
[PERSON: #I] ← (EXPR) ← [LIKE] → (PTNT) → [DOUGHNUT: #].
[PERSON: #I] → (STAT) → [OWN] → (PTNT) → [DOUGHNUT: #].

The symbol # marks context-dependent referents. By itself, # represents the definite article *the*. The symbol #I represents the speaker in the current context.

Not all events and processes have agents. Consider the sentences *John opened the door* and *The door opened*. In the first sentence, John is the agent of opening. But in the second, the door is not an animate being that could be an agent. In fact, the door seems to have the same semantic relationship to the concept OPEN in both sentences. To test for agents, the *on purpose* pattern may be used:

John opened the door on purpose.
The door opened on purpose.

Only agents can do things on purpose. Because the first sentence sounds normal, John is the agent of the concept OPEN. But the second sentence sounds odd. Therefore, the door cannot be the agent of OPEN. Instead, it would be the patient in both sentences:

[PERSON: John] ← (AGNT) ← [OPEN] → (PTNT) → [DOOR: #]
[DOOR: #] ← (PTNT) ← [OPEN].

For the verb *open*, the subject could be either the agent or the patient. For some verbs, such as *rise*, the subject is always the patient. Consider the sentence *If John raises the temperature, then the temperature will rise*. The *on purpose* test would show that John is the agent of RAISE, but the temperature is not the agent of RISE. Instead, the temperature is the patient of both RAISE and RISE.

The PTNT relation links a concept corresponding to a verb to a concept of something that is acted upon, experienced, or just passively involved. But consider the next pattern:

What I did to the doughnut was *eat* it.
What I did to the doughnut was *make* it.

There is something odd about the second sentence: Eating can be done to a preexisting object, but making cannot be done to the object because it didn't even exist before the act of making. The creation of an object as a result of an action must be indicated by the relation RSLT instead of PTNT:

[PERSON: #I] ← (AGNT) ← [MAKE] → (RSLT) → [DOUGHNUT: #].

Sometimes the distinction between patient and result may not be obvious, as in the sentence *I baked the doughnut*. That sentence has two possible interpretations: *I created the doughnut by baking it*, which would use the RSLT relation; or *I took a doughnut that already existed and baked it*, which would use the PTNT relation.

In the last example, the concept [BAKE] had different relations linked to the concept corresponding to the object. Sometimes a verb might have different relations linked to its subject. White (1975) discussed the distinction between thinking about an object and thinking that some proposition is true. Consider the following examples:

I was engaged in thinking about the problem.
I was interrupted at thinking about the problem.
I was tired out by thinking about the problem.

All of these sentences sound natural with the ending *about the problem*. But if that phrase were replaced with *that the sky is blue*, they would make no sense. The phrase *thinking about* passes the tests for an action with an agent:

I am thinking about the problem.
What I did was think about the problem.
I thought about the problem on purpose.

But all of these sentences would sound odd if the phrase *about the problem* were replaced with *that the sky is blue*. Therefore, *think that* cannot be an action with an agent, and it must be a state with an experiencer:

[PERSON: #I] ← (AGNT) ← [THINK] → (ABOUT) → [PROBLEM: #].
[PERSON: #] ← (EXPR) ← [THINK] → (PTNT) → [PROPOSITION: [SKY: #] → (ATTR) → [BLUE]].

In the second graph, the patient of [THINK] is a proposition whose referent is a nested graph that says that the sky has attribute blue.

An important classification of verbs is the system by Vendler (1967), which is a classic in conceptual analysis. He distinguished *achievements*, which are completed at one instant of time, from *accomplishments*, which have intermediate stages. Those two classes are distinguished by the following pattern:

John partially *wrecked* the house.
John partially *bought* the house.

The first sentence means that John did some damage, yet less than total destruction. But the second sentence is nonsense: Buying is an indivisible action that cannot be done partially. This test distinguishes achievement verbs like *buy* that refer to indivisible actions from accomplishment verbs like *wreck* that allow intermediate stages or degrees of completion. Other achievement verbs that cannot be done partially include *see*, *take*, *enter*, and *drop*. Some accomplishment verbs that can be done partially include *penetrate*, *cover*, *consume*, and *burn down*.

Some of the verb patterns test for shallow features that are on the borderline between syntax and semantics. But other patterns can detect deep-seated semantic properties as well. As an example, Katz (1966) used the following pattern:

The *razor blade* is good.
The *grain of sand* is good.

The first sentence is complete in itself and needs no further explanation. The second, however, immediately leads the reader to ask, *Good for what*? Terms that fit in the same co-occurrence class as *razor blade* include *aspirin*, *poker hand*, and *lung*. These words refer to things that have implicit purposes: A razor blade is supposed to be good for shaving, a lung for breathing, and a poker hand for winning a game. Words that fit in the same class as *grain of sand* include *molecule*, *integer*, *liquid*, and *planet*. These words have no implicit relations that provide a scale of goodness. A grain of sand may be good for causing an oyster to grow a pearl, and a planet may be good as a target for space exploration, but those uses are not part of the meaning of *sand* or *planet*. As a test to determine the implicit relationships for a concept type, consider the next example:

Here is a *razor blade*, but . . . .
Here is a *grain of sand*, but . . . .

In the first sentence, one would expect a continuation that denies some typical relationship of a razor blade, for example, *You can't shave with it* or *Its blade is chipped*. In the second, however, there is no obvious continuation. In a discussion about cultured pearls, it might make sense to say, You can't grow a pearl with it. Yet that continuation would be suggested by the previous context, not by any relationship intrinsic to the grain of sand itself. This example can be generalized to a co-ocurrence pattern known as the *but* test:

Macula is a dog, but she is not an animal.
Lucky is a pet, but he is not tame.
Muffy is a cat, but she is not black.
Yojo is a cat, but he is not a verb.

The first sentence is contradictory because the definition of dog implies animal. The second sounds normal because being tame is an expectation for a pet, but not a requirement. The third sounds odd with *but* because being black is not an expectation for a cat. That sentence, however, would sound normal with the conjunction *and* instead of *but*, because not being black is a possibility. The fourth is literally true, but it sounds odd because there is no way that anything could ever be both a cat and a verb. In general, for any concept type t and predicate $p(x)$, the *but* test has the following form:

x is an instance of t, {but | and} not $p(x)$.

A sentence of this form may be contradictory, natural, odd, or nonsensical with the conjunction *but* or *and*.

If contradictory, then either $p(x)$ is part of the definition of t, or it is implied by the definition.
If natural with *but* and odd with *and*, then $p(x)$ is implied by the expectations for t.
If odd with *but* and natural with *and*, then $p(x)$ is possible for t, but it is not an expectation.
If odd or nonsensical with both *but* and *and*, then $p(x)$ does not apply to instances of type t.

The collection of all the expectations for a concept type t forms the schema for t. The *but* test can be applied to each clause or predicate of a definition or schema to verify that it is well formed.

Each co-occurrence pattern divides a set of words into two classes: those that fit the pattern, and those that don't. A second test can subdivide either or both of those classes further. The class of transitive verbs that fit the pattern *I* ______ *the doughnut* were subdivided into the subclasses of actions versus states by the pattern *What I did was* ______ *the doughnut*. As a third test, the pattern *What I did to the doughnut was* ______ *it* distinguishes those actions that apply to a patient from those that generate a result. With n patterns, up to 2^n co-occurrence classes may be distinguished. Each pattern tests one *binary feature*, and each co-occurrence class is defined by the presence or absence of certain features. The ultimate meaning of labels like AGNT, PTNT, or RSLT is completely determined by the string of bits that specifies which tests succeeded or failed.

Explicit labels for concept and relation types are among the aspects that distinguish conceptual graphs from other systems of logic. In some notations, verbs are represented as predicates whose arguments are distinguished only by position. Transitive verbs would map to two-place predicates: eat(x, y), like(x, y), make(x, y), bake(x, y), and think(x, y). This form represents the syntactic distinction between subject and object, but it obscures the semantic distinction between agent and experiencer or patient and result. Distinctions between different kinds of baking and thinking could be shown with different predicates, such as bake1(x, y) versus bake2(x, y). But that notation fails to show the common action underlying bake1 and bake2, and it fails to show the difference between patient and result. With conceptual graphs, the concept type BAKE shows the common action, and the difference is shown by the relations (PTNT) or (RSLT). Although most people who use predicate calculus do not make those distinctions, the notation is general enough to represent them. In fact, the ϕ operator that translates a conceptual graph into a formula would generate the predicate bake(x) for BAKE and the predicate ptnt(x, y) or rslt(x, y) for the conceptual relation. The ϕ operator preserves the semantic distinctions expressed in conceptual graphs.

Individuals and Names

What is a name, and how does a name relate to an individual? Is a person's name more significant than his or her street address? If Jim and Tom both earn the same salary, does that mean that they share the same paycheck? Philosophical puzzles like these assume major practical significance with computer systems. It makes a big difference whether a computer represents Tom's salary with just a pointer to Jim's or whether it treats it as a separate entity and issues a separate paycheck. The possibility of several people living in the same house makes addresses inappropriate as unique identifiers. Yet names are not unique either. In database systems, names are not used as primary identifiers. Instead, each individual is assigned a unique internal identifier. Social security numbers are the identifiers used in the databases of the U.S. Government. Are those identifiers merely a computational convenience, or do they have some deeper philosophical significance?

With computers, things that can be stored directly in the machine must be distinguished from things that can be represented only by a symbol. Numbers and character strings are the easiest things to represent; they are called *lexical objects*. Names are lexical objects because they can be represented uniquely by a character string in a database. People, houses, and trees, however, cannot be flattened out and squeezed onto the surface of a disk; they are *nonlexical objects*. Because the computer cannot store nonlexical objects, it must represent them by lexical objects called *surrogates*, such as the internal identifiers used in databases. Although names are lexical objects, they are not suitable as surrogates because they are not unique. In conceptual graphs, lexical objects can be represented by quoted strings in the referent field of a concept. As surrogates for nonlexical objects, conceptual graphs use serial numbers marked with the symbol #.

[PERSON: #85437] → (NAME) → [WORD: "Tom Smith"].

This graph may be read *The person #85437 has a name, which is the word "Tom Smith."* The surrogate *#85437* is the primary identifier, and the name *Tom Smith* is demoted to the status of an entity of type WORD linked to the person by the NAME relation. If a name is unique within a context, it can be pulled into the referent field of a concept by a process of *name contraction*:

[PERSON: Tom Smith].

This is an abbreviated form that may be read *the person Tom Smith* or just *Tom Smith*. The quotes around the name are dropped to show that the person is not the name, but is being identified by the name. If quotes were written in the referent field, [PERSON: "Tom Smith"] would show a person who happened to be a character string.

A failure to distinguish proper names from common nouns is a frequent source of errors in knowledge bases. Students who are learning Prolog, for example, often represent the sentence *Tom likes Mary* by a predicate of the form like(tom, mary). They generalize that technique to represent *Cats like fish* by the predicate like(cats, fish). But since *cat* and *fish* are common nouns they should be represented by predicates, not by constants. The sentence *Cats like fish* becomes a rule:

like(X, Y) ← cat(X) & fish(Y).

This rule says that for all X and Y, X likes Y if X is a cat and Y is a fish. The rule that proper names map into Prolog constants whereas common nouns map into predicates works well for most applications. But sometimes, one would like to store types in a data base or write rules that reason about types. To do that, types would have to be represented by constants instead of predicates. That possibility could be represented in Prolog by a predicate kind (X, T), which would relate an entity X to its type T. The sentence about cats and fish would then become

like(X, Y) ← kind(X, cat) & kind(Y, fish).

This rule says that for all X and Y, X likes Y if X is of kind cat and Y is of kind fish. The Prolog predicate kind(X, Y) corresponds to a KIND relation in conceptual graphs. Following is the graph for the sentence *The animal Yojo is of type cat*:

[ANIMAL: Yojo] → (KIND) → [TYPE: cat].

The type label TYPE occurs on concepts whose referents are types rather than indicid-uals. The KIND relation kinks a concept of the individual animal named Yojo to a concept of the type *cat*.

The symmetry between proper names and common nouns can be restored in Prolog by using a predicate name (X, N), corresponding the NAME relation in conceptual graphs. Then the sentence *Tom likes Mary* would become,

like (X, Y) ← name(X, Tom) & name(Y, Mary).

This rule says that X likes Y if X is named Tom and Y is named Mary. The name predicate permits names to be demoted from the role of unique identifiers. With this representation, the variables X and Y must range over surrogates, similar to the referents used in conceptual graphs.

The distinction between surrogates and names, introduced as a computational convenience, is a general solution to the proper philosophical treatment of names. When surrogates are treated as the primary identifiers, a name becomes a characteristic that is no more fundamental than address, weight, or hair color. That approach solves various paradoxes in logic:

Sam believes that Dr. Jekyl is a gentleman.
Dr. Jekyl is Mr. Hyde.
Sam does not believe that Mr. Hyde is a gentleman.

The paradox arises from the rule of substituting equals for equals. If the name "Dr. Jekyl" is substituted for "Mr. Hyde" in the third sentence, then Sam would both believe and not believe that Dr. Jekyl is a gentleman. But with the name predicate, the above argument becomes

Sam believes $(\exists x)$(name(x, "Dr. Jekyl") $\wedge$ gentleman(x)).
$(\exists x)$(name(x, "Dr. Jekyl") $\wedge$ name(x, "Mr. Hyde")).
$\sim$ (Sam believes $(\exists x)$(name(x, "Mr. Hyde") $\wedge$ gentleman(x))).

The paradox disappears because "Dr. Jekyl" and "Mr. Hyde" are now treated as simple character strings that cannot be substituted for one another. Although the use of surrogates as primary identifiers resolves the paradoxes about names, the surrogates are not globally unique: Each conceptual system, either a database or a person's brain, must have its own set of surrogates. Names are externally printable character strings that are used to correlate the surrogates of one system with the surrogates of another.

Colors and Measures

Colors and measures also raise practical and philosophical issues. Is redness a thing in itself, or is it simply a property of something red? The color red could be represented by a one-place predicate that applies to something red, as in red(x), or it could be represented by a constant in a two-place predicate, color(x, red). With the predicate red(x), redness is a property that cannot exist apart from some red object x. With the predicate color(x, red), the word *red* becomes the name of some abstract entity that has the same ontological status as the physical object x. Measures, like colors, are abstract entities with a strange ontological status: You can point to instances of money that add up to \$5, but you can't point to the amount \$5 by itself. If Mary owes Tom \$5, she is not obligated to pay with a specific instance of money, because she could pay by check or with any combination of coins and bills whose measure is \$5.

The philosopher Willard van Orman Quine enunciated a famous test for determining the catalog of entities in an ontology: An entity type is a full-fledged member of the ontology if it is possible to quantify over instances of the type or to refer to an instance by a variable. That principle has been codified as Quine's dictum: *To be is to be the value of a quantified variable.* Because concepts in a conceptual graph correspond to variables in predicate calculus, Quine's dictum determines whether something should be represented as a concept type. It suggests the following criteria for a concept type t:

Can you say, "There exists an instance of t"?
Can you count instances of t or measure some amount of t?
Can you refer back to an instance of t?

Verbs and adjectives are often represented by predicates or relations instead of being considered as quantifiable entities. But consider the following sentences:

A cat chased a mouse. The chase lasted 39 seconds.
I have a red scarf. But the red faded to a dull orange.

Because it is possible to refer to instances of chasing or instances of redness, they must be represented by quantified variables in predicate calculus or by concepts in conceptual graphs. The verb *chase* is normally represented by the concept [CHASE], and the adjective *red* by the concept [RED]. When those concepts are mapped into predicate calculus, they are assigned existentially quantified variables: $(\exists x)$chase(x) or $(\exists y)$red(y). Then the instance of chasing or of redness can be referenced by the appropriate variable x or y.

The variables associated with concepts permit references to instances in the referent field, but not to type labels in the type field. The concept [RED], for example, has an implicit variable for the instance of redness, but no variable for the type RED itself. To permit references to types, the KIND relation in conceptual graphs or the kind predicate in Prolog may be used. As an example, the following is a graph for *a red pyramid*:

[PYRAMID] → (ATTR) 020 [RED].

In this graph, RED corresponds to a monadic predicate; it is not possible to treat it as a value in a database. The nest graph represents the phrase *a pyramid whose color is of type red*:

[PYRAMID] → (ATTR) → [COLOR] → (KIND) →
  [TYPE: red].

In this graph, *red* is now a value in the referent field of a concept. However, this graph is more complex than the previous one, because it now has three concepts and two relations. To simplify the graph, define a new relation COLR that links objects to their kind of color:

relation COLR(*x*, *y*) **is**
  [OBJECT: **x*] → (ATTR) → [COLOR] → (KIND)
    → [TYPE: **y*].

Using the COLR relation, the following graph represents the phrase *a pyramid of color red*:

[PYRAMID] → (COLR) → [TYPE: red].

This graph is identical in meaning to the graph for *a red pyramid*. Representing *red* in the referent field is more convenient for representing data in the database. Representing *red* as a type is more systematic for mapping to and from English. But either graph could be transformed into the other by expanding or contracting the definition.

Database designers often confuse entities with measures. For example, some databases treat salary as a lexical object, because the amount could be represented by a character string. But that character string only represents the measure of the salary, not the salary itself. If a salary were a character string, a person could double his or her salary just by going to the copying machine. The following graph shows a concept of a salary whose measure is an amount of $20,000.

[SALARY] → (MEAS) → [AMOUNT: $20,000].

For convenience, this graph may be abbreviated by the process of *measure contraction*. The symbol @ is placed in the referent field of a concept to indicate a measure:

[SALARY: @ $20,000].

In this concept, the symbol @ shows that $20,000 is a measure of SALARY. In the expanded graph, it is the name of the amount of the salary. Because "$20,000" is a name, it is also possible to expand this concept into the fullest form, which has quotes around the name:

[SALARY] → (MEAS) → [AMOUNT] → (NAME) →
  [WORD: "$20,000"].

This graph says that there exists a salary, which has as measure some amount, which has as name the word "$20,000." As this example illustrates, conceptual graphs can be compact and readable, or they can be pedantically explicit and detailed.

The distinction between entities and their measures clarifies some philosophical puzzles. A famous example is Barbara Partee's sentence that seems to imply that ninety is rising:

The temperature is ninety, and it is rising.

The logician Richard Montague (1973) proposed an ingenious, but rather complex construction to resolve the paradox. In effect, he interpreted it to mean that the extension of the temperature is 90° and the extension is rising. He interpreted the sentence *Ninety is rising* to mean that the intension of *ninety* is rising, a proposition that does not follow from the statement about extensions. Yet Montague's interpretation is not well motivated, especially because it is not reflected in the syntax or vocabulary of English.

Ideally, every feature of a semantic theory should be associated with syntactic cooccurrence patterns or common terms in the vocabulary. The distinction between quantities like temperature, salary, or weight and their measures is signaled in English and other languages by common words like *measure* and *amount*. Partee's sentence could be interpreted as an abbreviation for the longer form *The temperature, which now has a measure of 90°, is rising*. This sentence may be represented by the following conceptual graph:

[AMOUNT: 90°] ← (MEAS) ← [TEMPERATURE: #]
  ← (PTNT) ← [RISE].

In this graph, the temperature, but not the amount, is directly linked to [RISE]. After the temperature rises, its measure will be a different amount, but the original amount of 90° will remain the same. This interpretation relies on the common tendency of natural languages to abbreviate cumbersome phrases that can be understood from context. For convenience, conceptual graphs support similar abbreviations. By measure contraction, the above graph may be simplified to the following:

[TEMPERATURE: # @ 90°] ← (PTNT) ← [RISE].

This graph may be read *The temperature is 90° and it is rising*, but that reading is only an abbreviated way of expressing the longer form that explicitly mentions measures and amounts.

Role Types and Natural Types

Most logic texts map adjectives and common nouns into one-place predicates. That works well in many cases: The phrase *a happy boy* becomes $(\exists x)(\text{happy}(x) \land \text{boy}(x))$; that formula implies that x is happy and x is also a boy. But when the method is applied to all adjectives and nouns, it runs into serious paradoxes. One natural language processor started with the following two sentences:

Sam is a good musician.
Sam is a bad cook.

From them, it derived the assertions,

good(sam) ∧ musician(sam).
bad(sam) ∧ cook(sam).

It then answered yes to all of the following questions:

Is Sam a bad musician?
Is Sam a good cook?
Is Sam a good bad musician cook?

The adjectives *good* and *bad* do not apply directly to a person or thing, but to some role that defines a standard of goodness. Sam is not being considered good or bad as a human being, but only in the role of musician or cook. The paradox does not arise with *happy boy* because happiness is attributed directly to the individual, not to a role. This example illustrates two important distinctions: Some nouns describe an individual only with respect to a particular role; and some adjectives qualify the role, not the individual in isolation.

This issue is closely related to the Aristotelian distinction between *essence* and *accident*: Sam is by essence a man and by accident a musician or cook. Concept types like MAN, DOG, or TREE are called *natural types*; they refer to individuals by properties that cannot be changed without destroying them. But concept types like CUSTOMER, PET, or LANDMARK are called *role types*; they refer to entities by some external property or relationship that can change without affecting the individual. In general, an instance of a natural type can be recognized by examining the individual without considering any external relationships or circumstances: ANIMAL, MAN, WOMAN, DOG, BEAGLE, BUILDING, or NUMBER. But a role type can only be recognized in context, often by very indirect means: PET, COOK, MUSICIAN, LAWYER, FATHER, SISTER, FRIEND, WIFE, PEDESTRIAN, DWELLING, or QUOTIENT. A person may not even know that he or she has assumed a new role until some time after it happens; that is often true of becoming a grandfather or a prizewinner. A role type is always a subtype of some natural type:

TEACHER < PERSON
PET < ANIMAL
PEDESTRIAN < PERSON
QUOTIENT < NUMBER

A role type always has implicit relationships:

A teacher teaches a subject to students.
A pet is owned by some person.
A pedestrian walks on the street.
A quotient is the result of division.

Natural types tend to be arranged in a strict tree of types and subtypes: CAT and DOG, for example, have no common subtypes. But role types create tangles in the hierarchy. As soon as the role type PET is introduced, all animals become candidates for becoming pets, and the hierarchy must be enriched with all of the induced subtypes: PET-CAT, PET-DOG, PET-FROG, PET-GORILLA, PET-ELEPHANT, PET-SPIDER, and even PET-PARAMECIUM.

Like *good* and *bad*, many adjectives do not modify the entity itself but rather some role the entity plays:

former senator
current president
alleged thief
recent graduate
nuclear physicist
incorrect result

These adjectives apply only to the roles, not to the individuals directly. One can say *former pet*, but not *former cat*, except as a kind of joke; the possibility of making the joke indicates that *former* does not normally apply to *cat*. As a natural type, the number 4 could never be incorrect; but considered as a result of 2 + 3, it would be incorrect. Sometimes the word that determines the role is elsewhere in the context. In the sentence *Ivan is a poor choice for shortstop, but he's a good choice for catcher*, the same person is both a poor choice and a good choice. CHOICE is a role type, which indicates that Ivan may be chosen for another role, SHORTSTOP or CATCHER, for which he may be good or bad.

In conceptual graphs, an adjective like GOOD cannot be linked to the thing it modifies by a simple attribute (ATTR) relation. Instead, it must have an additional link to the role type according to which the thing is classified. The canonical graph for GOOD must show both links:

[GOOD]-
  (ATTR) ← [T: *x]
  (BASE) → [TYPE] ← (KIND) ← [*x].

This graph shows that GOOD may be an attribute of something of any type, since T is the top of the type lattice and imposes no constraints whatever. However, the cross reference, indicated by the variable *x, shows that the good thing must be considered a kind of some base type, which sets the standard for goodness. This representation allows a parser (or a person in a conversation) to interpret a sentence like *Ivan is good* by quickly linking [GOOD] to [PERSON: Ivan] by the ATTR relation, while leaving open the base type until further information becomes available. If the speaker never mentions the base type, the listener can always ask, *Good in what way*?

Role types for nouns are the easiest to recognize, but the same criteria for distinguishing role types from natural types can also be applied to verbs. Some verbs directly describe an action, but others describe the role that the action fills. The following are some examples by Marchetti and Morris (1989):

Tom hid the ball by placing it in a box.
Mary killed Tom by stabbing him.

Hiding and placing are two different ways of describing exactly the same action. An external observer could see Tom "placing" the ball. But how could one tell that he was "hiding" it? Similarly, killing and stabbing are two ways of describing the same act. But what if Tom died a day after the incident? An external observer could immediately tell that the act was a stabbing. But only after seeing the consequences could it be described as a killing. This analysis suggests that placing and stabbing are easier to recognize by externally observable attributes than hiding and killing. And killing is easier to recognize than murdering. By the criteria for distinguishing natural types from role types, PLACE and STAB should be considered natural types, with HIDE, KILL, and MURDER as role types. As another example, consider the next two sentences:

The mayor spent a long time outlining his proposals.
He spoke for nearly three hours.

This passage has three verbs: *spending time*, *outlining*, and *speaking*. Yet all three describe exactly the same activity. Which are natural types and which are role types? Speaking must refer to a natural type, because an external observer can recognize speaking even without being able to understand the language. Outlining proposals orally is a subtype of speaking that could be recognized by someone who understood the language; it would also be considered a natural type, but a more detailed or specific subtype. Spending time, however, requires some interpretation or point of view that is not intrinsic to the activity. All activities have a duration, but they are only considered as "spending time" from the perspective of someone who is considering what else might be done in that time. In conceptual graphs, the three sentences would be parsed separately; but afterward, coreference links could be drawn between the three actions to show that they all represented the same activity.

As a general criterion, a natural type, whether described by a noun or a verb, can be recognized as such by the characteristics it has by itself. A role type can only be recognized by relationships to something else. Often the conditions needed to determine whether something is a role type depend on something that cannot be observed directly:

1. The agent's intentions.
2. A result or effect that has not yet occurred.
3. A piece of paper, such as a license, deed, or contract, or some mental change that results from an oral promise.
4. Another person or thing that might be far away: A person could become a father, a cousin, or even a sibling without ever knowing it.

As an example of how this distinction affects program design, consider a car registration system that must deal with object types CAR, PERSON, and OWNER. The natural types CAR and PERSON should be represented by explicit object types in the program, but OWNER is a role type. Where should information about a car's owner be stored? If it goes with the CAR type, then it would be difficult to find what car a person owns without checking every car in existence. If the information goes with the PERSON type, it would be difficult to find the owner of a particular car without checking all persons. If the information goes in both CAR and PERSON objects, a possible inconsistency might arise where a car's owner could be updated without changing the owner's car. The most flexible solution is to have a special type of object called OWNERSHIP with a subtype CAR-OWNERSHIP. Each object of type OWNERSHIP would have a starting date and a slot for an ending date, which would be null for current ownerships; an ownership would only represent one car owned by one owner for a continuous period of time. When a car is sold, a new ownership object would be created, but the old ownership object should not be destroyed. Each PERSON would have a list of ownerships, to which new ones might be added, but old ones would never be removed. Similarly, each CAR would have a list of ownerships to indicate its past history of owners. This organization would make it easy to find all the owners of a given car or all the cars of a given owner, and it would reduce the danger of inconsistencies.

Philosophical Basis

The distinctions embodied in natural languages represent the collective wisdom of many generations of people. An analysis of those distinctions and their interrelationships is a valuable guide for linguistics, philosophy, and knowledge engineering. But because languages are used for different purposes in possibly conflicting ways, they contain a wealth of concepts that may be inconsistent or incompatible with one another. Therefore, the linguistic analysis must be supplemented with logic in order to detect and avoid inconsistencies. Furthermore, the folk wisdom underlying language may also conflict with the findings of modern science. Therefore, the linguistic and logical analysis must also be supplemented with information about the subject matter, either from books or from interviews with the experts. Yet natural languages have proved to be remarkably resilient: The syntactic structures that evolved in prehistoric times are flexible enough for discussions of computers, mathematics, legal systems, and nuclear physics. Co-occurrence patterns can be used to discover the natural representations that underlie those structures. A knowledge engineer may need some linguistic training to use those patterns effectively, but they can also be implemented in automated tools where the computer presents a series of patterns to help a person do the analysis.

References

Buzan, T. (1974). *Use both sides of your brain*. New York: E. P. Dutton.

Geertz, C. (1983). *Local knowledge*. New York: Basic Books.

Katz, J. J. (1966). *The philosophy of language*. New York: Harper & Row.

Lenat, D. B., & Guha, R. V. (1990). *Building large knowledge bases*. Reading,MA: Addison-Wesley.

Marchetti, J. M., & Morris, R. A. (1989). Representing event identity in semantic analysis. *Proceedings of the second Florida Artificial Intelligence Research Symposium* (pp. 193–196). Orlando, FL.

Montague, R. (1973). The proper treatment of quantification in English. In R. Montague, *Formal philosophy* (pp. 247–270). New Haven, CT: Yale University Press.

Novak, J. D., & Gowin, D. B. (1984). *Learning how to learn*. Cambridge: Cambridge University Press.

Sowa, J. F. (1984). *Conceptual structures: Information processing in mind and machine*. Reading, MA: Addison-Wesley.

Sowa, J. F. (1989). Knowledge acquisition by teachable systems. In J. P. Martins & E. M. Morgado (Eds.), *EPIA 89: Lecture Notes in Artificial Intelligence* (pp. 381–396). Berlin: Springer-Verlag.

Vendler, Z. (1967). *Linguistics in Philosophy*. Ithaca, NY: Cornell University Press.

White, A. R. (1975). Conceptual Analysis. In C. J. Bontempo & J. J. Odell (Eds.), *The Owl of Minerva* (pp. 103–117). New York: McGraw-Hill.

Part III
Knowledge Elicitation Methods

6
Implications of Cognitive Theory for Knowledge Acquisition

Sallie E. Gordon

Introduction

Knowledge acquisition for expert systems is a subclass of any instructional situation whereby knowledge must be externalized from a human expert and transferred to one or more "systems." These systems have historically been other people, as in education and training. However, with the advent of artificial intelligence, the systems are more frequently becoming computers. There are many issues that cut across all types of knowledge acquisition. However, two key questions are, (a) What is the nature of the knowledge and/or skills used by an expert? and (b) What are the implications of the nature of expertise for methods of transferring the knowledge or skill?

In knowledge acquisition for expert systems, the traditional approach has been to develop elicitation methods that are derived from, or strongly related to, the knowledge base being developed. However, a better means of developing an artificial knowledge base might be first to address the question of the basic nature of expert knowledge and skill, and then to develop techniques appropriate for each type of knowledge. We have learned from educational research that the nature of knowledge has important implications for the development of effective methods of instruction (e.g., Anderson, 1989; Dillon & Sternberg, 1986; Gagne & Briggs, 1979; Glaser, 1984, 1989; Resnick, 1981; Scandura, 1977). One might assume that this holds for transfer to artificial as well as biological systems.

The goal of this chapter is to present theory and research from cognitive psychology relevant to the two questions posed above. I will address the first question by describing a particular viewpoint whereby development of expertise revolves around changes in declarative and procedural knowledge, and then by presenting empirical evidence relevant to this view. I will address the second question by outlining some implications of the theoretical viewpoint presented for designing and choosing expert system knowledge acquisition methods.

Cognitive Models of Expertise

Expertise and Knowledge Structures

Expertise has become a popular research topic in recent years, resulting in a growing body of literature on this subject. Some researchers have focused on the content and form of expert knowledge structures. Much of this research has consisted of identifying differences between novices and experts. (See Chi, Glaser, & Farr, 1988, for reviews of some of this work.) For example, research has shown that experts have knowledge structures that are more detailed as well as more organized than those of novices (Chase, 1983; Chi, Glaser, & Rees, 1982; Larkin, 1979; Reif & Heller, 1982); that novices tend to organize and perceive problems at a more concrete level whereas experts rely more on abstract concepts (Chi, Glaser,

& Rees, 1982; de Jong & Ferguson-Hessler, 1986); that experts perceive problems in large, meaningful patterns due to a superior organization in their knowledge base (Akin, 1980; Egan & Schwartz, 1979; KcKeithen, Reitman, Rueter, & Hirtle, 1981); and that experts are much faster than novices because of automatized procedural knowledge that does not require a conscious, "controlled" thinking through of the problem (Gentner, 1988).

Stage Models of Skill Acquisition

Some researchers have suggested that the content and organization of "static" knowledge, or knowledge that is "what we know," is only part of the picture. Many in cognitive psychology subscribe to the view that there is a qualitative difference between static *declarative* knowledge about facts and a more dynamic *procedural* type of knowledge (e.g., Anderson, 1983; Squire, 1987; Tulving, 1985). Several models of expertise assume that the development of expert knowledge takes place in stages related to the declarative and procedural systems (Anderson, 1983, 1985, 1987a, 1987b; Fitts & Posner, 1967; Rasmussen, 1980, 1983, 1986). In this section, I will review the basic assumptions of these models, focusing in particular on John Anderson's (1983) ACT* model because it is relatively well known. However, the reader is also referred to Rasmussen (1986) for a similar theory and a very readable and interesting research program.

Declarative and Procedural Knowledge

Much of Anderson's (1983) model of memory and skill acquisition revolves around the distinction between declarative knowledge and procedural knowledge. This distinction was first suggested by Ryle (1949) and refers to the difference between knowledge *that* and knowledge *how*. That is, declarative knowledge consists of *what* we know about objects, events, static relationships between concepts, and so forth. It can also include knowledge about how to do things, such as the steps needed to start a car. It is commonly assumed that our declarative knowledge is represented in a propositional network form (Anderson, 1983; Collins & Loftus, 1975; Kintsch, 1988; Norman & Rumelhart, 1975); is relatively static; and is easy to verbalize. In addition, most researchers distinguish between episodic and semantic declarative memory (Anderson, 1985; Gagne, 1985; Squire, 1987; Tulving, 1983, 1985). Episodic memory refers to memory for specific past experiences or events in a person's life, including the environmental and temporal cues associated with those events. Semantic memory refers to general knowledge of the world—such as facts, concepts, and vocabulary—without reference to temporal landmarks or particular contextual events.

Procedural knowledge is knowledge about how to perform various cognitive activities, or the dynamic process of operating on knowledge. A person may have the declarative knowledge of "To divide fractions, invert the divisor and multiply," but the procedural knowledge must be demonstrated by actually performing the division of fractions. Because of the dynamic nature of procedural knowledge, it is often represented in theory or simulation by if-then production rules (e.g., Neves & Anderson, 1981; Singly & Anderson, 1988). These rules represent the direct association between *conditions* perceived by the person (as a result of external or internal stimuli) and system *response(s)*.

Although declarative knowledge is often described as knowledge *that*, and procedural knowledge as knowledge *how*, this distinction can be misleading. The primary reason is that declarative knowledge can include "knowledge about procedures," such as the proper steps necessary to fix a flat tire. This type of knowledge is simply a special type of declarative knowledge consisting of an *ordered* sequence of actions. This "knowledge about procedures" is consistent with the use of the term *scripts* in artificial intelligence fields.

Stages of Skill Acquisition

A three-stage model of skill acquisition was posited by Fitts and Posner (1967) and updated by Anderson (1983, 1987a). The theory relies on the declarative-procedural distinction because the types of knowledge are manifested differentially in the three stages.

In the first, *cognitive*, stage, people are primarily involved in accumulating declarative knowledge from various sources. If a person must perform a task, relevant sections of the declarative knowledge are retrieved from long-term memory and operated on by *domain-general* procedural knowledge (procedures that can be applied to declarative structures in any content area). An example of a domain-general procedure would be, "If goal is to transform current state into goal state, then match current state to goal state to find the most important difference" (Anderson, 1985). In the preliminary stage, decision making and problem solving tend to be slow, tedious, and prone to error.

As we become more competent in the domain, we gradually move into a second, *associative*, stage. The repeated use of declarative knowledge in given situations results in *domain-specific* procedures, that is, direct associations between specific conditions and the resultant action. The need for operating on declarative knowledge gradually becomes bypassed. The advantage to this process is that when conditions in the environment match the conditions of the procedural rule, the action is automatically invoked, circumventing the longer and more tedious process of retrieving declarative knowledge and applying general productions to it (Charness & Campbell, 1988).

Finally, in the third, *autonomous*, stage, the procedures become highly automatized. That is, the associations become strengthened and more highly specialized or tuned toward particular types of situations. Procedural knowledge at this stage operates in a very fast, automatic fashion (Anderson, 1983; Gagne, 1985; Charness & Campbell, 1988).

During the third stage, simple productions become composed into, or replaced by, more complex, inclusive productions. Because the latter type of productions compress a large number of instantiating conditions and resulting actions, they lose the ability to support verbalizable knowledge (Neves & Anderson, 1981). Thus, as procedures become composed and automatized, the ability to verbalize knowledge of the skill decreases (Anderson, 1985). When performance of a task has become completely automatized, processing requires virtually no cognitive resources, is autonomous, and is unavailable to conscious awareness (Anderson, 1983; Carr, McCauley, Sperber, & Parmalee, 1982; Logan, 1980, 1988; Marcel, 1983; Neely, 1977; Posner & Snyder, 1975; Shiffrin & Dumais, 1981).

In summary, the model described here suggests that people becoming competent in a given domain move away from the use of symbolic or declarative knowledge and toward a reliance on perceptual, nonverbalizable procedural knowledge. This can be viewed as an associative memory phenomenon, in that the stimulus components triggering the use of declarative knowledge come to be directly associated with system output or actions. This model is relevant to the development of expert systems in at least two ways. First, artificial knowledge bases can be structured to mirror this process of moving from declarative knowledge to a compiled procedural base (e.g., Altman & Buchanan, 1987). Second, if experts' performance is often based on nonverbalizable procedural knowledge, we will need to identify knowledge elicitation or acquisition methods that tap this implicit knowledge and do not rely specifically on overt verbalization (either spoken or written). Although I address both of these issues below, it is the second possibility that will be of primary focus.

There are two fundamental assumptions in the theory as outlined above: that knowledge structures can essentially be divided into declarative and procedural systems, and that expertise is acquired whereby initial use of declarative knowledge is later compiled into procedural knowledge. Before describing some of the key implications of the theory for knowledge acquisition, I will briefly review evidence that the theory is, in fact, a tenable one. Accordingly, in the next three sections I will review empirical findings relevant to the first (declarative/procedural) assumption; discuss alternative theoretical accounts of those data; and finally, review support for the second assumption regarding progression through stages.

Findings Relevant to the Declarative/Procedural Distinction

There is a general category of research that has been used as support for the distinction between declarative and procedural knowledge (Squire, 1987; Squire & Cohen, 1984; Tulving, 1985; Willingham, Nissen, & Bullemer, 1989). This consists of demonstrations of *dissociations* between different types of learning tests, such as direct memory tests versus performance measures. That is, variables that affect one type of learning measure do not affect another. (Specific instances will be described below.) Consistent patterns of dissociation between different measures of learning and memory have been found in amnesic patients and have also been experimentally created in normal research participants. Each of these categories will be reviewed below; for a more detailed review of the majority of the experimental findings see Richardson-Klavehn and Bjork, 1988; Schacter, 1987; and Shimamura, 1986.

Dissociations in Amnesics

Amnesics are generally characterized by profound deficits in learning and memory. Although they are often described as being unable to explicitly recall or recognize information from previous experiences (e.g., Schacter, 1987), research has shown that this description is too simplistic. That is, amnesics actually fail to learn two types of information: specific facts or relationships among concepts, and the contextual relationships concerning where and when the facts were learned (Evans & Thorn, 1966; Schacter, Harbluk, & McLachlan, 1984; Shimamura & Squire, 1987, 1988). It is only the latter type of deficit that is tapped by "explicit" memory tests such as recall and recognition.

Shimamura and Squire (1988) demonstrated that these two types of learning disability are separate and distinct impairments. In a series of studies, they found that all amnesics exhibited difficulty in remembering recently learned facts. In addition, *some* were able to remember certain facts but were unable to remember the episode in which the learning occurred. This second type of forgetting is referred to as source amnesia and is the type of amnesia typically measured with recall and recognition tests. The researchers interpreted their findings as indicative of a difference between forgetting in semantic memory (knowledge of newly acquired facts) and episodic memory (knowledge about the context in which the facts were learned). As Squire (1987) noted, "source amnesia . . . does not appear in all amnesic patients, but the ones who show it do so consistently. Moreover, the patients without source amnesia can have just as severe a memory impairment as patients with source amnesia" (p. 173).

While memory for newly acquired facts and specific learning episodes is grossly impaired in amnesics, certain kinds of learning are well preserved. These types of learning typically fall into two categories: (a) perceptual-motor and cognitive skills, and (b) priming effects.

As examples of the first category, patients are able to learn perceptual-motor skills such as mirror-tracing (Milner, Corkin, & Teuber, 1968); bimanual tracking (Corkin, 1968); reading of mirror-inverted script (Cohen & Squire, 1980); putting together a jigsaw puzzle (Brooks & Baddeley, 1976); and serial pattern learning requiring a motor response (Nissen & Bullemer, 1987). Given enough time, amnesics can also acquire complex cognitive skills such as computer programming (Glisky, Schacter, & Tulving, 1986). These patients all showed profound memory deficits for recall or recognition of the specific learning episodes responsible for the skill acquisition.

Many of the studies of memory dissociation in amnesics have demonstrated that while recall and recognition were impaired, amnesics showed normal "learning" as evidenced by repetition priming. Repetition priming is a general term indicating that the presentation of a particular stimulus (i.e., a picture or word) has a facilitative effect for later tasks requiring the naming or identification of the same stimulus. For example, in a word identification task, a person might see the word *horse* at Time 1. At a later time, the person might be presented with a word stem completion task (e.g., Complete the following word: hor__) or a fragment completion task (e.g., Complete the following word: h_r_e). The initial word presentation is said to have had a priming effect to the extent that the person says *horse* following the partial cue either more often or more quickly than under conditions where the word was never presented before the completion task.

Amnesics have been shown to exhibit normal learning as measured by priming tasks. For example, in a series of studies by Warrington and Weiskrantz (1970, 1974, 1982), participants showed both poor recall and recognition memory for familiar words from a previously presented word list. However, priming occurred for word-stem completion tasks and word fragment tasks at a level equal to that of nonamnesics. Other studies have obtained similar findings when participants were asked to identify figures (Milner et al., 1968); produce common associations after exposure to word pairs (Shimamura & Squire, 1984); generate category instances (Graf, Shimamura, & Squire, 1985); and produce idioms in a word association task (Schacter, 1985). All of these studies showed severe impairment when patients were asked to explicitly remember previous learning episodes and associated information; however, repetition priming was normal.

In summary, amnesics show memory loss of new information and/or the source of that new information. However, skills requiring direct performance of some task are preserved, as are priming effects. Both of these are relatively automatic types of performance and do not involve "strategic" or controlled remembering. The consistent dissociation found in amnesics between memory for specific learning episodes on the one hand, and skill acquisition or priming effects on the other hand, has contributed to an increasing interest in identifying corresponding types of dissociation in nonamnesic people.

Experimentally Created Dissociations

Findings of dissociations in normal or nonamnesic people fall into several general categories. These are listed below, along with a description of the standard types of finding for each category.

1. *Effects of experimental manipulations on explicit memory but not repetition priming.* Many studies have confirmed that certain initial study conditions, such as instructions to elaborate or instructions to generate the target words, will facilitate later recall and/or recognition memory, but not repetition priming (Graf & Mandler, 1984; Graf, Mandler, & Hayden, 1982; Graf & Schacter, 1987; Jacoby, 1983; Jacoby & Dallas, 1981; Schacter & Graf, 1986).

For example, Jacoby and Dallas (1981) had people read word lists, and half were given instructions to elaborate on the material by answering questions about the meaning of the words. Participants were later given a yes/no recognition test or a word identification test (subjects had to name briefly presented words). As has been traditionally found, elaboration facilitated recognition memory; however, it had no effect on word identification.

All of the studies in this category show that instructions that induce conceptual or semantic processing help explicit memory tests, such as recall or recognition, but do not facilitate repetition priming beyond simple presentation of the word or picture. In fact, some studies have shown that coceptual processing induced by instructions to generate the words reduced priming as compared with focusing on perceptual elements by simply reading the words (Blaxton, 1989; Clarke & Morton, 1983; Smith & Branscome, 1988; Winnick & Daniel, 1970).

2. *Effects of experimental manipulations on repetition priming but not explicit memory.* Several studies have shown that certain experimental manipulations tend to affect repetition priming but not recall or recognition. For example, some researchers have presented

stimuli in one modality, such as visual pictures, and then looked for priming effects in another modality, such as words. In this type of study, it has been consistently found that priming effects are much stronger within the *same* mode (e.g., stimuli presented as printed words and later tested as printed words) than when stimuli are presented in one mode and then later tested for priming in another mode (Clarke & Morton, 1983; Graf et al., 1985; Jacoby & Dallas, 1981; Kirsner, Milech, & Standen, 1983; Roediger & Blaxton, 1987; Tulving, Schacter, & Stark, 1982; Weldon & Roediger, 1987). This is known as *modality specificity* and has been found to strongly affect repetition priming but not tests of memory such as free recall and recognition (e.g., Graf et al., 1985; Weldon & Roediger, 1987).

3. *Temporal dissociations between explicit tests and repetition priming*. While memory for words presented in word lists typically declines very quickly, repetition priming has been shown to last for relatively long periods of time. For example, Sloman, Hayman, Ohta, Law, and Tulving (1988) showed a reliable amount of priming in a word fragment completion task for a period of 16 months. In a study of picture naming, Mitchell and Brown (1988) showed that repetition priming showed no decline whatsoever over a period of 6 weeks, while recognition memory for the pictures showed a progressive and rapid decline. Finally, it should be noted that in opposition to repetition priming, *semantic* priming (using a semantically related word to prime an associated word) has been shown to have very rapid decline (Dannenbring & Briand, 1982).

4. *Dissociations in elderly populations*. Dissociations similar to those found for amnesic patients have been identified in normal elderly populations (Light & Singh, 1987; Light, Singh, & Capps, 1986). For example, Light and Singh (1987) found that people over 60 exhibited much poorer recall and recognition than younger people, but showed no differences in repetition priming as measured by percent correct for both word-stem completion and word fragment identification tasks.

Similarly, Mitchell (1989) evaluated 11 measures of memory in young and old age groups. Mitchell found that elderly people exhibited lower levels of recall and recognition, but no differences on any other measures. More interestingly, Mitchell performed a factor analysis resulting in a three-factor solution, with the three factors strikingly similar to an episodic/semantic/procedural memory distinction.

5. *Perceptual learning*. It has recently been argued that a person can acquire basic cognitive skills or procedural knowledge in the form of rules of which the person is unaware (Lewicki, 1985, 1986; Lewicki, Czyzewska, & Hoffman, 1987). Anecdotal examples of this "learning without awareness" phenomenon include the idea that people are unable to articulate semantic and syntactic rules of the language they use, although they can generate correct speech acts and indicate when a particular example does not sound right (Lewicki et al., 1987).

An increasing number of studies have demonstrated that people are able to learn complex rules or stimulus constraints in perceptual-motor tasks without being aware of the rules or constraints (Cohen, Ivry, & Keele, 1988; Day & Marsden, 1982; Franks, 1982; Hayes & Broadbent, 1988; Kellogg, 1982; Kellogg, Robbins, & Bourne, 1978; Lewicki, 1986; Lewicki et al., 1987; Lewicki, Hill, & Bizot, 1988; Nissen & Bullemer, 1987; Stadler, 1989). In these studies, participants showed learning through some type of performance but were not aware of the nature of the learning and were not able to verbally report the rules or source of their behavior.

As an example, Lewicki and his colleagues (Lewicki et al., 1987) had people perform a visual search task that required them to indicate which of four quadrants on a display contained a target digit. Subjects pressed a button denoting one of the four quadrants as quickly as possible after seeing the target appear on the screen. For six trials, the target was the only stimulus on the screen. On the seventh trial, the target was embedded in a field of 35 distractor digits. The location of the target on the seventh trial was a function of the target location in four of the previous six trials. After 168 presentations of the same rule relating the sequence of easy trials to the location in the seventh trial, the rule was abandoned. When the rule was changed, reaction time for the seventh trial increased drastically, showing a large negative transfer effect. This indicates that learning had taken place. In addition, the participants were not able to detect the relational rule when questioned after the learning phase and, in fact, denied any awareness of such rules.

As a more stringent criterion of whether people are actually "aware" of the rules underlying performance, Stadler (1989) replicated the study by Lewicki (Lewicki et al., 1987). However, instead of testing awareness in a subjective manner by simply asking people if they could state the rules, he measured awareness in a more rigorous or objective manner. Subjects performed the same visual search task for a target digit, indicating the quadrant in which the target appeared. As in previous studies, learning of the rules was evidenced by a decreasing reaction time on the seventh trial over the course of the learning phase blocks and by a negative transfer when the task was changed to different rules. However, the interesting result is that when subjects were asked to indicate, *before* the seventh trial, which quadrant the target would be located within, they were not able to perform the task above chance. Note that the only difference between the two tests is whether subjects had to predict or simply respond to the target in the critical trial. This fact is important for two reasons: (a) The dissociation was within implicit memory, not between explicit and implicit memory, as was much of the research previously reviewed; and (b) The evidence seems quite compelling that rules were learned in some way that was not open to conscious introspection, even in a "nonverbal" way.

In a similar series of studies, Willingham and his colleagues (Willingham et al., 1989) recently demonstrated that learning can take place independently in two systems, one of which is not available to conscious introspection. Participants reacted to a light in one of four positions on a horizontal bar. The light position was determined by various repeating sequence rules. Results showed that people often learned the sequences without explicit declarative knowledge of the associations. However, occasionally participants *did* have an awareness of the sequence rules, facilitating performance above and beyond the effects of nonconscious learning. The researchers interpreted this effect as being due to enhanced attentional expectancies. In addition, their subjects appeared to be much more variable on the rate of declarative learning versus procedural learning. Willingham et al. concluded that for their task, "procedural and declarative knowledge may be acquired separately; one need not have knowledge of one type in order to build the other type of knowledge" (p. 1059).

Other studies have shown similar dissociations. For example, Reber and his colleagues (Reber, 1967, 1976; Reber & Lewis, 1977) presented people with letter strings organized according to rules of a synthetic grammar. These participants learned to identify grammatically correct strings even when they were not consciously aware of the rules.

Broadbent and colleagues (Broadbent & Aston, 1978; Berry & Broadbent, 1984) showed that people could learn through practice to make better decisions about controlling systems (such as a simulated city transportation system with multiple constraints). And yet the same people did not improve in their ability to answer verbal questions. Subsequently, Broadbent and Aston (1978) also showed that verbal instructions could significantly enhance question answering yet have no effect on decisional *performance* (a phenomemon that has been generally recognized in education for quite some time; see, e.g., Nickerson, Perkins, & Smith, 1985).

Hayes and Broadbent (1988) conducted three studies in which people had to interact with a simulated person on a computer in such a way as to shift the computerperson's behavior to "friendly." Performance and interview data indicated that varying the dynamics of the system could induce either conscious verbal processing or unconscious automatic processing. Hayes and Broadbent interpreted these results as being due to two separate learning modes, only one of which involves a verbal system. They also noted that when critical stimulus characteristics are relatively few and very salient, subjects were more likely to learn in the declarative system, or *s-mode*, as the researchers referred to it.

Several hypotheses or classification schemes have been offered to account for the results described in this section. In the next section, I will briefly describe some of these schemes. I will then describe a view that can adequately account for the types of data described earlier—essentially an extended version of the declarative/procedural systems view.

Theoretical Explanations

The Task View

It has been suggested that many of the previously reviewed research tasks can be described as either *explicit* or *implicit* tests of memory (e.g., Schacter, 1987). Explicit tests essentially involve the "conscious recollection of recently presented information, as expressed on traditional tests of free recall, cued recall, and recognition" (Schacter, 1987, p. 501). Implicit tests include any measure where learning is evidenced by a "facilitation or change in task performance that is attributable to information acquired during a previous study episode" (Schacter, 1987, p. 501). Examples of implicit tests include perceptual-motor skills and priming in various modalities.

The explicit/implicit distinction is really more of a categorical description of experimental tasks than a theory per se. As such, the description is commonly used by researchers in discussing effects found in dissociation studies. However, for our purposes the dichotomy of explicit versus implicit tasks is too limited, in the sense that the dichotomy does not adequately describe or "frame" the types of research just reviewed. For example, I reviewed work earlier showing that source amnesia can be dissociated from other types of amnesia, both of which are measured by explicit memory tests.

Another problem for this dichotomy is the fact that recent research has demonstrated dissociations *within* implicit tests. For example, Stadler (1989) had people press a button to either predict a target location or react to a target location. Neither of these measures could be termed an explicit memory test for a previous learning episode, yet clear dissociations were found.

The Process View

A somewhat similar approach has been suggested by Jacoby (1983) and Roediger and his colleagues (e.g., Roediger, Weldon, & Challis, 1989). Roediger et al. point out that most explicit tests tap into *conceptually driven* cognitive processes, whereas most implicit tests tap into *data-driven* cognitive processes. Conceptually driven processes tend to rely on the encoded meaning of concepts, or on semantic processing, whereas data-driven processes tend to tap the "perceptual record of past experience" (Roediger et al., 1989, p. 16). This view explains the dissociation data by arguing that conceptual processes used during encoding will affect only conceptually driven retrieval processes (as in explicit tests). On the other hand, implicit data-driven tests will only be affected to the extent that stimuli surface characteristics are different between original presentation and test materials.

Unfortunately, there is usually a confound such that explicit tests involve *intentional remembering* as well as being conceptually driven, whereas implicit tests do not involve intentional acts of remembering. Certain types of findings indicate that it is the intentional remembering component that can produce dissociations. For example, in studying amnesics, Graf, Squire, and Mandler (1984) found that the exact same word stem can result in dissociations between intentional remembering (as in cued recall) and word-stem completion.

Likewise, Schacter (1985) gave amnesic patients word pairs as stimuli and then gave one of the words as a test cue. Instructions accompanying the cue were either for explicit recall of the previously paired word or for free association (measuring repetition of the previously paired word). Given the exact same cue, dissociations were still found between the two types of test. This indicates that it is not the characteristics of the cue per se, but the "intentional remembering" aspect that is critical. In addition, this categorization scheme would have difficulty with other recent findings, such as those obtained by Stadler (1989).

The Memory Systems View

Another explanation for the dissociation data is that memory takes place within two or more separate systems. The most popular version of this view is the distinction between declarative and procedural memory. Proponents of this view include, but are not limited to, Anderson (1976, 1983, 1985; Singley & Anderson, 1989); Hayes and Broadbent (1988); Squire and Cohen (Cohen, 1984; Squire, 1987; Squire & Cohen, 1984); Lewicki (1986); Mitchell (1989); and Tulving (1983, 1985).

Squire and Cohen (Cohen, 1984; Squire, 1987; Squire & Cohen, 1984) have suggested that *procedural* memory is the type of memory spared in amnesics; it represents learning through changes in how pre-existing cognitive operations are carried out. Procedural memory includes motor skills, cognitive skills, simple classical conditioning, habituation, sensitization, perceptual after-effects, and other automatic associative phenomenon. Squire (1987) suggests that procedural memory is not a unitary system but more a "collection of different abilities, each dependent on its own specialized processing system" (p. 164). As such, procedural memory tends to be modality bound, whereas declarative memory is not. Perhaps most important, procedural memory cannot be reflected in verbal reports or judgments of familiarity. Rather, it can only be expressed directly in performance.

Squire (1987) suggests that *declarative* memory is the type of explicit memory that has been impaired in amnesia; it is "memory that is directly accessible to conscious recollection. It can be declared. It deals with facts and data" (p. 152). In addition, declarative memory is divided into episodic memory and semantic memory (as described earlier). Episodic memory is affected in patients with source amnesia, and the fact that only some amnesics exhibit source amnesia is taken as evidence for the distinction between episodic and semantic memory.

In discussing the underlying neuropsychological systems, Squire (1987) speculates that the declarative system developed after the procedural system in the evolution of intelligence, an idea also pointed out by Schwartz (1989) in discussing knowledge acquisition for expert systems. A related suggestion is that ontologically, declarative memory develops later than procedural memory (e.g., Nadel & Zola-Morgan, 1984).

The system view championed by Squire (1987) can adequately account for most of the data for amnesics and experimentally induced dissociations. However, some researchers have argued that the skills subsumed under procedural memory are too disparate to be accounted for by one system. For example, Schacter (1987) pointed out that skill learning and priming can be dissociated experimentally (Butters, 1987). However, this cannot be viewed as a critical point because Squire (1987) explicitly notes that there are diverse and specific components within the procedural memory system (this will be expanded below). Schacter (1987) also suggests that the declarative/procedural account is inadequate because amnesics can retrieve new facts and vocabulary even though they have no explicit memory for the learning episodes. Clearly, this can be accounted for by the distinction between episodic memory and semantic memory.

A systems view that makes no distinctions *beyond* declarative and procedural memory cannot account for some of the empirical data. However, given some modularity within the two subsystems, the approach is quite adequate in accounting for a very wide variety of findings. Squire makes a distinction between episodic memory and semantic memory within the declarative system. We will extend this approach by combining the declarative/procedural view with modular approaches such as those proposed by Schneider and Detweiler (1987) and Crowder (1989).

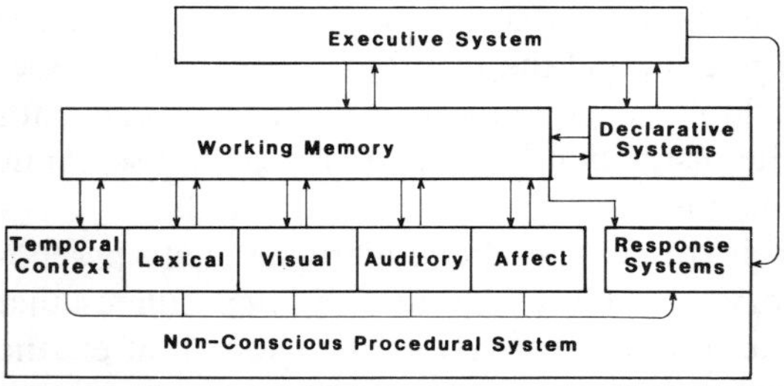

FIGURE 6.1. A system/mode model of learning and memory.[1]

[1] The content of modules can be processed relatively independently, associated with one another in the procedural/habit system, and/or transferred into working memory depending on threshold and attentional factors. Information from temporal context and visual or auditory modules are associated in working memory and become the basis for episodic memories. An executive control system mediates and organizes the information flow to working memory, the declarative systems, and response systems.

A System/Mode View

The "system/mode" view described here is very similar in nature to the models proposed by Schacter (1989) and Moscovitch (1989). The reader is referred to the chapters by Schacter and Moscovitch, as well as to *Memory and brain* by Squire (1987), for neuropsychological data supporting this type of model.

The fundamental assumptions of this view are that several systems operate in parallel and are characterized by different processing mechanisms. The basic form of the system is diagrammed in Figure 6.1. Information enters the system through basic perceptual modules, such as visual and auditory systems. Some additional modules have been included in the illustration, although it is not implied that there are no modules *other* than those represented. The experimental findings supportive of the various modules shown in this scheme are reviewed elsewhere (see Moscovitch, 1989; Schacter, 1989; Schneider & Detweiler, 1987).

The output of the modules can be brought into working memory by an executive system. The modules are important in that priming effects can be localized within these units. That is, repetition priming effects can be thought of as occurring in a somewhat isolated fashion without reference to interitem associations. The model outlined here can explain priming of seemingly declarative material (i.e., words) because of the existence of a separate lexical module, independent of conceptual or declarative memory.

Items or information in the modules can also be associated with one another, and with the response system modules, within the procedural/habit subsystem. These associations are not directly open to introspection. The procedural system is involved in incremental skill learning, and as Schacter (1989) notes, "the procedural/habit system does not have any connections with the conscious awareness system" (p. 364). He continues by pointing out that this fact does not mean that the system is isolated; "it is possible to voluntarily *initiate* various acquired skills or procedures" (p. 365).

Explicit or episodic memory occurs as a result of information being brought from context modules as well as visual, auditory, or other modules into working memory. This combined information is then stored in declarative/episodic memory and is potentially available for retrieval via recall or recognition tasks. The executive system essentially guides and organizes these retrieval functions. Thus, any processes that affect the combinations of information brought into conscious or working memory during the initial study phase will affect later "explicit" retrieval of the information from episodic memory.

Declarative memory also has a component based on semantic associations, without particular regard to the specific context of learning. It is not clear at this time whether semantic memory is stored separately from episodic memory from the outset or whether the contextual elements somehow become separated from episodic memory to create semantic memory. For the purposes of this analysis, the differences are not particularly critical.

A system characteristic that is important for the purposes of the current analysis is the fact that the executive control system can only bring certain types of system information into working memory and conscious awareness. This includes verbal or symbolic information, images, feelings, and so forth—anything from the individual modules or from declarative memory. It cannot bring procedural knowledge into conscious awareness. These are direct associative ties, usually between perceptual stimuli (either internal or external) and some type of response (either internal or external).

In the stage model of expertise described earlier, an important assumption was the idea that declarative processing always precedes procedural knowledge "developmentally." That is, a novice first performs a skill by accessing and using symbolic knowledge in the declarative system. This slowly evolves or becomes "compiled" into procedures. Notice that the system/mode view described does not account for this transition over the course of skill acquisition. The purpose of the next section is to review research relevant to the transitional assumption and determine whether further elaboration or modification of the system/mode view is necessary.

Empirical Support for Stages

There has been relatively little research that directly addresses the assumption that declarative knowledge always precedes procedural. Anderson (Neves & Anderson, 1981; Singley & Anderson, 1989) has presented some evidence that in certain domains, this is the normal sequence of activity. For example, Neves and Anderson (1981) describe data from verbal protocols of people learning to write geometry proofs. The protocols suggest that they first proceed through an effortful use of conceptual or background knowledge in working memory. Over time, protocols become much shorter, bypassing the original intermediate steps.

Sanderson (1989) attempted to directly test two of the assumptions in Anderson's (1983) model: (a) that declarative knowledge precedes procedural, and (b) that as procedural knowledge is developed, declarative knowledge becomes less accessible. Sanderson used a task similar to the city transportation problem used by Hayes and Broadbent (1988). She manipulated various features of the task and found important dissociations between task performance and verbal knowledge. More specifically, when performance hinged on visuospatial characteristics of the task, performance improved but verbalization did not. According to Sanderson, the task format induced a perceptual solution strategy that made verbal reasoning about the system irrelevant.

Sanderson's (1989) data are important because they bear on the stage model in several ways. First, they indicate that declarative and procedural learning can occur in parallel, and that declarative knowledge does not have to precede procedural knowledge. Second, the results indicated that the development of procedural knowledge does not necessarily lead to a decrease in the ability to verbalize declarative knowledge. Often, the development of declarative and procedural knowledge go hand in hand.

Several other studies have also indicated that procedural knowledge does not have to be predated by declarative knowledge. For example, the perceptual learning studies reviewed earlier demonstrated that procedural learning can occur for some tasks from the beginning (e.g., Lewicki et al., 1987; Stadler, 1989; Willingham et al., 1989). This process seems to occur when performance of the task depends on complex, perceptual characteristics of the stimuli and not on symbolic, conceptual activity.

In summary, the findings to date indicate that procedural knowledge does not necessarily have to be "compiled" as a result of using declarative knowledge. Rather, the two types of knowledge are acquired in parallel, with the relative amount of learning within each system highly dependent on the task being performed. This is entirely consistent with the systems/mode view of knowledge acquisition outlined above (refer back to Figure 6.1). In some tasks, people seem to rely heavily on "controlled" declarative processing during initial learning. In addition, procedural condition/action associations build up over time and can eventually take much of the workload from declarative processing. Although this shift from declarative to procedural dominance seems to occur for tasks relying predominantly on symbolic, conceptual knowledge, other, more "perceptual" tasks may rely heavily on the acquisition of procedural knowledge from the beginning.

Given a relatively strong body of research supporting the system/mode view described earlier, I will now proceed to a discussion of how the distinction between declarative and procedural knowledge holds certain implications for eliciting and representing knowledge for expert systems.

Implications for Knowledge Acquisition

The system/mode view presented earlier is one way of imposing structure on a wide body of research that can reasonably account for the data. In addition, it provides a useful starting point from which to discuss knowledge acquisition methods.

Some General Implications for Knowledge Acquisition

In the course of task performance, experts will use declarative knowledge, procedural knowledge, or both. In novel problem-solving or decision-making situations, experts may predominantly use declarative knowledge. This process will be manifested by verbal concepts coming into working memory at various points during the process. Subjectively, the process will feel slow and effortful. In familiar problem-solving tasks, domain-specific procedures will result in quick answers placed in working memory with virtually no intervening thoughts. Finally, for tasks that are more perceptual in nature, the process may be essentially procedural whether or not the stimuli are familiar.

In all of the above situations, the expert has verbal access only to the information that comes into working memory, and that information is declarative in nature. The major implication of this view is that by definition, procedural knowledge *cannot* be directly verbalized. It is therefore counterproductive to ask an expert *how* he or she made a decision or solved a problem. The best that the expert can do is verbalize the thoughts that came to working memory as a product of the procedures and use declarative knowledge to conjecture what those procedures must have been (e.g., Johnson, 1983).

At this point, there are several important questions relevant to knowledge acquisition for expert systems. Given that experts use both declarative and procedural knowledge:

1. Should we try to elicit one or both types of knowledge?
2. How can we elicit declarative knowledge?
3. How can we elicit procedural knowledge?
4. Should we develop knowledge representations that embody both types of knowledge?

The Case for Measuring Declarative and Procedural Knowledge

There are at least two reasons why we should try to capture and use both declarative and procedural knowledge. The first is based on a desire to accurately mirror or capture the "expertise" of the expert(s). There is an increasing amount of evidence that people use different types of processes to accomplish tasks and that at least some of these processes are not verbalizable (e.g., Lewicki et al., 1987; Lewicki et al., 1988; Stadler, 1989). If, as Anderson (1983), Rasmussen (1986), and others suggest, much of expert skill lies within the bounds of nonverbalizable procedural knowledge, then capturing only declarative knowledge would miss much of what experts know. As Schwartz (1989) has pointed out, much of what experts do involves complex pattern recognition and other associative processes. Similarly, capturing only procedural knowledge would miss the flexible conceptual verbalizable knowledge that is often critical for handling novel, difficult problem-solving situations.

In addition, many state-of-the-art expert systems are moving in the direction of a multilevel knowledge structure. These systems have been dubbed *second generation expert systems* or *multilevel expert systems* (Hart, 1986; Musen, Fagan, Combs, & Shortliffe, 1987; Steels, 1987). Some multilevel expert systems rely on compiled "production rules" (procedural knowledge) for the normal and familiar situations. However, they also use domain-general procedures on a "deep model" (declarative conceptual knowledge) as a backup for use with novel or unusual cases. The declarative knowledge subsystem also provides the second generation of expert systems with the basis for extensive explanation facilities. The concepts and ideas are accessed and used to explain the reasoning behind the production rules being fired (Clancey, 1983; Steels, 1987).

Finally, some expert systems are being modified to have pattern recognition components in the form of parallel connectionist systems (networks consisting of very large numbers of simple but highly interconnected "units" operating in a highly parallel nonsymbolic associative manner). These systems would be directly analogous to the associative procedural systems used by human experts (Gordon, 1991).

Given that expert systems are moving in the direction of having multiple systems similar to the knowledge systems employed by humans (Gordon, 1991), it would seem desirable to identify methods for transferring human expert declarative and procedural knowledge to corresponding expert system components.

Eliciting Declarative Knowledge

Any procedure requiring the expert to verbalize his or her knowledge would be appropriate for acquiring declarative knowledge. Many, if not most, methods for eliciting expert knowledge rely on some form of verbalization by the expert. The major types of knowledge elicitation methods falling within this category will be briefly reviewed.

Unstructured Interviews

Unstructered interviews usually consist of the knowledge engineer's asking whatever questions seem relevant and/or likely to elicit new information. Notes are taken, or a tape recording is made and later transcribed (Gordon & Gill, in press). According to the system/mode view, the information elicited using unstructured interview techniques is declarative in nature. That is, in passively describing his or her knowledge and actions, the expert is using domain-general search procedures to access and verbalize stored declarative knowledge (and perhaps to construct new inerences in addition to previously existing information).

Structured Interviews

Interviews can be structured using a number of methods. These include free association (Mitchell, 1987); laddering (Diederich, Ruhmann, & May, 1987; Grover, 1983); question probes (Gordon & Gill, 1989, in press); forward scenario simulation (Diederich et al., 1987; Grover, 1983); goal decomposition (Grover, 1983); and the repertory grid technique (Boose, 1986, 1988). While these methods vary greatly in the method used to structure the interview, they have one element in common: They require the expert to access relatively static knowledge and verbalize that knowledge in some form. As such, the methods are directly eliciting declarative knowledge.

Think-Aloud Verbal Protocols

Some knowledge acquisition methodologies consist of giving experts one or more problems to solve and asking them to think aloud as the problem is solved. (See Ericsson & Simon, 1984, for a review of this type of "concurrent" verbalization.) The verbalization is recorded, resulting in a verbal protocol that is then analyzed to identify knowledge or strategies used by the expert. Often, the experts are asked to verbalize any thoughts that occur to them during the process *and also* to explain their problem solving behavior.

What type of knowledge is being elicited? First, the expert is being asked to perform some task, and thus actual task behavior will be based on some mixture of declarative knowledge and procedural knowledge. The exact mix will depend on the type of task being performed (e.g., perceptual vs. symbolic analysis), as well as the familiarity of the task. Thus, behavior itself is a function of all types of knowledge.

However, in most verbal protocol analysis, the emphasis is predominantly or completely placed on the *verbalizations* of the expert, including an explanation of *why* he or she performed the specific behaviors. What is elicited in the verbal protocol *itself* will thus be declarative knowledge coming from two sources: declarative knowledge that is used during problem solving itself, and inferencing processes being performed on declarative structures to provide explanations.

To generate explanations, an expert will use two strategies. Sometimes search procedures will result in accessing previously learned declarative knowledge about appropriate "strategies" (Nisbett & Wilson, 1977). These can be directly verbalized. Lacking these, the expert may attempt to construct a plausible explanation. Explanation construction is accomplished by observing information such as situational cues and one's own behavior and then using domain-general skills to infer rules that could account for the behavior. This process is similar to Sanderson's (1989) description of *mental model building*. According to Sanderson, observation of system information during the course of task performance results in modifications to a semantic mental model. This model acts as the source of information for verbalizations of knowledge.

Retrospective Verbal Protocols

Retrospective protocol analysis is similar to traditional protocol analysis; however, it splits the problem-solving and explanation processes chronologically. The expert solves the problem and subsequently tries to explain his or her reasoning processes that occurred during the course of problem solving. (See Ericsson & Simon, 1984, and Stauffer, 1989, for advantages and disadvantages of this method.) The information contained in the retrospective protocol is then evaluated or analyzed in some manner. For example, Waldron and Waldron (1988) used a retrospective protocol technique to study how an engineer had designed a leg mechanism for a multilegged walking machine.

As with the "on-line" verbal protocols, our framework would lead us to conclude that the knowledge verbalized by experts in retrospective protocol includes memories of knowledge states in consciousness during problem solving and new declarative explanations (either retrieved from long-term memory or inferred).

Acquiring Procedural Knowledge

Procedural knowledge can, by previous definition, never be directly elicited from the expert. To acquire procedural knowledge, rules must be developed by first observing multiple examples of actual task performance under a variety of circumstances. Data collected from the observation are then used to infer the relationships between informational input and verbal and/or behavioral output. The associative rules can be inferred either by the knowledge engineer, the expert, or both working together. Another alternative for acquiring procedural knowledge is to give the input and output characteristics of the problems to a computer system to learn directly via an appropriate technique such as neural networks.

Task Performance

With any of these three methods, a variety of problems is presented to the expert. The variety spreads across at least two dimensions: situation or stimulus characteristics and task familiarity. The theoretical distinction between the use of domain-general and domain-procedural rules strengthens the argument for having the expert solve a variety of problems ranging from very familiar to very novel. The problems are preferably given to experts under real-world conditions, and the expert is asked to simply proceed as normal. Any naturally occurring verbalizations, such as those that would occur during a consultation, can be spoken aloud. For example, an X-ray technician would be given as wide a variety of films as possible, both common and rare.

In collecting task performance data, at least two alternative methods are possible. First, input characteristics and behavioral output can be measured "on-site" during job performance. If the expert performs a wide variety of problem solving or decision making during the normal course of events this can be an efficient method. However, sometimes the normal stream of instances includes too many "common" instances with correlated features and not enough variety. Under these circumstances, problems or stimuli must be created in order to orthogonalize input characteristics and get adequately unusual instances. To do this, it may be possible to identify critical attributes ahead of time and vary them orthogonally to the degree possible (and realistic). Orthogonality of input characteristics is necessary to correctly identify critical conditions of the various rules. (This rationale is similar to that of experimental design for identifying causal factors and interactions.)

Consider the example of a forester's helping a landowner in deciding what tree to plant in a given location for a given purpose. By evaluating the information that is typically requested and used during problem solving, it is possible to identify the various location characteristics and other "input" features that are part of the information used in decision making. New scenarios different from those used in previously solved problems can be created by combining characteristics. These scenarios are then given to the expert, and the results of the decision-making processes are recorded.

This method has the distinct advantage of allowing the expert to perform within, or in a manner similar to that used in, the real-world environment. He or she is under no stress to verbalize internal processes that are difficult to explain. Evidence for the importance of this unencumbered task environment comes from research on complex decision making. For example, Porter (1987) conducted an experiment in which people were asked to make complex decisions and estimates. More specifically, Air Force cadets were given information supposedly representing several officers' responses to a variety of questions. The cadets were asked to assign the officers to categories such as gender, rank, academic degree, and so forth. Half of the participants were aksed to make the judgments and also simultaneously to explicate their choices. Another group was allowed to simply guess without explicating the decisional processing. The second group made responses that were significantly more accurate in terms of actual base rate information than those of the group induced to verbalize their decisional choices.

As a final comment, it should be pointed out that acquiring procedural knowledge by using actual task performance is certainly no less time-consuming than eliciting declarative knowledge. Conservative estimates of the number of procedural rules held by master chess players have been on the order of 50,000 different pattern recognition productions (Anderson, 1985). By the time someone becomes an expert in a certain field, he or she has accumulated a very large repertoire of production rules. As of this time, there is no shortcut for "extracting" or acquiring these rules—for transfer to either human or computer.

Human Inferencing

To infer procedural rules, the system developer and expert "observe" the relationship between three types of information: the situational charateristics, the intermediate knowledge states verbalized during problem solving, and the final answer. This input/output information is then used to infer rules that could feasibly underlie the behavior.

The difficulty of inferring production rules in this manner will depend on a number of factors. First, the difficulty will increase rapidly with the number and complexity of the stimulus characteristics and task behaviors. A computer program could potentially facilitate the process when there are numerous combinations. In addition, inferencing will become more difficult as the task becomes more of a perceptual one, rather than a symbolic one.

Caution should be used in inferring rules from behavior, especially for tasks that have a large perceptual component. If we cannot infer our own internal rules governing behavior (e.g., Reber & Lewis, 1977), it is not (necessarily) more likely that someone else can perform this inference. It is for this reason that many tasks, especially perceptual ones, may be best acquired directly using machine learning, as discussed below. This obviates the need to make explicit or verbalize the rules underlying performance.

One obvious problem with inferring procedural rules based on performance is that we would not know the degree to which the inferred rules correspond to actual internal procedures. In confronting this problem, plausibility may prove to be one useful criterion. For example, a colleague recounted to me a story about a fire chief who attributed a particular case of problem solving to his "sixth sense" (see Klein, Calderwood, & Clinton-Cirocco, 1986). This inference was not considered particularly plausible by the knowledge engineers. The situational characteristics and behavior were evaluated and a new, more acceptable inference derived.

A more rigorous criterion for evaluating whether the procedural rules have been adequately captured would come from directly *testing* the validity of the rules. If the procedural knowledge of the expert has been captured, then subsequent problems given to both human expert and expert system should result in similar performance. Although one can never be sure that the procedural knowledge of the expert has been acquired, such validation through extensive testing would lend strong evidence. The obvious drawback is that this may impose additional time demands on the expert, unless behavioral measures could be collected in a relatively unobtrusive way on-site.

Machine Learning

As mentioned earlier, another method for inferring procedural rules is the use of machine learning. This is a method where an entire set of problem/solution examples is given to a computer to internalize or learn (Arciszewski, Mustafa, & Ziarko, 1987; Langley & Carbonell, 1984; Michalski, Carbonell, & Mitchel, 1986; Witten & McDonald, 1988). Some computer programs incorporate complex algorithms to learn the expertise represented in the examples provided by the human expert. Other advanced systems, such as neural networks, learn from the examples even if there is uncertainty, fuzziness, or even inconsistency in the examples.

Machine learning seems to hold promise for inferring the relationships between situational characteristics and behavior, especially in cases involving pattern recognition or highly "perceptual" stimuli (Schwartz, 1989). However, these systems still suffer from certain drawbacks, including the fact that people must still choose a large and relevant set of examples, and "massage" the features and behaviors into a form readable by the machine.

Conclusions

There is a great deal of evidence to show that people rely on at least two very different types of knowledge and, similarly, exhibit two different types of learning. The empirical findings relevant to these differences were framed within a system/mode view (see again Figure 6.1) that distinguishes between declarative and procedural learning systems. It was emphasized that although we possess static, verbalizable knowledge, much of our learning and skilled behavior is driven by a dynamic, procedural knowledge that is not verbalizable. And more, these systems may often behave independently. This means that just because someone can verbalize declarative knowledge, it doesn't mean that the person can go out and actually do the job. Conversely, experts may be very good at doing their job but unable to teach it to others except by task demonstration.

It follows from the theoretical analysis presented here that different knowledge acquisition methods are more suitable for measuring one type of knowledge than another. If the goal for knowledge acquisition is to acquire both types of knowledge, then certainly *at least* two different methods should be employed. For acquiring declarative knowledge, appropriate methods include those requiring introspection and verbalization, such as structured interviews. Although think-aloud problem-solving methods elicit declarative knowledge, the requirement to *explain* the behavior can be intrusive. For this reason, such think-aloud protocols can be used to elicit declarative knowledge; however, the problem-solving behavior accompanying verbal protocols cannot also be safely used to infer procedural knowledge.

Problem solving, when experts are not asked to explain their behavior, or are asked only to verbalize thoughts that naturally come to working memory, can be very useful for inferring procedural knowledge. If domain-general procedures are being used for task performance, intermediate declarative states of knowledge in conscious awareness can be used to infer those domain-general procedures (as is currently done by Rasmussen and others studying problem solving). If domain-specific procedures are used, the situational characteristics and behavioral outcomes can be analyzed to infer the specific and finely tuned procedural rules.

The theory and research reviewed here suggest that experts may often rely heavily on non-verbalizable types of knowledge. The fact that most of our knowledge elicitation methods rely on verbalization should be at least somewhat disconcerting. Given the existing results thus far, which indicate the parallel and often independent nature of learning systems, perhaps current practices in knowledge acquisition should be modified. In addition, researchers should progress toward identifying the divergence and convergence provided by various forms of knowledge elicitation.

References

Akin, O. (1980). *Models of architectural knowledge*. London: Pion.

Altman, R. B., & Buchanan, B. G. (1987). Partial compilation of strategic knowledge. *Proceedings of the AAAI Sixth National Conference on Artificial Intelligence* (pp. 399–404). Los Altos, CA: Morgan Kaufmann.

Anderson, J. R. (1976). *Language, memory, and thought*. Hillsdale, NJ: Erlbaum.

Anderson, J. R. (1983). *The architecture of cognition*. Cambridge, MA: Harvard University Press.

Anderson, J. R. (1985). *Cognitive psychology and its implications*. New York: W.H. Freeman.

Anderson, J. R. (1987a). Skill acquisition: Compilation of week-method problem solutions. *Psychological Review, 94*, 192–210.

Anderson, J. R. (1987b). Production systems, learning, and tutoring. In D. Klahr, P. Langley, & R. Neches (Eds.), *Production system models of learning and development* (pp. 437–458). Cambridge, MA: MIT Press.

Anderson, J. R. (1989). The analogical origins of errors in problem solving. In D. Klahr & K. Kotovsky (Eds.), *Complex information processes: The impact of Herbert A. Simon* (pp. 343–371). Hillsdale, NJ: Erlbaum.

Arciszewski, T., Mustafa, M., & Ziarko, W. (1987). A methodology of design knowledge acquisition for use in learning expert systems. *International Journal of Man-Machine Studies, 27*, 23–32.

Berry, D. C., & Broadbent, D. E. (1984). On the relationship between task performance and associated verbalizable knowledge. *Quarterly Journal of Experimental Psychology, 36A*, 209–231.

Blaxton, T. A. (1989). Investigating dissociations among memory measures: Support for a transfer-appropriate processing framework. *Journal of Experimental Psychology: Learning, Memory, and Cognition, 15*, 657–668.

Boose, J. H. (1986). *Expertise transfer for expert system design*. New York: Elsevier.

Boose, J. H. (1988). Uses of repertory grid-centred knowledge acquisition tools for knowledge-based systems. *International Journal of Man-Machine Studies, 29*, 287–310.

Broadbent, D. E., & Aston, B. (1978). Human control of a simulated economic system. *Ergonomics, 21*, 1035–1043.

Brooks, D. N., & Baddeley, A. D. (1976). What can amnesic patients learn? *Neurophsychologia, 14*, 111–122.

Butters, N. (1987, February). *Procedural learning in dementia: A double dissociation between Alzheimer and Huntington's disease patients on verbal priming and motor skill learning*. Paper presented at the meeting of the International Neuropsychological Society, Washington, DC.

Carr, T. H., McCauley, C., Sperber, R. D., & Parmalee, C. M. (1982). Words, pictures, and priming: On semantic activation, conscious identification, and the automaticity of information processing. *Journal of Experimental Psychology: Human Perception and Performance*, *8*, 757–777.

Charness, N., & Campbell, J. I. D. (1988). Acquiring skill at mental calculation in adulthood: A task decomposition. *Journal of Experimental Psychology: General*, *117*, 115–129.

Chase, W. G. (1983). Spatial representations of taxi drivers. In D. R. Rogers & J. H. Sloboda (Eds.), *Acquisition of symbolic skills* (pp. 391–405). New York: Plenum.

Chi, M. T. H., Glaser, R., & Farr, M. J. (Eds.). (1988). *The nature of expertise*. Hillsdale, NJ: Erlbaum.

Chi, M. T. H., Glaser, R., & Rees, E. (1982). Expertise in problem solving. In R. Sternberg (Ed.), *Advances in the psychology of human intelligence* (Vol. 1, pp. 17–76). Hillsdale, NJ: Lawrence Erlbaum Associates.

Clancey, W. J. (1983). The epistemology of a rule-based expert system: A framework for explanation. *Artificial Intelligence*, *20*, 215–251.

Clarke, R., & Morton, J. (1983). Cross-modality facilitation in tachistoscopic word recognition. *Quarterly Journal of Experimental Psychology*, *35A*, 79–96.

Cohen, A., Ivry, R., & Keele, S. W. (1988). *Attention and structure in sequence learning* (Technical Report No. 88–5). Institute of Cognitive and Decision Sciences, University of Oregon, Eugene, OR.

Cohen, N. J. (1984). Preserved learning capacity in amnesia: Evidence for multiple memory systems. In L. R. Squire & N. Butters (Eds.), *Neuropsychology of memory* (pp. 83–103). New York: Guilford Press.

Cohen, N. J., & Squire, L. R. (1980). Preserved learning and retention of pattern-analyzing skill in amnesia: Dissociation of "knowing how" and "knowing that." *Science*, *210*, 207–209.

Collins, A. M., & Loftus, E. F. (1975). A spreading-activation theory of semantic processing. *Psychological Review*, *82*, 407–428.

Corkin, S. (1968). Acquisition of motor skill after bilateral medial temporal excision. *Neuropsychologia*, *6*, 255–265.

Crowder, R. G. (1989). Modularity and dissociations in memory systems. In H. L. Roediger & F. I. M. Craik (Eds.), *Varieties of memory and consciousness: Essays in honour of Endel Tulving* (pp. 271–294). Hillsdale, NJ: Erlbaum.

Dannenbring, G. L., & Briand, K. (1982). Semantic priming and the word repetition effect in a lexical decision task. *Canadian Journal of Psychology*, *36*, 435–444.

Day, B. L., & Marsden, C. D. (1982). Two strategies for learning a visually guided motor task. *Perceptual and Motor Skills*, *55*, 1003–1016.

deJong, T., & Ferguson-Hessler, M. G. M. (1986). Cognitive structures of good and poor novice problem solvers in physics. *Journal of Educational Psychology*, *78*, 279–288.

Diederich, J., Ruhmann, I., & May, M. (1987). KRITON: A knowledge-acquisition tool for expert systems. *International Journal of Man-Machine Studies*, *26*, 29–40.

Dillon, R. F., & Sternberg, R. J. (Eds.). (1986). *Cognition and instruction*. Orlando, FL: Academic Press.

Egan, D. E., & Schwartz, B. J. (1979). Chunking in recall of symbolic drawings. *Memory and Cognition*, *7*, 149–158.

Ericsson, K. A., & Simon, H. A. (1984). *Protocol analysis: Verbal reports as data*. Cambridge, MA: MIT Press.

Evans, F. J., & Thorn, W. A. F. (1966). Two types of posthypnotic amnesia: Recall amnesia and source amnesia. *International Journal of Clinical and Experimental Hypnosis*, *14*, 162–179.

Fitts, P. M., & Posner, M. I. (1967). *Human performance*. Belmont, CA: Brooks Cole.

Franks, I. M. (1982). Rehearsal and learning of embedded movement sequences during a tracking task. *Perceptual and Motor Skills*, *55*, 615–622.

Gagne, E. D. (1985). *The cognitive psychology of school learning*. Boston, MA: Little, Brown & Company.

Gagne, R. M., & Briggs, L. J. (1979). *Principles of instructional design*. New York: Holt, Rinehart & Winston.

Gentner, D. R. (1988). Expertise in typewriting. In M. T. H. Chi, R. Glaser, & M. J. Farr (Eds.), *The nature of expertise* (pp. 1–21). Hillsdale, NJ: Erlbaum.

Glaser, R. (1984). Education and thinking: The role of knowledge. *American Psychologist*, *39*, 93–104.

Glaser, R. (1989). Expertise and learning: How do we think about instructional processes now that we have discovered knowledge structures? In D. Klahr & K. Kotovsky (Eds.), *Complex information processing: The impact of Herbert A. Simon* (pp. 269–282). Hillsdale, NJ: Erlbaum.

Glisky, E. L., Schacter, D. L., & Tulving, E. (1986). Computer learning by memory-impaired patients: Acquisition and retention of complex knowledge. *Neuropsychologia*, *24*, 313–328.

Gordon, S. E. (1991). Front-end analysis for expert system design. *Proceedings of the Human Factors Society 35th Annual Meeting* (pp. 278–282). Santa Monica, CA: Human Factors Society.

Gordon, S. E., & Gill, R. T. (1989). Question probes: A structured method for eliciting declarative knowledge. *AI Applications in Natural Resource Management*, *3*, 13–20.

Gordon, S. E., & Gill, R. T. (in press). Knowledge acquisition with question probes and conceptual graph structures. In T. Lauer, E. Peacock, & A. Graesser (Eds.), *Questions and information systems*. Hillsdale, NJ: Erlbaum.

Graf, P., & Mandler, G. (1984). Activation makes words more accessible, but not necessarily more retrievable. *Journal of Verbal Learning and Verbal Behavior*, *23*, 553–568.

Graf, P., Mandler, G., & Hayden, P. (1982). Simulating amnesic symptoms in normal subjects. *Science*, *218*, 1243–1244.

Graf, P., & Schacter, D. L. (1987). Selective effects of interference on implicit and explicit memory for new associations. *Journal of Experimental Psychology: Learning, Memory, and Cognition*, *13*, 45–53.

Graf, P., Shimamura, A. P., & Squire, L. R. (1985). Priming across modalities and priming across category levels: Extending the domain of preserved function in amnesia. *Journal of Experimental Psychology: Learning, Memory, and Cognition*, *11*, 385–395.

Graf, P., Squire, L. R., & Mandler, G. (1984). The information that amnesic patients do not forget. *Journal of Experimental Psychology: Learning, Memory, and Cognition*, *10*, 164–178.

Grover, M. D. (1983). A pragmatic knowledge acquisition methodology. *International Joint Conference on Artificial Intelligence*, *83*, 436–438.

Hart, A. (1986). *Knowledge acquisition for expert systems*. New York: McGraw-Hill.

Hayes, N. E., & Broadbent, D. (1988). Two modes of learning for interactive tasks. *Cognition*, *28*, 249–276.

Jacoby, L. L. (1983). Perceptual enhancement: Persistent effects of an experience. *Journal of Experimental Psychology: Learning, Memory, and Cognition*, *9*, 21–38.

Jacoby, L.'L., & Dallas, M. (1981). On the relationship between auto-biographical memory and perceptual learning. *Journal of Experimental Psychology: General*, *110*, 306–340.

Johnson, P. E. (1983). What kind of expert should a system be? *The Journal of Medicine and Philosophy*, *8*, 77–97.

Kellogg, R. T. (1982). Hypothesis recognition failure in conjunctive and disjunctive concept-identification tasks. *Bulletin of the Psychonomic Society*, *19*, 327–330.

Kellogg, R. T., Robbins, D. W., & Bourne, L. E. (1978). Memory for intratrial events in feature identification. *Journal of Experimental Psychology: Human Learning and Memory*, *4*, 256–265.

Kintsch, W. (1988). The role of knowledge in discourse comprehension: A construction-integration model. *Psychological Review*, *95*, 163–182.

Kirsner, K., Milech, D., & Standen, P. (1983). Common and modality-specific processes in the mental lexicon. *Memory & Cognition*, *11*, 621–630.

Klein, G. A., Calderwood, R., & Clinton-Cirocco, A. (1986). Rapid decision making on the fire ground. *Proceedings of the Human Factors Society 30th Annual Meeting* (pp. 576–580). Santa Monica, CA: Human Factors Society.

Langley, P., & Carbonell, J. G. (1984). Approaches to machine learning. *Journal of the American Society for Information Science*, *35*, 306–316.

Larkin, J. H. (1979, December). Processing information for effective problem solving. *Engineering Education*, pp. 285–288.

Lewicki, P. (1985). Nonconscious biasing effects of single instances on subsequent judgments. *Journal of Personality and Social Psychology*, *48*, 563–574.

Lewicki, P. (1986). Processing information about covariations that cannot be articulated. *Journal of Experimental Psychology: Learning, Memory, and Cognition*, *12*, 135–146.

Lewicki, P., Czyzewska, M., & Hoffman, H. (1987). Unconscious acquisition of complex procedural knowledge. *Journal of Experimental Psychology: Learning, Memory, and Cognition*, *13*, 523–530.

Lewicki, P., Hill, T., & Bizot, E. (1988). Acquisition of procedural knowledge about a pattern of stimuli that cannot be articulated. *Cognitive Psychology*, *20*, 24–37.

Light, L. L., & Singh, A. (1987). Implicit and explicit memory in young and older adults. *Journal of Experimental Psychology: Learning, Memory, and Cognition*, *13*, 531–541.

Light, L. L., Singh, A., & Capps, J. L. (1986). Dissociation of memory and awareness in young and older adults. *Journal of Clinical and Experimental Neuropsychology*, *8*, 62–74.

Logan, G. D. (1980). Attention and automaticity in Stroop and priming tasks: Theory and data. *Cognitive Psychology*, *12*, 523–553.

Logan, G. D. (1988). Automaticity, resources, and memory: Theoretical controversies and practical implications. *Human Factors*, *30*, 583–598.

Marcel, A. T. (1983). Conscious and unconscious perception: An approach to the relations between phenomenal experience and perceptual processes. *Cognitive Psychology*, *15*, 238–300.

McKeithen, K. B., Reitman, J. S., Rueter, H. H., & Hirtle, S. C. (1981). Knowledge organization and skill differences in computer programmers. *Cognitive Psychology*, *13*, 307–325.

Michalski, R. S., Carbonell, J. G., & Mitchell, T. M. (Eds.). (1986). *Machine learning: An artificial intelligence approach* (Vol. 2). Los Altos, CA: Morgan Kaufmann.

Milner, B., Corkin, S., & Teuber, H. L. (1968). Further analysis of the hippocampal amnesic syndrome: 14 year follow-up study of H. M. *Neuropsychologia*, *6*, 215–234.

Mitchell, A. A. (1987). The use of alternative knowledge-acquisition procedures in the development of a knowledge-based media planning system. *International Journal of Man-Machine Studies*, *26*, 399–411.

Mitchell, D. B. (1989). How many memory systems? Evidence from aging. *Journal of Experimental Psychology: Learning, Memory, and Cognition*, *15*, 31–49.

Mitchell, D. B., & Brown, A. S. (1988). Persistent repetition priming in picture naming and its dissociation from recognition memory. *Journal of Experimental Psychology: Learning, Memory, and Cognition*, *14*, 213–222.

Moscovitch, M. (1989). Confabulation and the frontal systems: Stretegic versus associated retrieval in neuropsychological theories of memory. In H. L. Roediger & F. I. M. Craik (Eds.), *Varieties of memory and consciousness: Essays in honour of Endel Tulving* (pp. 133–161). Hillsdale, NJ: Erlbaum.

Musen, M. A., Fagan, L. M., Combs, D. M., & Shortliffe, E. H. (1987). Use of a domain model to drive an interactive knowledge-editing tool. *International Journal of Man-Machine Studies*, *26*, 105–121.

Nadel, L., & Zola-Margan, S. (1984). Toward the understanding of infant memory: Contributions from animal neuropsychology. In M. Moscovitch (Ed.), *Infant memory* (pp. 145–172). New York: Plenum.

Neely, J. H. (1977). Semantic priming and retrieval from lexical memory: Roles of inhibitionless spreading activation and limited-capacity attention. *Journal of Experimental Psychology: General*, *106*, 226–254.

Neves, D. M., & Anderson, J. R. (1981). Knowledge compilation: Mechanisms for the automatization of cognitive skills. In J. A. Anderson (Ed.), *Cognitive skills and their acquisition* (pp. 57–84). Hillsdale, NJ: Erlbaum.

Nickerson, R. S., Perkins, D. N., & Smith, E. E. (1985). *The teaching of thinking*. Hillsdale, NJ: Erlbaum.

Nisbett, R. E., & Wilson, T. D. (1977). Telling more than we can know: Verbal reports on mental processes. *Psychological Review*, *84*, 231–259.

Nissen, M. J., & Bullemer, P. (1987). Attentional requirements of learning: Evidence from performance measures. *Cognitive Psychology*, *19*, 1–32.

Norman, D. A., & Rumelhart, D. E. (1975). *Explorations in cognition*. New York: W.H. Freeman.

Porter, D. B. (1987). Classroom teaching, implicit learning and the deleterious effects of inappropriate explication. *Proceedings of the Human Factors Society 31st Annual Meeting* (pp. 289–292). Santa Monica, CA: Human Factors Society.

Posner, M. I., & Snyder, C. R. R. (1975). Attention and cognitive control. In R. L. Solso (Ed.), *Information processing and cognition: The Loyola symposium* (pp. 55–85). Hillsdale, NJ: Erlbaum.

Rasmussen, J. (1980). The human as a system component. In H. Smith & T. Green (Eds.), *Human interaction with computers*. London: Academic Press.

Rasmussen, J. (1983). Skills, rules, knowledge: Signals, signs, and symbols and other distinctions in human performance models. *IEEE Transaction on Systems, Man, and Cybernetics*, *13*, 257–267.

Rasmussen, J. (1986). *Information processing and human-machine interaction: An approach to cognitive engineering*. New York: Elsevier Science.

Reber, A. S. (1967). Implicit learning of artificial grammars. *Journal of Verbal Learning and Verbal Behavior*, *5*, 855–863.

Reber, A. S. (1976). Implicit learning of synthetic languages: The role of instructional set. *Journal of Experimental Psychology: Human Learning and Memory*, *2*, 88–94.

Reber, A. S., & Lewis, S. (1977). Implicit learning: An analysis of the form and structure of a body of tacit knowledge. *Cognition*, *5*, 333–361.

Reif, F., & Heller, J. I. (1982). Knowledge structures and problem solving in physics. *Educational Psychologist*, *17*, 102–127.

Resnick, L. B. (1981). Instructional psychology. *Annual Review of Psychology*, *32*, 659–704.

Richardson-Klavehn, A., & Bjork, R. A. (1988). Measures of memory. *Annual Review of Psychology*, *39*, 475–543.

Roediger, H. L., & Blaxton, T. A. (1987). Effects of varying modality, surface features, and retention interval on priming in word fragment completion. *Memory and Cognition*, *15*, 379–388.

Roediger, H. L., Weldon, M. S., & Challis, B. H. (1989). Explaining dissociations between implicit and explicit measures of retention: A processing account. In H. L. Roediger & F. I. M. Craik (Eds.), *Varieties of memory and consciousness: Essays in honour of Endel Tulving* (pp. 3–41). Hillsdale, NJ: Erlbaum.

Ryle, G. (1949). *The concept of mind*. San Francisco, CA: Hutchinson.

Sanderson, P. M. (1989). Verbalizable knowledge and skilled task performance: Association, dissociation, and mental models. *Journal of Experimental Psychology: Learning, Memory, and Cognition*, *15*, 729–747.

Scandura, J. M. (1977). *Problem solving: A structural/process approach with instructional implications*. New York: Academic Press.

Schacter, D. L. (1985). Priming of old and new knowledge in amnesic patients and normal subjects. *Annals of the New York Academy of Sciences*, *444*, 41–53.

Schacter, D. L. (1987). Implicit memory: History and current status. *Journal of Experimental Psychology*, *13*, 501–518.

Schacter, D. L. (1989). On the relation between memory and consciousness: Dissociable interactions and conscious experience. In H. L. Roediger & F. I. M. Craik (Eds.), *Varieties of memory and consciousness: Essays in honour of Endel Tulving* (pp. 355–389). Hillsdale, NJ: Erlbaum.

Schacter, D. L., & Graf, P. (1986). Effects of elaborative processing on implicit and explicit memory for new associations. *Journal of Experimental Psychology: Learning, Memory, and Cognition*, *12*, 432–444.

Schacter, D. L., Harbluk, J. L., & McLachlan, D. R. (1984). Retrieval without rocollection: An experimental analysis of source amnesia. *Journal of Verbal Learning and Verbal Behavior*, *23*, 593–611.

Schneider, W., & Detweiler, M. (1987). Connectionist/control architecture for working memory. In G. H. Bower (Ed.), *The psychology of learning and motivation*, Vol. 21 (pp. 53–119). San Diego, CA: Academic Press.

Schwartz, T. J. (1989, December). Parables of neural networks. *AI Expert*, pp. 54–59.

Shiffrin, R. M., & Dumais, S. T. (1981). The development of automatism. In J. R. Anderson (Ed.), *Cognitive skills and their acquisition* (pp. 111–140). Hillsdale, NJ: Erlbaum.

Shimamura, A. P. (1986). Priming effects in amnesia: Evidence for a dissociable memory function. *Quarterly Journal of Experimental Psychology*, *38A*, 619–644.

Shimamura, A. P., & Squire, L. R. (1984). Paired-associate learning and priming effects in amnesia: A neuropsychological study. *Journal of Experimental Psychology: General*, *113*, 556–570.

Shimamura, A. P., & Squire, L. R. (1987). A neuropsychological study of fact learning and source amnesia. *Journal of Experimental Psychology: Learning, Memory, and Cognition*, *13*, 464–474.

Shimamura, A. P., & Squire, L. R. (1988). Long-term memory in amnesia: Cued recall, recognition memory, and confidence ratings. *Journal of Experimental Psychology: Learning, Memory, and Cognition*, *14*, 763–770.

Singley, M. K., & Anderson, J. R. (1988). A keystroke analysis of learning and transfer in text editing. *Human-Computer Interactions*, *3*, 223–274.

Singley, M. K., & Anderson, J. R. (1989). *The transfer of cognitive skill*. Cambridge, MA: Harvard University Press.

Sloman, S. A., Hayman, C. A., Ohta, N., Law, J., & Tulving, E. (1988). Forgetting in primed fragment completion. *Journal of Experimental Psychology: Learning, Memory, and Cognition*, *14*, 223–239.

Smith, E. R., & Branscombe, N. R. (1988). Category accessibility as implicit memory. *Journal of Experimental Social Psychology*, *24*, 490–504.

Squire, L. M. (1987). *Memory and brain*. New York: Oxford University Press.

Squire, L. R., & Cohen, N. J. (1984). Human memory and amnesia. In J. McGaugh, G. Lynch, & N. Weinberger (Eds.), *Proceedings of the conference on the neurobiology of learning and memory* (pp. 3–64). New York: Guilford Press.

Stadler, M. A. (1989). On learning complex procedural knowledge. *Journal of Experimental Psychology: Learning, Memory, and Cognition*, *15*, 1061–1069.

Stauffer, L. (1989). Eliciting and analyzing data about the engineering-design process. *Proceed-*

ings of the NSF Design Theory and Methodology Conference, University of Massachussetts, Amherst, MA.

Steels, L. (1987). Second generation expert systems. In M. A. Bramer (Ed.), *Research and development in expert systems*, Vol. 3 (pp. 175–183). Cambridge: Cambridge University Press.

Tulving, E. (1983). *Elements of episodic memory*. New York: Oxford University Press.

Tulving, E. (1985). How many memory systems are there? *American Psychologist*, *40*, 385–398.

Tulving, E., Schacter, D. L., & Stark, H. A. (1982). Priming effects in word-fragment completion are independent of recognition memory. *Journal of Experimental Psychology: Learning, Memory, and Cognition*, *8*, 336–342.

Waldron, M. B., & Waldron, K. L. (1988). A time sequence study of a complex mechanical system design. *Design Studies*, *9*, 95–106.

Warrington, E. K., & Weiskrantz, L. (1970). Amnesia: Consolidation or retrieval? *Nature*, *228*, 628–630.

Warrington, E. K., & Weiskrantz, L. (1974). The effect of prior learning on subsequent retention in amnesic patients. *Neuropsychologia*, *12*, 419–428.

Warrington, E. K., & Weiskrantz, L. (1982). Amnesia: A disconnection syndrome? *Neuropsychologia*, *20*, 233–248.

Weldon, M. S., & Roediger, H. L. (1987). Altering retrieval demands reverses the picture superiority effect. *Memory and Cognition*, *15*, 269–280.

Willingham, D. B., Nissen, M. J., & Bullemer, P. (1989). On the development of procedural knowledge. *Journal of Experimental Psychology: Learning, Memory, and Cognition*, *15*, 1047–1060.

Winnick, W. A., & Daniel, S. A. (1970). Two kinds of response priming in tachistoscopic word recognition. *Journal of Experimental Psychology*, *84*, 74–81.

Witten, I. H., & McDonald, B. A. (1988). Using concept learning for knowledge acquisition. *International Journal of Man-Machine Studies*, *29*, 171–196.

7
Knowledge Acquisition and Constructivist Epistemology

Kenneth M. Ford and Jack R. Adams-Webber

Introduction

The most fundamental step in the knowledge acquisition phase of the development of an expert system is the elicitation of knowledge from a skilled individual. The knowledge acquisition phase has typically involved the knowledge engineer's working closely with a specialist to elicit relevant knowledge from the latter's domain. This is typically a tedious and ad hoc cycle that consists of extensive verbal interviews followed by the construction of prototypes, testing, and more interviews. This approach has two significant drawbacks—it has been extremely laborious, and domain experts often have difficulty articulating their knowledge in forms useful to the knowledge engineer. Indeed, it has been suggested (Feigenbaum & McCorduck, 1983) that "the problem of knowledge acquisition is the critical bottleneck in artificial intelligence" (p. 80).

A commonly proposed partial solution to the knowledge acquisition bottleneck is the design and implementation of automated tools for the purposes of interacting with domain experts, acquiring and organizing knowledge, and automatically generating a prototype expert system. Among the expected benefits of automating at least a portion of the knowledge acquisition process are the following:

1. An automated approach may be more efficient than manual interviewing methods, thereby reducing the great expense presently incurred in the knowledge acquisition phase.
2. Automated approaches may prove able to elicit expertise not easily obtained by manual interviewing methods, thus producing systems with greater expertise.

To provide these desired benefits, automated approaches to knowledge acquisition must assist the knowledge engineer in avoiding the domain experts' cognitive defenses and reduce the representation mismatch—the difference between the manner in which the domain expert normally states knowledge and the way it is represented in the expert system knowledge base. The design and construction of knowledge acquisition tools have recently become areas of intense research and development.

As noted by Bradshaw and Boose (1990), a major difficulty has been that much of the aforementioned work lacks a plausible theoretical foundation:

> As a consequence of incomplete theory and a limited repertoire of practical approaches to the dynamics of the modeling process, knowledge engineers have had to rely on intuition and experience as the primary means of developing and testing effective procedures. (p. 129)

Many of those engaged in knowledge acquisition (as researcher or practitioner) may be classified as toolmakers and/or tool users. Toolmakers should exploit theory as a means of building their tools on a sound footing and as a framework in which to make explicit their epistemological assumptions. Furthermore, theory may offer toolmakers a useful infrastructure upon which to build highly integrated collections of tools and techniques (i.e., hybrid knowledge acquisition workbenches). Tool users may employ theory as the basis for a principled application of the myriad tools now coming available. Rarely do working knowledge engineers explicitly consider the epistemological underpinnings of the methods and tools they employ in their task. In fact, many knowledge engineering practices imply an epistemology that some practitioners would perhaps reject if considered explicitly.

Not surprisingly, the clinical and experimental psychology literatures contain much research that is relevant to the problem of improving communication between knowledge engineers and domain experts (Bainbridge, 1979; Ericsson & Simon, 1984; Hoffman, 1987). In particular, Kelly's (1955) personal construct theory has served as the basis for several recent approaches to the design and construction of automated (i.e., computer-based) knowledge acquisition tools (Boose, 1984; Boose & Bradshaw, 1987; Ford, Petry, Adams-Webber, & Chang, 1991; Ford, Stahl, Adams-Webber, Cañas, Novak, & Jones, 1991; Gaines & Shaw, 1986). In the remainder of this chapter we provide a brief summary of personal construct theory and discuss the implications of a constructivist epistemology for work in knowledge acquisition.

Personal Construct Theory

Personal construct theory, as formulated by Kelly (1955, 1969, 1970) and summarized by Adams-Webber (1987), is essentially a formal model of the organization of human cognitive processes. As such, it provides a comprehensive and systematic foundation for addressing epistemological issues (cf. Fischler & Firschein, 1987; p. 68). According to Kelly's theory, each individual constructs a personal model of the world in a manner commonly associated with the method of scientists. More precisely, Kelly's (1955) notion of the *personal scientist* assumes that each individual seeks to predict and control events by forming relevant hypotheses and then testing them against experience (cf. Mischel, 1964). In Kelly's (1955) own words, "The aspirations of the scientist are essentially the aspirations of all men" (p. 43).

Kelly (1970) argued that all of the basic assumptions of personal construct theory (a single postulate and 11 corollaries) can be derived logically from one explicit epistemological premise, his principle of "constructive alternativism." According to this principle, "reality" does not reveal itself to us directly but rather is subject to as many different constructions as we are able to invent (cf. Adams-Webber, 1989; Agnew & Brown, 1989a). Thus, any given event is open to a variety of alternative interpretations. This does not imply, however, that each interpretation is as good as any other. On the contrary, different ways of construing the same event can be evaluated by comparing them systematically in terms of their relative predictive utility (Mancuso & Adams-Webber, 1982). It is likely that some interpretations of an event will prove more useful than others for anticipating similar events in the future.

Kelly's (1955) fundamental postulate asserts that "a person's processes are psychologically channelized by the ways in which he anticipates events" (p. 46). As Adams-Webber (1989) notes, this postulate specifically implies that all of our representational processes are fundamentally anticipatory. Kelly (1955) elaborated the logical implications of this proposition in terms of 11 corollaries, 5 of which are directly relevant to the central thesis of this chapter.

1. Kelly's (1955) *dichotomy corollary* states: "A person's construction system is composed of a finite number of dichotomous constructs" (p. 59). The basic units of analysis in personal construct theory are bipolar dimensions termed *personal constructs*. These constructs can be viewed as *templets* that a person "creates and then attempts to fit over the realities of which the world is composed" (Kelly, 1955, p. 8). Kelly contended that the underlying distinction that lends each construct its meaning is dichotomous in form, for example, happy/sad, young/old, odd/even.

He argued further that the dichotomous nature of personal constructs is an essential feature of the way in which people organize their experience. For example, if the same event could be perceived simultaneously as equally pleasant and unpleasant in the same respect, then this distinction would have no definite meaning in relation to that particular event. The event in question could be said to lie outside the "range of convenience" of this particular construct.

2. Kelly's (1955) *construction corollary* states: "A person anticipates events by construing their replications" (p. 50). Each person employs personal constructs to forecast events and later to evaluate the predictive utility of the forecasts. This does not mean that the same event ever actually recurs but rather that we use our personal constructs to represent perceived similarities and differences among events; then we organize these representations into coherent patterns, or "schemata," within the framework of which we are able to detect certain recurrent themes in our experience over time and then feed these representations forward in the form of expectations about future events (cf. Ford, 1989).

3. Kelly's (1955) *experience corollary* states: "A person's construction system varies as he successively construes the replication of events" (p. 72). Kelly (1955) proposed that with the passage of time, the perception of new events constitute an ongoing validational process that serves to confirm or disconfirm an individual's anticipations. As a result, a person's constructs undergo continuous, progressive change as they are revised in the course of experience. Indeed, Kelly assumed that specific changes in either the structure or content of our personal constructs occur primarily in response to predictive failures (cf. Adams-Webber & Mancuso, 1983).

4. Kelly's (1955) *range corollary* states: "Each construct is convenient for the anticipation of a finite range of events only" (p. 68). Each of a person's constructs has a specific *range of convenience*, which comprises "all those things to which the user would find its application useful." Accordingly, the range of convenience of a construct defines its extension in terms of a single aspect of a limited domain of events (Mancuso & Eimer, 1982). Not only individual constructs but also, by implication, systems and subsystems of interrelated constructs have specific ranges of convenience. This suggests that some degree of functional differentiation between subsystems of constructs within the same system can enhance its overall range of convenience with respect to the variety of events that can be accommodated within its framework (Adams-Webber, 1979).

5. Kelly's *organization corollary* states: "Each person characteristically evolves, for his own convenience in anticipating events, a construction system embracing ordinal relationships between constructs" (p. 56). A particular construct seldom, if ever, stands alone in our experience, because it is usually deployed together with one or more other related constructs in interpreting and predicting events. Indeed, a necessary condition for organized thought is some degree of overlap between constructs in terms of their respective "ranges of convenience" (Adams-Webber, 1970). It is this overlap, or intersection, between the constructs' extensions that enables an individual to formulate hypotheses. That is, in interpreting an event we essentially categorize it in terms of one or more constructs, and then by reviewing our personal systems of related constructs we can derive predictive inferences from that initial categorization. For example, suppose that within the context of an individual's personal construct system, the relatively subordinate construct *polite/rude* were to be subsumed by the relatively superordinate construct *considerate/inconsiderate*. This individual would tend to expect considerate behavior on the part of any persons perceived as polite. It is this predictive function of a person's construct system that provides the logical rationale for the Kellyan view that human beings are characterized by an "anticipatory stance."

Like the prototype of the scientist that he is, man seeks prediction. His structured network of pathways leads toward the future so that he may anticipate it. This is the function that it serves. Anticipation is both the push and the pull of the psychology of personal constructs. (Kelly, 1955, p. 49)

The notion of "personal" constructs lends emphasis to the theme that individuals create and maintain unique models of the world.

In other words, as "personal scientists," we humans frequently anticipate the occurrence or nonoccurrence of future events based on our willingness to project observed uniformities into the future. Thus, we continually glide from the past into the future with our previous experience preceding us—illuminating and organizing the manner in which subsequent events will be manifest to us. (Ford, 1989, p. 190)

The process through which people continuously anticipate events and test the efficacy of their constructions has been termed the *experience cycle* (cf. Neimeyer, 1985):

Kelly believed that this cycle—anticipation, investment, encounter, confirmation/disconfirmation and constructive revision—represented a useful heuristic for conceptualizing human experience. (p. 278)

Supporting Empirical Research

Adams-Webber (1979) has reviewed research that provides empirical support for some of the central implications of personal construct theory, including the following hypotheses:

1. *Experience within a specific domain will produce gradual increases in the degree of interrelatedness between constructs that are typically applied to events in that domain* (Adams-Webber & Mirc, 1976; Baldwin, 1972; Bannister, 1963; Benjafield, Jordan, & Pomeroy, 1976; Bodden & James, 1976; Cochran, 1976; Fransella, 1972; Lemon, 1975). For example, Adams-Webber and Mirc (1976) found that student teachers repeatedly tested with repertory grids (cf. Adams-Webber, 1984) at regular intervals during their initial 6 weeks of supervised classroom experience showed systematic increases in the average degree of interrelationship among certain textbook "curricular role" (experimental) constructs that were used to define their professional roles, whereas there were no systematic changes in relationships among an equal number of "lay" (control) constructs not directly relevant to defining the role of teacher. At the outset, these student teachers showed significantly less interrelationship among the curricular role constructs than did a group of experienced teachers; however, after the former had completed 6 weeks of practice teaching in local schools there were no significant differences between the two groups with respect to the degree of overall relatedness among the curricular role constructs.

2. Although people can impose different interpretations on the same events, and it is highly unlikely that any two people develop construct systems with exactly the same pattern of relations between constructs (Kelly, 1970), *there seems to be considerable consensus within the "normal" population concerning the pattern of relationships among many constructs*. For example, Bannister (1962) found that when people categorized photographs of strangers on the basis of the same psychological constructs, there was a significant level of agreement concerning the overall pattern of interrelationship among these constructs, despite the fact that there was very little agreement in terms of how particular photographs were rated. Applebee (1975, 1976) also showed that the degree of consensus among children about the relationships between specific constructs increases gradually with age. Like Bannister (1962), Applebee observed more intersubject agreement concerning relationships between constructs than about the ratings assigned to particular elements.

3. *Personal construct systems tend to develop with age and experience*. As children mature, and presumably gain in social sophistication, they increasingly differentiate between themselves and others in terms of bipolar constructs (Adams-Webber, 1985a, 1985b). Scarlett, Press, and Crockett (1971) report that there is a steady increase with age in the number of different constructs that children use in describing people, which is Crockett's (1965, 1982) operational definition of "cognitive complexity."

In addition, older children use significantly more psychological constructs in describing persons than do younger children, and girls and women of all ages tend to use significantly more psychological constructs, and more different constructs, than do their male peers in describing persons (Barratt, 1977; Brierley, 1967; Duck, 1975; Little, 1968; Scarlett et al., 1971). Specifically, these studies report systematic decreases with age in the relative frequency of use of constructs that refer to people's appearance (e.g., "She is tall"); social roles ("He is captain of the football team"); and behavior ("Bill tells lies"), whereas there is a corresponding increase in the use of personality constructs ("Mary is shy") and intellectual constructs ("Lewis is stupid"). In the youngest age groups tested—8 and younger—appearance and role constructs are used more often than either behavior or personality constructs. At intermediate ages—from 8 to 10—behavior constructs are used most frequently. During puberty and adolescence—from 12 to 15—the use of personality constructs increases dramatically, and by midadolescence they are the most prevalent of all.

Thus, as Adams-Webber (1979) notes, from the standpoint of Kelly's theory, cognitive development involves systematic transformations in both the structure and content of personal construct systems. A progressive increase in terms of the number of different constructs that are used, that is, "cognitive complexity," is accompanied by a gradual shift of emphasis from a primary concern with concrete descriptions of appearance, social roles, and behavior to a predominant interest in the more abstract dimensions of personality and intelligence.

A Constructivist Epistemology

As we have seen, Kelly (1955) developed a constructivist model of human representational processes in which he assumed that their primary function is the anticipation of events. Several other psychological theories emphasize the anticipatory nature of cognitive processes, including those of Bartlett, Dewey, Neisser, Piaget, Rotter, and Tolman, among others. Indeed, since at least as far back as Dewey's (1896) critical analysis of the reflex arc concept, many cognitive psychologists have recognized that, although the *nominal* stimulus can often be identified as a physical event external to the person (i.e., it could, in principle, be described adequately by a physicist), the *functional* stimulus (i.e., what needs to be explained by a psychologist) is constituted by the anticipatory processes of the person. As Kelly (1955) noted, "Dewey emphasized the anticipatory nature of behavior and the person's use of hypotheses in thinking. The psychology of personal constructs follows Dewey in this respect" (p. 129).

Heidbreder (1933) points out that Dewey's insight led Dewey's student Watson (1919) to conclude that the scientific investigation of sensation and perception, let alone thinking, was impossible because the conscious experience of the person is not directly accessible to external observers. This dogma became part of the philosophical groundwork for the development of a radical behaviorism, which in its most extreme from included an attempt to explain all human behavior—even verbal behavior—in terms of external events and the organism's responses as related to its past history of inputs (Skinner, 1957). Nonetheless, as Deese (1969) has put the case, psychology is concerned not only with overt behavior, but also with experience, including our everyday conscious acts of interpreting and anticipating events.

Bartlett (1932) suggested that perception and memory involve not only the registration of sensory patterns but also the construction of these sensory data into something having significance that goes beyond their sensory character. He referred specifically to the process of connecting a given stimulus pattern with some preformed setting or "schema" (a term borrowed from Head, 1920) as an "effort after meaning." Bartlett also suggested that although a stimulus array may possess "reactive significance" at the level of reflex responses, as soon as the reacting persons become *aware* of the material with which their reactions deal, there is meaning. In this sense, even the most elementary perceptual processes involve inferential constructions that go beyond the given sensory data.

More precisely, according to Bartlett's theory, any perceived similarity between events must depend on underlying tendencies (active schemata) that lead to the grouping together of items of input that possess a welter of diverse sensory characteristics. It follows that whenever two or more events, separated by an interval of time, are perceived as the same, some kind of information has been retained during the interval.

From the specific perspective of personal construct theory, outlined above, Bartlett's most important contribution was to enhance our understanding of the fundamental problem of how we can recognize a given event as the same or as different from that which we had anticipated. For example, let us suppose that a person recognizes a currently perceived event as the same as one that was observed on a previous occasion. Because this kind of recognition is frequently highly detailed, there must be some way in which specific information is preserved in the perceiving system from the first to the second occasion.

The traditional solution to this important problem assumed that recognition of the same event on a subsequent occasion requires the reexcitation of a specific "trace" or comparison with a preserved "copy" of the previous sensory input. As Asch (1969), among others, pointed out, this so-called solution still leaves open the fundamental question of how the present stimulus input makes contact with the *correct* trace or copy without its *prior recognition*, which is exactly what needs to be explained in the first instance. It was Bartlett's (1932) crucial insight that "in all cases recognizing is rendered possible by the carrying over of orientation or attitude from the original presentation to the representation" (p. 193).

It also seems clear that in order for a person to recognize the same event on a second occasion, the new sensory data (input) must exert some control over the perception of similarity. That is, there must be some common properties in the two stimulus patterns that the processes of cognition are prepared to seize on and elaborate (cf. Neisser, 1976). Thus, even if perception on each occasion involves certain inferential constructions, the input information itself must also play a role in accurate recognition. In short, whenever there is repeated perception of the same event, the stimulus patterns that activate sensory processes are presumed to have something in common. Nonetheless, as Bartlett's profound analysis of the problem revealed, something more is required.

Following Bartlett, Neisser (1967) argued that stimuli do not simply impose their impressions on a passive receptor. For instance, we are able to "see" an object only after an elaborate process of construction, which typically makes use of both the available stimulus material and "traces" of previous acts of construction. It follows, according to Neisser (1967, 1976), that the whole conception of structured cognitive processes is fundamentally different from that of a simple response sequence.

Neisser (1967) pursued Bartlett's (1932) suggestion that experience leads to a gradual building up of cognitive structures that are nonspecific but organized representations of a great number of individual acts of construction. Neisser proposed further that a psychological system stores information about its own constructive processes rather than the products of those constructions. That is, the information that is retained consists of traces of similar acts of construction, and it is organized in ways that correspond to the structure of those acts. These cognitive structures (schemata) control the fate of information that is to be stored and are themselves information of the same kind. Thus, they are integral parts of all of our memories, and they also provide articulate patterns into which new material can be assimilated.

In a similar vein, Piaget (1960) referred to this sort of anticipatory schematization as a "gestalt with a history." More precisely, he contended that

> perception itself does not consist in a mere recording of sensorial data, but includes an active organization in which decisions and preinferences intervene and which is due to the influence of perception as such on this schematization of actions or of operations. (Piaget, 1971, pp. 86–87)

From the standpoint of the entire cognitive system as an operational whole, the activity of each schema can be viewed as "the part (i.e., the sector of activity or functioning sector) played by a substructure in relation to the functioning of the total structure and, by extension, the action of the total functioning on the functioning of the substructure" (Piaget, 1969, p. 165).

Within the specific framework of personal construct theory, the schemata that are used to assimilate sensory input consist of sets of interrelated bipolar distinctions, or constructs (Mancuso & Adams-Webber, 1982). Thus, Kelly (1955) went somewhat further than Bartlett, Neisser, and Piaget in specifying the formal structure of our cognitive processes. He posited that the same act of construction that establishes some basis of perceived similarity between two or more events also serves to simultaneously differentiate them from still other events. This implies that each construct is fundamentally an integrating and a differentiating operation whereby at least two events are perceived as similar to one another and, at the same time, are perceived as specifically different from at least one other event. In Kelly's (1969) own words,

> It must be understood that the personal construct abstracts similarity and difference simultaneously. One cannot be abstracted without implying the other. For a person to treat two incidents as different is to imply that one of them appears to be like another that he knows. Conversely, for a person to treat two incidents as similar is to imply that he contrasts both with at least one other incident he knows. We intend this to be considered as an essential feature of the personal construct by means of which we hope to understand human behavior. (pp. 102–103)

In more formal terms, given a set of three elements A, B and C (the minimum context of any construct), A and B may be perceived as similar to one another in the same respect in which C contrasts with both. For instance, Carol views adultery and bourbon as gratifying and crabgrass as a source of frustration. On the other hand, she regards dieting as neither gratifying nor frustrating; it simply lies outside the range of convenience of this distinction in the context of her personal experience. However, dieting may fall within the range of convenience of other constructs in her system, such as "safe versus dangerous."

Husain (1983) points out that Kelly also assumed that "human cognitive activity is continuous throughout its entire range from basic categorization right up to moral evaluation . . . (and that) all constructs are of the same type and hence continuous with one another" (p. 13). Husain also points out that

> The data (to which we apply our constructs) are given as a temporal flux in which each datum stands in fixed and irreversible order of temporal succession. In this flux there are recurrent patterns of similarity that mark off certain rhythms or segments (see Agnew & Brown, 1989b; Jones, 1976). . . . They do not depend on us or on our constructs but provide the empirical basis from which our constructs can be abstracted as empirical hypotheses. (p. 14)

An important epistemological issue is that of how such constructs can become adapted to the environment. To ignore this question would leave us in the position that any possible interpretation of an event is just as useful as any alternative interpretation. Thus, as Mancuso and Adams-Webber (1982) point out, the problem of adaptation in personal construct theory is essentially a matter of convenience in anticipating events, and therefore it is related logically to Kelly's range corollary, which

> states explicitly that a construct is convenient for the anticipation of a finite range of events only. If this were not the case, adaptation, as discussed above, could not occur. That is, if our constructions were not anticipatory in the sense that they are open to positive and negative feedback from events, then psychological development would not be constrained by the environmental parameters that are the source of sensory input. (p. 30)

Agnew and Brown (1989a) assert that Kelly's model of human cognition adequately addresses "the recurrent question arising in the history and philosophy of science and in artificial intelligence studies: How do we reduce the problem under investigation to manageable size?" (p. 153). They point out that

Kelly (1955) early recognized that individuals must possess mechanisms that automatically restrict their range of attention. He postulated that a construct, or a hierarchy or network of constructs, bounds our anticipations of particular experience, and selects abstractions from possible worlds, large or small, to serve the anticipations. Our constructs reflect our bounded rationality by limiting the number of events addressed, and by operating within a restricted or manageable frame of reference. (p. 154)

They note that, according to Kelly's range corollary, any single construct or system of constructs has a limited range of convenience that comprises all those events to which an individual would find its application predictively useful. It follows that the range of convenience of a construct, or system of interrelated constructs, will (by definition) delimit the specific search space that is relevant to evaluating that construct or system in terms of its predictive efficiency. This consideration would resolve in functional terms the problem of reducing the search space of a problem to manageable size in that, as Agnew and Brown (1989a) imply, the search space of any question is automatically constrained by the range of convenience of the constructs that we employ in trying to answer it (see Adams-Webber, 1989; Ford, 1989; Ford & Chang, 1989).

Nonetheless, this "solution" provides us with no guidance at all concerning the basic *pragmatic* issue of what particular constructs we should attempt to apply to a given problem. From a Kellyan standpoint, this question is entailed in Kelly's (1955) fundamental tenet of "constructive alternativism," which implies that events are, in principle, subject to as many alternative ways of construing them as we ourselves can invent (cf. Adams-Webber, 1979).

A related, larger problem that is also raised by Agnew and Brown (1989b) is that "if our knowledge relies on robust feedforward mechanisms, and highly selected abstracted feedback, then much of such knowledge *must be highly fallible* [italics added]" (p. 21). Mischel (1964) raises essentially the same issue in relation to Kelly's model; that is, How can our anticipations ever be invalidated if we assess all feedback from the environment in terms of the same set of constructs (i.e., schemata) that originally were used to formulate those anticipations? Warren (1964) has further elaborated Mischel's question in the following terms:

Taking a more general view, I consider the point Mischel raises here to be a basic problem for all psychological theories which attempt to take perception seriously. It is a matter of the veridicality of perception or construction and how it is checked by the perceiver. It crops up under headings like "the Selectivity of Perception" or the "Transformation of Information Input." All theories using the concept of "hypotheses" or "expectation" run into this issue sooner or later. (p. 11)

From a constructivist standpoint (whether Kantian or Kellyan), by definition there can be no independence of the thing cognized from the cognition of it (cf. Husain, 1983). For example, as Mischel (1969) puts the case, "Kant not only held that, since experience is not 'given' but is constructed by us according to rules that we prescribe for it, what we know is always things as they appear to us, never things in themselves" (p. 18). It follows that all of our experience is constituted by our own constructions. Our anticipations are themselves constructions of the same sort, only projected toward future events. As Agnew and Brown (1989a) have correctly inferred, it follows that the feedback in terms of which we evaluate our anticipations is also constructed by ourselves and does not necessarily reveal the real nature of events as they are in themselves, that is, independent of our own construing.

In direct contrast, epistemological realism assumes a crucial independence of the objects of our knowledge from our own construing processes; that is, "what we know are things existing independently of cognition" (Martin, 1957, p. 13). As Parker (1962) has explained, "the thesis of cognitive independence means only and exactly that an object of cognition is independent of and unrelated to the cognition *corresponding to it*" (p. 382). This thesis also presupposes, as Gilson (1940) pointed out, that there is "a *real* adequation of the intellect and the object informing it . . . [which is] a primitive ontological relation between intellect and object" (p. 237).

In Kellyan terms, insofar as the principle that explains human action lies in the mind (i.e., cognition) and not in external events, it consists of our intention to bring about a correspondence between our future experience and certain of our anticipatory representations (cf. Mancuso & Adams-Webber, 1982). On the other hand, as Agnew and Brown (1989a) note, "reality does not directly reveal itself to us" (p. 6). Consequently, the problem of how closely our personal constructs correspond with "events-in-themselves" simply cannot arise from the perspective of Kelly's (1955) constructive alternativism or any other form of constructivist epistemology (Husain, 1983; Mischel, 1969).

The question that does remain, however, is the fundamental epistemological issue of how we can evaluate the adequacy of our knowledge when we cannot step outside of the framework of the representations that we ourselves have constructed in order to compare them directly with an external "reality." Kelly (1955) was reaching toward a possible resolution of this problem when he adopted a strictly pragmatic approach to assessing the adequacy of our representations. He claimed that they should be evaluated in terms of their basic *function*, which, according to his fundamental postulate, is anticipation.

None of the explicit assumptions of personal construct theory logically precludes the possibility that the structure of reality may someday become fully apparent to us. As Agnew and Brown (1989a) point out, "Kelly's model does not rule out the possibility of isomorphism between subjective criteria, on the one hand, and domain structure, on the other" (p. 19).

Kelly (1955) did suggest, however, that the only criterion currently available for evaluating the adequacy of our construing is its predictive utility with respect to our own experience. He assumed that as we gradually improve our capacity to anticipate events, the overall pattern of our experience will become increasingly coherent, and consequently we shall encounter less confusion. In short, our confidence in our personal constructs tends to be enhanced by any new experience that seems to be consistent with our anticipations. As Warren (1964) explains,

> [Kelly] makes the business of validation of constructs also a matter of construing, either at a different level of construing from the original construing or by employing different but systematically related constructs . . . [The] criterion for a person's assessment of the outcome of his anticipations [is] the internal consistency of the personal constructions within the person's construction system . . . Truth becomes a matter of coherence within a system rather than of correspondence with reality. (p. 11)

On the one hand, we are not in a position to specify the relationship between events and our own representations. On the other hand, we are ready to infer that whenever we improve our capacity to anticipate events the overall pattern of our experience will become more coherent. Moreover, there is no specific reason for us to assume that our construing will not continue to accommodate gradually to whatever (unknown) parameters define reality. It could be the case that, as Agnew and Brown (1989b) suggest, "reality plays an indirect and approximate editing role for some of our perceptions and beliefs" (p. 17). Thus, we agree with Agnew and Brown (1989a) that

> Kelly's theory can provide for an optimism that some knowledge, through time and through intra- and inter-individual winnowing, achieves increased "external" and general validity, knowledge that represents more than disposable cultural myths, or highly local or personal empirical or symbolic fabrications. (p. 11)

Furthermore, their suggestion helps to explain how our anticipatory representations (hypotheses) can be functionally useful, despite our being unable to determine their so-called objective truth status. Even if all of our current representations are of indeterminate validity in terms of their degree of correspondence with an independent reality underlying events, they may still prove useful for anticipating new possibilities as we persevere in our efforts to improve the range of convenience of our constructions and to explore still unknown potentials of human experience. According to Kelly (1969), the relevant issue is not whether our constructions are true or false but rather the *pragmatic* question of which of them might serve as useful axes of reference for charting alternative courses of action in terms of their anticipated consequences and then making sense of feedback from experience (Adams-Webber & Mancuso, 1983).

Implications of the Constructivist Epistemology for Prominent Issues in Knowledge Elicitation

In the next sections we examine the implications of the constructivist epistemology with regard to the construction of classical diagnostic expert systems—not necessarily knowledge-based systems in general.

The Mining Analogy

The mining analogy, which pervades much of the current knowledge acquisition literature, reveals the underlying realist epistemological assumptions that are fundamentally at odds with the constructivist epistemology developed in this chapter. Specifically, the mining analogy implies that the act of eliciting an expert's knowledge consists of "mining those jewels of knowledge out of their heads one by one" (Feigenbaum & McCorduck, 1983, p. 2). The assumptions are that there exists some "gold standard" or reality of knowledge and that a domain expert has captured a discrete (presumably large) part of this existent knowledge.

We feel that in some important respects this mining analogy is quite misleading. Expertise is not like a natural resource that may be harvested, transferred, captured, or hoarded but is instead largely constructed by the expert and reconstructed by the knowledge engineer. A pivotal aspect of the knowledge acquisition process is the knowledge engineer's act of construction in which the utterances/tool interactions of the domain expert are transformed into an implementation formalism. Expertise (and knowledge in general) is not a zero-sum game. The issue of the appropriateness of the mining analogy for work in knowledge acquisition reflects the long-standing "realist" versus "constructivist" debate found in the literatures of philosophy and psychology—as discussed earlier.

Inability to Verbalize Expertise

Several studies have reported that when domain experts are asked to explain how they reach a given conclusion, they often construct plausible lines of reasoning that have little or no relevance to their actual problem-solving methods (Johnson, 1983). Waterman (1986) has described a particularly troublesome knowledge engineering paradox; that is, *the more competent domain experts become, the less able they are to describe the knowledge they use to solve problems*. For example, although novices (e.g., intern radiologists) may frequently best their more experienced mentors (i.e., radiology faculty) on standardized written tests (radiology textbook knowledge), they are usually unable to approach expert level performance at the actual job task (e.g., interpreting a mammograph) (Chang, Ford, & Petry, 1987). It appears that in many domains, the "experts" are largely unable to communicate to their students the specific knowledge that makes them experts. It stands to reason that if domain experts are unable to transfer (over years) a significant portion of their expertise to their top students, then traditional verbal interview-based knowledge acquisition methods cannot be expected to do substantially better. As a partial consequence, many of today's expert systems may more appropriately be referred to as "novice" or "advanced novice" systems (Dreyfus & Dreyfus, 1986).

In most interesting domains (e.g., diagnostic medicine), true expertise is more than the successful accumulation of "book knowledge." In fact, much of an expert's unique collection of knowledge and skills are of his or her own construction. Human experts acquire their expertise not only from explicit knowledge of the sort found in textbooks (i.e., widely shared consensual beliefs) but also from personal experience. They consequently construct a repertory of working hypotheses, or rules of thumb (i.e., functional but fallible anticipations held with high confidence and uncertain validity; cf. Agnew & Brown, 1989b) that, combined with their fund of book knowledge, make them expert practitioners.

Thus, the greater their expertise, the further the experts' schemata or construct systems deviate from those of typical practitioners, and the greater the importance of personally constructed knowledge. As a consequence, experts may not be able to verbalize the incremental knowledge responsible for evolving them beyond a competent practitioner, because there often is not a societally (or even a domain-specific) shared method of expressing it. In other words, the expert has developed a collection of functional but fallible hypotheses (i.e., a personal construct system) that in some important respects does not coincide with publicly available domain knowledge, thus rendering it difficult to state explicitly. But perhaps herein lies the most significant facet of expertise.

From this perspective, a critical task in knowledge acquisition research is the development of adequate tools and techniques for the purpose of assisting the knowledge engineer in bringing the experts' self-constructed knowledge to the surface, thus making explicit the valuable heuristic knowledge that experts possess but are frequently unable to articulate.

Multiple Experts in the Same Domain

In response to several perceived shortcomings associated with the current generation of expert systems, many researchers and practitioners have suggested that new systems be based on the knowledge of multiple domain experts. There are typically two reasons offered for the rush toward multiple domain experts: (1) narrow expertise, which is also known as the fragility problem (referring to the case in which an expert system fails because its knowledge is too specific or limited), and (2) that no single expert may be capable of adequately solving the problem of interest. These may also be regarded as problems of breadth and depth, respectively. This view is reflected even in lay publications such as the *New York Times* (March 29, 1984):

> Many say the possibility of aggregating the knowledge and insights of several experts in the same field opens the prospect of computer-aided decisions based on more wisdom than any one person can contain. (p. A16)

In our view, the aforementioned problems do not so much argue either for or against multiple domain experts but rather counsel caution against unreasonable expectations and in the inappropriate use of expert systems.

Regarding expert systems, the breadth problem is in the nature of the beast. Our goal in the construction of expert systems is not to create a broadly intelligent system (e.g., a Renaissance man) but rather to focus on replicating a very high degree of human performance in some limited area of expertise (an idiot savant). The sudden collapse of competence associated with many expert systems is less a function of limited (i.e., narrow) expertise than it is a by-product of the representational and computational paradigms employed in their design and construction. Although the addition of multiple experts may shift the point of collapse somewhat outward, it will not change the abrupt nature in which an expert system degrades when operating outside its area of competence.

The depth problem implies that for the domain in question there are no true experts available (i.e., no single practitioner capable of excellent performance). Such domains are not well suited to the application of expert system technology. It would seem to be a basic tenet that in order to build an expert system one must have access to an expert. Given the constructed nature of expertise, assembling a roomful of competent practitioners will not necessarily lead to expertise either collectively or individually. We are reminded of Dijkstra's (1982) warning, "It is impossible to sharpen a pencil with a blunt axe. It is equally vain to try to do it with ten blunt axes instead" (p. 130).

The issue of knowledge acquisition from multiple experts is a line of clear demarcation between "realists" and "constructivists." As noted earlier, many workers in knowledge acquisition act as though there exists some "gold standard" or reality of knowledge, of which experts discover various parts. They argue that two well-known experts may disagree because they each have access to different chunks of the total knowledge. However, from a constructivist perspective we would expect that experts in the same domain are likely to agree about the vast majority of their knowledge (i.e., widely shared consensual beliefs) and yet have major differences in their largely self-constructed expertise.

If a single expert represents the pinnacle of his or her profession and the knowledge acquisition phase is reasonably successful, it is unlikely that adding more experts will improve performance. In fact, it is our experience (in medical expert systems) that adding multiple experts tends to cause a "regression to the mean" and that the resulting system is less expert than any of the true experts. And if we incrementally added more experts (eventually including entire species), we would build (by definition) an "average system," not an expert system.

Consider the unhappy circumstance of a trip to the hospital for diagnosis of an apparently serious difficulty: You want the physician to speak from his or her expertise, and then perhaps you will seek another well-known physician for another (i.e., second) opinion. You do not want a homogenized, "averaged" opinion. Thus, given the unusually luxurious scenario in which the knowledge engineer has multiple bona fide domain experts at his or her disposal, it is often preferable to build separate knowledge bases for each expert rather than attempting to incorporate their expertise into a single knowledge base.

> The method of arriving at and presenting dissenting opinions seems to be very useful. By examining the differences between consensus and dissenting opinions, the end user can readily see the range of acceptable solution sets with their justifications, rather than just one solution set. (Boose, 1986, p. 214)

It is our experience that in many situations only this notion of "running the experts in parallel" (i.e., independent expert systems) seems to work—either theoretically or practically speaking.

It should be stressed that in the foregoing discussion of the implications of the constructivist epistemology we are not painting all efforts at knowledge acquisition from multiple experts with the same dark brush. Specifically, several researchers are in the process of developing tools and techniques from within the constructivist framework for addressing this thorny problem (Shaw & Gaines, 1988).

Summary

> Theories without tools are empty, tools without theories are blind.
>
> —with apologies to Kant (1902)

Many knowledge engineers are currently involved in the development, application, and validation of expert systems. The most fundamental step in the knowledge acquisition phase of the construction of an expert system is the elicitation of knowledge from a skilled individual. Like experts in many other fields, knowledge engineers typically rely on intuition and personal experience, and rarely examine the theoretical implications of the assumptions underlying the various methods and tools that they employ. In contrast, we have suggested that an explicit model of human cognitive processes—including perception, memory, representation, anticipation, and reasoning—might serve to help us simplify some of the central problems in this field.

Integration is the battle cry as toolmakers rush to produce hybrid knowledge acquisition tools. However, it is becoming widely realized that ad hoc combinations of techniques and tools—a sort of Occam's hell—may not contribute much to ameliorating the knowledge acquisition bottleneck.[1] As we noted earlier, theory can offer an infrastructure upon which to build highly integrated hybrid knowledge acquisition tools in a principled way. In addition, this same theory may serve as a valuable reference and theoretical guide for the tool's users.

In this chapter we have explored some specific ways in which personal construct theory, in particular, may prove useful in describing and modeling cognitive processes involved in the acquisition of knowledge by domain experts. Personal construct theory was formulated originally by Kelly (1955, 1969, 1970) to explicate the general principles that govern the evolution of complex patterns of organization within an individual's personal experience. It is essentially a constructivist model of the human representational processes and assumes that the primary function of these processes is the anticipation of events (Adams-Webber & Mancuso, 1983). We observed basic logical parallels between some of the central assumptions of this model and certain principles of cognitive structure developed by Asch (1969); Bartlett (1932); Deese (1969); Dewey (1896); Neisser (1967, 1976); and Piaget (1969, 1971).

We explained further that this approach, as well as alternative constructivist models, explicitly rejects the fundamental thesis of cognitive independence that is entailed in epistemological realism (and underlies the so-called mining analogy in knowledge engineering); that is, "What we know are things existing independently of cognition" (Martin, 1957, p. 13). In contrast, all forms of constructivism assume that "what we know is always things as they appear to us, never things in themselves" (Mischel, 1969, p. 18). We have argued, however, that there is no specific reason for us to assume that a cognitive system cannot continue to accommodate gradually to whatever (unknown) parameters define reality (Adams-Webber, 1989) and that "some knowledge, through time and through intra- and inter-individual winnowing, achieves 'external' and general validity" (Agnew & Brown, 1989a, p. 11).

We have proposed specifically that personal construct theory is a potentially rich field of ideas for those interested in constructing computational models of human cognition and developing related systems for the assessment and representation of knowledge. For example, we have shown that Kelly's (1955) range corollary implies that the progressive differentiation of function between separate subsystems of constructs (schemata) will gradually enhance the overall deployability of any cognitive system in terms of the variety of events that can be anticipated within its framework and, at the same time, will narrow the specific focus (or search space) of any particular problem. We also reviewed empirical research findings that support some of the central implications of this theory, such as the hypotheses (a) that experience within a specific domain will produce gradual increases in the degree of interrelatedness between constructs that are routinely applied to events in that domain (e.g., Adams-Webber & Mirc, 1976), and (b) that human cognitive development involves systematic transformations in both the structure and content of personal construct systems (Adams-Webber, 1979).

Finally, we hypothesized specifically that domain experts acquire their expertise not only from explicit knowledge of the sort found in textbooks (i.e., widely shared consensual beliefs) but also from a fund of personal experience essentially consisting of functional but fallible anticipations held with high confidence and uncertain validity (cf. Agnew & Brown, 1989b), which, combined with their book knowledge, make them expert practitioners. It would seem to follow that the greater their expertise, the further the experts' construct systems deviate from those of typical practitioners and the greater the importance of personally constructed knowledge. In short, experts have developed personal construct systems that in some important respects do not coincide with publicly available domain knowledge, rendering it extremely difficult for them to articulate their knowledge explicitly to either colleagues, students, or knowledge engineers. From this perspective, the critical task in knowledge acquisition research is the development of adequate tools and techniques for assisting the knowledge engineer in bringing the experts' self-constructed knowledge to the surface, thus making explicit the valuable heuristic knowledge that experts possess but are frequently unable to articulate.

Acknowledgments. Many people have influenced the development of this work. We owe particular thanks to Neil Agnew, Jeff Bradshaw, Robert Hoffman, Joseph Novak, Howard Stahl, and Jeff Yerkes. This work was supported in part by Florida High Technology Enhancement Funds.

Note

1. General concensus expressed in the group meeting of the knowledge acquisition tools special interest group at fourth Knowledge Acquisition for Knowledge-Based Systems Workshop, Banff, Alberta, Canada.

References

Adams-Webber, J. R. (1970). Actual structure and potential chaos. In D. Bannister (Ed.), *Perspectives in personal construct theory* (pp. 30–45). London: Academic Press.

Adams-Webber, J. R. (1979). *Personal construct theory: Concepts and applications*. New York: Wiley.

Adams-Webber, J. R. (1984). Repertory grid technique. In R. Corsini (Ed.), *Encyclopedia of psychology* (p. 225). New York: Wiley Interscience.

Adams-Webber, J. R. (1985a). Self-other contrast and the development of personal constructs. *Canadian Journal of Behavioural Science*, *17*, 303–314.

Adams-Webber, J. R. (1985b). Construing self and others. In F. Epting & A. W. Landifield (Eds.), *Anticipating personal construct psychology* (pp. 58–69). Lincoln, NE: University of Nebraska Press.

Adams-Webber, J. R. (1987). Personal construct theory. In R. Corsini (Ed.), *Concise encyclopedia of psychology* (pp. 824–825). New York: Wiley Interscience.

Adams-Webber, J. R. (1989). Kelly's pragmatic constructivism. *Canadian Psychology*, *30*, 190–193.

Adams-Webber, J. R., & Mancuso, J. C. (1983). The pragmatic logic of personal construct psychology. In J. R. Adams-Webber & J. C. Mancuso (Eds.), *Applications of personal construct psychology* (pp. 1–10). New York: Academic Press.

Adams-Webber, J. R., & Mirc, E. (1976). Assessing the development of student teachers' role conceptions. *British Journal of Educational Psychology*, *46*, 338–340.

Agnew, N. M., & Brown, J. L. (1989a). Foundations for a theory of knowing: 1. Construing reality. *Canadian Psychology*, *30*, 152–167.

Agnew, N. M., & Brown, J. L. (1989b). Foundations for a theory of knowing: 2. Fallible but functional knowledge. *Canadian Psychology*, *30*, 168–183.

Applebee, N. (1975). Developmental changes in consensus in construing within a specified domain. *British Journal of Psychology*, *66*, 473–480.

Applebee, N. (1976). The development of children's responses to repertory grids. *British Journal of Social and Clinical Psychology*, *15*, 101–102.

Asch, S. E. (1969). A reformulation of the problem of association. *American Psychologist*, *24*, 92–102.

Bainbridge, L. (1979). Verbal reports as evidence of the process operator's knowledge. *International Journal of Man-Machine Studies*, *11*, 411–436.

Baldwin, B. (1972). Change in interpersonal cognitive complexity as a function of a training group experience. *Psychological Reports*, *30*, 935–940.

Bannister, D. (1962). Personal construct theory: A summary and experimental paradigm. *Acta Psychologica*, *20*, 104–120.

Bannister, D. (1963). The genesis of schizophrenic thought disorder: A serial invalidation hypothesis. *British Journal of Psychiatry*, *109*, 680–686.

Barratt, B. B. (1977). The development of peer perception systems in childhood and early adolescence. *Social Behavior and Personality*, *5*, 351–360.

Bartlett, F. (1932) *Remembering: A study in experimental and social psychology*. London: Cambridge University Press.

Benjafield, J., Jordan, D., & Pomeroy, E. (1976). Encounter groups: A return to the fundamental. *Psychotherapy: Theory, Research and Practice*, *13*, 387–389.

Bodden, J., & James, L. E. (1976). Influence of occupational information giving on cognitive complexity. *Journal of Counseling Psychology*, *23*, 280–282.

Boose, J. H. (1984). Personal construct theory and the transfer of human expertise. *Proceedings AAAI–84* (pp. 27–33). Los Altos, CA: William Kaufmann.

Boose, J. H. (1986). Rapid acquisition and combination of knowledge from multiple domain experts in the same domain. *Future Computing Systems*, *1*, 191–216.

Boose, J. H., & Bradshaw, J. M. (1987). Expertise transfer and complex problems: Using Aquinas as a knowledge-acquisition workbench for knowledge-based systems. *International Journal of Man-Machine Studies*, *26*, 3–28.

Bradshaw, J. M., & Boose, J. H. (1990). Decision analysis techniques for knowledge acquisition: Combining information and preferences using Aquinas and Axotl. *International Journal of Man-Machine Studies*, *32*, 121–186.

Brierley, D. W. (1967). *The use of personality constructs by children of three different ages*. Unpublished doctoral dissertation, University of London.

Chang, P. J., Ford, K. M., & Petry, F. E. (1987). *PA 31: The production of expert system rules from repertory grid data based on a logic of confirmation*. Paper presented at the Seventh International Congress on Personal Construct Psychology, Memphis, TN.

Cochran, L. R. (1976). Categorization and change in conceptual relatedness. *Canadian Journal of Behavioural Science*, *8*, 275–286.

Crockett, W. H. (1965). Cognitive complexity and impression formation. In B. A. Maher (Ed.), *Progress in experimental personality research* (Vol. 2, pp. 47–90). New York: Academic Press.

Crockett, W. H. (1982). The organization of construct systems. In J. C. Mancuso & J. R. Adams-Webber (Eds.), *The construing person* (pp. 62–95). New York: Praeger.

Deese, J. (1969). Behavior and fact. *American Psychologist*, *24*, 515–522.

Dewey, J. (1986). The reflex arc concept in psychology. *Psychological Review*, *3*, 357–370.

Dijkstra, E. W. (1982). *Selected writings on computing: A personal perspective*. New York: Springer-Verlag.

Dreyfus, H. L., & Dreyfus, S. E. (1986). *Mind over machine*. New York: Free Press.

Duck, S. W. (1975). Personality similarity and friendship choices by adolescents. *European Journal of Social Psychology*, *5*, 351–365.

Ericsson, K. A., & Simon, H. A. (1984). *Protocol analysis: Verbal reports as data*. Cambridge, MA: MIT Press.

Feigenbaum, E. A., & McCorduck, P. (1983). *The fifth generation*. New York: Addison-Wesley.

Fischler, M. A., & Firschein, O. (1987). *Intelligence: The eye, the brain, the computer*. Reading, MA: Addison-Wesley.

Ford, K. M. (1989). A constructivist view of the frame problem in artificial intelligence. *Canadian Psychology*, *30*, 188–190.

Ford, K. M., & Chang, P. J. (1989). An approach to automated knowledge acquisition founded on personal construct theory. In M. Fishman (Ed.), *Advances in Artificial Intelligence Research* (Vol. 1, pp. 83–132). Greenwich, CT: JAI Press.

Ford, K. M., Petry, F. E., Adams-Webber, J. R., & Chang, P. J. (1991). An approach to knowledge acquisition based on the structure of personal construct systems. *IEEE Transactions on Knowledge and Data Engineering*, *3*(1), 78–88.

Ford, K. M., Stahl, H., Adams-Webber, J. R., Cañas, A. J., Novak, J., & Jones, J. C. (1991). ICONKAT: An integrated constructivist knowledge acquisition tool. *Knowledge Acquisition*, *3*, 215–236.

Fransella, F. (1972). *Personal change and reconstruction*. London: Academic Press.

Gaines, B. R., & Shaw, M. L. G. (1986). Induction of inference rules for expert systems. *Fuzzy Sets and Systems*, *18*, 315–328. Amsterdam: Elsevier Science Publishers.

Gilson, E. (1940). *The spirit of medieval philosophy: 1931–1932 Gifford Lectures*. New York: Scribners.

Head, H. (1920). *Studies in neurology*. London: Froede, Hodder, & Stoughton.

Heidbreder, E. (1933). *Seven psychologies*. New York: Appleton-Century-Crofts.

Hoffman, R. R. (1987, Summer). The problem of extracting the knowledge of experts from the perspective of experimental psychology. *The AI Magazine*, pp. 53–67.

Husain, M. (1983). To *what* can one apply a construct? In J. R. Adams-Webber & J. C. Mancuso (Eds.), *Applications of personal construct psychology* (pp. 11–28). New York: Academic Press.

Johnson, P. E. (1983). What kind of expert should a system be? *The Journal of Medicine and Philosophy*, *8*, 77–97.

Jones, R. (1976). Time, our lost dimension: Toward a new theory of perception, attention and memory. *Psychological Review*, *85*, 323–355.

Kant, I. (1787/1902). *The critique of pure reason* (2nd ed.). New York: Macmillan.

Kelly, G. A. (1955). *The psychology of personal constructs*. New York: Norton.

Kelly, G. A. (1969). A mathematical approach to psychology. In B. A. Maher (Ed.), *Clinical psychology and personality: The selected papers of George Kelly* (pp. 7–45). New York: Wiley.

Kelly, G. A. (1970). A brief introduction to personal construct theory. In D. Bannister (Ed.), *Perspectives in personal construct theory* (pp. 1–29). London: Academic Press.

Lemon, N. (1975). Linguistic development and conceptualization. *Journal of Cross-Cultural Psychology*, *6*, 173–188.

Little, B. (1968). Factors affecting the use of psychological versus non-psychological constructs on the rep test. *Bulletin of the British Psychological Society*, *21*, 34.

Mancuso, J. C., & Adams-Webber, J. R. (1982). Anticipation as a constructive process. In J. C. Mancuso & J. R. Adams-Webber (Eds.), *The construing person* (pp. 8–32). New York: Praeger.

Mancuso, J. C., & Eimer, B. N. (1982). Fitting things into sorts: The range corollary. In J. C. Mancuso & J. R. Adams-Webber (Eds.), *The construing person* (pp. 130–151). New York: Praeger.

Martin, W. O. (1957). *The order and integration of knowledge*. Ann Arbor, MI: University of Michigan Press.

Mischel, T. (1964). Personal constructs, rules, and the logic of clinical activity. *Psychological Review*, *71*, 180–192.

Mischel, T. (1969). Scientific and philosophical psychology: A historical introduction. In T. Mischel (Ed.), *Human action: Conceptual and empirical issues* (pp. 1–40). New York: Academic Press.

Neimeyer, R. A. (1985). *The development of personal construct psychology*. Lincoln, NE: University of Nebraska Press.

Neisser, U. (1967). *Cognitive psychology*. New York: Appleton-Century-Crofts.

Neisser, U. (1976). *Cognition and reality: Principles and implications of cognitive psychology*. San Francisco, CA: Freeman.

Parker, F. (1962). A demonstration of epistemological realism. *International Philosophical Quarterly*, *2*, 367–393.

Piaget, J. (1960). *The psychology of intelligence*. New York: Harcourt Brace.

Piaget, J. (1969). The problem of common mechanisms in the human sciences. *Human Context*, *1*, 163–185.

Piaget, J. (1971). *Psychology and epistemology*. New York: Viking.

Reinhold, R. (1984, March 29). Reasoning ability of experts is codified for computer use. The New York Times, pp. A1, A16.

Scarlett, H. H., Press, A. N., & Crockett, W. H. (1971). Children's descriptions of their peers. *Child Development*, *42*, 439–453.

Shaw, M. L. G., & Gaines, B. R. (1988). A methodology for recognizing consensus, correspondence, conflict, and contrast in a knowledge acquisition system. *Proceedings of the 3rd Knowledge Acquisition for Knowledge-Based Systems Workshop* (pp. 30–1–30–19). Banff, Alberta, Canada: SRDG Publications.

Skinner, B. F. (1957). *Verbal behavior*. New York: Appleton-Century-Crofts.

Warren, N. (1964). *Constructs, rules, and the explanation of behavior*. Paper presented to the Symposium on Construct Theory and Repertory Grid Methodology, Brunel University, London, England.

Waterman, D. A. (1986). *A Guide to Expert Systems*. Reading, MA: Addison-Wesley.

Watson, J. B. (1919). *Psychology from the standpoint of a behaviorist*. Philadelphia: J. B. Lippincott.

8
Eliciting and Using Experiential Knowledge and General Expertise

David S. Prerau, Mark R. Adler, and Alan S. Gunderson

Expert systems generally contain the knowledge of human domain experts in the form of heuristic rules. The heuristics encapsulate the rules of thumb that the experts have gained through extensive experience in the field. These experience-proven heuristics are used by the experts as a major part of their problem-solving approach. When confronted with rare and novel problems to which the heuristics may not apply, human experts can apply, in addition, their fundamental understanding—or "deep knowledge"—of the domain. In GTE's COMPASS and PROPHET expert systems, we have elicited and used expert knowledge that is at an intermediate level, lying between knowledge based on direct experience and knowledge based on fundamental domain principles. We call this intermediate level *expertise knowledge*. Use of this type of knowledge allowed us to extend the range of both the resulting systems beyond the heuristics gained from the experts' field experience.

This chapter discusses the elicitation and use of expertise knowledge in expert systems. We utilized these techniques in the development of both the COMPASS and PROPHET expert systems, which contain knowledge based on both the experience and the expertise of our domain experts. In this chapter, we use the COMPASS expert system as our primary example. In the first section of the chapter, we define and describe three levels of knowledge: experiential knowledge, deep knowledge, and expertise knowledge. The second section describes the COMPASS expert system and discusses its task, the analysis of GTE's No. 2 EAX switching system. The following section describes how we elicited and used the expert knowledge. Next, we discuss the differences between performing knowledge elicitation for knowledge based on experience and for knowledge based on the general expertise of our domain expert. The last major section of the chapter discusses the risks involved in basing an expert system's knowledge on the general expertise of an expert.

Experience, Expertise, and Deep Knowledge

Most of the expert systems being produced today are based on the day-to-day experiences of domain experts. Experts have a thorough understanding of the fundamental principles that are involved, based on lengthy experience and often on extensive training. Through their familiarity with the domain, their training, and their experiences in solving problems within that domain, the experts develop heuristics, or rules of thumb. We call this knowledge based on extensive field applications *experiential knowledge*.

Experiential knowledge is associational, providing a direct relationship between situational data and conclusions, without the explicit use of the principles of the domain. The knowledge is usually related to a narrowly defined problem. For example, an expert automobile mechanic may hear a certain noise in a car and conclude that a certain problem exists. His conclusion is based on his having experienced the noise/problem correlation numerous times and may not include any consideration of the mechanical principles that link the two. Such associational, experiential knowledge has been referred to as shallow knowledge (Davis, 1989; Scarl, 1989).

At the other end of the scale, problem solving relies on knowledge of the fundamental principles of the domain, which is often called *deep knowledge*. (Chandrasekaran & Mittal, 1982; Sticklen & Chandrasekaran, 1985). This type of problem-solving ability corresponds to a researcher's attempts to solve a problem never before encountered or a novice's attempts to solve problems in a new domain. For example, a physicist investigating new areas relies on the fundamental principles of physics. The novice or researcher has not yet had enough experience to develop heuristics and must derive a solution from fundamental theorems or formulas.

For many situations, deep knowledge cannot be used, and experiential knowledge is essential to solve a problem effectively. The underlying domain models are not well understood, are inappropriate, or do not exist. For example, one may know a model of domain functionality, but not a model to cover failure modes—many components may be possible causes of a failure, but experience guides an expert to the most likely one. Alternatively, a model of the domain might exist, but it might be so complex or inefficient that it is not useful for practical purposes. For these reasons, these classes of problems are most appropriately solved by expert systems.

There is an intermediate level of knowledge between direct experience and deep knowledge. We call this level *expertise knowledge*. This level is characterized by knowledge that is shallow—however, this shallow knowledge is based on some combination of the deep knowledge of the experts and their related experiential heuristics, and not *directly* on their practical experience. Unlike the researcher or novice, the domain expert has been exposed to a variety of problems that may be related to the problem at hand. Heuristics that have proven successful in solving problems may be extended or modified by the expert to apply to a related set of problems. These extensions are derived from the domain expert's deep knowledge of the overall domain; however, they lack the benefit of the expert's practical experience on the problem at hand and are therefore likely to be less refined and less certain.

Nevertheless, if it would be beneficial for an expert system's domain to extend beyond the direct experience of any expert, it is possible to utilize the expertise of an expert to develop those parts of the system. If this is done successfully, the resulting system could have capabilities greater than those of the expert.

The COMPASS and PROPHET expert systems, developed by GTE Laboratories, utilize both the experience and the expertise of our domain experts. To illustrate our ideas on the elicitation and use of experience and expertise in expert systems, we will discuss how these concepts were applied to COMPASS. As mentioned, we utilized a similar approach in the development of PROPHET (Prerau, Gunderson, & Levine, 1988).

The No. 2 EAX Switch and the COMPASS Expert System

GTE's COMPASS, the Central Office Maintenance Printout Analysis and Suggestion System (Prerau, 1990; Prerau, Gunderson, Reinke, & Adler, 1990), is a large telecommunications expert system. It analyzes hardware fault messages from telephone company central office switching equipment and suggests maintenance actions to be performed. A large telephone "switch" produces thousands of maintenance messages daily, such as those shown in Figure 8.1. The analysis of such a collection of data is nonobvious and is often overwhelming to the neophyte. However, telephone maintenance personnel must analyze the printouts to determine actions required to maintain the switch. The manual analysis procedure requires a high level of expertise and a good deal of time.

```
*SOR* SORT TALLY REPORT  06/14       13:37:34
FLT NU  PM                IM               FM                FRM  ICNF         TIM
(HISTOGRAPH SCALING: * = 001)

01  8   0,3,1,3,0,2,0,1   3,0,1,1,3,3,0    1,3,0,0,3,0,3     2.0  NUC8.0 ,PC1  13/,19:43:52    1 *
0E  0   0,1,1,3,2,2,3,0   1,3,0,1,3,2,2    0,2,2,3,1,0,2     1.0  NUC0.1 ,PC0  14/,11:04:46    1 *
0E  3   2,1,1,3,3,1,2,1   1,2,1,1,3,1,2    1,1,2,2,1,0,2     1.2  NUC3.1 ,PC1  14/,09:00:10    1 *
0E  3   3,1,3,3,3,2,3,1   1,3,1,3,3,0,0    1,0,0,3,1,3,3     1.3  NUC3.0 ,PC1  14/,11:02:54    1 *
0E  3   3,3,2,2,3,3,0,1   3,0,1,2,2,0,0    1,0,0,0,3,2,0     1.3  NUC3.0 ,PC1  14/,09:00:10    1 *
0E  3   3,3,3,3,1,2,0,0   3,0,0,3,3,3,1    0,3,1,0,3,1,2     1.3  NUC3.0 ,PC0  13/,23:10:13    1 *
12  0   0,1,0,3,0,3,0,3   1,0,3,0,3,1,0    3,1,0,0,1,3,2     2.0  NUC0.0 ,PC0  13/,23:42:05    1 *
12  0   0,1,1,0,1,3,3,3   1,3,3,1,0,1,3    3,1,3,3,1,1,1     2.0  NUC0.0 ,PC1  13/,19:27:17    1 *
20  10  0,0,0,3,3,1,0,0   0,0,0,0,3,0,2    0,0,2,0,0,1,0     0.0  NUC10.0,PC1  14/,11:46:25    1 *
20  10  0,1,1,0,0,0,2,2   1,2,2,1,0,0,3    2,0,3,2,1,2,2     0.0  NUC10.0,PC1  13/,21:09:24    1 *
20  11  0,1,2,2,0,1,1,3   1,1,3,2,2,0,3    3,0,3,1,1,0,1     0.0  NUC11.0,PC1  14/,13:30:49    1 *
20  11  0,3,0,2,2,1,2,2   3,2,2,0,2,3,1    2,3,1,2,3,2,2     0.0  NUC11.1,PC0  13/,14:45:56    1 *
20  11  0,3,0,3,0,0,2,3   3,2,3,0,3,0,3    3,0,3,2,3,0,1     0.0  NUC11.0,PC0  13/,17:16:24    1 *
20  12  3,1,1,0,2,1,0,0   1,0,0,1,0,1,3    0,1,3,0,1,0,1     0.0  NUC12.0,PC0  14/,09:05:25    1 *
20  12  3,1,1,0,3,2,0,0   1,0,0,1,0,1,2    0,1,2,0,1,1,2     0.0  NUC12.0,PC0  13/,19:35:46    1 *
20  13  0,1,3,3,3,1,0,1   1,0,1,3,3,3,2    1,3,2,0,1,3,3     0.0  NUC13.0,PC1  14/,11:04:19    1 *
20  13  1,1,3,0,0,0,1,1   1,1,1,3,0,1,0    1,1,0,1,1,3,2     0.0  NUC13.1,PC1  14/,10:38:47    1 *
20  13  2,1,1,2,1,2,1,1   1,1,1,1,2,1,3    1,1,3,1,1,1,0     0.0  NUC13.0,PC1  14/,12:42:27    1 *
20  13  2,2,0,1,0,1,0,2   2,0,2,0,1,1,0    2,1,0,0,2,1,0     0.0  NUC13.1,PC1  14/,11:04:22    1 *
20  14  0,2,1,1,2,0,1,2   2,1,2,1,1,3,1    2,3,1,1,2,1,1     0.0  NUC14.0,PC1  14/,13:34:44    1 *
20  16  0,2,0,1,3,0,1,1   2,1,1,0,1,2,2    1,2,2,1,2,1,2     0.0  NUC16.1,PC1  14/,11:17:50    1 *
20  16  0,2,3,1,3,3,3,1   2,3,1,3,1,2,2    1,2,2,3,2,1,2     0.0  NUC16.0,PC0  13/,17:51:50    1 *
20  16  0,3,3,1,0,1,2,1   3,2,1,3,1,2,2    1,2,2,2,3,1,2     0.0  NUC16.1,PC0  13/,17:39:41    1 *
20  16  1,0,0,1,2,1,3,1   0,3,1,0,1,3,2    1,3,2,3,0,1,2     0.0  NUC16.1,PC1  14/,10:05:39    1 *
20  16  1,0,3,2,3,0,1,1   0,1,1,3,2,0,0    1,0,0,1,0,2,1     0.0  NUC16.1,PC1  14/,09:40:53    1 *
20  16  1,0,3,2,3,0,1,1   0,1,1,3,2,2,0    1,2,0,1,0,2,0     0.0  NUC16.0,PC1  14/,09:40:55    1 *
20  16  1,0,3,2,3,0,1,2   0,1,2,3,2,0,1    2,0,1,1,0,3,3     0.0  NUC16.0,PC1  14/,09:40:51    1 *
20  16  1,0,3,2,3,2,1,2   0,1,2,3,2,0,1    2,0,1,1,0,3,2     0.0  NUC16.0,PC1  13/,16:37:47    1 *
20  16  1,3,1,2,2,1,2,1   3,2,1,1,2,3,2    1,3,2,2,3,1,2     0.0  NUC16.0,PC1  14/,10:16:08    1 *
20  16  2,1,2,2,3,3,1,2   1,1,2,2,2,0,1    2,0,1,1,1,3,3     0.0  NUC16.0,PC1  14/,12:10:35    1 *
20  17  0,0,3,3,1,0,2,0   0,2,0,3,3,3,2    0,3,2,2,0,1,0     0.0  NUC17.1,PC0  14/,03:15:26    1 *
20  18  0,1,0,2,2,0,2,0   1,2,0,0,2,3,3    0,3,3,2,1,0,1     0.0  NUC18.1,PC0  13/,21:24:26    1 *
20  19  0,0,0,1,3,2,3,1   0,3,1,0,1,2,3    1,2,3,3,0,2,3     0.0  NUC19.0,PC1  14/,12:28:47    1 *
20  19  0,0,0,1,3,2,3,1   0,3,1,0,1,3,2    1,3,2,3,0,1,0     0.0  NUC19.1,PC1  14/,12:27:01    1 *
20  20  0,2,2,1,0,2,2,0   2,2,0,2,1,2,0    0,2,0,2,2,2,3     0.0  NUC20.0,PC1  13/,15:58:58    1 *
20  5   0,1,0,2,1,1,2,0   1,2,0,0,2,3,0    0,3,0,2,1,3,3     0.0  NUC5.1 ,PC0  13/,16:53:39    1 *
20  5   0,2,3,0,3,1,2,0   2,2,0,3,0,3,0    0,3,0,2,2,3,3     0.0  NUC5.0 ,PC1  14/,10:07:33    1 *
20  5   1,3,0,3,0,1,0,0   3,0,0,0,3,3,0    0,3,0,0,3,3,3     0.0  NUC5.0 ,PC0  13/,16:14:12    1 *
20  5   2,2,0,1,2,0,1,0   2,1,0,0,1,3,0    0,3,0,1,2,3,3     0.0  NUC5.0 ,PC0  13/,16:54:04    1 *
20  5   2,2,3,3,0,0,3,3   2,3,3,3,3,3,3    3,3,3,3,2,3,2     0.0  NUC5.1 ,PC0  14/,01:25:50    1 *
20  5   3,0,2,1,2,0,0,0   0,0,0,2,1,3,0    0,3,0,0,0,3,3     0.0  NUC5.1 ,PC1  14/,10:04:53    1 *
20  5   3,2,1,2,0,0,2,2   2,2,2,1,2,2,1    2,2,1,2,2,1,0     0.0  NUC5.0 ,PC0  14/,01:25:50    1 *
20  5   3,2,1,2,0,0,2,3   2,2,3,1,2,2,0    3,2,0,2,2,0,3     0.0  NUC5.0 ,PC0  14/,01:25:55    1 *
20  5   3,2,1,2,0,0,3,0   2,3,0,1,2,1,0    0,1,0,3,2,2,2     0.0  NUC5.1 ,PC0  14/,01:25:57    1 *
20  5   3,2,1,2,0,0,3,1   2,3,1,1,2,0,1    1,0,1,3,2,3,2     0.0  NUC5.0 ,PC0  14/,01:25:59    1 *
20  7   1,2,2,3,3,0,3,1   2,3,1,2,3,1,3    1,1,3,3,2,2,3     0.0  NUC7.0 ,PC1  14/,12:40:31    1 *
20  8   0,0,3,2,0,3,0,2   0,0,2,3,2,2,1    2,2,1,0,0,2,0     0.0  NUC8.1 ,PC1  13/,14:13:15    1 *
21  1   0,1,0,0,3,3,3,3   1,3,3,0,0,3,0    3,3,0,3,1,3,3     0.0  NUC1.1 ,PC1  14/,03:15:26    1 *
21  10  0,3,0,1,0,3,1,0   3,1,0,0,1,3,1    0,3,1,1,3,2,0     0.0  NUC10.0,PC0  13/,16:22:58    1 *
21  2   0,1,1,1,1,0,1,3   1,1,3,1,1,1,2    3,1,2,1,1,3,1     0.0  NUC2.0 ,PC1  14/,11:04:21    1 *
21  2   0,1,1,1,1,0,1,3   1,1,3,1,1,2,0    3,2,0,1,1,2,0     0.0  NUC2.1 ,PC1  14/,11:03:54    1 *
21  2   0,2,0,1,1,0,1,2   2,1,2,0,1,2,2    2,2,2,1,2,1,2     0.0  NUC2.0 ,PC0  13/,23:19:04    1 *
21  2   0,2,3,1,0,3,0,2   2,0,2,3,1,3,1    2,3,1,0,2,3,0     0.0  NUC2.0 ,PC0  13/,23:19:04    1 *
```

FIGURE 8.1. An example of a small part of the COMPASS input.

COMPASS captures the knowledge utilized by a top switch expert in order to perform the maintenance analysis task for GTE's No. 2 Electronic Automatic Exchange (No. 2 EAX).

COMPASS is implemented using the KEE system from IntelliCorp. The COMPASS implementation (Prerau, 1990, chap. 11) utilizes multiple AI paradigms, including rules, frame hierarchies, demon mechanisms, object-oriented programming, and Lisp code accessed through KEE. COMPASS is a large expert system, consisting of approximately 1,000 frames (containing a total of about 15,000 slots), 400 rules, and 500 Lisp functions.

COMPASS has been deployed for use to aid maintenance at 46 No. 2 EAX sites, serving about 500,000 GTE telephone customer lines. The initial prototype version, COMPASS-I, was based entirely on our expert's *experience*, whereas a substantial portion of COMPASS-II, an upgraded, expanded, redesigned, and reimplemented version, was developed based on our expert's *expertise*.

The No. 2 EAX Switch

The No. 2 EAX switch can interconnect up to 40,000 telephone lines. It contains thousands of relays that open and close to complete a path between two telephones. A path consists of two half-paths: one into the switch and one back out. At any one time, thousands of such paths exist in a switch. The switch is continually setting up new paths as new telephone calls are being made. The switch is designed with redundancy—there are several different possible paths through the switch network that can be used to connect two particular telephone lines. If an isolated relay or other component fails and the switch attempts to use that component as part of a connection path, then that path will fail and the call will be aborted. When the call is redialed, the switch will utilize a different path to complete the call. Thus, the failure of a single component has little immediate impact on the overall performance of the switch. However, if the switch is neglected, more and more components will fail. Eventually the failure of enough components will prevent alternate routing to succeed for some portion of the switch.

The existence of a hardware failure in the No. 2 EAX system is normally detected by the system itself. As part of the process of selecting the path for a call through the switch, the path is tested electrically. If there is a short, an open, or some other problem on the path, the connection is aborted and a maintenance message is generated. The maintenance message specifies the general problem type and the complete path that was attempted through the switch. The switch's failure detection system only determines that a problem exists on the path—it does not determine the cause of the failure, that is, which of the many relays, wires, and other components of the path were faulty. Only by collecting maintenance messages and detecting common subpatterns can an analyst determine clues to the cause of the failure. In the simplest example, several faulty paths all have a single relay in common and have no other commonalties. In this situation, it is likely that the common relay caused all of the maintenance messages. A switchperson would look for such commonalties as evidence of the causes of the switch problems. But even in the simple case there can be complexities. It is also possible that each path had its own individual fault and that the appearance of the common relay in each path was coincidental. To analyze this and more complex situations, No. 2 EAX analysts must rely on their expertise and experience.

The Maintenance Task

The tasks of the No. 2 EAX maintenance person, and thus COMPASS, are to analyze the hundreds, or often thousands, of maintenance and warning messages that a No. 2 EAX puts out daily; determine the possible specific faults in the switch; and determine a series of maintenance actions (in priority order) to correct the faults. Successful performance of these tasks requires: (a) a thorough knowledge of the switch's architecture; (b) problem identification experience with the switch to determine which of hundreds of messages refer to a common switch problem; (c) message analysis experience to determine the most likely cause of the problem by examining the set of messages that correspond to it (due to ambiguity the maintenance messages do not indicate a single specific fault as the cause of a switch problem); and (d) repair experience to evaluate solution techniques.

We will now discuss the elicitation of No. 2 EAX maintenance expertise and experience from a top domain expert.

Knowledge Elicitation in COMPASS

COMPASS was developed at GTE Laboratories with the participation of one of the country's top maintenance experts for the No. 2 EAX switch. For one week each month for about two years, the COMPASS development team worked with him to acquire the knowledge needed for COMPASS. The knowledge elicitation was based primarily on intensive structured and unstructured interviews with the expert utilizing detailed test-case analysis.

The COMPASS domain expert's experience included working in telephone switching for more than 16 years, with about five years of experience working on the No. 2 EAX switch. During those years of maintaining the No. 2 EAX, he developed a thorough understanding of the switch architecture. He also gained a great deal of experience maintaining the switch, and from this experience he developed an excellent set of heuristics for identifying and solving problems.

Each month the expert would fly to our site for an intensive one-week knowledge elicitation session. Our primary concern was to use the time we had to capture as much expert domain knowledge as possible. We developed a pragmatic approach (Prerau, 1987; Prerau, 1990, chap. 9) that has proven successful in acquiring the knowledge for large-scale expert systems.

Our interactions with the expert might be broken into three phases. The first phase could be called the get-acquainted phase. The major goal was for the knowledge engineers to become familiar with the problem domain, its terminology (including jargon and acronyms), and the structure of the task. We primarily utilized unstructured and structured interviews. We felt very strongly that this phase was necessary to allow the knowledge engineers to begin to form an overall picture of the task at hand and to begin to form a design, not only of the expert system but also for the areas of knowledge that must be acquired. By learning the jargon of the expert in this phase, we were able to allow the expert to relate his knowledge in familiar terms as we progressed to the next phase.

During the initial phase, the knowledge engineers also read available background material, including training course notes and relevant published documents. As the knowledge engineers were learning about the domain, the expert was also learning something about expert system technology and was getting familiar with the research environment at our laboratories, which is quite different from the field conditions in the part of the company where he worked. We introduced the expert to English if-then-else rules so that he could understand the major representational technique we would be using to encode his knowledge. We felt it important that the expert learn at this point that we would be asking probing and often difficult questions, examining his answers for inconsistencies, and questioning his decisions. We tried to make the expert realize that these were not personal attacks but were the way we capture the information. Finally, the knowledge engineers and the domain expert tried to learn to work well together as a group for a week at a time in an enclosed conference room.

During the second phase, we began to capture and encode the expert domain information, primarily by the use of structured interviews and detailed test-case analyses. We found that the most valuable technique was to use real data and follow the steps that the expert took in his analysis. We recorded the knowledge found as we went along, primarily using English if-then rules. At first we wrote the rules on a whiteboard and photographed the board to preserve the rules. After the session was completed, the photographs were subsequently transcribed into a knowledge document, which was reviewed with the expert during the following session. After some knowledge was captured, we would "run" the rules manually on a different set of data. In this way we both checked our understanding of each rule (i.e., Did the rule capture the expert's knowledge?) and examined the rule's application (i.e., Did it work on a slightly different set of data?). During this phase, knowledge rules were added, edited, and removed. The application of the rules to the data served to structure the sessions with the expert. Usually rules needed to be made more specific as more and more examples were examined; occasionally, more general principles emerged.

For the more technical portions of the application, we would use a variety of techniques to capture the information. Sometimes hours, or even days, were required to develop the knowledge for a small part of the analysis. Venn diagrams proved useful in analyzing situations that involved sets of complex overlapping data.

The third phase involved implementing the knowledge and verifying and validating the resulting program. By utilizing the rapid-prototyping techniques available with advanced AI hardware and software, we were able to encode the knowledge of the expert in the time between visits. Then, we would ask the expert to analyze a set of data and compare his results to that of the expert system program. This accomplished three things:

1. We verified that implementation of the knowledge worked the same way as our manual "running" of the English rules.
2. When the program agreed with the expert's analysis, we gained more confidence in the knowledge we had captured.
3. When we found discrepencies between the expert and the program, this would focus our attention for subsequent knowledge elicitation sessions.

In addition to using actual data, we also found it useful to create hypothetical data to push the expert to define the limits of the rules. Most often this was done by starting with real data and modifying it until the expert felt that a rule was no longer applicable.

As the project progressed, we spent almost all of our time in this third phase. We had captured the overall functionality, but we were working on handling more and more special or unusual cases. During this phase, we were also adding rules to fill in the "gaps" between the rules we had.

The COMPASS knowledge elicitation effort is described in more detail in works by Prerau (1987; 1990, chap. 9).

Experiential Knowledge and General Expertise in COMPASS

The development of COMPASS consisted of two main stages: the development of an initial prototype system (COMPASS-I), and the development of an extended system (COMPASS-II). The initial system was constructed to demonstrate that expert system techniques were applicable to this domain. This version of COMPASS analyzed one type of maintenance message, the Network Recovery 20 (NR20) message. These NR20 messages are caused by a shorted relay, and—very important to maintenance analysis—the switch can determine which half-path through the switch contains the fault. The NR20 message analysis was singled out as the most appropriate subdomain in the No. 2 EAX switch for three reasons: First, these messages indicate the most critical problem. An expert system that was only capable of analyzing these messages would still have a significant payoff for the company, because each NR20 message indicates an aborted call. Second, our expert was most experienced in analyzing these messages. Because these messages indicate critical problems, the analysis is a high-priority task. There is a steady volume of these messages requiring analysis on a daily basis. Our expert had spent a significant portion of his time gaining experience with this class of message. Third, the analysis of NR20 messages covered a complete subset of the domain. A prototype system that analyzed only these messages would provide a good basis for expansion to cover all of the message types that require analysis.

The overall structure of the COMPASS-I expert system is based on the analysis steps applied by the expert and therefore is based on time-tested techniques developed in the field

(Prerau, 1990). The basic steps include the following: identifying commonalities among patterns of paths in the messages; generating a list of switch problems by associating sets of patterns; determining a set of possible faults that might cause each problem; determining a set of possible maintenance actions that might repair each fault; and finally, generating a prioritized list of maintenance actions to be taken to correct each problem.

After developing, testing, and demonstrating the COMPASS-I expert system for this one message type, COMPASS was expanded to include all 10 No. 2 EAX message types that require detailed analysis. These message types included System Malfunction Analysis (SMA) 110, 111, and 112 messages. SMA 110–112 messages are caused by an open circuit in the path. The opens are typically caused by a relay that is still functioning but is too slow to close in the allotted time. These messages are early warning signs that a relay is about to fail.

Although SMA 110–112 problems are less critical than NR20 problems, there is a big payoff in being able to identify and repair these problems. Because SMA 110–112s are an early warning of a soon-to-be faulty component (which would later be recognized as an NR20), the ability to find all SMA problems would have the major benefit of finding the faulty components before any calls were aborted, thus improving service quality. If all SMA problems were found, very few, if any, of the critical NR20 problems would occur. This analysis can result in maintenance activity advancing from a reactive to a proactive mode.

There are several important differences between the SMA messages and NR messages that affect message analysis, including the following:

1. *The SMA messages indicate both half-paths that generated the fault.* Only one half-path contains the faulty relay, whereas the other half-path in each message describes a path with no faults. However, the switch diagnostics cannot determine which of the two half-paths contains the faulty element. Therefore *half* of the data are spurious. Naturally, this large amount of spurious data makes analysis much more difficult, even for an expert. More messages are required to provide enough confirming messages to determine a pattern.

2. *There is a much greater volume of SMA messages than NR messages.* SMA errors are less critical yet more common. The number of error messages makes the analysis more difficult for a human analyst, even an expert. There are often no more than 25 to 50 NR20 messages on a given day, but there might be several hundred SMA 110–112 messages. Each SMA message contains almost twice the data of an NR message because it specifies two half-paths, whereas an NR message specifies a single half-path. Therefore, the total amount of SMA data could be 10 to 20 times greater than the NR data. As mentioned, a major step in the analysis involves determining certain half-path commonalities among the messages. Such a process increases significantly in complexity as the number of messages increases.

3. *The expert had very little field experience in doing a detailed analysis of SMA messages.* Switch maintainers spend most of their time attending to the NR problems. These problems are more critical than the SMAs, and the SMAs were much more difficult to analyze (see the above two points). Thus, the maintainers, even experts, do a superficial analysis of SMA data and often, when other tasks must be done, skip the SMA analysis completely.

Because the SMA 110–112 data are either ignored or partially analyzed, our expert had far less experience in analyzing SMA data than he had in analyzing NR20 data. He did possess knowledge pertaining to the meaning of the SMA messages and the likely causes of the problems, but he lacked the experience in solving them. Therefore, when it came to eliciting our expert's knowledge for this portion of COMPASS, we had to rely on (a) his deep knowledge of the switch architecture, and (b) his ability to modify the shallow heuristics that he developed for NR domain to the SMA domain, that is, his expertise. Because the COMPASS knowledge acquisition had already developed the knowledge for the NR analysis, we attempted to model the SMA analysis to coincide with this as much as possible. Often, parallels were drawn to the SMA analysis from the more firmly based NR analysis. The overall

Table 8.1. NR and SMA Knowledge Rules That Are Similar.

NR 20-22 BC DUAL EXPANSION ONE PGA DOMINANT 2 MESSAGES ANALYSIS RULE

IF There exists a BC Dual Expansion One PGA Dominant Problem
  AND
    The number of messages is 2
THEN
    Fault is in the PGA in the indicated expansion (.35)
  AND
    Fault is in the IGA (.35)
  AND
    Fault is in the PGA in the silent expansion (.2)
  AND
    Fault is in the Backplane (.1)
BECAUSE
    Two messages are not enough to justify the PGA fault being more likely than the IGA fault

SMA 110-112 BC DUAL EXPANSION ONE PGA DOMINANT 2 MESSAGES ANALYSIS RULE

IF There exists a BC Dual Expansion One PGA Dominant Problem
  AND
    The number of messages is 2
THEN
    Fault is in the PGA in the indicated expansion (.45)
  AND
    Fault is in the IGA (.45)
  AND
    Fault is in the Backplane (.1)
BECAUSE
    Two messages are not enough to justify the PGA fault being more likely than the IGA fault

Table 8.2. NR and SMA Knowledge Rules That Differ Significantly.

NR 20-22 NESTED CLUSTER POST-CLUSTERING RULE

IF Cluster Y is contained in Cluster X
THEN
    Delete Cluster Y
BECAUSE
    This eliminates coincidental clusters that occur within a problem

SMA 110-112 NESTED CLUSTER POST-CLUSTERING RULE

IF Cluster Y is contained in Cluster X
THEN
  IF Cluster Y is of higher precedence than Cluster X
    AND
      The number of messages in Cluster X that are not in Cluster Y is less than a sufficient number
    AND
      The number of messages in Cluster X that are not in Cluster Y < = 0.5 × the number of messages in Cluster Y
    AND
      Cluster X is not a reduced cluster
    AND
      None of the messages in X that are not in Y are SMA 110 messages with 8 repetitions
  THEN
      Delete Cluster X
    AND
      Cluster Y is a reduced cluster
  ELSE
    Delete Cluster Y
BECAUSE
  Normally, this rule is used to eliminate coincidental clusters that occur within a problem. However, since spurious matches are likely to occur in SMA messages, it is possible that the larger cluster is the spurious one. To paraphrase the rule: Take the larger cluster unless the smaller one within it is almost as big and has higher precedence.

SMA analysis could follow the same basic steps as the NR analysis. NR rules were examined for seemingly related cases in the SMA analysis, and a modified form of the NR knowledge was used for the SMAs where that could be done to maintain the analogy as much as possible. Table 8.1 shows sample NR and SMA knowledge rules that are similar. However, many times the SMA rules were found to be significantly different from the NR rules, as shown in Table 8.2. Thus, although the SMA portion of COMPASS is not based on lengthy experience of a domain expert, it represents the best available knowledge that could have been be applied to the problem. The SMA knowledge, primarily based on our expert's general expertise, required some techniques for knowledge eliciation different from those used for the NR analysis, which is based on the direct experience of our expert. We will discuss these differences in the next section.

Differences in Knowledge Elicitation for Experiential Knowledge Versus General Expertise

Our initial knowledge elicitation sessions concerned only the NR messages. During these sessions we, the knowledge engineers, not only learned the heuristics that later became COMPASS rules but also gained an understanding of the No. 2 EAX switch. In effect, we became miniexperts in the domain ourselves.

Through all of these discussions, our domain expert was speaking from his direct experience gained over years of servicing the No. 2 EAX switch. Our questioning of the expert might direct him to more and more obscure possible causes and rarer conditions, but his experience in the field made him an unquestioned authority. When he said, for example, that a certain message pattern indicated a particular malfunction about four times as often as a different malfunction, there was no room for argument—certainly not from the knowledge engineers. The expert had seen that particular pattern innumerable times during his years of analyzing switches, and there was essentially no possibility that he could be far wrong—his heuristics had been exhaustively tested during his years of maintaining the No. 2 EAX. Another expert might indicate a slightly different ratio of, say, 75:25 instead of 80:20. We did have several other experts look at the knowledge base to confirm it. However, because the COMPASS expert's knowledge was firmly based on maintenance experience and had proven successful in the field, no other expert could (or did) find significant errors in the knowledge, though the other experts might use a slightly different approach to an analysis. Also, by the nature of the COMPASS domain, minor differences in heuristics usually produced little effect, if any, on the results of the analysis. If COMPASS could embody most of our expert's NR20 heuristics, it could be almost as good an NR20 analyst as he was.

As mentioned, in our extension of COMPASS, we added (among other things) knowledge to analyze the SMA 110–112 messages. These SMA problems were rarely analyzed in the field, so the COMPASS expert relied on his basic knowledge of the switch to derive the knowledge for the expert system. It was his expertise, not his experience, that formed the basis of the discussions. He posited rules for analyzing SMA messages based on this expertise, such as the SMA rules shown in Tables 8.1 and 8.2. Sometimes these rules were developed using parallelism with some experiential heuristics identified for NR20 problems or other parts of COMPASS, as shown in Table 8.1. His domain expertise included his detailed knowledge of the switch architecture, the meaning of the SMA 110–112 maintenance messages, and knowledge of the possible malfunctions that could generate the errors. However, his rules for SMA 110–112 were not proven in the field and thus did not have the same authority as his NR20 heuristics.

The major difference in our interviewing techniques between eliciting knowledge based on experience and knowledge based on general expertise was due to a lower confidence in the latter knowledge. Our domain expert was less sure of the information he conveyed, using fuzzy terms rather than specific numeric ranges. Also, the knowledge engineers were more skeptical of the rules they received. In part this was because of the recognition of the expert's limitations in this particular area, but it was also because these rules were dependent on mathematical concepts (probability, set theory, etc.) that were familiar concepts to the knowledge engineers. Thus, the SMA analysis rules stated by the expert were often questioned by the knowledge engineers rather than accepted as absolute truths, whereas the expert's NR knowledge was generally accepted without challenge. By the time the project got to the point of eliciting knowledge about SMA analysis, the knowledge engineering group had gained sufficient knowledge of the domain to question the expert's nonexperiential heuristics. In some of the detailed analysis of message commonality, the mathematics and statistics background of the knowledge engineers was at least as relevant as the field experience of the domain expert. For example, in certain cases, we used Venn diagrams to describe intersecting sets of maintenance messages, a technique familiar to the knowledge engineers but not to the domain expert.

The interviews for this portion of COMPASS were much longer and more tedious. There was much more discussion of probability theory (more often among the knowledge engineers than between them and the expert) and much more reliance on fabricated examples to test the boundaries of a rules applicability. The knowledge elicitation techniques we used were equally suited for both areas of knowledge; the difference was more in the intensity and the emphasis of the interviewing process.

Another difference between the NR knowledge acquisition and SMA knowledge acquisition was that the NR analysis started from scratch, whereas the SMA knowledge acquisition attempted throughout to find and make use of parallels between the analyses of the NR and SMA messages. As knowledge engineers, we had captured a large measure of both the expertise and experience of our expert in the rules relevant to the NR analysis. We could compare and contrast the NR and SMA analyses in the development of the new SMA rules. In some cases, the knowledge proved to be virtually identical. In cases where we found strong differences, we would reexamine the original NR knowledge to see if it possibly contained an error. Although usually the differences were caused by real differences between the SMA and NR analyses, occasionally we would find that there was an error in the original NR knowledge.

The development of SMA knowledge, as well as the reexamination of the NR knowledge, would often lead to lively debates between the expert and the knowledge engineering team. Our expert would base his arguments on both his thorough understanding of the switch and his years of field experience. The knowledge engineering team would call on their recently gained knowledge of the switch functionality and their statistical and mathematical expertise. These debates would lead to more detailed explanations of the switch functionality to explain the often major differences in the somewhat parallel knowledge. These discussions led to the refinement of both the NR and the SMA knowledge.

In the end, both sets of knowledge were developed. The NR knowledge, based on the expert's expertise and experience, fared better in our field tests, as could be expected. However, the SMA knowledge, based on expertise alone, proved to be almost as accurate. The SMA knowledge was generally as reliable as the NR knowledge in determining the cause of the problems and generating the corrective maintenance actions. The principal difficulty in the SMA analysis was in the initial recognition of problems in the data, and this problem was caused by weakness in determining the proper pattern size thresholds used to recognize the problems. The problem-recognition difficulty was clearly a result of the lack of experience our expert had in accurately setting these thresholds. This was in direct contrast to similar types of thresholds that he was able to specify accurately for the NR analysis.

Though we developed both sets of knowledge, we realized that there were more risks related to using knowledge based on general experience than there were in utilizing experiential knowledge. We will discuss these risks in the next section.

Risks of Expertise-Based System

There are several risks to using an expert system—such as the SMA analysis part of COMPASS—that are not based on solid field experience. For example:

1. *False positives*: The expert system might identify problems that do not really exist, such as COMPASS's finding a problem in the switch that is not really there.
2. *False negatives*: The expert system may not identify problems that do exist, such as COMPASS's failing to find an important switch problem.
3. *Nonoptimal heuristics*: The expert system's heuristics may find a usable result but not find the best possible result, such as if the priority order of COMPASS's recommended maintenance actions is incorrect; field experience provides knowledge about which techniques offer a more likely solution.

False positives in COMPASS mean that COMPASS will recommend that a switch maintainer perform maintenance actions for a problem that does not exist. Performing maintenance actions for nonexistent problems might involve a significant waste of time and resources. In addition, at least in a small percentage of the cases, the extra work on unneeded maintenance actions might itself introduce additional problems in the switch. For both of these reasons, false positives in COMPASS, though not disastrous, can be costly.

False negatives in COMPASS are less of a problem. There is considerable redundancy in a central office switch, so it is not critical to find every fault immediately. Most of these network problems will get worse over time. A problem that is not caught today will probably show up with more evidence tomorrow. Therefore, throughout COMPASS we chose to minimize false positives. With this in mind, all of our heuristics were based on eliminating the consideration of any possible switch problems that were based on inconclusive evidence. Because we knew that the expertise-based SMA analysis was less solid than the experience-based NR analysis, the slanting of heuristics away from false positives was emphasized to a significantly greater extent in the SMA analysis than in the NR analysis. As mentioned, in field performance the SMA analysis proved to be generally as reliable as the NR analysis in determining the cause of the problems and in generating the corrective maintenance actions, but it performed less well (though very acceptably) in the initial recognition of problems in the data.

Conclusions

In COMPASS, both the experiential knowledge and the general expertise of our expert were captured in the expert system. Although the initial system was based on direct experience of an expert, major portions of the expanded COMPASS system utilized the expert's general expertise. As we have discussed, the different kinds of knowledge required some differences in the techniques used for elicitation. Although the system did not use deep knowledge of the electrical switch components, the expertise knowledge fell into the middle ground between deep knowledge and the surface knowledge that typifies most practical applications of expert systems today.

The payoff in the case of COMPASS is a system that provides a better overall analysis than the expert who provided the knowledge to the system. It combines a computer's speed and completeness of analysis with the human expert's expertise in understanding the switch architecture. Generally, the portion of the system that provides analysis based on the expert's experience will never surpass the expert. However, the portion of COMPASS that analyzes the SMA messages provides a much better analysis than that of expert maintenance personnel. (Note that due to time pressure experts sometimes skip analysis steps that are usually unproductive. In such cases,

COMPASS is more complete in its analysis and thus may outperform the experts even for the experience-based portion of the system.)

There are some risks in basing the expert system's analysis on unproven knowledge, that is, system knowledge that is not based on the experience of the expert in the field. The resulting analyses will almost certainly contain some inaccuracies, and these may be very hard to detect. In our domain, maintenance of a highly redundant switching system, there was little risk of any major disaster. Any undetected problem can only get worse until it is eventually obvious enough to be detected. Even the least optimal ordering of maintenance actions will eventually solve a detected problem if the set of actions on the list is correct. The advantage of using the portion of COMPASS founded on expertise rather than experience is that there is a strong possibility that problems will be detected earlier—in the SMA rather than the NR stage. The number of customer-affecting problems present in the switch is thus reduced, resulting in improved quality of service. So in this particular case, the payoff far outstrips the risk.

References

Chandrasekaran, B., & Mittal, S. (1982). Deep versus compiled knowledge approaches to diagnostic problem solving. *Proceedings of the National Conference on Artificial Intelligence* (AAAI–82), pp. 349–354.

Davis, R. (1989). *Form and content in model-based reasoning*. Paper presented at 1989 Workshop on Model Based Reasoning, 11th International Joint Conference on Artificial Intelligence, Detroit, MI.

Prerau, D. (1987). Knowledge acquisition in the development of a large expert system. *AI Magazine*, *8*(2), 43–51.

Prerau, D. (1990). *Developing and managing expert systems: Proven techniques for business and industry*. Reading, MA: Addison-Wesley.

Prerau, D., Gunderson, A., & Levine, S. (1988). The PROPHET expert system: Proactive maintenance of telephone company outside plant. *Proceedings of the Fourth Annual Artificial Intelligence and Advanced Computer Technology Conference* (AI'88), Long Beach, CA.

Prerau, D., Gunderson, A., Reinke, R., & Adler, M. (1990). Maintainability techniques in developing large expert systems. *IEEE Expert*, 5(2), pp. 71–80.

Scarl, E. (1989). *Sensor failure and missing data: Further inducements for reasoning with models*. Paper presented at 1989 Workshop on Model Based Reasoning, 11th International Joint Conference on Artificial Intelligence, Detroit, MI.

Sticklen, J., & Chandrasekaran, B. (1985). Use of deep-level reasoning in medical diagnosis. *Proceedings of the Expert Systems in Government Symposium*, pp. 496–502, McLean, VA.

9
Managing and Documenting the Knowledge Acquisition Process

Karen L. McGraw

Introduction

As increasing numbers of knowledge-based systems are developed for operational use, the importance of effective knowledge acquisition becomes more evident. Knowledge acquisition, the process of eliciting and representing (i.e., in computers) expertise from domain experts into a system, consumes the single largest block of development time (Feigenbaum, 1977; Hayes-Roth, Waterman, & Lenat, 1983; McGraw & Harbison-Briggs, 1989). To reduce the time required for this process, numerous tools have been developed that automate or systematize the elicitation and transfer of knowledge from expert to knowledge base. Among these advances are developments in machine learning that enable an expert system to acquire knowledge from "experience" (Michalski, 1980; Michie, 1982); expert system shells that allow domain experts to write their own rules (e.g., MacSMARTS, NEXPERT); and knowledge engineering tools that assist the knowledge engineer (e.g., McGraw & Harbison-Briggs, 1989, chap. 10).

The process of eliciting knowledge during the development of operational knowledge-based systems also often requires that developers apply a method and/or facility for tracking and documenting the information. It is expected that knowledge engineers could benefit from a tool that would enable them to not only acquire and manage knowledge acquisition sessions but also to develop a knowledge document that reflects domain experts' definitions of major concepts, acronyms, and vocabulary. This information can be valuable to current system development and evaluation efforts. After it is acquired and organized, the knowledge document can be made available to groups that are developing systems in similar domain areas.

This chapter describes an easy-to-use, standalone tool that can be used to manage the knowledge acquisition process. As designed, the tool provides vital program documentation, enables knowledge base traceability, and includes embedded instruction in knowledge acquisition techniques.

A Statement of Need

To develop effective knowledge-based systems, practitioners must use methods and techniques that will help ensure that the system is well organized, consistent with an expert's reasoning, and reasonably complete. Early efforts in knowledge engineering were modeled on small-scale projects in which the domain expert and a single developer interacted to build the final project (Hayes-Roth, Waterman, & Lenat, 1983). The project size allowed developers to approach the task informally and select knowledge acquisition techniques in an ad hoc manner. Early methods for knowledge engineering were often based on the five-step cycle of identification, conceptualization, formalization, implementation, and testing (Hayes-Roth, Waterman, & Lenat, 1983). As testing revealed inadequacies in the system, developers evaluated the deficiency and revisited a phase of the cycle. Subsequent system development entailed working in the area of presumed weakness, moving again through cycle phases, and retesting the developing prototype. The purpose of this approach was to provide a general "path" for development. The path was purposefully nonrestrictive to allow the system to be enhanced continually as knowledge acquisition enabled the developers to discover more about the domain. While this tactic provided a structure within which a system architecture and basic domain structure could be defined, it did not provide any control or guidance on the knowledge acquisition techniques that were used or the techniques by which the system's content could be deemed valid. Additionally, it provided no guidance on how to plan for and document knowledge acquisition during system development.

Current expert system development efforts tend to be larger and more complex (i.e., higher cost, more functional modules, multiple knowledge bases, etc.) than their predecessors. In this environment the ad hoc nature of early knowledge engineering methods has proven to be less than desirable. State-of-the-art methods tend to depict the knowledge engineering process in a more managed phase approach, drawing on approaches popularized in the software engineering field.[1] Some methods, like the adapted waterfall model for knowledge acquisition (McGraw & Harbison-Briggs, 1989), suggest not only steps or phases within the cycle but also techniques appropriate for each phase and guidelines for decision points and iteration. Within this model, developers first work through the feasibility and requirements phases, in which they analyze customer requirements to determine functional objectives. Using these objectives as guidelines, they begin to converge on the overall software development plans that help ensure the desired functionality. At a finer level, software development plans yield a core set of goals that, when met, determine project completion.

Using the functionally derived goals, developers continue to traverse the model, moving through the following phases: conceptual analysis, domain analysis, solution analysis, structured elicitation, implementation and prototyping, and evaluation. Within each phase they select specific knowledge acquisition and implementation techniques to enable them to focus on specific needs and type or level of knowledge required. Although the phases within the cycle are depicted as steps, the model does not bind the developer to a specific sequence. More accurately, it enables what could be termed a *managed iteration*. When a prototype review or test reveals an inadequacy, developers determine which type of deficiency has been noted, move back into the phase that focuses on that type of knowledge, select specific knowledge acquisition techniques, and work with experts to remove the deficiency. For example, development efforts may iterate to the conceptual analysis stage if developers determine that important conceptual definitions or relationships are missing or incomplete. Techniques they might use at this stage include concept sorting or taxonomy development.

Although this model helps developers plan their approach to knowledge acquisition, select technique types according to developmental phase, and refine knowledge acquisition efforts, problems remain. The following sections briefly describe typical problems developers encounter even when using a method that provides more control.

Knowledge Acquisition Management

Although a systematic approach helps the knowledge engineer determine the various types of information that must be acquired and techniques that can help him or her do so at each phase, it does not assist the knowledge engineer in knowledge acquisition management. To document the cycles, track the knowledge acquisition plan, and record the information elicited, the developer could implement a paper-based management system. McGraw and Seale (1987) implemented a paper-based system that included knowledge acquisition planning forms, knowledge acquisition recording forms, and knowledge acquisition session scheduling.[2] (For more information on this system, refer to McGraw & Harbison-Briggs, 1989, chap. 3.) This required that the knowledge engineer document knowledge in such a way to enable later verification and validation of the system. On a large system with multiple knowledge engineers, this requirement can result in documentation that varies greatly in quality, for whereas some knowledge engineers will follow management guidelines, others will be less prompt in recording their information or will record it a manner that is not easily understood by others. Additionally, the author encountered reluctance on the part of some programmers to work with a paper-based system.

Technique Selection

Knowledge engineering is still largely a "manual" activity. That is, although knowledge acquisition tools exist, most developers still acquire the majority of knowledge using manual techniques such as structured interviewing and process tracing. The time and labor demands of manual techniques are high, even when attempts are made to structure, plan, and manage knowledge acquisition sessions (McGraw & Harbison-Briggs, 1989; Prerau, 1986). The process of planning a session, selecting an appropriate technique, conducting the session, and summarizing session output (i.e., through protocol analysis or transcription analysis) can be tedious. For example, to plan, conduct, and review a 1-hour session a knowledge engineer can expect to expend, on the average, 9 hours (McGraw & Harbison-Briggs, 1989). Of this total, planning can take 40 percent of the time, whereas reviewing and summarizing the session can take as much as 45 percent.

Not only is structured knowledge acquisition time-consuming, but also the knowledge engineer's efforts are often hampered by the selection of an inappropriate technique. For example, recent surveys indicate that most knowledge engineers rely on interview sessions almost exclusively (Shadbolt & Burton, 1989; McGraw & Westphal, 1990), even when they might not be appropriate for current knowledge acquisition plans and the status of the prototype. The majority of expert system developers have relied on unstructured interviewing (Hart, 1986) because it appears easy and because they may not know how to use other, more specialized techniques. Other work suggests that knowledge engineers should combine structured interviews with other techniques for the best results (Kidd, 1985; Rolandi, 1986; Prerau, 1986). For example, Hoffman (1984, 1987) combined task analysis with structured interviewing to elicit numerous useful facts and heuristics. Additionally, McGraw and Seale (1988) suggest combining the interview with decision-making analysis techniques with small groups of experts.

"Lost" Knowledge

Finally, even if knowledge acquisition sessions are well planned, consist of a variety of effective techniques, and are well documented, development teams rarely develop what may be termed a *knowledge document*.[3] Consequently, both the developer and the customer miss an opportunity to capture and "package" basic domain knowledge. A knowledge document is an accumulation of the basic knowledge and vocabulary of the domain in a single document that is easy to add to, review, and reference. Thus, it is an organized set of critical domain information of use not only to knowledge engineers on the *current* project but also to other projects in the same domain and to the customer's organization. During its development the knowledge document can be shared among team members and can provide domain familiarization for new knowledge engineers; after the project is over it can be shared with other contractors or developers of systems in related domains.

To summarize, knowledge acquisition can be thought of as a process that requires not only AI expertise but also organization and management skills. Experience on large programs, such as the Force Requirements Expert System-Hawaii (FRESH) and Pilot's Associate, has shown that developers often find that it is useful to have some sort of data base to handle knowledge as it is acquired.[4] If the project is large and the development team is made up of multiple knowledge engineers and domain experts, developers also need some way to keep track of and use key information on domain experts and knowledge engineers (e.g., background and training, types of experience, problems, and sessions in which they participated). Customers and management personnel are interested in knowing what knowledge acquisition plans have been set, how many knowledge acquisition sessions are under way, and from which session or expert specific knowledge was elicited. Regardless of the techniques that are used, many manual-based knowledge acquisition efforts fail to adequately document knowledge in such a way that enables later verification and validation of the system (Liebowitz & DeSalvo, 1989; McGraw & Harbison-Briggs, 1989). Even fewer address the issue of training in knowledge acquisition techniques.

The following section provides background in the development and use of knowledge acquisition tools to solve some of the aforementioned problems and sets the stage for a discussion of a system developed specifically to enhance the management and documentation of knowledge acquisition efforts.

Knowledge Acquisition Tooling

To alleviate some of the perceived deficiencies of manual knowledge acquisition, developers have attempted to automate the knowledge acquisition process. These automation efforts range from the use of simple problem simulations (McGraw & Seale, 1987) to fully automated knowledge engineering environments or workbenches (Boose & Bradshaw, 1987). Most of these tools were developed to serve the needs of a specific system, a class of applications, or the program in which they were to be used. Additionally, the focus of these tools tends to be on the representation and/or implementation of domain expertise, as opposed to the logistics of handling the information. Specifically, none of these tools provides the knowledge engineer with an integrated system that links domain experts and areas of expertise with a data-base-like system for planning and managing knowledge acquisition sessions, and with embedded information describing knowledge acquisition techniques that might be appropriate. Furthermore, none of the existing tools is capable of recording or producing a knowledge document. Yet the development of current expert systems demands an increasing focus on the professional management of acquired information. In our experience with large development projects, not only are contracting agencies asking for knowledge traceability or system auditability (i.e., where the knowledge came from), but many of them are also beginning to recognize the value of the "condensed" domain knowledge that is acquired and organized during the development of a knowledge-based system.

Management and organization of the knowledge acquisition process could be streamlined by the use of an integrated, computer-based knowledge acquisition tool kit. The Hypertext Knowledge Acquisition Tool (HyperKAT) was designed as a tool kit to address these needs. HyperKAT is similar to computer-aided software engineering (CASE) tools in many respects. CASE tools provide software engineering support to developers working in the construction of more traditional (i.e., not expert systems) software systems. HyperKAT provides a management tool kit for developers of knowledge-based systems, intelligent tutoring systems, or intelligent computer-managed instruction systems. Unlike knowledge acquisition tools that are intended to *build* a knowledge base, are domain-specific, or require a symbolic processing computer, HyperKAT is a stand-alone management and documentation tool. It can be used across domains, regardless of the platform (e.g., type of expert system hardware) that has been selected for knowledge-base development. Currently, HyperKAT runs in the HyperCard environment on a Macintosh II, IIcx, or IIci computer with one megabyte of memory and System 6.0 or higher.

In designing HyperKAT, we considered the aforementioned problems and requirements to determine the characteristics that would be desirable in a knowledge acquisition management system. Specific design goals involved making the knowledge engineer and project manager's tasks simpler:

1. The tool should reduce the requirements for paper-based record keeping.
2. The tool should provide an integrated environment for planning and recording knowledge acquisition sessions or interactions.
3. The tool should enable a knowledge engineer to search through the data base of information for a specific session number, domain expert, or session topic.
4. The tool should provide a mechanism for selecting appropriate elicitation techniques.
5. The tool should enable a knowledge engineer to capture and depict not only verbal knowledge but also graphical knowledge.
6. The tool should make it easy to create a knowledge document as each session is reviewed, without leaving the environment.
7. The tool should help the knowledge engineer track time-on-task; total time expended on a session; whether or not the session has been reviewed; the topic, domain expert, and environment; whether or not the session was taped; and the technique used.
8. The tool should be able to provide printouts of session-related information, such as total sessions, sessions by specific knowledge engineers or domain experts, sessions on a specific topic, and so forth.

To meet these design goals, we investigated the feasibility of numerous software programs and related technologies. A key factor in the design of this tool was that the technology on which it was to be based had to be able to handle interrelated information and integrate it, allowing dynamic input and access. Over time, we began to view HyperKAT as less of a knowledge acquisition tool and more as an information processing tool for the knowledge acquisition process. The following section discusses knowledge acquisition as an information processing problem and lays the foundation for the design of HyperKAT and a more detailed discussion of its components and use.

Knowledge Acquisition as Information Processing

If developers require that knowledge engineers plan, record, and manage information related to knowledge acquisitions sessions, they sometimes encounter resistance because of the paperwork involved in adhering to such a system. On the Pilot's Associate program, the primary author and Mary Seale (McGraw & Seale, 1986) designed knowledge acquisition procedures and forms, and implemented a VAX-based knowledge acquisition system that stored key information (e.g., knowledge acquisition plans, session summaries, and domain expert files). Although it did provide documentation on the knowledge acquisition process, it proved to be less than adequate because the system did not provide links among important entities (e.g., between key domain terms and sessions in which they were discussed). Recent advancements in information processing technology such as hypertext, however, provide a feasible vehicle that can be used to manage and document knowledge acquisition activities. The following section describes hypertext and differentiates it from standard software implementation technologies to help the reader understand its selection and use in the HyperKAT system.

A Brief Background on Hypertext

Hypertext is a dynamic technology for building "nonlinear" systems for information management around a network of objects connected together by links (Halasz, 1988). Nonlinear systems are those that enable users to work with them in any "order" that meets their needs, rather than being forced into a step-by-step use that the system developer derived. In a typical hypertext system, windows that the user sees on the screen are associated with objects in a data base. Graphic links, depicting relationships among these objects, are provided among these objects in the data base

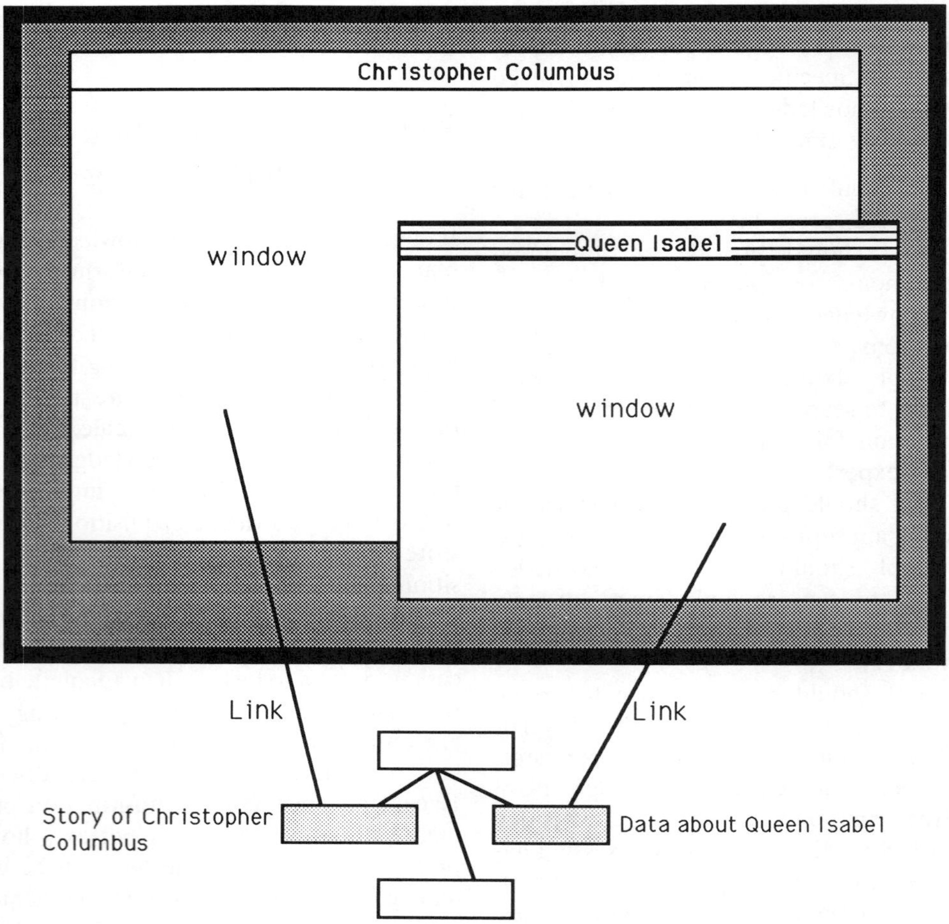

FIGURE 9.1. Basic features of a hypertext system.

(Conklin, 1987). Information is grouped, or chunked, into small units called nodes. This enables the user to store and access only the information in which he or she is interested and to move quickly from it to other information. As the information is retrieved (upon user demand), a single unit of information appears in the window on the screen. The user determines navigation through the data base by (a) selecting words or phrases for further investigation, for example, or (b) moving to a "bird's-eye view" in which all nodes of the system are visible and then selecting one for more detailed investigation. Figure 9.1 illustrates the hypertext concept.

Current hypertext systems support both graphics and text facilities (Conklin, 1987). Users can browse contents within the "data base" by navigating through the system. Not only is user browsing enabled, but it is also one of the most important components of hypertext systems. As he or she is browsing the system, the user can display some or all of the hyperdocument (i.e., a hypertext-based system of information). For example, the user can view the system at a detail level (i.e., a single window of information) or at a global level (i.e., the system appearing in the window as a multipronged graph). The detailed view helps the user get specific information from the system. The global graph enables the user to see all or part of the nodes in the system. This can provide contextual and spatial cues that can aid the user in getting the "big picture" of

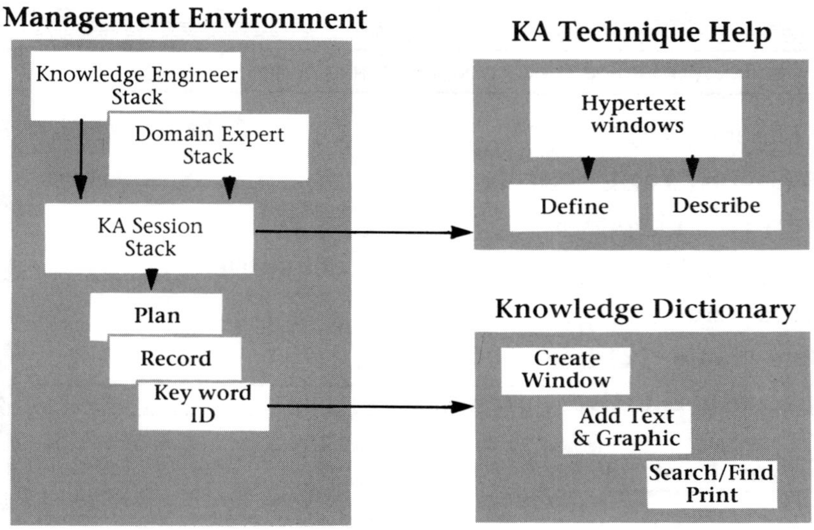

FIGURE 9.2. HyperKAT top-level design.

the nodes (i.e., pieces of information) being viewed and their relationship to other nodes in the system. As soon as the desired node is located, it can be accessed and displayed in the window. Then the user can navigate through the system, moving from one "file" or node directly to another file or node.

As categorized by Conklin (1987), hypertext systems fall into four general application areas. *Macro literary systems* include those that support massive on-line libraries. *Problem exploration tools* support early, unstructured thinking about a problem, such as systems for outline processing, programming, and design. *Browsing systems* are easy-to-use small-scale systems for teaching and public information. *General hypertext systems* enable "experimentation" with a broad range of applications. Applied hypertext technology includes such diverse tools as NoteCards, a development tool for hypertext systems (Halasz, Moran, & Trigg, 1987); KMS, a system designed to help organizations manage knowledge (Akscyn, Yoder, & McCracken, 1988); the Intermedia project, a tool for professors to use in organizing and presenting lesson material via computer (Garrett, Smith, & Meyrowitz, 1986); and HyperCard,™ a development tool for Macintosh systems that has been used for over 1,000 applications in the management and training areas.[5]

Specific applications of hypertext technology include electronic publishing; user interfaces; project management (e.g., Focal Point II by Mediagenics); embedded help systems for software applications (e.g., Claris SmartForm); financial modeling; electronic mail; and on-line user documentation (e.g., Microsoft Word 4.0).

HyperKAT as an Application of Hypertext for Knowledge Acquisition

Using HyperKAT, individual knowledge engineers and/or knowledge-based system managers can accomplish the following:

1. Organize plans for knowledge acquisition.
2. Organize and compile summaries of knowledge acquisition sessions.
3. Compile management-related information, such as number of sessions, types of sessions, typical session time, sessions per subsystem, and so forth.
4. Create and manage "data bases" of domain experts and knowledge engineers.
5. Link elicited domain knowledge to the session and expert from which it was acquired.

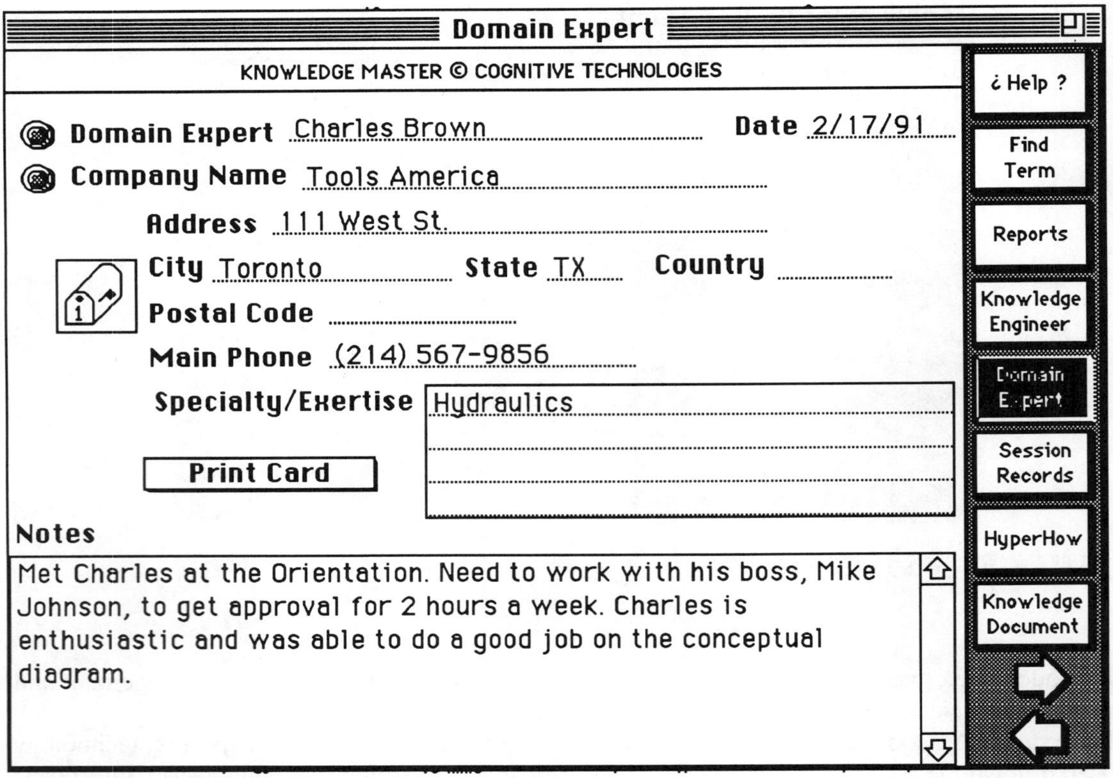

FIGURE 9.3. A domain expert card.

6. Receive assistance in selecting appropriate knowledge acquisition techniques for a specific session.
7. Enable the creation of a knowledge document from key words, concepts, or acronyms appearing in the knowledge acquisition session record.

Figure 9.2 depicts a top-level view of HyperKAT, its major components, and their integration.

Components of HyperKAT

As Figure 9.2 illustrates, HyperKAT consists of a number of interrelated components, including the knowledge acquisition planning and recording form or card, the domain expert and knowledge engineer stack, the embedded help for session techniques stack, and the knowledge document file. (Note that a *stack* is a file of cards based on a single format.) The reporting component enables printouts of data within these components, formatted according to user specifications. The following sections describe these components in more detail.

The Domain Expert/Knowledge Engineer Stack

One of the major components of management is the administration of resources. In the knowledge acquisition process, these resources are usually the domain experts and knowledge engineers. Each expert or knowledge engineer differs in terms of his or her experience, communication abilities, time commitment, current assignment, and previous performance. Having information on hand that describes some of these individual differences can help a knowledge engineering coordinator or system manager assign the appropriate knowledge engineer and/or domain expert to a specific system component or knowledge acquisition

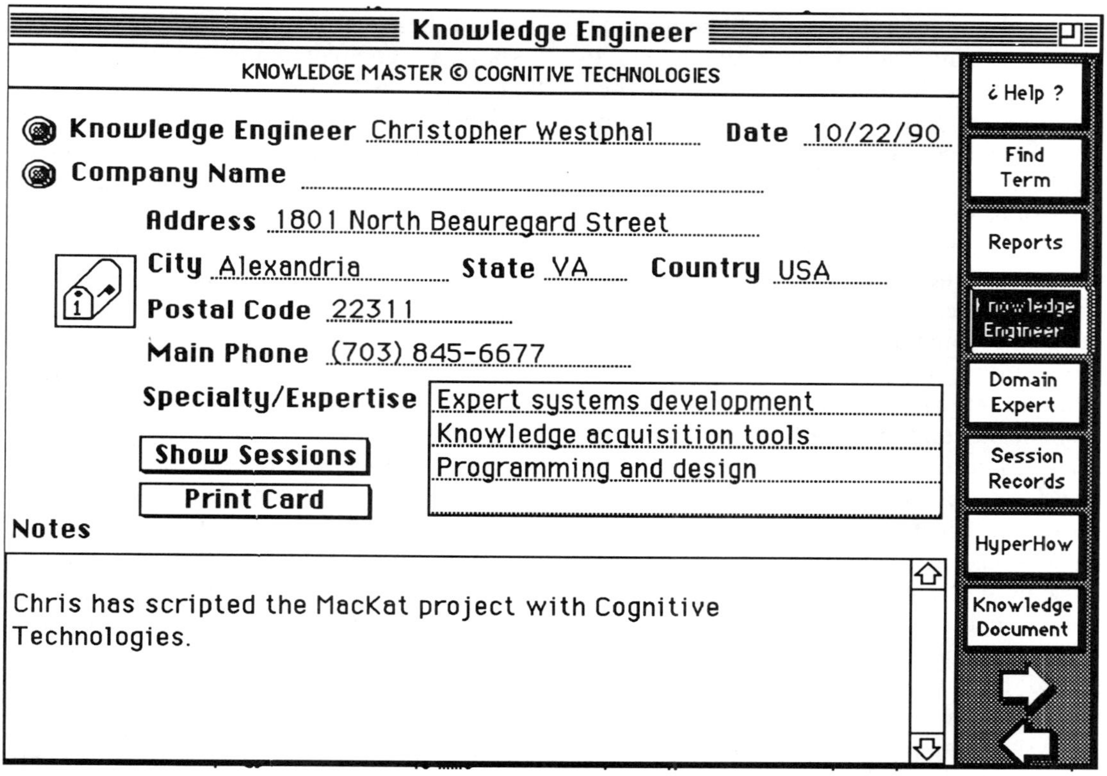

FIGURE 9.4. A knowledge engineer card.

session. For example, if the system being developed has a mission planning component and the files show that one of the knowledge engineers has had experience in mission planning, it is logical that he or she be selected to acquire mission planning knowledge from a domain expert. Similarly, if one of the domain experts requested work on the tactical planning component because he or she is interested in it, but has direct experience in route planning, he or she may not be the most appropriate choice for tactical planning knowledge acquisition.

In HyperKAT, resource management has been addressed by the use of a domain expert stack and a knowledge engineer stack. Each stack consists of a master "form" or card (similar to a frame), into which specific information about an individual can be listed, and individual forms or cards that represent specific resources. This information then can be used to aid in the selection of an individual, to document ongoing work, to keep track of sessions attended or missed, and to document any problems and concerns that could impact system validity. Figure 9.3 illustrates a domain expert card. Figure 9.4 shows a knowledge engineer card. Notice that each includes not only locational information such as name, address, and telephone number but also selection information such as availability, experience, and special expertise. The field names in each of the cards (e.g., address, phone, zip, etc.) all have associated preferences (i.e., suggested entry formats) that may be examined by "mousing" on the field name. Doing so will trigger a pop-up window that provides tips on what type and form of input is expected.

Note that each card is similar. Each contains identifying information slots, such as name, company name, addresses, telephone, and specialty or expertise. Each contains an area in which the developer can make free-form notes. The card itself is a window on the screen. At the top of the card is a menu command line that enables to user to save, quit, cut information, paste information, and gain access to object-creation facilities and a toolbox (i.e., selectable icons that enable text and graphic creation and editing). The user clicks on the radial button beside the name or company name slots to gain access to a pop-up menu of known names, from which a name can be selected or a new one added to the list. This facility enables ease of use, decreases the amount of typing users do, and thus decreases type-in errors that affect the ability of the system to search and sort based on a name.

By clicking on the Show Session button, the user gains access to a coded list of knowledge acquisition sessions (e.g., MP101, MP102, TP107) in which the expert or knowledge engineer has participated. If the user clicks the mouse in the Notes section, he or she can type in additional information, such as availability, previous problems, and so forth. As it is entered, the text wraps from one margin to the next. When more information is entered than can be shown at any one time, the scroll bars at the lower right of the screen become active (i.e., filled in with a pattern). The user can either click on an arrow to scroll up or down through the notes, or can drag the open box symbol up or down on the scroll bar to view a specific section of the notes.

The Knowledge Acquisition Planning/ Recording Card

The knowledge acquisition planning and recording stack is the heart of HyperKAT. The master card provides an integration of the knowledge engineer's plans and a summary of a knowledge acquisition session. A card is created for each knowledge acquisition session. Taken together, the cards form the planning and recording stack.

To enable program documentation and future searching and report generation tasks, this card contains certain identifying information, such as domain area, knowledge engineer, domain expert, and session type. Additionally, it provides an area for free-form typed input of knowledge acquisition plans. In this area general plans can be recorded, often in the form of simple outlines. Later, as sessions are completed and the information is "digested," the knowledge engineer selects the Review Session button to access a new type-in window and types in a record of the session, including general areas of discussion, goals met, and specific heuristics or information. A sample session record card for a tactical planning (as denoted by the TP prefix in Session #) module of an expert system is shown in Figure 9.5.

Because the information that is recorded on this card is very likely to be key data that will be searched for and reported on at a later date, the manner in which it is entered is important. For example, by clicking on the symbol beside Subsystem, the user selects the component of the knowledge-base system (e.g., the Tactical Planner) for which the knowledge will be acquired. After the subsection has been selected, the user clicks on the button beside Topic and chooses an existing topic from a pop-up menu or selects New to add a topic to the subsystem's topic list.[6] Finally, the knowledge engineer clicks on the button beside Session, which causes the slot to be filled with the next session number for that selected knowledge-base system component. This helps keep the numbering system sequential and nonredundant (i.e., two sessions in the same subsystem area will not have the same identification number). By clicking on the symbol beside either the domain expert field or the knowledge engineer field, the user can view a pop-up window listing the names of all known experts or engineers, and add or remove a name. If desired, each domain expert's card can be reviewed or all domain expert cards searched for a word describing necessary expertise or a specific feature. For example, if work on a knowledge-based system project for the insurance domain calls for expertise in annuities as a form of life insurance, the knowledge engineer could use the menu commands to search for "annuities" and identify experts with experience in that area. When an appropriate domain expert has been found, that name can be selected from the listing in the pop-up window. After it is selected, the name appears in the blank without necessitat-

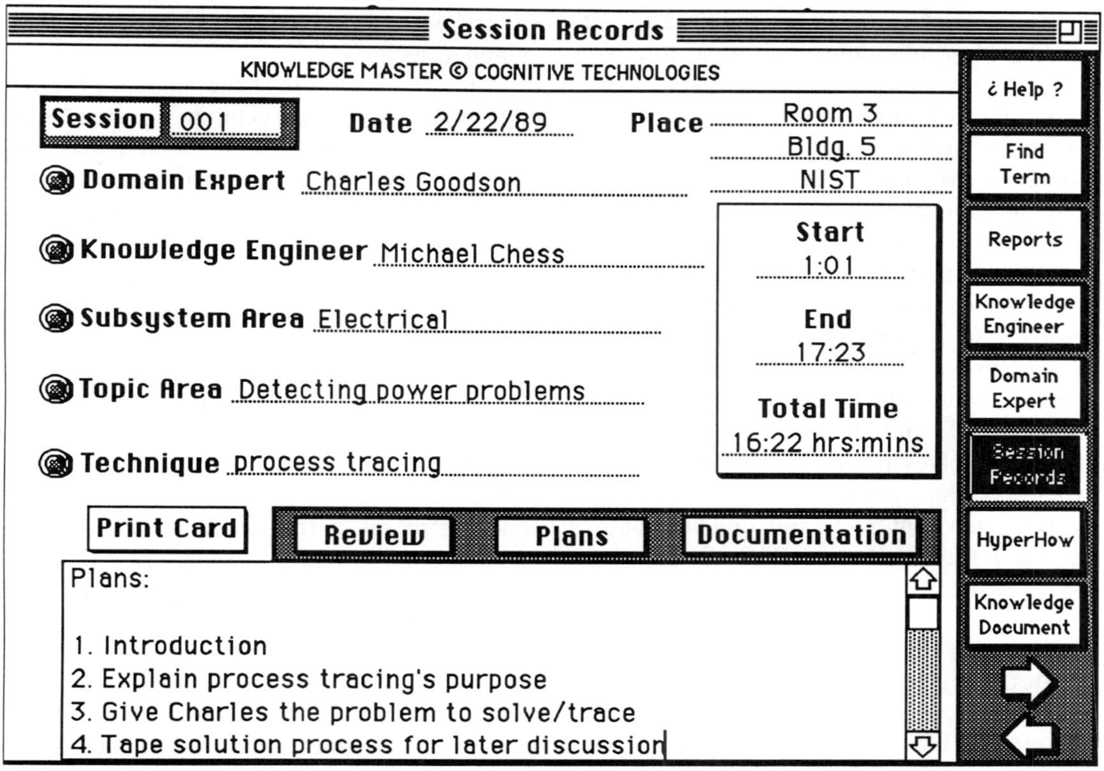

FIGURE 9.5. Sample knowledge acquisition session record card.

ing that the engineer type it. The process of selecting specific knowledge acquisition session techniques is similar. When the knowledge engineer clicks the symbol beside Technique, a pop-up menu appears listing the name of each technique that has already been entered and enabling the user to select one or more, or to add a new technique to the list. The embedded training component of this menu is discussed in a later section of the present chapter (see Embedded Help for Techniques).

The user types in the Place and Plans information. After a knowledge acquisition session has been completed the knowledge engineer can call up the Session Record and enter final data, such as start/end times and/or a summary, session outline, or heuristics derived from the session. When start time and end time values are entered, HyperKAT calculates total elapsed time. This information can be used by project management to track time-on-task and provide one measure of efficiency. If the session was taped, the knowledge engineer can enter the videotape or audiotape number to access the information in the tape library at a later time. Additionally, HyperKAT keeps a record on each card as to whether or not the knowledge engineer has entered session notes or summaries (representing a review of the session notes or tape). A printout of sessions that have been completed but not yet transcribed or recorded can stimulate management or the knowledge engineer to complete the recording task while the information is still fresh and easy to recall. (See the Reporting section for more information about HyperKAT printouts.)

Each session record card provides access (i.e., by clicking on the Review button) to a new window that includes not only a NotePad, which the knowledge engineer can use to record textual descriptions of facts, heuristics, and other pertinent data, but also a ScratchPad, which provides facilities (a window and primitive graphic tools) for the creation and labeling of graphic information, such as flowcharts or taxonomies. Thus, graphics can be used for conveying conceptual (Sowa, 1984); categorical and temporal (Lancaster, 1990); or functional (Means & Gott, 1989) views of the domain—information that is difficult if not impossible to record with text alone. After it is input, important graphical data that the expert provides need not be filed away, lost, or divorced from the session from which they came. This encourages the use of such data and enhances auditability or traceability of the knowledge. Finally, if the system under development involves the use of multiple experts, experience has shown that they can more easily review, interact, and enhance data in graphical forms (Boose, 1985; Stefik et al., 1987).

Embedded Help for Techniques

As noted earlier, surveys in the field of knowledge acquisition identify the interview as the most often selected knowledge acquisition technique (Gammack & Young, 1985). However, the interview is often selected inappropriately and managed incorrectly. For example, in the author's experience with over a dozen large knowledge-based systems, knowledge engineers tended to select the interview as the knowledge acquisition technique even when a technique such as a simulated problem scenario or task decomposition exercise would have been more appropriate. After this technique was selected, knowledge engineers frequently failed to manage and structure the interview through the use of primary and secondary questions; open and closed questions; and reflective (e.g., paraphrasing, prompting) and nonreflective (e.g., paralingual cues) techniques. In a review of over 50 hours of knowledge acquisition session tapes, "interviews" often had no introduction or formal closing (McGraw, 1989). Furthermore, even when a technique other than the interview was selected, knowledge engineers often ended up conducting an unstructured interview anyway—often to exclusion of the planned technique.

It may be that part of the tendency to select the interview as a technique is that some knowledge engineers feel that it is easy and that they do not know how to conduct sessions based on other techniques that are appropriate for knowledge acquisition. The Technique section of the HyperKAT knowledge acquisition session card not only enables the display of a menu of useful, general knowledge acquisition techniques (e.g., task analysis, structured interview, process tracing, etc.) but also lets the user access information (e.g., description, guidelines) on the use of these techniques. "Hyper-help" can be accessed by clicking to select the technique and selecting Help from the menu commands. This action causes the display of a pop-up, scrollable window in which text is presented that defines the technique, gives guidelines for successful use of the technique, and alerts the knowledge engineer to possible problems in its use. Figure 9.6 shows a sample help window displaying information on top-level knowledge acquisition techniques and terms. A small graphic is presented on each technique help card as a locational or navigational cue. The user scrolls through the text if it is not completely displayed in the window. After it is in a help window, the cursor becomes a pointing hand that enables the user to point to and click on the arrows at the base of the window or the menu commands beneath the window. The arrows let the user move either back to the card from which help was accessed (the "return" arrow) or to the previous or next set of help (i.e., on another technique).

Knowledge Document Creation and Reporting

Previous sections have described the actual components of HyperKAT. The following sections describe how HyperKAT enables users to create knowledge documents from knowledge acquisition session information and to print out user-defined reports.

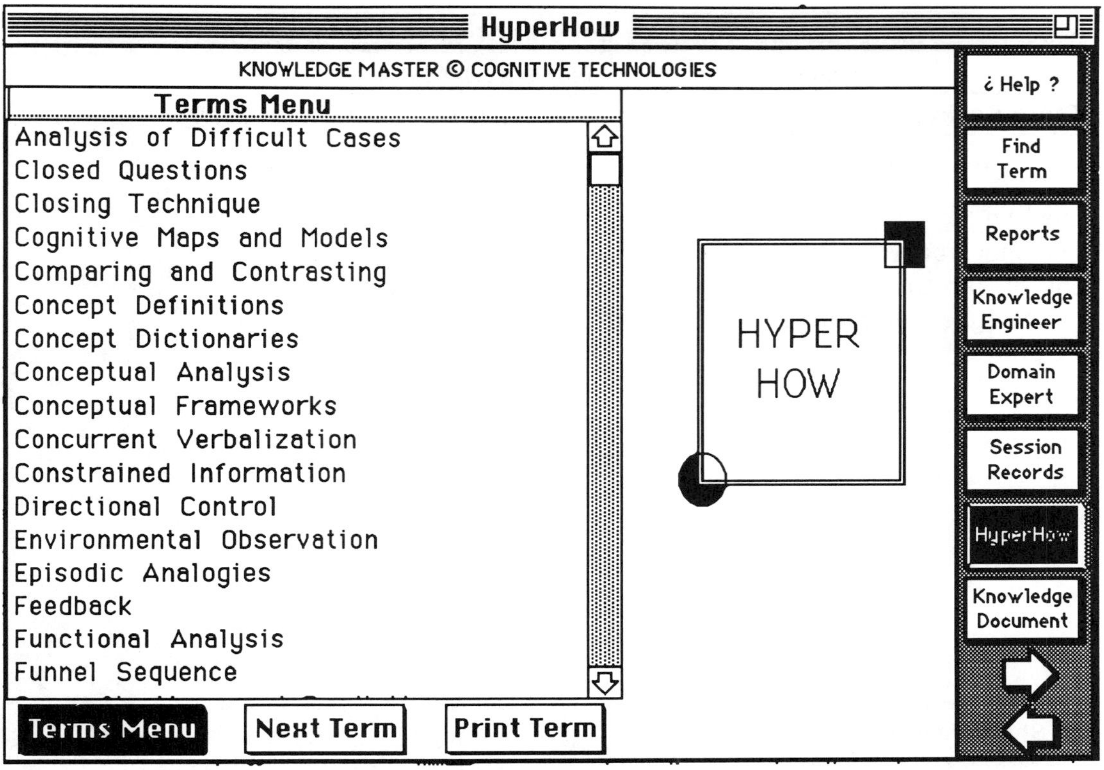

FIGURE 9.6. A sample "Hyper-help" window.

Knowledge Document

Traditional software engineering texts and practice suggest the use of data dictionaries for software development projects (Jensen & Tonies, 1979). A "dictionary" of domain knowledge (or on a larger scale, a knowledge document) is a valuable tool for knowledge-based systems. A knowledge document typically defines or depicts not only the terminology and acronyms within a domain but also the relationships among important concepts or functional components. For example, in the domain of military avionics a knowledge document would provide definitions and descriptions of aircraft (both "ownship" and the enemy's), weaponry, and maneuvers, among other basic information. It could also include diagrams depicting the categorical relationships of aircraft to each other, a taxonomy of weapon systems on different aircraft, or a set of heuristics governing strategies and tactics.

To determine accurately how big the system should be, define its functionality, set knowledge acquisition plans, and work effectively with domain experts, knowledge engineers must understand and have access to the type of information that is found in a knowledge document. Once the document has been developed, it can be used in many ways. For example, it can be used as a training aid, as new knowledge engineers come onto the program. It can be used as a reference document throughout the project, reflecting a common base of understanding and term usage for all team members. Later, the knowledge document can be delivered to the customer or contractor, providing an important reference tool for future development teams.

Knowledge documents, when developed at all, are generally created and stored as

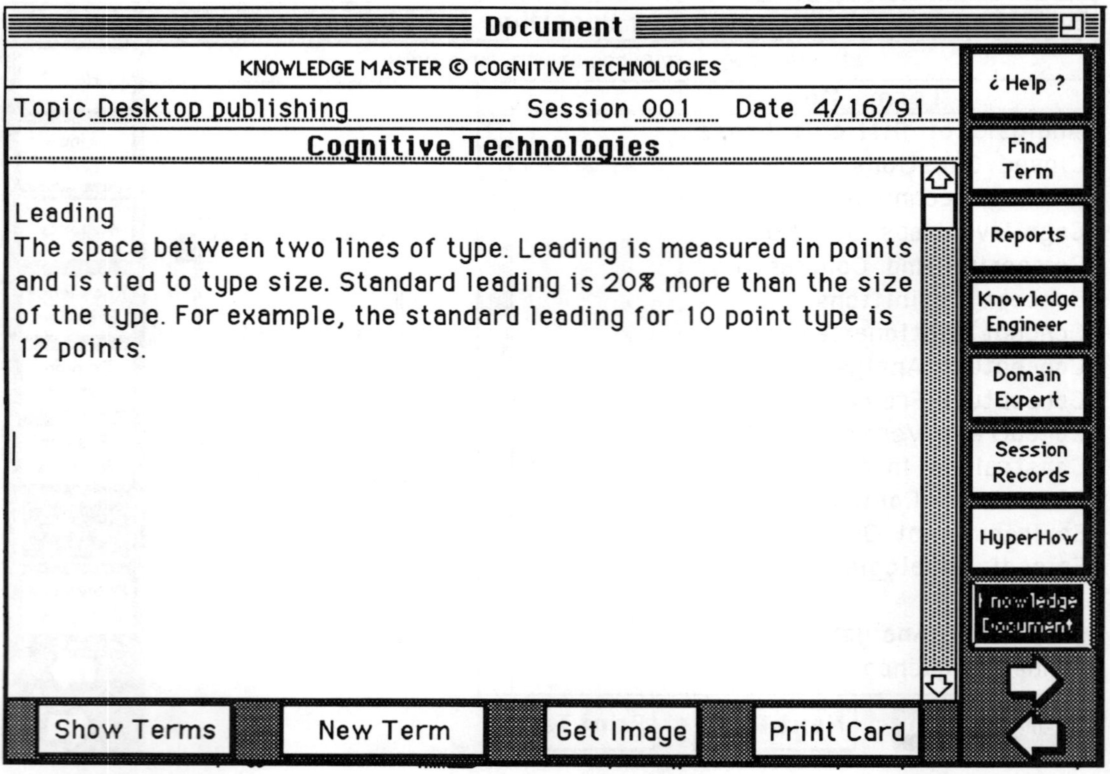

FIGURE 9.7. The knowledge document card.

numerous pieces of paper within a folder. In HyperKAT, the knowledge document is developed from information provided on the knowledge acquisition card or from free-form input. For example, the knowledge engineer can use selected words in the session review as entries in a knowledge document. By simply selecting a word, acronym, or phrase from the session review and selecting the Extract Text command, the knowledge engineer can define the selected item as an entry in the knowledge document. Figure 9.7 shows a sample knowledge documemt card, which consists of a pop-up window with a NotePad and a SketchPad area. The NotePad displays the text that was initially extracted by the user from the knowledge acquisition record card (in this figure, "leading"). The knowledge engineer can type a definition or description of the phrase, as shown in the window, or he or she can use the primitive graphic tools (selected from the Tools menu) to create a simple graphic on the SketchPad that represents information elicited on that topic. The knowledge document card contains menu commands and navigational icons (e.g., the arrows) like those found on other cards within HyperKAT.

Knowledge document cards can be searched for key words, updated, or printed out at any time for use by knowledge engineers or domain experts. After a word has been defined, it will appear in the record card with a special tag symbol. Clicking on a defined term activates its knowledge document card, complete with its definition and any graphic that might have been created to further describe it.

Reporting

Information that a knowledge engineer enters on the knowledge acquisition session record card can be extracted, compiled, and printed out in any of several report structures. Typical report structures that provide useful information for knowledge acquisition management include a tabular listing of knowledge engineers or domain experts by session number and topic, a listing of all sessions and their dates, and a listing of elapsed time for each session in a list of sessions. Data also can be printed out in user-defined formats. For example, the user could specify that the report be structured as a table in which columns include SESSION #, DATE, KE, DE, and TOPIC.

The reporting component of HyperKAT also enables users to do the following:

1. Print out a complete list of sessions in a table format that is specified by the user, as described above.
2. List all sessions that have not yet been annotated or "reviewed."
3. List sessions conducted by a particular knowledge engineer.
4. List sessions in which a particular domain expert participated.
5. List sessions dealing with a specific topic.
6. List sessions conducted using a specific knowledge acquisition technique.

Information of this type has many uses. For example, system developers could use it to simply review what has been accomplished during a specific stage or phase. Additionally, it could be used to monitor time expended per session, percentages of sessions in which specific techniques were used, average time expended, and so forth. This type of information is valuable in planning for future sessions and training knowledge engineering staff. Finally, such a system provides the first level of system traceability or auditability and can be useful in monitoring ongoing validation of the acquired knowledge.

A Sample Scenario With HyperKAT

Because HyperKAT is a hypertext-based product, users can enter it and move through it in the fashion that best meets their needs. Any single session might be vastly different from the session before or after it. For example, one might enter HyperKAT and work through existing knowledge acquisition record cards, extracting terms and phrases for the knowledge document. On the other hand, one might enter HyperKAT with the express purpose of planning or reviewing and writing up a knowledge acquisition session, or formatting and printing out a report of specific data. Therefore, it is difficult to provide an example of what a "typical" session with HyperKAT might look like. However, to provide the reader with concrete examples of how a user might interact with the system and work through its components, we provide the following scenario.

The domain for which knowledge is being acquired in this case is investments. The goal is to develop an expert system that would help an advisor suggest financial investments to clients, based on client factors such as investment amount, client age, family situation, other investments, interests, T-bill rates, and so forth. The developed system will be made available to advisors, who will access it from desktop computer systems.

Initially, the knowledge engineering coordinator for the project double-clicks on the HyperKAT icon, which displays a standard Macintosh window. In the window there are "stack" icons for each of the major components of HyperKAT (e.g., a domain expert stack, knowledge engineer stack, knowledge acquisition card stack, etc.). From this window she double-clicks the icon that represents the domain expert stack icon. A blank domain expert card appears, similar to that shown in Figure 9.3. She clicks on the radial button beside Domain Expert, the first slot on the card, and receives a pop-up menu listing all known (i.e., already entered) domain experts. The last choice on the menu is *New*. She selects the final option to create a card on a domain expert who is not already a part of the file. A type-in box appears, along with a prompt for her to type in the expert's name in the format FIRST/LAST and to press RETURN when the name is complete and correct. After she types the name and presses RETURN, she receives a verification message, which enables her to correct any typing mistakes or click YES to confirm that the name is correct. The message box disappears, and the new domain expert card can now be seen in total. The expert's name that the coordinator just entered appears in the slot beside Domain Expert. Next, the coordinator clicks on the radial button beside Company Name, and a pop-up appears from which she can select a company name from an existing list or type in information about a company that is not in the data base. In this case the company is already in the data base, so she clicks to select it from the list. As she does so, the Address, City, State, Country, Postal Code, and Main Phone slots are filled in automatically. The cursor appears beside the Specialty/Expertise slot. The knowledge engineer coordinator types in the information that the domain expert had provided concerning his specialty. Next, she clicks in the Notes area on the card and types in additional information, such as the expert's direct phone number and availability. After the final domain expert's card has been created, the knowledge engineering coordinator creates a similar "data base" of knowledge engineers on the program.

At this point the knowledge engineering coordinator exits the personnel card stacks and double-clicks on the icon that represents Subsystems. Subsystems can be thought of as components or modules that have been identified as being required for final functionality of the knowledge-based system. For example, an investment advisor might require information on IRAs, KEOGHs, CDs, savings accounts, savings bonds, annuities, and so forth. Each of these can be identified as a separate area or subsystem. Later, when they fill in the knowledge acquisition session card, knowledge engineers can click on the radial button beside *Subsystem* (refer to Figure 9.5) and select one that has been defined. This method speeds overall input, reduces typing-related errors, and enables the system to sort and locate knowledge acquisition sessions related to each of the system's subareas.

As an individual knowledge engineer works with HyperKAT, he or she might first call up a knowledge acquisition card by double-clicking on the knowledge acquisition card stack. (This card was shown in Figure 9.5.) Instead of being required to type in critical information, such as domain expert name or system component, the knowledge engineer clicks on the radial button beside appropriate entries (e.g., Session, Subsystem, Domain Expert, etc.) and selects an entry from pop-up menus. To reduce errors related to misnumbering, the number of a new session card for a subsystem area is automatically incremented and is placed in the Session # slot. Session numbers within each subsystem depict a sequential numbering scheme. Numbers are preceded by names or acronyms for system components as defined by the knowledge engineering coordinator (as noted in the previous section). Next, the knowledge engineer clicks on the Topic button and views a list of topics that have been identified as knowledge acquisition targets for this subsystem. For example, topics for the Annuities area include General, Costs, Constraints, Prerequisites, and New. After a topic has been selected, it appears in the slot. Next, the knowledge engineer selects the Technique button and is presented with a pop-up window of primary knowledge acquisition techniques. From this list the knowledge engineer double-clicks to select the planned technique, which is then placed in the slot.

After identifying information has been entered, the knowledge engineer can type in goals or plans for the session in a scrollable window using standard word processing features (e.g., type, backspace, delete, cut, copy, or paste). For example, for an introductory session in the annuity area, a knowledge engineer would fill in the first part of the card by selecting the appropriate subsystem (annuities), topic, domain expert (someone with expertise in this area), and technique (e.g., process tracing with case analysis). Then he or she would click in the Plan area and type in sample goals, such as the following:

1. Differentiate annuities from standard life insurance policies.
2. Identify assumptions from which an advisor works when suggesting this investment opportunity.
3. Determine constraints and attribute values for the selection of annuities.

When the card is filled in, the knowledge engineer can print and distribute it or electronically distribute it to other knowledge engineers on the team, the knowledge engineering coordinator, and/or a domain expert.

After the knowledge acquisition session is over, the knowledge engineer reenters the system by double-clicking on the knowledge acquisition card stack and uses the menus to find and display the card for the particular session number. First, he or she clicks in the slot beside Start, types in the beginning time, and presses RETURN. If the user clicks once on the Start text itself, instructions as to the format of the time entry, plus an example (e.g., "Please type the time, followed by AM or PM—8:00 AM.") appears in a pop-up window. The knowledge engineer repeats the process for the End time. As soon as both times are accepted, HyperKAT calculates elapsed session time and places the value in the Total Time slot. Next, the knowledge engineer clicks in the slot beside the appropriate recording mechanism (e.g., videotape or audiotape) and identifies any supporting taped material. Specific notes about the session can be input by typing in the Notes section. For example, the knowledge engineer might want to note that the domain expert will be sending him or her a copy of one of major firm's informational brochure, "All About Annuities."

The knowledge engineer now reviews the original session plans. Plans that were met can be noted (e.g., by typing in a symbol such as a +). Plans that were not met can be cut and pasted onto cards for upcoming sessions. Information that summarizes the session and reviews major concepts, terms, and heuristics can be entered by clicking on the Review button at the base of the card. (Refer back to Figure 9.5). This opens up a small window (which is linked in the data base to this session card) with scroll bars, navigational arrows, and menus, like many of the other windows in HyperKAT. From this window the knowledge engineer types in notes about the session, including points of discussion and specific heuristics mentioned by the expert (e.g., "IF savings is for a child's college education AND the child is in good health and less than 5 years old, an interest-bearing life policy is very desirable"). He or she can use the format that best fits the data. For example, the knowledge engineer can select and copy the plans into the review window and can then add information beneath each goal statement that describes accomplishments in that area. Alternately, review information can look like an outline, can be set up with numbered lines, or can look like a transcript. It can also be formatted as a functional record, with a general summary, followed by key words (to enable searching later) and specific heuristics or critical information.

After the session has been reviewed and documented, the knowledge engineer uses the navigational arrows or menus to exit the knowledge acquisition card component. The information on the recently completed knowledge acquisition card can be printed and distributed to others on the team and/or sent to the domain expert for review and verification or clarification. At a later date the knowledge engineer can call the card back up; click the Review button; and select and extract (i.e., copy) words, phrases, or acronyms for inclusion in the knowledge document. As was shown in Figure 9.6, the knowledge engineer can ask the expert to help him or her provide descriptions, definitions, or graphics for each term in the knowledge document. In our example the expert might be called on to help define and differentiate terms such as *whole life*, *term life*, *cash value*, and *annuity*, among others.

At the end of a development (i.e., prototyping) cycle, or whenever management desires, the knowledge engineering coordinator can use HyperKAT to compile, format, and print out reports of knowledge acquisition activity and knowledge document status. These printouts may be a hard copy of each domain expert card, knowledge engineer card, knowledge acquisition card, or knowledge document entry. Or, if the knowledge engineer sets it up as a table, the printout can show selected information that is sorted and compiled from the cards within the stacks.

FEATURES	BENEFITS
Stand-alone tool	HyperKAT is not bound to the knowledge base itself or directly to the developmental platform.
Links knowledge from source through plans and heuristics derived from sessions	Developers can track knowledge, enabling auditability or traceability.
User-defined links between key terms and concepts to definitions and descriptions	Hypertext technology enables the creation of a domain "knowledge document" that may have multiple uses.
Pop-up window for entry of key terms such as expert and engineer names, subsystem names, and knowledge acquisition techniques	Item selection reduces type-in errors, enhances entry speed, and enables more efficient database searching on key words.
Accommodates multiple users	One package can be used on a program-wide basis.
Stores multiple types of information in separate stacks	HyperCard structure enhances modularity and reusability, while decreasing storage/memory problems.
Icon-based and menu-driven commands	HyperKAT uses standard Macintosh user interface standards and is easy to use.
Printing Feature	HyperKAT provides both hardcopy and electronic versions of cards and data for review and program documentation.
Searching Feature	Knowledge engineers can search and quickly find target words.
On-line descriptions and guidelines for knowledge acquisition techniques	HyperKAT provides on-the-job assistance.

FIGURE 9.8. Features and benefits summary.

Conclusion

HyperKAT is a hypertext-based knowledge acquisition management tool. Figure 9.8 summarizes its features and benefits. As opposed to many knowledge acquisition aids or tools, HyperKAT's primary purpose is to manage and provide dynamic records of the process of acquiring and assembling knowledge about a domain. As such, it can be valuable for documenting knowledge acquisition activity, providing a means to trace where knowledge came from, and assigning individual knowledge engineers and domain experts to particular system components. If a critical knowledge engineer (or domain expert) leaves the project, at least a summary of his or her plans, session accomplishments, and session support information (i.e., videotape or audiotape identification) exists to help retain project momentum. Additionally, HyperKAT's reporting structure enables knowledge engineer coordinators to provide printouts of information that customers often request during phase or prototype reviews: number of knowledge acquisitions held during the time frame, participating personnel, topics discussed, techniques used, and so forth. This same information can help the project's management team review its efforts and use that data to plan the next cycle's (or project's) plans more realistically.

During the verification and validation phase of a project's development, the knowledge acquisition card stack can provide a thread between the resultant expert system and the original knowledge source, domain expert, knowledge engineer, and material. This information can help build face validity for the system, because it enables developers to identify where they got knowledge and verify knowledge they have acquired with other knowledge sources. For example, during a panel review of the development of a large piloting expert system one of the panel members asked where a specific piece of information came from. Similarly, another panel member requested that he be allowed to review the information that we had acquired about mission planning phases. A tracking system like HyperKAT enables developers to answer questions and respond to requests that affect system validity (and perhaps continuing project funding).

In a lesser sense HyperKAT can be used during a knowledge acquisition session. For example, after one of the initial knowledge acquisition sessions it may be prudent for the knowledge engineer and domain expert to work together and define, illustrate, and discuss the key terms and concepts that are pulled from the session review into the developing knowledge document. Such sessions not only save the knowledge engineer time and research but also can spark clarification of knowledge or the offering of new knowledge that may be stimulated by the reflection that this interactive task requires.

Furthermore, HyperKAT enables the ongoing development of a knowledge document that can be used to consolidate basic domain information. In turn, the knowledge document can be used as a training tool. For example, as knowledge engineers leave the project and are replaced by new personnel, the knowledge document provides a wealth of information that has proven to be important to the project. Reviewing knowledge document material can help new knowledge engineers come up to speed in the domain faster than if they were left to research, request information from multiple sources, and digest information on their own. Customers or clients may also recognize the value of a knowledge document. It represents a hard copy of vast amounts of domain expertise that they might want to share as a reference document for new employees.

Acknowledgments. The author is indebted to Christopher Westphal of the Institute for Defense Analyses, who as a member of the HyperKAT design team was responsible for programming HyperKAT and providing the screens for the figures in this chapter. Special thanks go to Robert Hoffman, who has worn his editor's hat so well during the chapter development process. The FRESH and Pilot's Associate program referred to in this chapter are DARPA-sponsored contracts and part of the Strategic Computing Program. The views expressed in this chapter about those programs are those of the author and do not reflect the views or opinions of DARPA, Texas Instruments, the Navy, or the Air Force.

Notes

1. Developers of AI systems can learn many lessons from the decades of work in the software engineering field. For more information on the adaptation and use of software engineering techniques in expert system development the reader may consult Partridge (1986) and McGraw and Harbison-Briggs (1989).
2. A system like this was used to document knowledge acquisition efforts on the DARPA-sponsored Pilot's Associate program within Texas Instruments.
3. Knowledge acquisition texts and/or articles (e.g., McGraw & Harbison-Briggs, 1989; Prerau, 1986) suggest that system developers should keep such a document. (See Regoczei & Hirst, this volume.)
4. FRESH is one of the application areas in DARPA's Strategic Computing Program and is operational at CINCPACFLT, Hawaii. The author participated in these programs while part of the Texas Instruments Artificial Intelligence Lab.
5. HyperCard is a trademark of Apple Computer.
6. Topics are specific areas of knowledge acquisition interest within each subsystem. For example, topics for the TP subsystem would include defensive tactics, offensive tactics, armaments, and so forth.

References

Akscyn, R., Yoder, E., & McCracken, D. (1988). The data model is the heart of interface design. *Proceedings of CHI '88* (pp. 138–244). New York: Association for Computing Machinery.

Boose, J. (1985). A knowledge acquisition program for expert systems based on personal construct psychology. *International Journal of Man-Machine Studies*, *23*, 495–525.

Boose, J., & Bradshaw, J. (1987). Expertise transfer and Complex Problems: Using AQUINAS as a knowledge acquisition workbench for expert systems. *International Journal of Man-Machine Studies*, *26*, 495–525.

Conklin, J. (1987). Hypertext: An introduction and survey. *Computer*, *20*, 17–41.

Feigenbaum, E. (1977). The art of AI. In *Proceedings of the Fifth International Conference on Artificial Intelligence* (pp. 281–285). Los Altos, CA: William Kaufmann.

Gammack, J., & Young, R. (1985). Psychological techniques for eliciting expert knowledge. In M. Bramer (Ed.), *Research and development in expert systems* (pp. 57–69). London: Cambridge University Press.

Garrett, N. L., Smith, K. E., & Meyrowitz, N. (1986). Intermedia: Issues, strategies, and tactics in the design of a hypermedia document system. In *Proceedings of the 1986 Conference on Computer-Supported Cooperative Work*. (pp. 163–174), Austin TX: MCC Software Technology Program.

Halasz, F. G. (1988, July). Reflections on NoteCards: Seven issues for the next generation of Hypermedia systems. *Communications of the ACM*, *31*, 10–18.

Halasz, F. G., Moran, T. P., & Trigg, T. (1987). NoteCards in a nutshell. In *Proceedings of the 1987 ACM Conference on Human Factors in Computing Systems* (pp. 45–52). Toronto, Canada: Association for Computing Machinery.

Hart, A. (1986). *Knowledge acquisition for expert systems*. New York: McGraw-Hill.

Hayes-Roth, F., Waterman, F. D., & Lenat, D. (Eds.). (1983). *Building expert systems*. Reading, MA: Addison-Wesley.

Hoffman, R. R. (1984). *Methodological preliminaries to the development of an expert system for aerial photo interpretation* (Report No. ETL–0342). Ft. Belvoir, VA: U.S. Army Corps of Engineers, Engineer Topographic Laboratories.

Hoffman, R. R. (1987, Summer). The problem of extracting the knowledge of experts from the perspective of experimental psychology. *AI Magazine*, *8*, 53–64.

Jensen, R., & Tonies, C. (1979). *Software engineering*. Englewood Cliffs, NJ: Prentice-Hall.

Kidd, A. (1985). Human factors in the design and use of expert systems. In A. Monk (Ed.), *Fundamentals of human-computer interaction* (pp. 210–233). New York: Academic Press.

Lancaster, J. (1990). Cognitively based knowledge acquisition: Capturing categorical, temporal, and causal knowledge. In K. McGraw & C. Westphal (Eds.), *Readings in knowledge acquisition: Current practices and trends*. Chichester, England: Ellis Horwood.

Liebowitz, J., & DeSalvo, D. (Eds.). (1989). *Structuring expert systems*. Englewood Cliffs, NJ: Yourdan Press.

McGraw, K. (1989). *Communication techniques for software support specialists*. Training seminar material. Annapolis, MD: Cognitive Technologies.

McGraw, K., & Harbison-Briggs, K. (1989). *Knowledge acquisition: Principles and guidelines*. Englewood Cliffs, NJ: Prentice-Hall.

McGraw, K., & Seale, M. (1986). *Pilot's Associate program knowledge acquisition guidelines and procedures*. DSEG Artificial Intelligence Lab document. Dallas, TX: Texas Instruments.

McGraw, K., & Seale, M. (1987). Structured knowledge acquisition techniques for combat aviation. In *Proceedings of NAECON '87*, *4*, 1340–1348. Dayton, OH.

McGraw, K., & Seale, M. (1988). Knowledge elicitation with multiple experts: Considerations and techniques. *Artificial Intelligence Review*, *2*, 24–31.

McGraw, K., & Westphal, C. (Eds.). (1990). *Readings in knowledge acquisition: Current practices and trends*. Chichester, England: Ellis Horwood.

Means, B., & Gott, S. (1988). Cognitive task analysis as a basis for tutor development: Articulating abstract knowledge representations. In J. Psotka, L. D. Massey, & S. A. Mutter (Eds.), *Intelligent tutoring systems: Lessons learned* (pp. 35–58). Hillsdale, NJ: Erlbaum.

Michalski, R. (1980). Knowledge acquisition through conceptual clustering: A theoretical framework and algorithm for partitioning data into conjunctive concepts. *International Journal of Policy Analysis and Information Systems*, *4*(3), 219–243.

Michie, D. (1982). The state of the art in machine learning. In D. Michie (Ed.), *Introductory readings in expert systems* (pp. 208–228). Edinburgh, Scotland: Edinburgh University Press.

Partridge, D. (1986). *Artificial intelligence: Applications in the future of software engineering*. New York: Ellis Horwood.

Prerau, D. (1986). *Knowledge acquisition in the development of a large expert system*. (Report, Computer and Intelligent Systems Laboratory). Waltham, MA: GT&E Laboratories.

Rolandi, W. G. (1986). Knowledge engineering in practice. *AI Expert*, *1*, 58–62.

Shadbolt, N. R., & Burton, A. M. (1989, April). A survey of methods for eliciting the knowledge of experts. In C. Westphal & K. McGraw (Eds.), *SIGART Newsletter* (pp. 15–18). New York:

Special Interest Research Group on Artificial Intelligence of the Association for Computing Machinery.

Sowa, J. (1984). Conceptual Structures: Information Processing in Mind and Machine. Reading, MA: Addison Wesley.

Stefik, M., Foster, G., Bobrow, D., Kahn, K., Lanning, S., & Suchman, L. (1987). Beyond the chalkboard: Computer support for collaboration and problem solving in meetings. *Communications of the ACM*, *30*(1), 32–47.

10
Using Knowledge Engineering to Preserve Corporate Memory

Gary A. Klein

The Value of Expertise

Expertise is a key resource in any organization, but it is usually not treated with the same care as other resources. Few organizations have any methods for preserving or expanding their experience, or even taking stock of their current expertise. In contrast, other corporate resources, such as financial assets and equipment, are monitored with great attention. Ironically, many corporations are proud of their experience. They boast of "being in business for over 75 years." But if you enter their offices you may have trouble finding anyone who has worked in the same job more than 2 or 3 years. When staff members retire, the organization does little to preserve their expertise; if an exit interview is performed it is usually directed at learning how the person feels about the job and the organization, rather than trying to elicit the accumulated tricks of the trade. Supervisors acknowledge the person's expertise through a small party or a gift, perhaps a plaque.

Because most organizations do not really understand how to value their own expertise, they tend to lose it. They fail to develop a workable corporate memory. The result is that they repeat errors because they fail to take advantage of lessons learned. The thesis of this article is that organizational experience is a resource to be managed, akin to cash, equipment, goodwill, and inventory. Imagine the outrage *you* would feel if your organization decided to upgrade to a new computer system and all your software files became unusable. Yet if the head of the computer department retires or leaves, the discussion centers on who will be the replacement. This article will examine expertise as a resource that can be identified, valued, and preserved. Knowledge engineering is the activity of finding, eliciting, and applying expertise.

First, I will examine the nature of expertise. Then the strategy of "bottling" expertise, as in expert systems, will be described, followed by a more general look at what it means to engineer knowledge. Knowledge engineering will be defined as a broad domain that subsumes intelligent system development. Finally, some low-technology applications of knowledge engineering will be presented, including technology transfer and training.

The Nature of Expertise

What is expertise? A simple answer would be that expertise is the growth of knowledge and skills through experience. It may be more helpful, however, to look at some of the specific aspects of expertise anchored within a concrete domain. One domain with which I am familiar is urban fire fighting (Klein, Calderwood, & Clinton-Cirocco, 1986), so the following discussion will contrast the expertise of a rookie fire fighter with that of an experienced fireground commander.

The rookie tries to learn the basic procedures. One fire fighter told us about his first

TABLE 10.1. Forms of knowledge.

1. Procedures
2. Specific details
3. Declarative knowledge
4. Physical relationships
5. Interpersonal knowledge
6. Perceptual/cognitive skills
7. Perceptual/motor skills
8. Goals
9. Precedents
10. Cultural knowledge

day: When an alarm came in he rode out on a fire engine, and as it turned a corner he saw a building engulfed in flames. "Look at that," he said to himself. "I hope someone called the fire department." Then he remembered that *he* was the fire department. Someone put a hose in his hands and told him where to spray the water, and in what pattern, and that was all he did until the water was turned off. He didn't have any clue about what the other fire fighters were doing that day.

Contrast this to a fireground commander with 20 years of experience who is used to allocating people and resources; risking lives in the process; and making decisions in less than a minute, sometimes in just a few seconds. There are a number of ways that the rookie and the experienced fire fighter differ, listed in Table 10.1.

With regard to procedures, the rookie doesn't know what all of the procedures are and has not yet even begun to master (i.e., make automatic) the procedures that he or she does know.

With regard to specific details, the rookie may never have dealt with a specific type of building before. The experienced fireground commander has done so, knows how to place engines and trucks, and may have even conducted a survey inside the very building that is now on fire.

With regard to declarative knowledge, the experienced fireground commander knows many facts about fires, about materials, and about buildings. One commander told us he was successful because he had been a builder before joining the fire department. He could visualize the construction of a home or office building and thus could tell how a fire might be weakening the roof supports.

The experienced fireground commander has learned to be sensitive to spatial, causal, and temporal relationships. The rookie has difficulty orienting the layout of rooms on one floor to the floor below, anticipating the consequences of actions, or recognizing the best time to intervene.

On an interpersonal level, the experienced fireground commander knows the crew. He or she can anticipate the crew members' reactions, their strengths, and their weaknesses. He or she can sense when someone doesn't understand an order or believes that a plan is stupid but doesn't want to say so. In an emergency, lives can depend on knowing how others will respond. One fire fighter described the frustration he felt in working with a new supervisor; during a difficult moment, when it was time to pull a crew off a spongy roof, the fire fighter had to waste time by coming down from the roof to argue in person for his belief. He felt that this interruption wouldn't have been needed if he had been working with his usual supervisor because they both knew how to read each other.

In terms of perceptual and cognitive skills, the experienced fireground commander has learned how to "see." The experienced commander notices patterns quickly and detects distinctions that are invisible to the novice. One commander described how he believed he knew where the seat of the fire was and ordered the members of his crew of train their hoses on it. After about 20 seconds he realized he was mistaken. He had expected the smoke to change color and intensity, and when it didn't he was forced to reassess. When he found the seat of the fire, he was able to estimate its temperature by the color of the flames. This type of knowledge is hard to articulate, even though it is vital to performance. It is part of what Polanyi (1966) has called *tacit knowledge*. (In our interview, we did not elicit the perceptual knowledge but merely determined how the fire fighter consistently used such knowledge.)

A simple example of a perceptual motor ability is the ability of a fire fighter to direct a stream of water at a window that has been identified as the seat of the fire. The hose itself is heavy, water pressure may be uneven, and the distance to the window must be calibrated to set up the best arc. The effects of the wind and of fire-related updrafts and downdrafts must also be taken into account. If the water supply is uncertain it is important to waste as little water as possible. Because the fire may be spreading exponentially it is also important to get a high volume of water on the seat of the fire as quickly as possible.

Another important part of expertise is setting appropriate goals. Fire fighting seems a simple domain, with a well-defined goal: to put fires out. But a fireground commander may have been given the wrong address. In arriving at the scene it may not be clear whether there is smoke (which is dangerous itself) or flames. If there are flames, should the commander try to search for occupants, put the fire out, or prevent the fire from spreading? If there are not enough resources to put the fire out, should the commander request more resources? It turns out that the fire fighter's goals are quite complex, so the ultimate goal is to do the best possible job. If more resources are needed, ask for them, but be careful not to ask for too many, because that may create a shortage elsewhere. In one fire, the commander erred in trying to extinguish the blaze and called in a second alarm. Several fire fighters were injured. The building had been condemned, and it was unsafe for operations. He should have just let it burn down. The commander should have had enough experience to recognize what the attainable goals were.

The expert's knowledge also includes precedents from past experience. Someone who has worked in an organization knows some of the corporate history—what is usually done in a given type of situation. This history provides a sense of typicality, which guides the recognition of appropriate actions and goals. It is also a source of specific analogues. One fireground commander told about trying to extinguish a blaze in a four-story apartment building. He noticed a billboard on the roof and ordered the crowd back because he remembered a previous fire in which the flames burned through the billboard supports, sending the billboard crashing down into the street.

There is a family of practices (cf. Dreyfus, 1972; Heidegger, 1962) surrounding everyone's actions. Many of the practices are tacit and context-specific. More than that, they *are* the context that defines the rest of what is going on. In the case of fire fighting, these include practices about when to argue and when to obey orders to section a roof, when to make suggestions to a commander or to a subordinate, how to signal disagreement without saying anything, how to appraise the confidence of fire fighters on a roof who have reported that the roof on which they are standing feels spongy.

The forms of knowledge in Table 10.1 help to describe what it means to have expertise. I am not claiming that they have any philosophical significance, and surely other people would identify some different dimensions. The purpose of this list is merely to provide a sense of the *range* of capabilities that are acquired when someone works in an organization for a number of years.

The different types of knowledge obviously interact. For example, if a fire fighter drives a fire truck into the parking lot of an apartment building he or she has previously investigated, he or she can make sure to turn sharply to park as close as possible to the available fire hydrant rather than having to waste time backing up and pulling in again. Those extra 20 seconds can be important. Thus, knowledge about goals and about specific details of the parking lot enhances perceptual/motor performance. I could spend a fair amount of space tracing all of these interrelations, but that is not the point of this chapter. The conclusion is that expertise is not a unitary quality. Expertise has many facets, most involving concrete knowledge about specific tasks, objects, and people. Expertise is not the same as being smart or even the same as possessing a set of reasoning abilities that transfer easily between domains. Expertise is not "general intelligence." (See Ceci & Liker, 1986, for a nice demonstration of this point.) A useful construct of general intelligence may facilitate the acquisition of expertise, but it does not substitute for expertise (see Sternberg & Frensch, this volume).

The purpose of examining the concept of expertise is to provide a basis for evaluating strategies for capturing expertise by representing it in different ways. The next section will turn to such strategies.

Bottling Expertise

The development of expert systems was important for helping cognitive psychologists appreciate the importance of content knowledge. The technology of expert systems has shown that it is possible to specify, elicit, and preserve some aspects of expertise. This technology serves as an existence proof for the viability of the domain of knowledge engineering.

To briefly describe the field, intelligent systems are computer programs that use a limited set of reasoning processes to manipulate a knowledge base. The knowledge base is the set of content rules generated by experts. The focus is usually on content knowledge rather than on process knowledge of general reasoning strategies. The unique power of intelligent systems resides primarily in the set of content rules. The term *knowledge engineering* (Waterman, 1986) was coined to describe the activities of computer programmers eliciting rules from experts and coding them in LISP or Prolog in a knowledge base.

Early on, the intelligent systems paradigm also revealed an interesting fact: It is hard to elicit knowledge rules from domain experts (Lehner & Adelman, 1989). Unfortunately, behavioral science (circa 1975) had not developed unique methods for finding out from experts what they know. There were no generally accepted techniques for eliciting implementable rules and other forms of knowledge, and it was necessary to jury-rig methods designed for basic research. After 100 years of claiming to be a science, after several decades of cognitive simulation and information processing and information science, experimental psychology had not addressed the problem of how to efficiently elicit or organize knowledge in general, let alone the knowledge of experts.

Why had expertise been so ignored in American psychology? One reason for avoiding expertise was that psychologists prefer to study *process* rather than *content*. From the beginning of the 20th century, American researchers have been fascinated by fundamental processes, whether in the form of classical and operant conditioning principles or 4-stage models of problem solving or heuristics. For this reason, behavioral scientists went out of their way to study nonverbal organisms (rats, pigeons). When studying people, psychologists used artificial stimuli and laboratory tasks to make sure there would be no contamination of the results. *Contamination* here means that the results might reflect prior experience. Researchers tried to avoid studying expertise, even inadvertently, because of the fascination with process. European researchers such as Claparède (1902), Duncker (1945), and de Groot (1965/1978) did not fall prey to this trap. (Even in American research there are obvious exceptions to this trend, e.g., the work of Munsterberg, 1913, in industrial psychology and Shanteau, 1984, on expert decision making.) The functionalist school would have undoubtedly studied expertise, but this program was cut short by the emphasis on behaviorism that held sway since the 1930s. The emphasis on process was justified because of the potential for high payoff. As behavioral scientists learned about the processes of memory and learning, the hope was to develop general theories of cognition that could be applied to fields such as instruction.

Starting with Larkin, McDermott, Simon, and Simon (1980) and Chi, Glaser, and Farr (1988), there has been a recent focus on expertise. (See Chi et al., 1988, for an excellent summary.) But there is a difficulty in studying domains such as physics; chess (e.g., Chase & Simon, 1973; Simon & Barenfeld, 1969); and bridge (Charness, 1979). These are relatively context-free tasks, omitting some factors listed in Table 10.1, such as perceptual/motor skills, interpersonal knowledge, and contextual details. The picture of expertise that emerges from this research must be limited.

While in many ways this has been interesting research, it has not found process differences between novices and experts. In fact, there may not be process differences. Chi et al. have nicely demonstrated that additional years of experience do *not* enable an expert to think differently, that is, to use different reasoning strategies. The available data do suggest that the additional experience provides more knowledge, enabling a more powerful use of strategies that are part of the repertoire of both experts and novices. But see Prawat (1991) for a dissenting opinion. Researchers' strategy of looking only at process has not yet paid off.

We see the same development in the field of artificial intelligence (AI). One of the clearest expressions of a focus on process is the General Problem Solver (GPS) (Newell & Simon, 1972). This is an AI program that consists of a set of heuristics (e.g., means/ends analyses) that were intended to be applicable to a wide set of tasks, ranging from chess to cryptarithmetic. One of the important lessons of the GPS was that heuristics alone were not sufficient to do useful work. (In Table 10.1, heuristics and other analytical methods would be classed as procedures.) Today we do not hear about a GPS 9.8. GPS had process knowledge but no content knowledge. If computer programs able to represent and store content knowledge could be developed, perhaps AI could become useful in applied settings. This was the premise that led to intelligent systems.

Intelligent systems do not assume that skilled personnel are smarter or use better reasoning strategies. Instead, the premise of an intelligent system is that the content knowledge makes the difference. The content knowledge of a retired expert could be elicited and represented as rules or frames, and inferential procedures could access the relevant rules. Thus, the expert's knowledge base could be applied to a decision even when the expert was no longer present. The intelligent system could be considered a bottle containing the elicited knowledge.

With the advent of intelligent systems, organizations may start to look differently at their senior employees. Now that there is potential for capturing expertise, organizations can worry more about losing it. And the employees themselves can regard their experience differently. Before, they may have been happy to pass on their tricks to the next generation. Now they can wonder whether to give away their knowledge or sell it back to the company. Expertise has taken on a new value.

There is a parallel between the use of intelligent systems to engineer knowledge and the development of petroleum engineering. Through the early 19th century, petroleum had always been a nuisance. It fouled the crops and fouled the drinking water. Sometimes it was sold as a folk remedy. Usually farmers just scraped it off their boots. Then in 1855, a Yale chemist discovered how to extract kerosene from petroleum, and immediately the picture changed. Headlines read "Good News for Whales," for here was a plentiful and inexpensive alternative source of fuel. Now there was an incentive to find, extract, process, and utilize this new resource, petroleum.

Because of the advent of intelligent systems, there is now an incentive to find, extract, process, and utilize knowledge. Whether we rely on intelligent systems or find alternative delivery strategies, there may be a new discipline emerging for engineering knowledge and expertise.

Knowledge Engineering as a Discipline

At present, *knowledge engineering* is a phrase used to describe the process of getting rules out of the heads of experts and into the innards of intelligent systems. There may be a broader field here: applied cognitive psychology or applied experimental epistemology (Hoffman & Deffenbacher, in press). Intelligent systems are just one means of "engineering" knowledge, and even if those systems were to suddenly disappear we would still be interested in knowledge engineering.

To continue the parallel of knowledge engineering to petroleum engineering, in treating petroleum as a resource, there is a need to do the following:

1. Locate sources of petroleum.
2. Assay the quality of the petroleum.
3. Extract the petroleum.
4. Process and refine the petroleum.

The tasks of systems engineering are planning, constructing, and guiding. Building on the analogy to petroleum engineering, we can suggest five aspects of knowledge engineering:

1. Locating sources of expertise in an organization.
2. Assaying the cost/benefits of engineering the expertise.
3. Acquiring the knowledge.
4. Codifying the knowledge.
5. Applying the engineered knowledge.

The remainder of this chapter examines each of these aspects.[1]

Locating Sources of Expertise in an Organization

Usually the first question asked is, "Who is an expert?" Sometimes this is an innocent question, but often it is motivated by skepticism that anyone can tell who the experts are. The skepticism is justified. There do not appear to be any simple procedures for distinguishing experts from nonexperts. There are elaborate criteria such as speed, errors, and recall scores, all of which are laborious to arrange and impractical to use in an applied setting. This is a real problem.

One way around the problem is to search for people with many years of experience. Yet we know that having years of experience is not the same as having expertise. In our research with fireground commanders we soon learned that a person with 12 years of command experience in a rural or suburban station that rarely has a two-alarm fire may be less proficient than someone with 2 years of experience in a decaying inner-city fire station that handles several multialarm fires a month.

Our current operating procedure is to ask the experts' peers—people at the head of an organization, if they appear credible and if they are familiar with the personnel—to identify the domain experts. This enables us to duck the question of who is an expert. But this is the operational definition most often used in intelligent systems work.

For the remainder of this article, the term *expert* will be defined as a person who has worked at a task for a significant period, usually several years, with a relatively broader range of experiences than other workers, and with mastery of a wider set of skills than most other workers. Dreyfus and Dreyfus (1986) describe how people who have mastered a skill can perform it without conscious analysis, and so an expert would be a person who had achieved such mastery for a greater number of skills in a given domain.

The question of who is an expert does not easily go away. I assume that there are experts who are consistently knowledgeable across different aspects of the job. In fact, we have not found many people who fit this description. In any domain there are some people who have had a set of experiences that makes them familiar with a subset of the domain's tasks and problems, x, others familiar with set y or set z, some who have some familiarity with both x and y, and so forth. Rarely is there anyone who is familiar with it all. To call someone a domain expert is to pretend that he or she is familiar with x, y, and z. It would be good for knowledge engineering if such experts existed, but we must not delude ourselves. Instead, we must acknowledge that there is expertise in the organization and that from different individuals we can learn about tasks and problems x, y, and z.

Therefore, we can replace the question, "Who is an expert?" with the question, "Where is the expertise in a given organization?" The job of the knowledge engineer is to get a sense of who knows what and to elicit from selected personnel the knowledge that they have acquired.

In locating people with expertise, guidance comes from the obvious fact that it is partially based on having direct and indirect experience.

Thus, we will search out individuals who have had more experience, especially if their experience was varied and challenging.

A disadvantage to searching for expertise from different sources is the problem of reconciling different judgments. Often the inconsistency will disappear if one probes about the underlying context. For example, several fireground commanders assured us that upon arriving at an occupied structure such as an apartment building, the first thing to do is "search and rescue." But one fireground commander described a time when he found a fire in the basement of an apartment building and tried to extinguish it before initiating search and rescue procedures. Is there a mismatch here? No, because the conditions had to be taken into account—the fire was just starting, and there was a reasonably good chance of putting it out, whereas diverting the crews to search and rescue would have ensured that the fire got a good start, running the risk of losing the whole building.

Of course, there will sometimes be disagreements among experts. Instead of ignoring these we can try to represent differences of opinion. In general, we have found a fairly high consistency in the judgments of different people with experience in the same domain.

An example might be useful here. We interviewed a fireground commander who described the importance of extrasensory perception. He claimed that there were times when being able to see into the future was essential for success, and he gave us a prime example. He once led his crew into a house that was on fire. It was a simple structure, not much out of the ordinary, and yet he felt that something was wrong, so without any evidence or justification he ordered his men to get outside. No sooner had they exited than the floor where they had been standing collapsed. Had they waited even another few seconds, they would have been plunged into the fire beneath them.

Not believing in such a strong version of ESP, we carefully probed the incident. We discovered first that the fireground commander had been slightly suspicious from the beginning. The crew had gone into the building (not noticing a basement), had taken the hoses into the back where the kitchen was, and had aimed the water at the supposed seat of the fire, but there was no reduction in flames. They tried again, with no result. So there was some initial uncertainty. We also discovered that the fireground commander remembered that he and his crew drew back into the living room and that it had been extremely hot, especially because the fire was supposed to be in the kitchen. (They didn't realize it was in the basement, right below them.) Further, the living room was very quiet, and fires are loud and noisy, so the combination of heat and quiet was unsettling. By the end of the interview, the fireground commander realized that the three cues and their interrelation were enough to signal mismatch, like a tilt sign that made him too uneasy to stay in the building any longer. He reluctantly relinquished his claim to having ESP.

Assaying the Cost/Benefits of Engineering Expertise

Not every valuable commodity should be engineered. Shale is a source of petroleum, but the current costs of processing shale often are great, so the price of petrochemicals has rarely justified the processing of shale. If the price of petrochemicals increases, and if more efficient means are developed for processing shale, then it would become economically feasible to do so. The usual cost/benefit analysis is applied in calculating the economics of the decision: the cost of processing the shale and the price of petroleum.

If knowledge is a resource, its valuation is also driven by the cost of the elicitation and the benefits it confers.

Cost

The cost of the elicitation is determined by how scattered the expertise is and by the knowledge elicitation procedures needed. If several interviewers are required, tape recordings must be transcribed, videotapes must be reviewed and coded, and special tasks must be developed. Or if other costly strategies are required, the economies will be reduced.

Benefits

Knowledge has different values at different times. We may feel that knowledge is always good, but unless it can be put to work and confer direct benefits on an organization, it is not worth acquiring. The expertise of a fireground commander is of lesser value when limited opportunities for promotion result in a backlog of highly qualified people waiting for their chance to move up. The knowledge of a skilled automotive worker is of greater value if layoffs have eliminated the apprentices and assistants who had been developing mastery of the job. A further complication is that the price of a new product may be unrelated to its subsequent price after the product matures and acquires competitors. A general guide to secretaries' tricks and shortcuts for word processing might have been very useful in 1984, but by 1988 it was not as exciting.

At present there are few techniques for assessing the costs and benefits of knowledge engineering. As more and more knowledge engineering studies are performed, the field will gain additional experience in knowledge engineering projects, and a set of typical cases will be built. Eventually the prior cases will be developed into more formal and abstract methods for performing knowledge assays.

Acquiring the Knowledge

There has been a great deal of interest in knowledge acquisition methods. Starting in 1985, there have been annual international meetings in Italy and in Banff, Alberta, Canada, to describe and evaluate knowledge acquisition. In addition, there have been books (Hart, 1988) and special issues of journals (IEEE, 1989) devoted to knowledge acquisition.

There has been much less work specifically on evaluating knowledge elicitation strategies. It is important to understand the efficiency and effectiveness of such strategies in order to work out the economies discussed above. Hoffman (1987) has performed pioneering research in this area. He examined several different strategies—unstructured interviews, structured interviews, analysis of familiar tasks, limited information tasks (using familiar tasks but omitting information that is typically available), constrained processing tasks (watching how an expert performs a familiar task under time pressure or other constraint), and the method of tough cases (analysis of how experts handle difficult tasks).

In evaluating these methods, Hoffman found that the analysis of familiar tasks enables the expert to feel comfortable but is fairly time-consuming. Structured and unstructured interviews are useful for initial data but are also time-consuming. Limited information tasks and constrained processing tasks can be tailored to achieve greater power but leave experts uncomfortable. The analysis of tough cases can yield valuable information, but such cases occur unpredictably.

Hoffman tried to quantify the efficiency of the different methods and found that the limited information and constrained processing tasks and the method of tough cases yielded the highest rates of new observation-related propositions and new inference-related propositions, approximately two to four propositions per total task minute (including preparation time). Structured interviews were somewhat less efficient, and unstructured interviews, the most common form of knowledge elicitation used to build expert systems, were the least efficient, generating only 0.13 propositions per total task minute.

Burton, Shadbolt, Hedgecock, and Rugg (1987) performed similar pioneering research and got a similar pattern of results. Protocol analysis of think-aloud tasks are the least efficient in generating new propositions. In contrast, card sort and laddered grid tasks were disliked by domain experts but were more efficient in generating knowledge, and about as efficient as a structured interview method.

Critical Decision Method

Hoffman (1987) has noted that the approach he called *the method of tough cases* is relatively efficient but has the shortcoming that such cases occur unpredictably. One way around this shortcoming is to query domain experts

TABLE 10.2. Critical decision interview probes.

Probe Type	Probe Content
Cues	What were you seeing, hearing, smelling . . .?
Knowledge	What information did you use in making this decision, and how was it obtained?
Analogues	Were you reminded of any previous experience?
Goals	What were your specific goals at this time?
Options	What other courses of action were considered or were available to you?
Basis	How was this option selected/ other options rejected? What rule was being followed?
Experience	What specific training or experience was necessary or helpful in making this decision?
Aiding	If the decision was not the best, what training, knowledge, or information could have helped?
Time pressure	How much time pressure was involved in making this decision? (Scales varied.)
Situation assessment	Imagine that you were asked to describe the situation to a relief officer at this point, how would you summarize the situation?
Hypotheticals	If a key feature of the situation had been different, what difference would it have made in your decision?

about nonroutine incidents that have occurred in the recent past. Flanagan (1954) first popularized this strategy as the critical incident method for describing what a job really demanded.

Klein, Calderwood, and Clinton-Cirocco (1986) adapted the method to study decision strategies of fireground commanders. This critical decision method (CDM) focuses on nonroutine cases and applies a set of question probes to elicit the domain experts' decision strategies, perceptual discriminations, pattern recognition, and so forth. These are illustrated in Table 10.2. The rationale for the CDM is that expertise does not stand out in routine cases, which may be performed by mediocre and skilled personnel using the same strategies. Expertise does not emerge for novel cases, because the novelty may limit the use of experience. Nonroutine cases, however, are a rich source for observations about expertise. Such tasks have atypical aspects but are still linked to prior experiences. Following the petroleum engineering analogy, nonroutine tasks represent a rich deposit that is most economical to exploit. Such tasks cannot be performed automatically. Domain experts have to "stretch," and as they do so their knowledge becomes more accessible. Moreover, by probing domain experts about specific cases we get at concrete aspects of their knowledge; in contrast, general questions often result in vague and abstract responses that are not very helpful. (Senjen, 1988, has shown the value of nonroutine cases for AI research.)

The CDM is described in greater detail elsewhere (Klein, Calderwood, & MacGregor, 1989). Basically, the interviewer performs several sweeps through the same incident. The first sweep acquires the story itself. The second sweep fixes the events to a timeline. The third sweep elaborates on decision points, goals, and situation assessment. A fourth sweep examines opportunities for errors. The method has been shown to have high levels of reliability as assessed by interrater reliability between coders (Taynor, Crandall, & Wiggins, 1987). We have applied it in a number of domains to study the expertise of nurses in a Neonatal Intensive Care Unit, software programmers, program managers, urban fireground commanders, incident commanders in charge of forest fire operations, tank platoon leaders, battle managers, design engineers, paramedics, and consumers. Some of the applications are described later.

The CDM is just one example of an efficient knowledge acquisition strategy. As we learn to perform better assays of knowledge in an area, and as we develop more powerful techniques for applying the knowledge, we will be able to improve the knowledge acquisition methods at our disposal.

Codifying the Knowledge

After knowledge is elicited, the task is to codify it, which usually means finding a way to represent it for efficient storage, retrieval, and use. Two codification techniques will be described: (a) the technologically sophisticated intelligent system approach and (b) the low-technology codification in the form of incident accounts. Doubtless there are other forms of codification for the knowledge of domain experts. The contrast between intelligent systems and incident accounts is useful for illustrating strengths and weaknesses of formal versus informal strategies and for illustrating the value of knowledge engineering apart from intelligent systems work.

Intelligent Systems

The best-known strategy of codification involves intelligent systems. For a number of years AI researchers and computer programmers have been developing approaches for codifying knowledge, and there is a voluminous literature on this subject. Some of the best-known sources are Waterman (1986) and Hayes-Roth, Waterman, and Lenat (1983). Intelligent systems can be used to express, and codify into rules, the declarative knowledge of experienced domain experts. Inference strategies are embedded within these complex rule bases to manage the handling of tasks and problems. Many efforts are currently under way for using intelligent systems to capture knowledge.

Researchers are beginning to learn the boundary conditions for applying intelligent systems. The initial response was uncritical enthusiasm, but after a number of failures it has become clear that intelligent systems are not always appropriate. As Huang (1989) has put it, "most . . . systems are brittle in the sense that they are not immune to even minor flaws in their encoded knowledge or slight changes in the environment" (p. 489). Holland (1986) and Huang have described some strategies for limiting the problems of brittleness. There is a definite role for intelligent systems, and we are gradually learning where they apply best—relatively simple domains where a manageable set of straightforward rules will provide useful guidance.

Of perhaps greater value than pure intelligent systems are hybrid knowledge-based systems, which blend a small intelligent system rule base with more powerful algorithms to attack calculational problems that are too complex for people. For example, Automated Air Load Planning System (AALPS) has been very helpful for configuring military cargo aircraft (Anderson & Ortiz, 1987). AALPS has a small rule base to ensure that trailers are loaded in the same aircraft as the trucks that will be hauling them and to ensure survivability by making sure that not all helicopters are loaded on the same aircraft. AALPS also relies on algorithms for calculating weight distribution so that the operators can determine if the aircraft will become dangerously unbalanced on takeoff or even during the flight when some cargo items are extracted for parachute landings.

The use of intelligent system technology in hypertext also will be interesting to pursue. It appears that the large-scale intelligent system, such as XCON, that was once touted as the sign of things to come is not very practical. We may see more applications that use limited intelligent system components, perhaps small rule bases of 300 to 400 rules, to handle limited tasks embedded within larger systems.

Case-based reasoning systems (Kolodner, 1988) are a type of intelligent system. They have a special appeal because they are structured around analogues rather than rules. Therefore, explanation is in terms of the analogue selected, the criteria for selecting it, the adjustments made to the analogue, and the reasons for those adjustments. Such an explanation is much more understandable than a presentation of the 200 rules fired during the session. Barletta and Mark (1988) have developed a case-based reasoning system to assist production workers in operating an autoclave. The types and configuration of parts placed in the autoclave can affect the failure rate, and the problem is too complex to be modeled, so the system identifies previous successful configurations and enables the workers to use these prior cases to select the sets of new parts to be autoclaved together. Recent efforts developed a case-based reasoning system to assist in estimating the costs of manufacturing parts (Reed & Klinger, 1991) and explored the use of previous aircraft designs to efficiently and rapidly generate new conceptual designs (Thordsen, Henke, King, & Stottler, 1989).

The work on case-based reasoning appears to transcend some of the limitations of traditional intelligent systems, but only in certain types of domains and problems. The general limitations of intelligent systems remain and make such systems imperfect vehicles for storing and representing many types of expertise.

Therefore, the goal of capturing the expert in a bottle does *not* seem to be realistic anymore. It is not so much that there is a knowledge engineering bottleneck that causes it to take too long to get the expert's knowledge into the bottle; the problem is more that there are important aspects of expertise that don't fit into the bottle. Dreyfus (1972; Dreyfus & Dreyfus, 1986) has covered at length the inherent limitations of conventional artificial intelligence. Developers are realizing how context-sensitive so many tasks and subtasks are. The job of modeling the entire world is required in order to build a flexible and responsive system for handling most important tasks, and so developers are learning how to restrict the domain and select fairly context-free tasks to increase the probability of success. Building an intelligent system to screen loan applicants is a good application because the initial screening should not require deep domain knowledge. Building an intelligent system to assess welfare eligibility of potential clients of a social service agency is also straightforward because the criteria are usually well established and the rule interpretations are available. But it would not be a good idea to build an intelligent system to help a new nurse diagnose infections in newborns because most of the diagnoses will depend on perceptual discriminations that the new nurse will not be able to make reliably for many months. In other words, if-then-else types of rules will be inadequate whenever the interpretation of the antecedent conditions require sophisticated human judgment.

Therefore, it is not feasible to build a rule base that covers many of the factors presented in Table 10.1. Intelligent systems can do an excellent job for capturing procedures, specific details, physical relationships, and declarative knowledge. They will do an uneven job of dealing with perceptual/cognitive skills, goals, and precedents; and they will do a poor job of managing perceptual/motor skills, interpersonal knowledge, and cultural knowledge.

One way to describe this problem is that the value of the expert is the *difference* between what he or she knows and what the less skilled people in the organization know. It is this difference, this knowledge delta between experts and others, that is important. An intelligent system project would have to spend most of its time on the foundation knowledge for this knowledge delta, but such knowledge is commonly available and hence of relatively low value. Most of the effort would be wasted on trivial procedures that are already well known by advanced beginners. Fortunately, there is a low-technology method—incident accounts—that efficiently codifies knowledge deltas.

Incident Accounts

Expertise is often packaged in the form of incident accounts—context-rich accounts of nonroutine incidents. These accounts illustrate a variety of forms of expertise and link these forms to the contextual features of the situation. They capture interpersonal knowledge, specific details, goals, and precedents, and they illustrate types of cultural knowledge. They define the types of perceptual/cognitive and perceptual/motor skills needed, and they usually allow an inference of the types of procedures, declarative knowledge, and physical relationships that are required. We have found that concept maps (Novak & Gowin, 1984) are a useful adjunct to incident accounts, because the concept maps provide an abstract structure to balance the concrete stories.

Accounts of nonroutine incidents, commonly referred to as *stories*, focus on the most advanced aspects of experience, not on the trivial procedures. In an organization, stories are records of lessons learned, analogues, and key decisions, stored in a form that is easy to call up when needed. They function like a voice of experience so that even when the people with expertise are not present the stories of their accomplishments are still available to remind the decision makers that "This is like the time when __ __ __."

If corporate memory is the sum of experiences and lessons learned by the people in an organization, it does not simply reside in the memory of these people, to be lost as they leave the organization. Their incident accounts serve as vicarious experience for the rest of the organization. Looking at it this way, the most experienced personnel in an organization are repositories of stories. They can size up a situation and remember another like it, remembering what was tried and how it worked out. They can leave the story behind when they are no longer working in the organization. In more traditional cultures, the role of storyteller can become formalized and ritualized. In modern organizations, storytelling is less formalized; there are no vice presidents in charge of corporate stories.

Schank (1990) has distinguished between several types of stories: official stories (the way the organization wants the incident described), paradigmatic stories (incidents showing how typical cases are handled), and exceptions (stories about nonroutine incidents). Each type of story serves different functions. As research using the CDM has shown, the stories about exceptions would appear to be the richest and most efficient communications about lessons learned. Another aspect of corporate memory is the history of major decisions. These specialized stories help to identify the way the organization works. How were financial crises handled? How was the company started? How were budgets set? What happened when key executives disagreed? How was dissension treated? These stories help to transmit cultural knowledge about the organization. In the next section, some examples will be presented to show applications of incident accounts.

Who would qualify as knowledge engineers? For the type of perceptual probing we have used in studying incident accounts, and for informal techniques of knowledge engineering, there would be little value to relying on personnel training to use LISP and Prolog. The types of probes we have used with the CDM have emphasized perceptual and cognitive cues, therefore suggesting the importance of a background in experimental psychology. Nevertheless, we have obtained good results from interviewers with a background in clinical psychology as well as anthropology. It remains to be seen how much knowledge engineers will need to be aware of issues in epistemology and basic cognitive theory. On the one hand, it seems obvious that if one is going to engineer knowledge, one should have a clear idea of what *knowledge* refers to. On the other hand, it is not obvious that theoretical questions are going to be resolved in a way that will facilitate knowledge engineering. It seems more critical that the interviewers be able to maintain an orientation of interest, curiosity, and respect regarding the expertise they are studying.

Applying the Engineered Knowledge

After expert knowledge has been codified it must be further engineered to help workers perform tasks. Three applications for engineered knowledge are: decision making, technology transfer, and training.

Decision Making

The goal of intelligent systems—including knowledge-based systems, case-based reasoning, and other related techniques—is to build systems that assist decision makers. The structure of the system is to represent and link the rules, frames, and cases so that all relevant knowledge is called upon to handle a given query. The system user merely has to answer a set of relevant questions, which describe the features of the decision task, and the system will generate a recommended action. For domains where intelligent system technology is appropriate (e.g., routine banking and insurance decisions), the goal of a decision-making system can be realized. In other domains, intelligent systems and knowledge-based systems will serve in more of an advisory role, recommending a course of action but leaving the decision to the user. (See Wiener & Nagel, 1988, for an overview.) Therefore, intelligent decision support systems can be used even where the more ambitious decision-making role is not accepted.

Technology Transfer

The introduction of new technology requires background knowledge for users, but such knowledge is rarely provided. The pace of technological change is far more rapid now than it was 40 years ago, which means that expertise becomes obsolete more quickly. When a technological change is introduced, there is likely to be little experience available to guide its use, and knowledge engineering can help to improve this situation.

Crandall and Calderwood (1989) used the CDM to study nurses in a Neonatal Intensive Care Unit of a hospital that is 1 of 14 sites in the country where a new surfactant drug is being studied. A surfactant drug assists respiration. Informal observations are that with surfactant and collateral technologies, babies born 24 to 28 weeks after conception are now being saved, whereas a few years ago they would have lived but a few hours. The use of these new techniques raises some problems, however. Neonatal Intensive Care Units are having to treat extremely small "microbabies," sometimes weighing less than 2 pounds—they even lack functional livers, kidneys, and immune systems!

Infection is always a problem in a Neonatal Intensive Care Unit, but for microbabies it is a much more severe problem because the spread is so fast and the babies are so vulnerable. It takes 24 to 48 hours to perform tests to detect infection, and in this time the infection can spread too far to be controlled. Crandall and Calderwood (1989) used the CDM to determine that the nurses had learned subtle perceptual cues alerting them to infection and other problems before these problems were identified by the standard tests. Moreover, some of these cues were not in the medical or nursing literature, and some of them were opposite to the cues found in full-term babies.

In this situation, the knowledge engineering was effective in obtaining crucial information in the form of stories supplemented with a list of important cues. Under ordinary procedures, if a surfactant proved effective it would be approved for general use in hospitals but without any special efforts to increase user awareness of the implication: the need to care for a newly created patient type, microbabies. The research of Crandall and Calderwood (1989) suggests that the knowledge engineering product, the stories, could be disseminated along with the drug itself.

For many new medical technologies, it can be valuable to apply knowledge engineering by compiling the experience of the nurses or other domain experts who have spent time building up expertise during the testing phase. In the case of drugs like surfactants, knowledge engineering may be helpful to ensure that some of the lives saved by the technology are not lost through inexperience in handling technological by-products.

The nurses who were interviewed tended to discount their capabilities, often arguing that "intuition" was what had helped them through difficult cases. When the interviewers probed for critical incidents where their experience had been important, a different picture emerged. For example, one nurse related an incident involving a neonate who was being cared for by another nurse. The neonate had rapidly turned dark blue, and its blood pressure had suddenly fallen. The nurse we were interviewing had noticed these symptoms and realized that it was a rare pneumopericardium problem (i.e., air had entered the membrane surrounding the heart, setting up a counter-pressure). The nurse in charge of the neonate misunderstood the problem, because the heart rate monitor showed a steady 80 beats a minute. The nurse relating the incident had seen a pneumopericardium episode in the past and insisted that the rest of the team was not treating the problem correctly. She pushed a syringe into the physician's hand so that he could release the trapped air around the heart. At that moment, an x-ray confirmed that she was correct, and the physician promptly carried out her directive and saved the baby. Later, they figured out that the mechanical heart monitor was only recording the electrical impulses sent to the heart, 80 per minute. In fact, the trapped air was preventing the heart from contracting at all!

Accounts such as this convey different types of domain knowledge—the importance of prior experience, the perceptual cues that were important, the need to carefully interpret data from monitoring apparatus, and permission to risk looking foolish in front of other nurses and physicians.

The initial incident accounts were generally less informative about perceptual cues than the enriched accounts after the interviewers had a chance to probe about perceptual and cognitive issues. Furthermore, the incident account data were found to be effective for training of nursing students about problems as well as the context surrounding those problems. In this way, knowledge engineering can support the introduction of new technology and can help reduce the resistance to accept such technology.

Training

Intelligent systems could support training if they let system users study the explanations to become familiar with the rule base and upgrade their skills to the point where they may not need the intelligent system.

A direct use of knowledge engineering is to provide training by letting personnel with low or medium skill levels get inside the head of the domain experts. Whether the delivery system is an intelligent system or a paper-based collection of incident accounts or training materials based on some other type of knowledge elicitation procedure, the users can go through a process of cognitive modeling. The knowledge engineering captures some of the thought processes and content knowledge of the experts. When this is presented to novices, or even to people showing adequate performance, it can be useful. The trainee can see how the expert is thinking, not just what the expert did, so there is a chance to model thinking processes (e.g., the cues that were noticed, the types of goals selected, reasons for rejecting certain options).

Stories are often a form of ritualized instruction, even informal stories. Consider parents who get together to watch their children at playgrounds or at little league games and describe unusual events, near disasters, or surprising opportunities. These stories are a form of vicarious learning, helping each listener to recognize symptoms of diseases or of fractured arms. Consider also pilots who gather around an officers' club bar at the end of a day and describe radar sets that suddenly malfunctioned and how they worked around the problem. The topic is different, but the form of instruction is largely the same.

The CDM has been used to identify training requirements and generate training materials in several applications, particularly in the area of software development and debugging skills (Thordsen et al., 1989; Weitzenfeld, Riedl Freeman, Klein, & Musa 1991). The CDM has also been used to capture expertise of skilled government program managers to accelerate the learning curve of new scientists and engineers (Crandall & Klein, 1989). In addition, in the example cited above, the CDM was used with nurses in a Neonatal Intensive Care Unit, and the resulting stories were evaluated with nursing students (Crandall & Calderwood, 1989). A list of critical cues for identifying sepsis were extracted from the stories. Student nurses were presented with the stories and then tested for recall of the sepsis cues. Recall was maintained at a high rate after several weeks.

Other knowledge-engineering methods are being used for training. Campbell (1989) described a coaching technique that is a variant of protocol analysis. IBM wanted to improve users' ability to learn new software applications. Instead of collecting protocol data as users worked through standard exercises, Campbell applied a different method. Tasks were presented, and the users were told they could ask for help at various points; these help requests were treated as data to find out where confusion was entering and what the nature of the confusion was. In other words, the natural coaching help was the basis of understanding what the users did not understand.

The use of knowledge engineering for training can be traced back to Bloom and Broder (1950), who obtained protocols from good and poor students who were attempting to solve multiple-choice problems. Bloom and Broder allowed the poor students to contrast their own protocols with those of the good students and to identify any important differences they could. This technique proved very effective in teaching the poor students how they could be approaching problems. Multiple-choice test scores for the poor students were significantly increased. Rather than treating each question as one for which the answer was remembered or not, the poor students learned that the good students did not quit when they didn't know the answer—instead they began to problem solve to improve their chance for a successful guess.

All of these knowledge-engineering methods show how to let trainees explore what was in the mind of the expert during a nonroutine task—what goals were selected and why certain goals were rejected, what options were chosen, what factors were noticed, and what cues were monitored. By using knowledge engineering to facilitate cognitive modeling, it should be possible to bring trainees up to speed more quickly.

Stories and Corporate Memory

There are several ways that organizations transmit knowledge: through formal training (courses and self-study programs); procedures (standard operating procedures, often detailed in manuals); and lessons learned. It is this last category, the lessons learned, that offers the greatest opportunity for knowledge engineering. The formal training and procedures are oriented to introductory training of new workers. The lessons learned can uniquely be used to upgrade the skills of workers with moderate experience. And lessons learned are best transmitted through stories.

Every time a worker faces an unfamiliar task, something is learned. If the worker is new the task may be familiar to everyone else, and the worker is the only one to do any learning. The best lessons come from tasks that no one has seen before—a unique type of customer complaint, a type of equipment malfunction that reveals a design flaw, or a computer-programming job that calls for some creative uses of files. Notice that these are concrete experiences in handling the problems faced by the organization. They are usually nonroutine incidents that serve as guidelines for future problems. Often the best lessons learned come from interesting failures that should not be repeated.

Corporate memory can be defined as the lessons learned from concrete on-the-job problem solving and decision making. When workers share their experiences, organizations can accelerate their learning curve. Often, the experiences will illustrate typical ways of reacting, but sometimes they will capture nonroutine events that show the exceptions to applying standard procedures. Organizations that can build up vicarious experience in this way are developing a true corporate memory. Even if the experiences aren't always shared, it is important that there be someone in the organization who can be consulted about what to do in a difficult or perplexing case. Organizations need people who have been around for years and can remember, "Oh yes, we tried that once, but it made the problem even worse."

The magazine *Consumer Reports* is an example of how this can work. Whenever I want to make a purchase of a product such as a microwave oven, I read through *Consumer Reports* and pick the model I want. But almost invariably I find that that model is no longer being sold. Yet it doesn't matter. The article has made me a more sophisticated buyer; it has alerted me to features I want as well as features to avoid. And it does this by telling me stories about what happened when a microwave oven was used in a certain way. These are often stories about nonroutine outcomes, and in reading them I become more expert myself.

Chi, Glaser, and Farr (1988) have shown that trainees like to use examples but do not always benefit from the examples if they don't have guidelines for their use, or if they lack experience to know how to select or modify the examples. This is a problem with using analogues and stories. Such materials may need to be supplemented with abstract principles to drive home the point of the story and to help the trainee appreciate what is relevant. That is why *Consumer Reports* frames its stories with the abstract dimensions of the product evaluations.

When will knowledge engineering be most useful? The greatest benefit from knowledge engineering will occur during organizational transitions and times of change. Under stable conditions, organizations do a fairly good job of preserving their corporate memory. They intend to do the same sort of work, maintain low turnover rates, and provide ample opportunities for training through apprenticeship. Lessons learned and stories can be directly shared between one generation of workers and another.

Corporate memory breaks down when conditions become unstable. If an organization starts growing rapidly, then there won't be the same chance to learn through apprenticeships. If a company downsizes, then it may lose experienced people who may not have an opportunity to pass on their lessons learned. If an organization has a turnover problem, then the normal routes for corporate memory transmission may be closed. If an organization acquires another company and loses that company's top management, then critical pieces of corporate memory may vanish.

When an organization changes its product line, or a product enters a new phase of its life cycle or is introduced to a new market, or an organization changes its internal production procedures, then the nature of jobs will be altered and information about intricacies of the new environment must be rapidly disseminated. It is such points of transition that create vulnerability to the loss of expertise.

If instability, transition, and turnover cause a deterioration of corporate memory, then we find the same symptoms of memory loss as in Alzheimer's victims. We see projects that have begun in good faith and that have been turned over to managers who don't understand what the project was intended to accomplish, just as an Alzheimer's victim who goes to a room and can't remember why. We see mistakes being repeated with no sign that the organization is taking steps to eliminate their causes, just as Alzheimer's patients have trouble changing their lifestyle to make things simpler. We don't see growth in sophistication and expertise but rather a reduction of skilled performance. We see more crises from problems that should have been anticipated and averted long before they grew out of control. We see increasing exhaustion from fighting a constant barrage of short-term crises.

Conclusions

Whether knowledge engineering is used for individual training or for organizational development and maintenance of a corporate memory, the issue is the same—how to elicit the different types of knowledge listed in Table 10.1 and make these available at some level so that it is possible to attain skill progression rather than an endless cycle of people coming up to speed, disappearing, and being followed by replacements coming up to speed. In order to continually increase the proportion of people achieving high degrees of job proficiency and mastery it will be necessary to foster an organizational culture in which expertise is valued and active steps are taken to preserve and build on prior experiences.

It has been asserted that knowledge engineering is a general discipline of acquiring knowledge from domain experts and making it available for a variety of functions, including decision making, technology transfer, and training.

The various methods of knowledge engineering have the potential for maintaining corporate memory and for preserving organizational expertise. In this chapter, I have tried to identify a general discipline, knowledge engineering. I have also presented some methods, starting with intelligent systems but including some alternatives. These alternatives should continue to expand and develop as we become more sophisticated in our abilities to capture and apply expertise.

Note

1. Knowledge engineering, as discussed in this chapter, is distinct from the concept of *cognitive engineering* (e.g., Mancini, Woods, & Hollnagel, 1987). Cognitive engineering is the attempt to design systems that are better adapted to the thought processes of users, whereas knowledge engineering is the attempt to capture the expertise of skilled performers.

Acknowledgments. This research was conducted under contracts MDA903-85-C-0099, MDA903-86-C-0170, and MDA903-89-C-0032 from the U.S. Army Research Institute for the Behavioral and Social Sciences, and we thank Dennis Leedom, Rex Michel, Stan Halpin, and Judith Orasanu for their support. The views, opinions, and/or findings contained in this paper, however, are those of the author and should not be construed as an official Department of the Army position, policy, or decision. I would also like to thank Beth Crandall, Paula John, and Buzz Reed for reviewing and criticizing earlier drafts of this chapter.

References

Anderson, D., & Ortiz, C. (1987). AALPS: A knowledge-based system for aircraft loading. *IEEE Expert*, *2*, 71–79.

Barletta R., & Mark, W. (1988). Explanation-based indexing of cases. *Proceedings Case-Based Reasoning Workshop* (pp. 50–60). San Mateo, CA: Morgan Kaufmann.

Bloom, B. S., & Broder, L. J. (1950). *Problem solving processes of college students*. Chicago: University of Chicago Press.

Burton, A. M., Shadbolt, N. R., Hedgecock, A. P., & Rugg, G. (1987). A formal evaluation of knowledge elicitation techniques for expert systems: Domain 1. *Proceedings of the First European Workshop on Knowledge Acquisition for Knowledge-Based Systems*, *D3*, 1–21.

Campbell, R. (1989, May). *Expertise in programming: The beginning of a developmental approach*. Paper presented at the Seventh Annual Adelphi University Conference on Applied Experimental Psychology, New York.

Ceci, S. J., & Liker, J. K. (1986). A day at the races: A study of IQ, expertise, and cognitive complexity. *Journal of Experimental Psychology, 115*, 255–266.

Charness, N. (1979). Components of skill in bridge. *Canadian Journal of Psychology, 33*, 1–16.

Chase, W. G., & Simon, H. A. (1973). The mind's eye in chess. In W. G. Chase (Ed.), *Visual information processing*. New York: Academic Press.

Chi, M. T. H., Glaser, R., & Farr, M. J. (Eds.). (1988). *The nature of expertise*. Hillsdale, NJ: Erlbaum.

Claparède, E. (1902). *L'association des idées*. Paris: Doin.

Crandall, B., & Calderwood, R. (1989). *Clinical assessment skills of experienced neonatal intensive care nurses* (Final report prepared for the National Center for Nursing, NIH, under Contract No. 1 R43 NR01911 01). Yellow Springs, OH: Klein Associates Inc.

Crandall, B., & Klein, G. A. (1989). *Organizational expertise inside the government*. (Final report under subcontract to MacAulay Brown, Inc., for F33615–88–C–1707). Yellow Springs, OH: Klein Associates Inc.

de Groot, A. D. (1965/1978). *Thought and choice in chess* (2nd ed.). New York: Mouton.

Dreyfus, H. L. (1972). *What computers can't do: A critique of artificial reason*. New York: Harper & Row.

Dreyfus, H. L., & Dreyfus, S. E. (1986). *Mind over machine: The power of human intuitive expertise in the era of the computer*. New York: Free Press.

Duncker, K. (1945). On problem solving. *Psychological Monographs*, *58*(5, Whole No. 270).

Flanagan, J. C. (1954). The critical incident technique. *Psychological Bulletin*, *51*, 327–358.

Hart, A. (1988). *Expert systems: An introduction for managers*. London: Kogan Page.

Hayes-Roth, F., Waterman, D. A., & Lenat, D. B. (1983). *Building expert systems*. Reading, MA: Addison-Wesley.

Heidegger, M. (1962). *Being and time*. New York: Harper & Row.

Hoffman, R. R. (1987 Summer). The problem of extracting the knowledge of experts from the perspective of experimental psychology. *AI Magazine*, *8*, 53–67.

Hoffman, R. R., & Deffenbacher, K. A. (in press). An ecological sortie into the relations of basic and applied science: Recent turf wars in human factors and applied cognitive psychology. *Ecological Psychology*.

Holland, J. (1986). Escaping brittleness: The possibilities of general-purpose learning algorithms applied to parallel rule-based systems.

In R. Michalsik, J. Carbonell, & T. Mitchell (Eds.), *Machine learning: An aritficial intelligence approach*. Los Altos, CA: Morgan Kaufmann.

Huang, D. (1989). A framework for the credit-apportionment process in rule-based systems. *IEEE Transactions on Systems, Man, and Cybernetics*, *19*(3), 489–498.

IEEE. (1989). Knowledge elicitation. [Special issue, May/June.] International Journal of Man-Machine Studies. (1989).

Klein, G. A., Calderwood, R., & Clinton-Cirocco, A. (1986). Rapid decision making on the fire ground. *Proceedings of the Human Factors Society 30th Annual Meeting*, *1*, 576–580. Dayton, OH: Human Factors Society.

Klein, G. A., Calderwood, R., & MacGregor, D. (1989). Critical decision method for eliciting knowledge. *IEEE Transactions on Systems, Man, and Cybernetics, Special Issue*, 462–472.

Kolodner, J. (Ed.). (1988). *Proceedings Case-Based Reasoning Workshop*. San Mateo, CA: Morgan Kaufmann.

Larkin, J., McDermott, J., Simon, D. P., & Simon, H. A. (1980). Expert and novice performance in solving physics problems. *Science*, 20(208), 135-148.

Lehner, P. E., & Adelman, L. (1989). Perspectives in knowledge engineering. *IEEE Transactions on Systems, Man, and Cybernetics*, *19*(3), 433–434.

Mancini, G., Woods, D. D., & Hollnagel, E. (Eds.). (1987). Cognitive engineering in dynamic worlds [Special issue]. *International Journal of Man-Machine Studies*, *27*. London: Academic Press.

Munsterberg, H. (1913). *The psychology of industrial efficiency*. Boston, MA: Houghton Mifflin.

Newell, A., & Simon, H. A. (1972). *Human problem solving*. Englewood Cliffs, NJ: Prentice-Hall.

Novak, J. D., & Gowin, D. B. (1984). *Learning how to learn*. New York: Cambridge University Press.

Polanyi, M. (1966). *The tacit dimension*. Garden City, NY: Doubleday.

Prawat, R. S. (1991). The value of ideas: The immersion approach to the development of thinking. *Educational Research*, *20*(2), 3–10.

Reed, F. D., & Klinger, D. W. (1991). A case-based reasoning system for performing manufacturing assessments: Phase II. (Prepared under Contract F33615-89-C-5702 to Wright Laboratory, Wright-Patterson AFB.) Yellow Springs, OH: Klein Associates Inc.

Schank, R. C. (1990). *Tell me a story: A new look at real and artificial memory*. New York: MacMillan.

Senjen, R. (1988). Knowledge acquisition by experiment: Developing test cases for an expert system. *AI Applications in Environmental Science*, *12*, 52–55.

Shanteau, J. (1984). Some unasked questions about the psychology of decision makers. In M. E. El-Hawary (Ed.), *Proceedings of the IEEE Conference on Systems, Man, and Cybernetics*. New York: IEEE.

Simon, H. A., & Barenfeld, M. (1969). Information-processing analysis of perceptual processes in problem solving. *Psychological Review*, *76*, 473–483.

Taynor, J., Crandall, B., & Wiggins, S. (1987). *The reliability of the critical decision method* (KATR-863(B)-87-07F; Prepared under contract MDA903–86–C–0170 for the U.S. Army Research Institute Field Unit, Leavenworth, KS). Yellow Springs, OH: Klein Associates Inc.

Thordsen, M. L., Henke, A., King, J. A., & Stottler, R. H. (1989). *A case-based reasoning system for conceptual design and software modelling*. (Prepared under contract F33657–88–C–2145 for USAF/AFSC, Aeronautical Systems Division/XRX, Wright Patterson AFB, OH.) Yellow Springs, OH: Klein Associates Inc.

Waterman, D. A. (1986). *A guide to expert systems*. Reading, MA: Addison-Wesley.

Weitzenfeld, J. S., Riedl, T. R., Freeman, J. T., Klein, G., & Musa, J. (1991). Knowledge elicitation for software engineering expertise. *Proceedings Fifth Software Engineering Institute Conference on Software Engineering Education*, Pittsburgh, PA, October.

Wiener, E. L., & Nagel, D. C. (Eds.). (1988). *Human factors in aviation*. San Diego, CA: Academic Press.

Part IV
Psychological Research on Expertise

11
On Being an Expert: A Cost-Benefit Analysis

Robert J. Sternberg and Peter A. Frensch

Everyone knows, or at least believes, that it is a good thing to be an expert. Experts within a given field are the ones who are respected, cited, and sought out. They can attain money as well as renown for their expertise, and some experts bask almost indefinitely in the fame and fortune their expertise brings them.

A large literature exists to show the benefits of expertise. Indeed, a recent volume (Chi, Glaser, & Farr, 1989) is devoted exclusively to the nature of expertise. Many studies have been conducted that attempt to show just what it is that experts do better than novices in a wide variety of domains, such as chess (Charness, 1981; Chase & Simon, 1973; de Groot, 1965); bridge (Charness, 1979; Engle & Bukstel, 1978); medicine (Elstein, Shulman, & Sprafka, 1978); typewriting (Gentner, 1989; Rabbitt, 1978); physics (Larkin, McDermott, Simon, & Simon, 1980); and mathematics (Schoenfeld & Herrmann, 1982; Siegler & Shrager, 1984). These studies represent only a handful of the many recent investigations of expertise in a wide variety of domains.

The modern studies that probably were most responsible for the burgeoning literature on expertise were the chess studies of deGroot (1965) and Chase and Simon (1973). They showed that expertise can be understood in large part in terms of access to enormous amounts of knowledge stored in long-term memory. When experts and novices were briefly presented with random positions of chess pieces on a chess board, the experts did no better than the novices in recalling the positions of the pieces. But when the positions made sense in terms of the game of chess, experts performed much better in recalling the positions. In other words, their superiority depended on (a) their being able to bring superior past knowledge to bear on the recall task and (b) stimuli that followed the rules.

In this chapter, we shall consider just what an expert is, looking first at cognitive competence and then at the attribution of cognitive competence, that is, what people believe an expert to be. We shall then consider the benefits and, finally, the costs of being an expert. Our account of expertise will draw in part on Sternberg's (1985a) triarchic theory of human intelligence and on our own recent research on expertise.

The Nature of Expertise

In order to understand expertise, it is necessary to consider two distinct and not always related aspects of it: competence, and the attribution of competence. We will consider each of these aspects in turn.

Expertise as Cognitive Competence

Experts can do "automatically" things that nonexperts can do only with great effort or not at all. In other words, what comes easily to an expert comes only with great difficulty or does not come at all to the novice. In explaining how

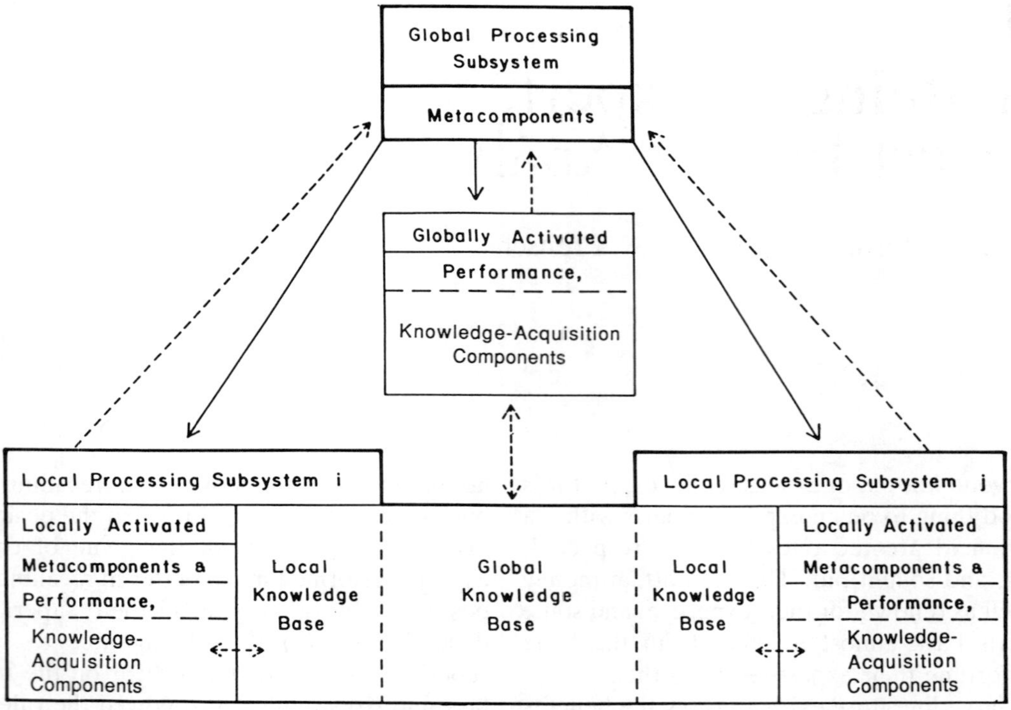

FIGURE 11.1. Sketch of the proposed control system for controlled and automatic processes.[1]

[1]The designations *i* and *j* represent two arbitrary local processing systems. Global processing is controlled; local processing is automatic. Solid arrows represent passage of control (activation). Broken arrows represent passage of feedback. Metacomponents—processes used in planning, monitoring, and evaluating—in the global processing system are able to instantiate themselves and other kinds of components in each of the local processing systems, and local processing systems are able to return control to the global processing system when their productions are unable to handle a problem. Performance components implement the instructions of the metacomponents, and knowledge acquisition components learn how to process information within a given domain in the first place. (From "The evolution of theories of intelligence," by Robert J. Sternberg, 1981, *Intelligence*, *5*, p. 224. Copyright 1981 by Ablex Publishing Corporation. Reprinted by permission of the publisher.)

automatization works, we shall refer to Figure 11.1, which is based on Sternberg's (1985a) triarchic theory of intelligence and the research Sternberg and others have done that is relevant to the theory.

Our model of automatization of information processing (based on Sternberg, 1985a) proposes that controlled information processing—the kind we do when we are thinking about what we do—is under the conscious direction of the individual and is hierarchical in nature, with executive processes (processes used to plan, monitor, and revise strategies of information processing) directing nonexecutive processes (processes used actually to carry out the strategies that the executive processes select). Automatic information processing is preconscious and not under the conscious direction of the individual, and is nonhierarchical in nature: There is no functional distinction within automatic information processing between executive and nonexecutive processing. Instead, task execution occurs through a production system—a set of condition-action sequences—where all kinds of processes function at a single level of analysis.

In processing information from new domains, and especially novel ones, the individual relies primarily on controlled, global processing. In such domains, the individual is not yet expert. A central executive directory activates nonexecutive processes and receives direct feedback from them. Information processing is of strictly limited capacity, and attention is focused on the task at hand. The total knowledge base stored in long-term memory is available for access by the processes used in the given tasks and situations.

In processing information from old domains or domains that are within the expert's realm of expertise, the individual relies primarily on automatic, local processing. A central executive initially activates a system consisting of locally applicable processes and a locally applicable knowledge base. Multiple local systems can operate in parallel. Performance in these systems is automatic and of almost unlimited capacity; attention is not focused on the task at hand. Only knowledge that has been transferred to the local knowledge base is available for access by the processes utilized in a given task or situation. A critical point is that activation is by executive processes in the global system to the local system as a whole. The executive processes can instantiate themselves as part of this local system; when used in this instantiation, they do not differ functionally from processes of any other kind.

In domains where one has little expertise, processing is largely focused in the global processing and knowledge system. As expertise develops, greater and greater portions of processing are transferred to (or packed into) a given local processing system. The advantage of using the local system is that activation is of the system as a whole, rather than of individual processes within the system, so that the amount of attention that needs to be devoted to use of the system is much less than it is under global control. Indeed, attention allocation for a whole local system is comparable to that for a single lower order process activated by the global system as part of the global system's functioning. The disadvantages of using the local system are that it is able to call on only a limited knowledge base—particularly, the knowledge base that has been packed into that local system—and that the local system is able to call on only those processes that have been packed into the local system. Experts are able to handle a wide variety of situations through the use of the local system, because they have packed tremendous amounts of information into the local system. Novices can hardly use local systems at all, because these systems have as yet acquired relatively few processes and relatively little knowledge.

Control passes to a local processing system when an executive process recognizes a given situation as one for which there is no potentially relevant local system. The local system is presumed to be of the nature of a production system, with a set of productions ready to act on the problem at hand. When a task is performed, an individual responds to specific conditions with specific actions. Each action is tied to a particular eliciting condition. The productions comprise functions that are executive in nature, as well as functions that are not. But all of these functions are integrated into a single, nonhierarchical system. Control is passed back to the global processing system when, during task performance, none of the productions in a system is able to satisfy a presented condition. When the bottom of the production list is reached but no given condition is satisfied, global processing is necessary to decide how to handle the new task or situation. As soon as this task or situation is successfully handled, it is possible to pack what is learned from the global processing of the new experience into a given local processing system, so that the next time such a situation is encountered there will be no need to exit from the local processing system.

According to this view, the extent to which one develops expertise in a given domain will actually depend on one's ability to pack new information, in a usable way, into local processing systems and to gain access to this information as needed. This view contrasts with that of certain theorists who seem to believe in the primacy of knowledge itself in intelligent functioning (e.g., Chi, Glaser, & Rees, 1982; Keil, 1984). We believe these investigators may place too much stress on the knowledge base itself and not enough on the ability of the individual to pack this knowledge into an effective and efficient local processing system. According to the present view, the reason that, of two people who play chess a great deal, one may become an expert and the other remain a duffer is that the first has been able to exploit information in a highly efficacious way, whereas the latter has not. The greater knowledge base of the expert is at least as much the result as the cause of the chess player's expertise, which derives from the expert's ability to organize effectively the information he or she has encountered in many, many hours of play.

Consider an example of how this view of expertise would apply to a concrete situation, such as becoming an expert automobile driver. When one first learns to drive, the task can seem very difficult, especially if one is learning to drive a vehicle with manual transmission. One needs virtually simultaneously to attend to many different aspects of a complex task, such as watching the road, manipulating the steering wheel, manipulating the clutch, pressing the gas pedal or the brake, watching the instrument panel, monitoring the traffic to the rear as well as that in front, and somehow putting together all of the facets of task performance that are needed in order to make the car run. Driving seems so difficult in the early stages because almost everything one is doing is conscious and controlled. Because conscious information processing is serial, one has to move back and forth rapidly among the different aspects of a task, making sure that none of them are unattended for long enough to allow a dangerous situation to arise. Novice drivers are especially susceptible to accidents largely because they find it difficult to attend to all of the aspects of driving, so that if one of these aspects quickly needs attention the novice may not get to that one until it is too late.

Eventually, driving becomes very easy to most people. They can attend to and coordinate all of the different aspects of the task, plus listen to the radio and even carry on a conversation in their automobile, all seemingly at the same time. Their greater ease in driving is due to their having automatized many aspects of the driving task so that they are not even consciously aware of what they are doing or of how they are coordinating the different aspects of the task. For the most part, they can rely on local processing systems that are executed automatically and in parallel. They therefore are free to devote their attention elsewhere, such as to a conversation they may be having with someone else in the car. But they can never devote their full conscious attention to that conversation or to any other distraction, because there is always the possibility that they will need to revert with split-second timing from their local processing systems to the global one. If they detect an abnormality—such as another car coming too close or an accident up ahead or construction in their path—they have to be ready to leave automatic information processing and to divert their full attention back to the driving task. When an emergency situation occurs, they become more like the novice again, having to devote their full attentional resources to the situation at hand. And anyone who has been with a driver confronting an anomalous situation knows that it is not the right time to engage the driver in a conversation that is irrelevant to the driving task.

Expertise as an Attribution of Cognitive Competence

Although expertise can be understood in part by the kind of cognitive account given above and by other, related accounts given by various researchers in the field (see, e.g., the essays in Chi, Glaser, & Farr, 1989), we believe that these accounts miss an important aspect of expertise, namely, its attributed aspect. In the real world—as opposed to many psychological laboratories—expertise is, in large part, an attribution. A person is an expert because she is regarded as such by others.

Consider, for example, the shaman, or medicine man, in what one might call a primitive tribe. Is the shaman genuinely a medical expert? People in the tribe would almost certainly say so. We would be more likely to hedge our bets. We might say that the shaman is not really a medical expert, or that if he is, it is only with respect to the views of the tribe and others like it. It is unlikely that any of us would go to the shaman if we needed emergency medical care, because the shaman is not an expert by our own standards of medical practice. Yet, the shaman has automatized a large number of procedures for handling medical emergencies and is hence in possession of the cognitive competence that the tribe believes necessary for medical practice. Moreover, the tribe attributes expertise to the shaman. Why, then, won't we acknowledge the shaman as an expert?

We are unwilling to attribute expertise in medicine to the shaman because the knowledge and automatized routines available to the shaman are not those of contemporary accepted medical practice. Thus, merely having knowledge and automatized routines does not qualify an individual as an expert. The knowledge and routines have to be ones that we label as relevant—as bestowing expertise. The attributional nature of this bestowal is shown by the fact that there are those who have medical systems and beliefs different from our own. For example, Chinese medicine is quite different from ours, and whereas many of us would view, say, acupuncture as not very relevant to medical expertise, there are many Chinese who would view our form of medicine as not very relevant. Besides, it will no doubt be the case in the future that doctors will look back and marvel at the primitivism of our attempts to treat medical ailments. Indeed, one need only watch "Star Trek," the television science-fiction series, to see Dr. Leonard McCoy, an inhabitant of a future time, watching in horror the "primitive" medicine that we now practice.

The case of medicine shows the extent to which expertise is, in part, an attribution. The experts in one place or one time are not necessarily considered to be experts in another place or another time, because the knowledge and automatic routines considered to constitute expertise can differ from one time and place to another. The attributional component of expertise can be illustrated even more dramatically in the case of people in another profession, namely, stockbrokers, people to whom some of us regularly turn for advice on investments. We turn to stockbrokers and other investment counselors because we attribute to them an expertise that we believe that we do not have. Most of us cannot possibly spend the time that would be required to follow all of the companies in which one might buy stock or to know all of the various attributes of the companies that would be relevant for making a decision regarding a stock purchase or sale. We leave it to the brokers, and the research staff of the firms for which they work, to use their expertise to help us make wise and informed decisions. This is not to say, of course, that every stockbroker is an expert. It may take some shopping around to find an expert, just as it would to find an expert in medicine or law, for that matter.

A problem arises because it is not clear that any stockbroker at all can do significantly better than chance in predicting stock trends over an extended period of time. Economic analysts such as Malkiel (1985) have investigated the market predictions of various classes of market analysts, such as fundamental analysts, who look at attributes of a company in making decisions about whether to buy or sell stock, and technical analysts, who look at past trends in the value of that stock. Malkiel and others have concluded that the model that best characterizes the stock market is a random walk: In other words, one cannot do significantly better than chance in predicting market trends. Indeed, even a slight edge in prediction over the long run would make a very wealthy individual out of any consistently correct analyst or her clients. Editors at *Forbes* magazine actually threw darts at the *New York Times* stock page and constructed a model portfolio built on the essentially random selection that the darts gave. The random portfolio did substantially better than did most mutual-fund managers. Indeed, it has long been known that very few mutual-fund managers are able consistently to outperform the Standard and Poor's 500 Index—an index of performance of representative companies with large capitalizations—over any extended length of time. Thus, one might argue that expertise in stock prediction is sometimes almost exclusively attributional—that there is no base of information that allows anyone to predict particularly well over an extended period of time. Malkiel (1985), incidentally, admits that there are exceptions, but they appear to be few and far between.

Relativity is not limited to expertise: There is an attributed side to intelligence and other ability constructs as well (see Sternberg, 1985b; Sternberg, Conway, Ketron, & Bernstein, 1981). What people in one culture see as intelligent, people in another culture may see as unintelligent. Such differences can even exist within varying walks of life within a single culture. For example, the emphasis on quickness that pervades our notions of intelligence is not shared by cultures in most of the world (see Sternberg, 1985a). Indeed, in Hispanic cultures, there is very little premium on quickness. In much of the world, the intelligent person is believed to be reflective, deliberative, and measured in making judgments, rather than quick.

The point of this section is not to undermine the study of cognitive competence as a basis of expertise, but rather to point out that a full account of expertise needs to look at its social-attributional as well as its cognitive side. Expertise, like intelligence, is in part contextually determined. Let us now consider some of the benefits of expertise.

Benefits of Expertise

Expertise has several benefits. Let us consider what some of these are, both from the standpoint of cognitive competence and from that of the attribution of cognitive competence.

Cognitive Competence

An expert is competent in handling familiar tasks within the domain of her expertise. The expert is at an advantage in the domain of her expertise because her ability to stay for longer amounts of time in the better developed local processing system relevant to her expertise enables her to free global processing resources for dealing with new situations. The novice is overwhelmed with new information and must engage global resources so frequently that much of the new information that is encountered is quickly lost.

An expert is also proficient at learning new tasks because global processing resources are more readily available for the intricacies of the task or situation confronted. In essence, a loop is set up whereby packing more information and processes into the local system enables one to automate more processing and thus to have global resources more available for what is new in a given task or situation. The expert is also able to perform more distinct kinds of tasks in parallel, because whereas the global processing system is conscious and serial in its processing, multiple local processing systems can operate in parallel. For a novice, for example, driving a car consumes almost all of his available global resources. For an expert, driving a car consumes local resources and leaves central resources available for other tasks, unless a new situation (such as a road block) is confronted that is unfamiliar and requires redirection of control to one's global resources.

The advantages of expertise can be readily seen in a typical interaction between a faculty member and a graduate student, say, in psychology or some other science. Because the faculty member has automatized many of the facets of research that need to be discussed, such as experimental design and analysis, the faculty member is able to free global processing resources for dealing with the basic ideas being discussed and for dealing with larger issues, in general. The graduate student is more apt to get bogged down in local details such as counterbalancing groups or designing an experiment so that the data will be analyzable, with the result that the student may have relatively few resources left over to deal with what is conceptually interesting at a higher level. Thus, the expert not only knows more, but also potentially can learn more and see more into a problem because more global processing resources are free for dealing with novelty.

The Attribution of Cognitive Competence

Attributions also play a role in the benefits of expertise, such as those attributions exhibited in the so-called Matthew effect (Merton, 1968). The basic idea is that to those who already have a lot will come more and that to those who have relatively little will come less. Indeed, they may find themselves losing what little they have. In science, for example, individuals who have already gotten grants generally find it easier to get new grants because they have established a track record. Individuals at more prestigious institutions find it easier to get grants and to get articles accepted because it is assumed that they will be able to do good research. Thus, in a certain sense, the attribution of expertise feeds on itself. People who are viewed as experts are given more resources so that they will be better able to enhance the expertise they already have. Scientists who are at relatively smaller institutions and who are viewed as less expert (often without regard to their true credentials), may find themselves unable to compete successfully for the resources that would enable them to develop both their expertise and the attribution of it. In effect, they are caught in a rut that society does not let them escape. Of course, this principle applies beyond science. In virtually any domain, the attribution of expertise puts one in a position to acquire more of the resources that will enable one to accrue yet further expertise, or the labeling thereof.

There are other benefits to being viewed as an expert, among them, greater respect from others, greater potential power and influence within one's field of expertise, the ability to make more money, and so on. But expertise can also have costs, as considered below.

The Costs of Expertise

It may seem odd to refer to any expertise as involving costs as well as benefits, but we are convinced that certain costs do indeed inhere in being an expert. Again, one set of costs stems from cognitive considerations, the other from attributional ones.

Cognitive Competence: Background Literature

Consider again the expert who has developed a number of local processing routines for handling a large number and variety of situations she encounters in the course of her regular interactions with tasks in her domain of expertise. Suppose that, somehow, the nature of the domain changes. In science, for example, paradigms change, as do the scientific methodologies used to do research in various paradigms. In business, technology changes, as do people's tastes. The very same products or lines of products that may be appealing to people at one time may be unappealing or even seem distasteful at another time. The cachet of tobacco, for example, is gone and has been replaced by an aversive reaction on the part of many members of society. Models who at one time appear attractive, at another time appear too fat or too thin. Today's technologies are tomorrow's dinosaurs. A frequent problem for experts is how to remain an expert in the face of a rapidly changing world; and the problem at a cognitive level is that it is exceedingly difficult to break up and reorganize an automatized local processing system to which one in all likelihood no longer even has conscious access. In other words, the expert's gain in automatized information processing may be at the expense of flexibility. Paradoxically, therefore, although the expert may have more processing resources available to deal with novelty than does the novice, it may be the novice who is better able to deal with this novelty, because of lesser entrenchment in terms of the novice's approach to new kinds of problems.

Just as the modern literature on the advantages of expertise can probably be traced back to the work of deGroot (1965) and Chase and Simon (1973) on chess, the literature on the costs of expertise can probably most easily be traced back to work done by Gestalt psychologists on the effect of Einstellung, or "set" (e.g., Luchins, 1942; Luchins & Luchins, 1959). In his original study, Luchins (1942) presented people with so-called water-jug problems. In solving these problems, one has to figure out how to use jugs of particular capacity to obtain some specified amount of water. The finding of particular interest in Luchins's research was that if people were presented a series of problems, all of which could be solved by the same rule, they persisted in using this rule for subsequent problems, even if there was a simpler rule that could be used to solve the subsequent problems. Even more impressively, the effect of Einstellung varied with the number of times the people had used the same rule. Thus, people who had solved five problems by using the same rule were more likely to persist in using the rule when they could have used a simpler one than were people who had solved fewer problems using a particular solution method. In other words, the cost of expertise was reduced flexibility.

Other, more recent studies have shown similar outcomes. For example, Lewis, McAllister, and Adams (1951) showed that the amount of practice on a motor task was negatively related to the ability to learn a new, incompatible motor task. Shiffrin and Schneider (1977, Experiment 1) reported that people trained for several thousand trials to detect visual targets among distractors in a consistent-mapping condition found themselves in trouble when both the target and the distractor sets were reversed. Detection accuracy fell far below that found at the very beginning of training under such conditions. In fact, people needed about 2,400 trials of the reversal training to reach the level of performance equal to that originally obtained after 1,500 trials of learning.

Some recent research on inflexibility of thinking that is associated with automatic processing has been concerned with the "lack of control" criterion of automaticity, specifically with the ability or inability of an individual to stop ongoing automatic processing (e.g., Jonides, 1981; Logan, 1985; Logan & Cowan, 1984). The major evidence for the hypothesis that automatic processing is uncontrolled and unstoppable comes from experimental studies of the Stroop effect (Stroop, 1935). In the Stroop effect, unattended dimensions of a stimulus interfere with the processing of attended dimensions, presumably because the processing of the unattended dimensions is somehow initiated by the presence of a certain stimulus situation and, once started, cannot be stopped by conscious control. When the automatic (unattended) and the controlled (attended) processes are antagonistic, as in most experimental treatments of the Stroop effect, they interfere with each other, thus decreasing the reaction time and the quality of performance in the controlled primary task (e.g., Dyer, 1973; Kahneman & Chajczyk, 1983; MacLeod & Dunbar, 1988).

Recent Research From our Laboratory

We have conducted some research relevant to the notion that a cost of expertise may be reduced flexibility in information processing. Consider a recent series of three experiments on real-world experts (Frensch & Sternberg, 1989), in which we related knowledge base, structures, and automatization of information processing to flexibility in problem solving.

Experiment 1

We tested the general hypotheses that (a) the structure of a knowledge base and (b) its degree of automatization both affect the flexibility with which a problem solver can respond to novel task demands. We expected that problem solvers who had automatized their solution strategies to a great extent would be more affected by changes in task demands than would problem solvers who responded to problems in a less automatized manner. Thus, experts in any given domain would generally be more vulnerable to task-demand modifications than nonexperts would be. Furthermore, we assumed that the effects of changes in task demands were larger for those demands that were incompatible with the structure of the knowledge base than for demands that were compatible. Therefore, experts would be more affected when new task demands called for deep, abstract principles to be changed than when surface features were to be changed. For nonexperts, the reverse was expected. That is, they would be more affected by surface changes than by deep, abstract changes.

The specific domain of knowledge chosen to test these ideas was the game of bridge. There have been only a few studies on skill differences in bridge (e.g., Charness, 1979; Engle & Bukstel, 1978). These few studies generally obtained results that are compatible with the findings reported in other domains that were described above. It appears that master bridge players encode abstract distributional features of bridge hands that automatically evoke strategies containing long lines of play, like "create an end play." Nonexperts, on the other hand, seem to encode hands primarily in terms of surface features (honor-cards, for instance), which are associated with smaller lines of play, like "take a trick."

In the first experiment, bridge players of varying levels of skill played 12 simulated bridge games on a computer. Half of the games were played under normal conditions. In the other half, players were instructed to play slightly different versions of bridge. Version 1 introduced new nonsense names for honor-cards and suits; Version 2 rearranged the order of honor-cards and suits; and Version 3, the lead-rule change, modified the rule determining who began each play. Instead of the player who won the last trick, which is the common rule in bridge, the player with the lowest card in the last trick led into the next trick. The different versions were intended to tap different levels of subjects' information processing. Versions 1 and 2 were considered surface modifications; Version 3 was expected to exert its effect on a deeper, more abstract, and more strategic level.

The main results of this study can be summarized as follows: First, experts played faster and won more games than did nonexperts when regular bridge rules applied, which indicates that experts had automatized their problem-solving strategies to a greater extent than had nonexperts. Second, changes in task demands generally affected people who had highly automatized their problem-solving strategies more than those who relied on less automatized strategies. Experts were generally more affected by the rule modifications than were nonexperts. Third, changes in task demands that were incompatible with the structure of the knowledge representation had more pronounced effects on people's problem-solving processes than had changes that were compatible. Experts were more affected by the deep-structural change than by the two surface changes, whereas nonexperts were more affected by the surface changes. And finally, both expertise groups seemed to be able to adapt their existing knowledge bases quite rapidly. In all conditions, with the exception of the deep-structural-rule-change condition for experts, people did reach approximately their original performance levels after only a few games.

Generally, Experiment 1 provides support for the hypothesis that the degree of automatization and the structure of the knowledge base affect the inflexibility of a cognitive system. One might object, however, that it is not sufficient to demonstrate that groups that organize their knowledge bases differently and rely on different degrees of automatization also differ in how inflexible they are. Rather, one must demonstrate that subject groups that do *not* differ on the relevant variables are also equally flexible (or inflexible). Experiment 2 was designed to test the latter prediction.

Experiment 2

Thirty-four bridge players of different levels of skill were instructed to generate opening bids to given bridge hands as quickly as possible. As Charness (1979) pointed out, even novice bridge players are able to generate a reasonable opening bid. Furthermore, the choice of opening bids is based primarily on the distribution of honor-cards and on the number of total cards per suit in a given hand. Although there are a large number of different bidding systems to determine a bid, these systems are based on the same properties of hands and, generally, do not arrive at different opening bids. The selection of an opening bid is a rather complex computational process that, for most players, does not involve any strategic considerations. With regard to opening bids, more and less experienced bridge players are not expected to differ in their classification of hands into bid categories. Thus, expert and nonexpert bridge players do not use differently structured knowledge bases. In addition, because of the absence of strategic considerations, the computational process of selecting an opening bid can be expected to be automatized rather quickly.

Experiment 2 tested the following three hypotheses: First, experts and nonexperts will select identical opening bids in the same amount of time when the original bridge rules are in effect. Second, experts will generally not be more affected in terms of response time to bid by the rule changes regarding opening bids than will nonexperts. And third, response times of experts and nonexperts will not be differentially affected by the rule modifications. Because the selection of opening bids is based primarily on surface features of hands, we expected both expertise groups to be more affected by the surface changes than by the deep strategic change.

All of our hypotheses were confirmed. Experiment 2 demonstrated that groups that do *not* differ in degree of automatization and structure of knowledge base also do not differ in how flexibly they can use their existing knowledge, measured in terms of response times. The results of Experiment 2 are consistent with our claim that the degree of automatization and the structure of a knowledge base affect the flexibility of effective problem solving.

In Experiments 1 and 2, we were primarily concerned with the differential impact of surface and deep rule modifications on expert and nonexpert bridge players and with the mechanisms of interference in the lead-change condition. We did not discuss the mechanisms of interference for the two surface rule modifications, however. Whereas people's problems with the lead change can only be explained in terms of conceptual levels of processing (e.g., restructuring), people's difficulties with the name and rank-order changes might be attributed to perceptual phenomena, either in addition to interference on the conceptual level or alone. It might be argued, for instance, that the surface modifications affected all people primarily at the level of encoding and only to a minor degree at a conceptual level. Alternatively, it might be argued that experts were most affected on a perceptual level, whereas nonexperts were most affected on a conceptual level or vice versa. Experiment 3 was designed to test the effects of the name-change and rank-order-change conditions at the level of encoding.

Experiment 3

Players of differing skill levels were shown slides of bridge hands (13 cards displayed in a fanned position) in exactly the same way they usually perceived bridge hands. Slides were visible for only 5 seconds. After slide-offset, players were asked to write down as many of the cards as they could remember. They had the option of watching the same slide for as many times as they wanted to until they had remembered and written down all 13 cards of the seen hand. Half of the bridge hands shown were structured in the same way players usually structure their hands; for the other half, features of the visual display were changed so as to correspond to the name-change and rank-order-change conditions used in Experiments 1 and 2. In the name-change condition, the letters on honor-cards were changed to new ones. In the rank-order condition, the familiar rank orders of suits and high cards were changed; consequently, hand displays mirrored the new rank orders. Reaction times, number of cards written down per trial, and number of cards correctly identified per trial were recorded.

The three major results of Experiment 3 can be summarized as follows: First, replicating Charness's (1979) earlier findings for the domain of bridge, we found that experts encoded meaningful information faster and with fewer errors than did nonexperts. Second, the two types of display changes did not differ in their effects on people's encoding abilities. And finally, experts and nonexperts were not differentially affected by the two types of changes of the visual display.

The fact that experts and nonexperts were not differentially affected by the two types of changes has important implications for the discussion of Experiments 1 and 2, suggesting that the effects of the surface rule modifications obtained in these studies might, for both expertise groups, be partly due to the disruption of normally occurring encoding processes; however, the findings of Experiment 3 do not support the argument that the surface changes affected experts primarily on a perceptual level and nonexperts on a conceptual level, or vice versa.

Summary

In this section we have presented empirical evidence for our claim that two major characteristics of expertise—namely, (a) the structure of the knowledge base and (b) the automatization of knowledge—can lead to inflexible problem solving. The overall results of the experiments can be summarized as follows. First, problem solvers who have automatized their solution strategies are less flexible than ones who have not. Experts were generally more affected than nonexperts when task demands changed. Second, problem solvers are less flexible when dealing with task modifications that are incompatible with the structure of their knowledge representations than when dealing with changes that are compatible. We found that experts were more affected by deep strategic rule changes, whereas nonexperts were more affected by surface changes when the two groups structured their knowledge representations differently. When both groups operated on the same knowledge base, they were not differentially affected. Third, all people were able to adjust to new task demands. In most conditions, they returned to their original levels of performance quite quickly. And finally, the effects of the two surface-rule manipulations might be partly due to perceptual, rather than conceptual, phenomena.

These findings can be discussed, for instance, with regard to recent theoretical developments in the field of intelligence. It has long been a traditional belief that intelligence can be characterized, in part, as the ability to acquire new information. Several theorists have proposed that the ability of people to deal with novel and particularly partially novel information is an integral aspect of intelligence (e.g., Berg & Sternberg, 1985; Cattell, 1971; Horn, 1968; Raaheim, 1984; Snow, 1981; Sternberg, 1985). The present findings suggest that this ability can reflect individual differences in domain-specific strategies and categorization of knowledge.

Our view on expertise also lends itself very easily to explaining a phenomenon that has been reported frequently in the developmental literature: the decline of fluid intelligence with age (Horn, 1968). Our data suggest that this decline might be due, at least in part, to the automatization of responses to highly recognizable and familiar situations and to the organization of existing knowledge bases. In other words, by being more "expert" at performing a range of familiar tasks, older individuals might experience an impairment in performance when a novel task competes with what is already known.

The Attribution of Cognitive Competence

Up to now, we have dealt with the cognitive costs of expertise. But, perhaps oddly, there are attributional costs as well. One such cost is that after an individual has been labeled as an expert in a given domain he may become pegged: People may be unable or unwilling to view him as expert in some other domain. A stunning but sad example of this is the case of George Reeves, who successfully played Superman in the television series by that name. After the Superman series ended, Reeves found himself unable to obtain employment in other roles, because directors believed that he was so typecast as the Superman figure. Despondent and out of work, Reeves eventually committed suicide. Other actors, of course, have found themselves repeatedly cast in certain kinds of roles to the exclusion of other roles that they may in fact desire more.

The phenomenon is not limited to acting. We can see the same in the scientific world. It is often assumed that someone who is successful as an experimental physicist will not make it as a successful theoretical physicist, and vice versa. Within psychology, we often seem to assume in hiring that people who have done applied work will no longer be suitable for positions as academics, because somehow the applied work will have spoiled them for an academic setting. We sometimes find people within one area of psychology suspicious of those who started off in another area, as though their expertise in the first area will hinder rather than help them in the new area in which they choose to do research. Examples abound, but the point is that the attribution of expertise in one domain may increase reluctance of people to make the same attribution in another domain.

The attribution problem can become an internal one as well as an external one. When people succeed in a given domain of endeavor, especially if the success is a creative one, they are reinforced in multiple ways for their success. In at least some cases, they will have worked hard for this success and will be eager to enjoy the fruits that it has borne. The problem is that the reinforcements they receive may render them reluctant to strive for success in a new project or a new area of endeavor. They may find themselves afraid to try for success in another domain, because they reason that now that they are near the top, the only way left for them is likely to be down. Their fear, then, is that by trying something new, they will expose themselves as has-beens. It is safer for them to rest on their laurels than to show themselves up as foolish or incompetent in a new project or domain.

Their fear is not an irrational one. By regression effects alone, it is reasonable to suppose that an unusual success in one domain is likely to be followed by somewhat lesser success in another domain, or even the same domain. For example, many of us are familiar with the now well-known "rookie-of-the-year" phenomenon, whereby the rookie of the year in baseball virtually never does as well his second year as he did his first. Such a phenomenon is consistent with the concept of statistical regression, and it applies in any domain. One need not take a statistics course to have an intuitive feeling for what will happen if one extends one's reach into new domains. This is not to say that even greater success is not possible. It is merely to say that some people will be afraid to keep trying. Thus, the attribution of expertise to a person in a given domain may hinder that person from developing expertise in a new domain or even from further developing expertise in the first domain.

Conclusion

We have argued that expertise has two aspects: a cognitive one and an attributional one. The two may, but need not, correspond. In fact, someone with very considerable skills may not be recognized as an expert, whereas someone else with much lesser skills may in fact bear this designation. We have proposed a cognitive model that includes a mechanism whereby expertise is acquired, and we have suggested that the acquisition of expertise entails costs as well as benefits. In order to develop our understanding of expertise more fully, we need to look at its cognitive and attributional sides, and also at its costs and benefits.

Acknowledgments. Preparation of this article was supported by Contract MDA90385K0305 from the Army Research Institute.

References

Berg, C. A., & Sternberg, R. J. (1985). Response to novelty: Continuity versus discontinuity in the developmental course of intelligence. In H. W. Reese & L. P. Lipsitt (Eds.), *Advances in child development and behavior* (Vol. 19, pp. 2–47). New York: Academic Press.

Cattell, R. B. (1971). *Abilities: Their structure, growth, and action*. Boston, MA: Houghton Mifflin.

Charness, N. (1979). Components of skill in bridge. *Canadian Journal of Psychology*, *33*, 1–16.

Charness, N. (1981). Search in chess: Age and skill differences. *Journal of Experimental Psychology: Human Perception and Performance*, *7*, 467–476.

Chase, W. G., & Simon, H. A. (1973). Perception in chess. *Cognitive Psychology*, *4*, 55–81.

Chi, M. T. H., Glaser, R., & Farr, M. J. (Eds.). (1989). *The nature of expertise*. Hillsdale, NJ: Erlbaum.

Chi, M. T. H., Glaser, R., & Rees, E. (1982). Expertise in problem solving. In R. J. Sternberg

(Ed.), *Advances in the psychology of human intelligence* (Vol. 1, pp. 17–76). Hillsdale, NJ: Erlbaum.

deGroot, A. D. (1965). *Thought and choice in chess*. Paris: Mouton.

Dyer, F. N. (1973). The Stroop phenomenon and its use in the study of perceptual, cognitive, and response processes. *Memory and Cognition, 1*, 106–120.

Elstein, A. S., Shulman, L. S., & Sprafka, S. A. (1978). *Medical problem solving*. Cambridge, MA: Harvard University Press.

Engle, R. W., & Bukstel, L. (1978). Memory processes among bridge players of differing expertise. *American Journal of Psychology*, *91*, 673–689.

Frensch, P. A., & Sternberg, R. J. (1989). *Expertise and flexibility: The costs of expertise*. Unpublished manuscript.

Gentner, D. R. (1989). Expertise in typewriting. In M. T. H. Chi, R. Glaser, & M. J. Farr (Eds.), *The nature of expertise*. Hillsdale, NJ: Erlbaum.

Horn, J. L. (1968). Organization of abilities and the development of intelligence. *Psychological Review*, *75*, 242–259.

Jonides, J. (1981). Voluntary versus automatic control over the mind's eye movement. In J. Long & A. D. Baddeley (Eds.), *Attention and performance* (Vol. 9, pp. 187–203). Hillsdale, NJ: Erlbaum.

Kahneman, D., & Chajczyk, D. (1983). Tests of the automaticity of reading: Dilution of Stroop effects by color-irrelevant stimuli. *Journal of Experimental Psychology: Human Perception and Performance*, *9*, 497–509.

Keil, F. C. (1984). Transition mechanisms in cognitive development and the structure of knowledge. In R. J. Sternberg (Ed.), *Mechanisms of cognitive development* (pp. 81–99). San Francisco, CA: Freeman.

Larkin, J. H., McDermott, J., Simon, D. P., & Simon H. A. (1980). Expert and novice performance in solving physics problems. *Science*, *208*, 1335–1342.

Lewis, D., McAllister, D. E., & Adams, J. A. (1951). Facilitation and interference in performance on the modified Washburn apparatus: I. The effects of varying the amount of original learning. *Journal of Experimental Psychology*, *41*, 247–260.

Logan, G. D. (1985). Skill and automaticity: Relations, implications, and future directions. *Canadian Journal of Psychology*, *39*, 367–386.

Logan, G. D., & Cowan, W. B. (1984). On the ability to inhibit thought and action: A theory of an act of control. *Psychological Review*, *91*, 295–327.

Luchins, A. S. (1942). Mechanization in problem solving. *Psychological Monographs*, *54* (6, Whole No. 248).

Luchins, A. S., & Luchins, E. H. (1959). *Rigidity of behavior: A variational approach to the effects of Einstellung*. Eugene, OR: University of Oregon Books.

MacLeod, C. M., & Dunbar, K. (1988). Training and Stroop-like interference: Evidence for a continuum of automaticity. *Journal of Experimental Psychology: Learning, Memory, and Cognition*, *14*, 126–135.

Malkiel, B. G. (1985). *A random walk down Wall Street* (4th ed.). New York: Norton.

Merton, R. (1968). The Matthew effect in science. *Science*, *159*, 56–63.

Raaheim, K. (1984). *Why intelligence is not enough*. Bergen, Norway: Sigma Forlag.

Rabbitt, P. M. A. (1978). Detection of errors by skilled typists. *Ergonomics*, *21*, 945–958.

Schoenfeld, A. H., & Herrmann, D. J. (1982). Problem perception and knowledge structure in expert and novice mathematical problem solvers. *Journal of Experimental Psychology: Learning, Memory, and Cognition*, *8*, 484–494.

Shiffrin, R. M., & Schneider, W. (1977). Controlled and automatic human information processing: II. Perceptual learning, automatic attending, and a general theory. *Psychological Review*, *84*, 127–190.

Siegler, R. S., & Shrager, J. (1984). A model of strategy choice. In C. Sophian (Ed.), *Origins of cognitive skills* (pp. 229–293). Hillsdale, NJ: Erlbaum.

Snow, R. E. (1981). Toward a theory of aptitude for learning: I. Fluid and crystallized abilities and their correlates. In M. Friedman, J. P. Das, & N. O'Conner (Eds.), *Intelligence and learning* (pp. 345–362). New York: Plenum.

Sternberg, R. J. (1985a). *Beyond IQ: A triarchic theory of human intelligence*. New York: Cambridge University Press.

Sternberg, R. J. (1985b). Implicit theories of intelligence, creativity, and wisdom. *Journal of Personality and Social Psychology*, *49*, 607–627.

Sternberg, R. J., Conway, B. E., Ketron, J. L., & Bernstein, M. (1981). People's conceptions of intelligence. *Journal of Personality and Social Psychology*, *41*, 37–55.

Stroop, J. R. (1935). Studies of interference in serial verbal reactions. *Journal of Experimental Psychology*, *18*, 643–662.

12
Mnemonics and Expert Knowledge: Mental Cuing

Francis S. Bellezza

I am particularly interested in how special knowledge facilitates (or sometimes interferes with) the learning and recall of new information. This is reflected in my research on mnemonic devices, which are special techniques used to memorize information (Bellezza, 1981, 1983). Psychologists have long known that the particular knowledge already in memory somehow mediates the learning of new information, but just how this mediation process takes place is not yet fully understood. The research I discuss here addresses the question of how expert knowledge representations in memory influence learning and recall. The study of the relation between expert knowledge and learning, however, is not as common in psychological research as the study of the role of expert knowledge in decision making and problem solving. But as our understanding of expertise increases, the study of learning by experts will, I believe, contribute to the understanding of how experts use their knowledge.

Declarative Knowledge

In my research, the knowledge extracted from experts is declarative knowledge, that is, knowledge that is verbal and factual rather than procedural (Anderson, 1976; Gordon, this volume). This distinction goes back to Ryle (1949) and is the distinction between knowing *what* (declarative knowledge) and knowing *how* (procedural knowledge). The expert knowledge typically studied by knowledge engineers is some mixture of these two types. Both types of knowledge are important. Experts solve problems by using knowledge of procedures, by using factual knowledge, and by making inferences from their factual knowledge. They do not simply remember facts. Knowledge engineers typically focus on declarative knowledge in the early stages of the knowledge-elicitation process and use a variety of techniques to construct representations of the declarative knowledge of their experts (Hart, 1986; Hoffman, 1987, 1989; Olson & Rueter, 1987). However, I have limited my investigations to declarative knowledge only. I ignore, at least temporarily, the problem of how knowledge is used by the expert in a procedural manner. Because of this orientation, the issue of declarative knowledge representation and organization in memory become important in my work. Other psychological researchers have also concentrated on the declarative knowledge of experts. Chi and Koeske (1983), for example, studied children's expert knowledge of dinosaurs and how that knowledge was organized in memory. McKeithen, Reitman, Rueter, and Hirtle (1981) found that experts in the ALGOL W computer language organized reserved words in memory according to their functional meaning, whereas nonexperts organized them using common-language associations. In addition, early expert computer systems, such as systems involving knowledge about baseball (Green, Wolf, Chomsky, & Laughery, 1963), dealt primarily with declarative knowledge. Thus, research on expert knowledge that is primarily verbal knowledge is not unprecedented.

The problem of knowledge representation is a difficult problem for knowledge engineering, as it is for cognitive psychology (e.g., Anderson & Bower, 1973; Anderson, 1983; Norman & Rumelhart, 1975; Rumelhart & McClelland, 1986) and for much the same reasons. At some point in the study of any area of expert knowledge a theory of knowledge representation must be proposed (Hart, 1986). Many psychologists believe that knowledge can be best represented in memory using memory schemas.

Memory Schemas

The most influential idea in cognitive psychology regarding the organization of information in memory is the notion of a memory schema. (See Alba & Hasher, 1983; Bower, Black, & Turner, 1979; Graesser & Nakamura, 1982; and Norman, 1982, for discussions of contemporary usage of the term.) Memory schemas are special knowledge structures in memory that all normal adults in our society possess. They represent activities about which most of us are very knowledgeable, if not expert. Memory schemas allow us to comprehend, perform, and remember many activities that are part of our culture, such as eating at restaurants, going to movies, shopping for groceries, and driving an automobile. Furthermore, expert knowledge can be characterized as memory schemas that only a few people in our culture share. But in addition to their importance for representing knowledge in memory, schemas are also important as a source of mental cues. This feature is important to psychologists who are interested in learning.

Mental Cuing

Schematic knowledge can be a repository of expert knowledge and the basis for expert performance in decision making and problem solving. In the theory of mental cuing (Bellezza, 1986, 1987a), however, emphasis is placed on the schematic or expert structure as a source of mental cues to which new information becomes associated. Psychologists have given mental cues other names, such as secondary cues, mental cues, mnemonic cues, and cognitive cues, all of which distinguish them from objects and other stimuli functioning as cues in the physical environment. A good way to describe the theory of mental cuing is that it combines ideas characteristic of both schema theory and the theory of mnemonic devices (Bellezza, 1987a). The notion of schemas as sources of mental cues is in contradistinction to the notion, as discussed by Alba and Hasher (1983), of the schematic structure acting purely as a repository of general information not specific to any particular event.

The Organization of Mental Cues

The organization of expert knowledge becomes an important issue when studying how experts learn, because learning is often measured using a free-recall test. In a free-recall test very few recall cues are provided to the rememberer, who must therefore use a strategy of generating mental cues to assist in the recall of specific information. Learning and recall based on mental cuing seems to occur in the following way: The expert being questioned tries to retrieve from memory about a particular event the specific information that is desired by the questioner and that is immediately available. In addition, the expert also retrieves schematic knowledge from expert knowledge structures in memory to which the new information may have become associated.

For example, if an expert in football experiences some new event, such as a football game, and later has to recall the details of that event, how does this learning and recall take place? The retrieval process includes retrieving knowledge about football in general as well as retrieving information regarding the specific event. The expert's knowledge of football acts as a set of memory cues for the particular football game to be described. The expert may think about the various player positions on a football team, the ways in which points can be scored, the various plays that can be executed,

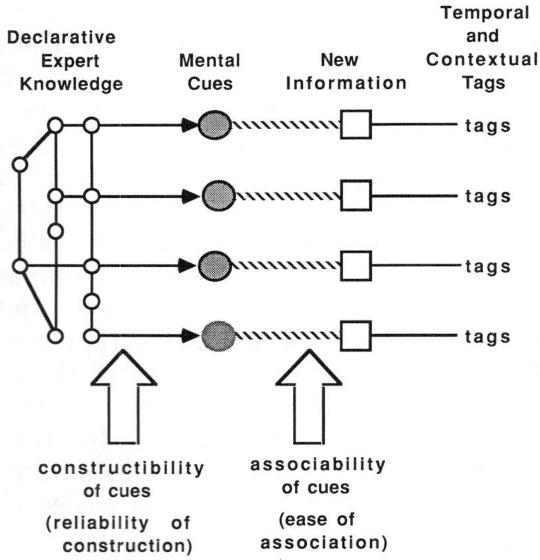

FIGURE 12.1. Schematic representation of the properties of constructibility and associability characteristic of the mental cues generated by an expert knowledge structure.[1]

[1] The interconnected open circles represent the network of expert knowledge in memory. The filled circles represent the mental cues that can be generated from the expert knowledge. The open squares indicate the new information from the environment that must become associated to the mental cues for learning to occur. The large arrow on the left emphasizes that the expert knowledge must reliably generate the same set of mental cues for the mental cues to have the property of constructibility. The large arrow on the right indicates that the mental cues and new information must be easily associated for the mental cues to have the property of associability. The context tags help to keep newly learned events discriminable from existing expert knowledge. (Reprinted with permission from Bellezza & Buck, 1988; copyright 1988, by John Wiley & Sons, Ltd.)

and so on. These all act as knowledge-based recall cues. In order to be effective, however, mental cues must have certain properties.

Properties of Mental Cues

Any body of expert declarative knowledge in memory can act as a source of memory cues to which new information can be associated. Figure 12.1 shows in a simplified way how the concepts, words, and images comprising expert knowledge can provide memory cues to which new information can become associated. Because the activated concepts and words are imageable and meaningful to the expert, they easily become associated to new verbal information (Paivio, 1971). This is the *associability* property of mental cues. Furthermore, because the mental cues are part of the expert's organized knowledge in memory, they should be generated in a reliable manner on different occasions (Olson & Rueter, 1987); that is, the same set of mental cues can be generated when learning occurs and when recall of the learned information is necessary. This reliability of generation of mental cues is the property of *constructibility*. Constructibility and associability of mental cues, as well as the two other properties of discriminability and bidirectional association, are discussed in more detail elsewhere (Bellezza 1981, 1986, 1987a).

Mental Cuing and Knowledge Engineering

What does mental cuing have to do with the theme of this volume, which is the problem of extracting knowledge from human experts in the process of building nonhuman expert systems? Both involve retrieval from memory and the problem of representation knowledge in memory. Indeed, the recall paradigm used in free-recall research, involving recall of an event, is similar to the interviews that occur when eliciting knowledge from an expert (Hart, 1986; Hoffman 1987, 1989; Olson & Rueter, 1987). One of the characteristics of both the free-recall task and some interview procedures is that the respondent is expected to provide relatively large amounts of information when provided with a relatively simple prompts. The minimal prompt used in our example of mental cues is, "Tell me about the football game." In a knowledge-elicitation procedure this is analogous to a prompt such as "Tell me about that problem you solved yesterday." Both involve the recall of events using minimal recall prompts and therefore minimal constraints on the information requested. The respondent may be encouraged to talk out loud without interruption, and his or her thoughts may be recorded as a verbal protocol.

Is using mental cues in the study of learning a good way to find out about how expert knowledge is represented in memory? Can the newly learned information act as a "tracer" to tell us more about the organization of knowledge structures in memory? Does the use of minimal recall prompts force the expert to reveal the organization of his or her expert knowledge in memory? Before trying to answer these questions, I will first describe some research on expert knowledge and mental cuing.

Free Recall of Word Pairs Related to Expert Knowledge

In the research described here, some of which was conducted in collaboration with Debra Buck (Bellezza & Buck, 1988), we tried to utilize college students' areas of expertise as sources of well-organized mental cues. We used the two areas of clothing and college football because we believed we would find some college students who were very knowledgeable about cloth and clothing but not knowledgeable about college football and would find other students very knowledgeable about college football but not knowledgeable about clothing.

In our first study, we devised two 50-item multiple-choice tests. One was a test of knowledge of cloth and clothing, and the second was a test of knowledge of college football. Both of these tests had good internal reliability, with coefficient alpha values of .80 and .92, respectively. Eighty-two college students were administered the clothing and football tests. The next day, the students attended a second session and were presented some verbal material. They were told to try to remember the material they were to read. First, they read a passage about the Aztec Indians and rated the visual imagery of each of its main ideas. They next went through two lists of 24 word pairs. The first list of word pairs involved clothing terms paired with common nouns. Examples of these pairs are *serge*–bomb, lemonade–*dicky*, *pinafore*–drum, ink–*chemise*. The second list involved football terms paired with other common nouns. Examples of these pairs are *quarterback*–lawyer, grapes–*punt*, *scrimmage*–nest, carrot–*touchdown*. As with the Aztec passage, the students rated the visual imagery of each pair.

On the third day the participants were asked to recall the Aztec passage; that is, they were asked to write down as much of the passage as they could without being given recall prompts other than the title. They were then asked to recall the two lists of pairs: the list containing clothing terms and the list containing football terms. These tests with minimal cues are called free-recall tests.

What was the rationale behind this study? We thought that those students who were knowledgeable about clothing would be able to recall many clothing terms because of the constructibility of clothing information in memory. Similarly, students who were knowledgeable about football would recall many football terms because of the constructibility of their football knowledge. A multiple-regression analysis was performed to determine how well recall of the expert terms could be explained using the score on the football test, the score on the clothing test, recall of the Aztec passage, and the participant's gender. Table 12.1 lists the beta weight for each predictor in the multiple-regression equation and indicates

TABLE 12.1. Multiple regression analysis of the free recall of expert terms from the pairs presented in Experiment 1.[1]

Predictor variable	Beta weight	Significance
Football terms (multiple R = .52)		
Aztec passage	.34	.01
Football test	.33	.02
Clothing test	−.03	—
Sex of participant	.45	.01
Clothing terms (multiple R = .52)		
Aztec passage	.16	—
Football test	.09	—
Clothing test	.21	.05
Sex of participant	.42	.01

[1] The significance value represents the probability that the corresponding beta weight value differs from zero by chance. If a significance level is missing, as indicated by the sign,—, then the beta weight cannot be considered different from zero using procedures based on statistical inference.

whether the beta weight was significantly different from zero when tested using a t test of statistical inference.

It can be seen from Table 12.1 that the score on the clothing test was significantly related to how many clothing terms were recalled. Its beta weight was significantly different from zero, but the beta weight for the football test was not. People who were experts in clothing recalled more clothing terms than did the nonexperts. Similarly, the score on the football test was significantly related to how many football terms were freely recalled, but the score on the clothing test was not. However, expertise was not the only factor that influenced recall. The amount recalled from the Aztec passage was positively related to recall of both types of expert terms because it acted as a test of the ability of each participant to recall any type of verbal information. Also, females tended to recall more than males regardless of the material presented.

TABLE 12.2. Multiple regression analysis of the free recall of neutral concrete nouns from the pairs presented in Experiment 1.[1]

Predictor variable	Beta weight	Significance
Paired with football terms (multiple R = .51)		
Aztec passage	.26	.02
Football test	.42	.01
Clothing test	−.06	—
Sex of participant	.54	.01
Paired with clothing terms (multiple R = .47)		
Aztec passage	.21	.05
Football test	.17	—
Clothing test	.22	.05
Sex of participant	.31	.04

[1] The significance value represents that probability that the corresponding beta weight value differs from zero by chance. If a significance level is missing, as indicated by the sign,—, then the beta weight cannot be considered different from zero using procedures based on statistical inference.

The free-recall test results showed that the clothing and football terms presented in the pairs were more constructible for the experts than for the nonexperts. This is what we expected. But did these expert terms act as mental cues for the information associated with them? That is, were the common nouns associated with the expert terms also better recalled? The answer is yes. In Table 12.2 we can see that scores on the football test were positively related to recall of the common nouns paired with the football terms. The football experts first recalled the football terms and then recalled the common nouns associated with them. Scores on the clothing test were not related to recall of these common nouns. The same point can be made regarding the lower half of Table 12.2. Scores on the clothing test were significantly related to recall of common nouns associated with the clothing terms, but scores on the football test were not.

As a final test in this study, we provided the students with the first word from each of the 24 pairs in the two lists and asked them to recall the word that went with it. Because the cues were presented, the need to generate the mental cues was eliminated. Therefore, this cued-recall test was a direct test of associability, that is, a test of how strongly each of the expert terms had been associated to its corresponding common noun. As in the free-recall test, clothing experts were able to recall significantly more words in the test of the clothing pairs, and football experts were able to recall significantly more words in the test of the football pairs. Bellezza and Buck (1988) provide a more detailed discussion of these results.

Implications for Expert Knowledge as Mental Cues

The results of this study support two of our notions regarding mental cues. First, expert terms acting as mental cues are more easily and reliably generated by experts than by nonexperts. That is, these mental cues are more constructible for experts that for nonexperts. Experts were able to recall more of the expert terms than were nonexperts. Second, expert terms are more associable for experts than for nonexperts. Experts were better able to associate common nouns to the expert terms and later use the expert terms as recall cues for the common nouns.

But word pairs are not the kind of information experts, or any other people, usually have to learn. In the next study we presented experts' descriptions of complete events related to, but contradicting, their expertise.

Free Recall of Passages Contradicting Expert Knowledge

In the research on human learning there is much evidence indicating that under certain conditions new information highly similar to existing knowledge is difficult to learn (Ausubel & Blake, 1958; Underwood, 1957). In terms of learning theories based on associative interference, this difficulty is called the paradox of expertise (Osgood, 1940; Smith, Adams, & Schorr, 1978), because experts seem to have less trouble than nonexperts absorbing information related to what they already know. The advantage the experts have in learning cannot be explained by interference theory, but it can be explained by mental cuing theory. Experts have well-organized knowledge structures in memory to which new information can be associated, and these knowledge structures aid experts' learning of new information, even if that information is similar to what the experts already know.

Debra Buck and I wanted our experts to learn information similar to what they already knew, but similar in a special way: The new information was to contradict their expert knowledge. Our question was, Do experts learn more poorly than nonexperts when confronted with information that is related to their expertise but in details contradicts it? In addition, we wanted to know how well experts performed in recalling textual information of the kind that people may have to deal with in school, work, and in other aspects of everyday life. Not all of the new information that is encountered in everyday life is consistent with what one already knows. It is not difficult to imagine an expert in any field remembering quite well an event or set of circumstances that contradicts what he or she expects. In fact, experts are often quick to recall past cases that were especially tough or salient and that deviated from the typical. This process may be a necessary one when problems are solved by experts using case-based reasoning (Lenat, 1984).

To answer our question we made up two passages called Crazy Wedding and Crazy Football. In the Crazy Wedding passage a wedding was described in which clothing usage contradicted all of the normal rules of how clothing should be worn. Some of the statements in the Crazy Wedding passage were, "The groom's mother wore a full-length dicky over a pink tunic"; "The bridesmaids wore their best black lingerie edged with pink piping"; and "As the featured soloist began to sing, she fingered the pleat in her pinafore."

Similarly, in the Crazy Football passage a football game was described that used many of the technical terms from college football but changed many of the rules. Some of the statements in this passage were, "The crazy football is round and covered with fur"; "Each team gets only one down and must advance the ball 5 yards"; and "The ball is passed only if the passer is beyond the line of scrimmage."

As in the first study, the experiment was spread over a 3-day period. On the first day, 107 college students were administered the football-knowledge and clothing-knowledge tests. On the second day, all the students had to read three passages and rate visual imagery of the ideas in them. The passages were the Aztec passage, the Crazy Football passage, and the Crazy Wedding passage. Finally, on the third day, the students had to freely recall all three passages.

As can be seen from the regression analyses shown in Table 12.3, recall of the Crazy Football passage was significantly related to scores on the football test. Those participants who scored high on the football test were better able to recall the Crazy Football passage than those who secored low. Scores on the clothing test were not significantly related to how well the Crazy Football passage was recalled. Similar results were obtained for the Crazy Wedding passage. Students who scored high on the clothing test were better able to recall the Crazy Wedding passage than those who scored low. Scores on the football test were unrelated to recall of the Crazy Wedding passage.

TABLE 12.3. Multiple regression analysis of the free recall of Crazy Football and Crazy Wedding passages presented in Experiment 2.[1]

Predictor variable	Beta weight	Significance
Crazy Football passage (multiple R = .61)		
Aztec passage	.44	.01
Football test	.47	.01
Clothing test	−.10	—
Sex of participant	.19	—
Crazy Wedding passage (multiple R = .55)		
Aztec passage	.40	.01
Football test	.16	—
Clothing test	.23	.02
Sex of participant	.35	.01

[1] The significance value represents that probability that the corresponding beta weight value differs from zero by chance. If a significance level is missing, as indicated by the sign,—, then the beta weight cannot be considered different from zero using procedures based on statistical inference.

The information in the clothing and football passages contradicted what the experts already knew, yet the experts remembered this information better than did the nonexperts. Why did this occur? Figure 12.1 shows that expert knowledge provides constructible and associable cues for the learning of new information. Yet, what is also necessary in mental cuing is some mechanism, such as context tagging (Anderson & Bower, 1972; Graesser, Woll, Kowalski, & Smith, 1980; see also Tulving, 1983), to prevent confusion in the memory system between the permanent expert knowledge about clothing or football and the new information stored in memory representing the Crazy Wedding and the Crazy Football stories. The experts remember the new information better than the nonexperts, but they remember it as an exception to what is usually the case.

Implications for Experts as Learners

Let me briefly summarize the implications of these two studies. Knowledge structures are highly organized sets of information in memory that contain conceptual components that are meaningful and imageable. Experts have knowledge structures that contain information about their area of expertise. The components of these knowledge structures can be used as mental cues to which new information can be associated. In the first study, we demonstrated that expertise helped the expert to recall common nouns unrelated to his or her area of expertise. Because the mental cues generated by the expert knowledge structure were both constructible and associable, new information was more easily acquired and recalled by experts than by nonexperts. The second study showed that experts were better than nonexperts at recalling information related to their area of expertise, even if that information contradicted what they already knew. We speculate that context tags in memory are able to keep separate newly acquired knowledge in memory from past knowledge.

Implications for Knowledge Elicitation

What do these two studies have to do with the problem of efficiently extracting expert knowledge from human experts? They suggest that expert declarative knowledge is more meaningful and imageable to the expert than to the nonexpert; that is, it is associable. They also suggest that expert declarative knowledge is well organized in memory; that is, it is constructible. Each time the expert knowledge is generated, it should have the same content and structure. When nonexperts try to generate expert information, a lot of guesswork and ad hoc assembling of information are probably involved. This means that when nonexperts try to generate expert information on two widely separated occasions, the information should vary greatly in content. That is, nonexperts are not reliable sources of expert information (McKeithen et al., 1981; Olson & Rueter, 1987). In our first study the better organization of expert knowledge in the memory of experts was demonstrated indirectly by the better recall of information that was neutral with regard to the area of expertise and in our second study by the better recall of information that actually contradicted the expert knowledge. In both studies, the superior organization of the expert information provided a constructible and associable set of mental cues that supported the new learning.

An implication of these two studies for the problem of knowledge extraction is that interview procedures should provide the knowledge engineer with valuable information. If expert knowledge is well organized, as it should be, then allowing the expert to think aloud to create verbal protocols should provide important information regarding the organization and content of the expert knowledge. Some investigators (Hart, 1986; Hoffman 1987, 1989), however, suggest that using interviews that are not highly structured can be a particularly inefficient way to extract expert knowledge. The knowledge generated may not be completely different from information generated in previous sessions, nor is it identical. The expert knowledge appears to be generated in a haphazard manner. The next study attempts to estimate how reliably (i.e., systematically), experts can provide information in a simple knowledge-generation task.

Reliability of Generating Expert Knowledge

In this third study, 65 college students were administered the previously developed clothing and football tests. The clothing and football tests are *reactive tests* because each question is answered by the examinee by choosing one of four answers. The next day, the students had to generate information for five clothing categories and for five football categories. This was a test of how much expert knowledge they could produce given minimal cues that acted as minimum constraints and so was a *generative test*. The five clothing categories used were (a) articles of clothing, (b) rules of correct dress, (c) brands of clothing, (d) types of cloth, and (e) occasions needing special dress. The five football categories were (a) names of famous football players, (b) rules of college football, (c) names of professional football teams, (d) types of football equipment, and (e) names of college football teams. For each category, the participants were given 3 minutes to generate as many items as they could. When they returned a week later for the final part of the experiment, they were surprised to learn that they again had to generate items for the same five clothing categories and the five football categories. They were told not to think about what they had done the previous week but simply to write down the appropriate information as they thought of it.

From these 65 students, 7 students were chosen as clothing experts. They scored among the top 10% of the students on the clothing test but scored below the mean on the football test. Another 7 students were chosen as football experts. They scored among the top 10% on the football test but below the mean on the clothing test.

Figure 12.2 displays the number of items generated for each category for the football experts and the clothing experts. Taking the mean number of items recalled in the two sessions, the experts always generated more information in each category that did the nonexperts. For the five clothing categories, the clothing experts generated a mean of 18.7 items, and the clothing nonexperts generated 13.5 items ($t(12) = 4.10$, $p < .002$). For the five football categories, the football experts generated a mean of 18.8 items, and the football nonexperts generated 11.3 items ($t(12) = 5.23$, $p < .001$). The experts have more information in their expert knowledge structures than do nonexperts. These results are also important because they show that generative measures of knowledge correlate with reactive measures, such as those used in the clothing and football multiple-choice tests.

But how well was expert knowledge organized in memory? The property of constructibility requires more than just verbal fluency in a particular content area. The element of organization is important. If the information the experts are generating really represents constructible knowledge, then each week the information generated should be pretty much the same. That is, if the expert information is organized, then this organization should be reflected by the unvarying structure of the information generated. To assess the degree of the organization of the expert knowledge generated, a measure of overlap was computed from the information generated in the two sessions. This overlap measure indicates

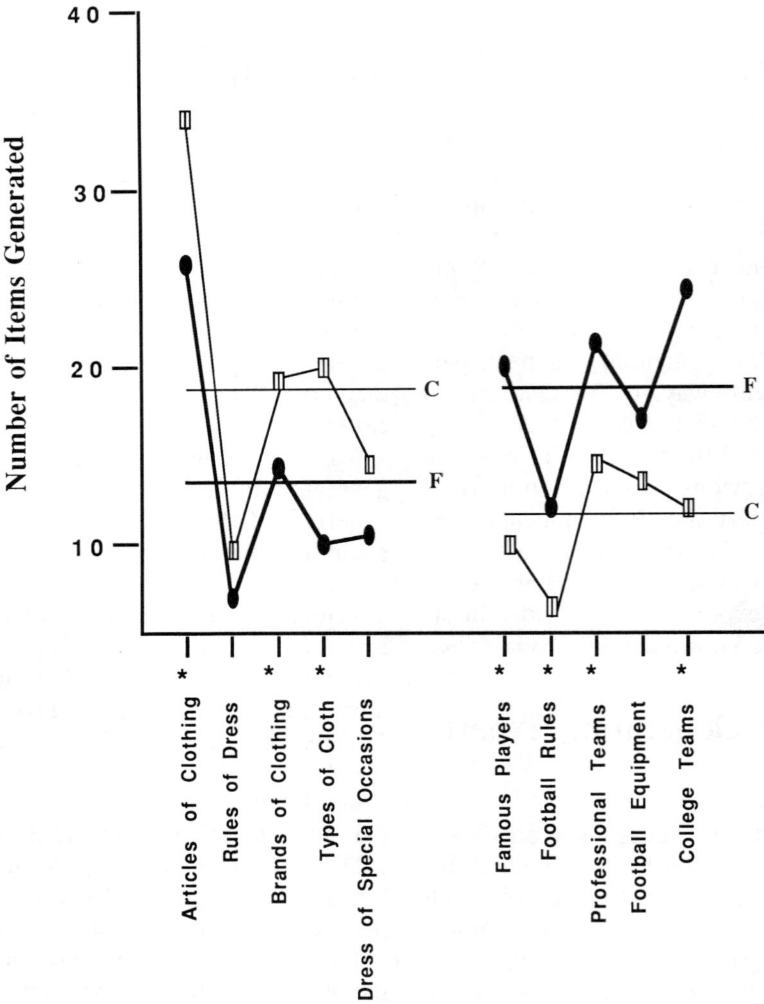

FIGURE 12.2. Mean number of items generated in the clothing and football categories by the clothing an football experts.[1]

[1] The light lines represent performance of the clothing experts and the heavy lines represent performance of the football experts. The horizontal lines represent mean performance over the five categories. The asterisks indicate on which categories performance for the clothing and football experts was significantly different.

what proportion of the items generated each week was the same in each category for each participant. The measure indicates how reliably information is retrieved from memory (Bellezza, 1984a, 1984b, 1984c, 1987b, 1988).

The results of the reliability analysis were surprising. Expert knowledge is supposed to be more constructible than nonexpert knowledge. Yet, as shown in Figure 12.3, there was slightly greater overlap of clothing information from one week to the next among the clothing nonexperts (.60) than among the clothing experts (.57). Although the superiority of the clothing nonexperts was not statistically significant ($t(12) = .52$, $p > .50$), we had expected the experts to be more reliable than the nonexperts. Equally surprising was the result that there was greater overlap of football infor-

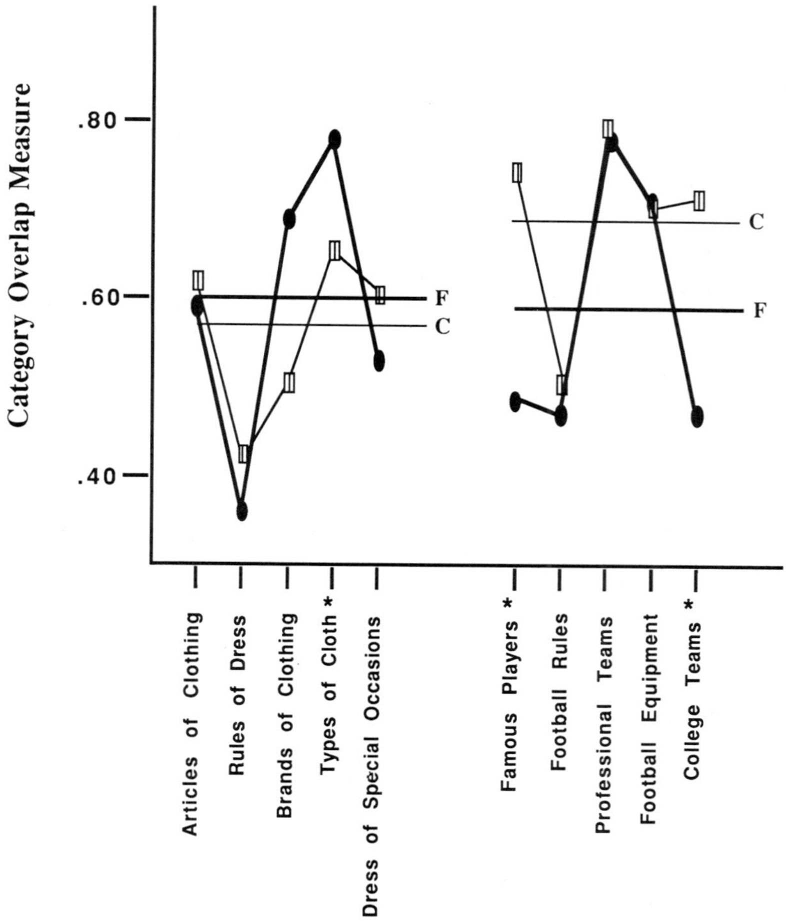

FIGURE 12.3. Proportion of overlap among the items generated in the two sessions for the clothing and football categories by the clothing and football experts.[1]

[1] The light lines represent performance of the clothing experts and the heavy lines represent performance of the football experts. The horizontal lines represent mean performance over the five categories. The asterisks indicate on which categories performance for the clothing and football experts was significantly different.

mation from one week to the next among the football nonexperts (.69) than among the football experts (.59), and this difference was statistically significant ($t(12) = 2.31, p < .04$).

What are we to make of these results? The overlap measure indicates that declarative knowledge about clothing is no better organized in memory for clothing experts than for clothing nonexperts, and declarative knowledge about football is better organized in memory for football nonexperts than for football experts! In the first session, low-knowledge participants should be doing more guessing and making use of information about which they are not certain. In the second session a week later, they should again guess and again use uncertain information. Hence, their overlap should be smaller. But this did not occur. Alternatively, it may be that the experts have so much information at their disposal that when tested each week they come up with appropriate but quite different information. Previous research (Bellezza, 1984a), using a random sample of college students and common categories such as the names of fruits, vehicles, and furniture, has shown that the larger the category tested, the lower the overlap from one generating session to the next. In a similar manner the experts may suffer from having too much information appropriate to the task.

Implications for Mental Cuing

Some of the clothing and football categories tested may be very large for experts compared to nonexperts. So, some expert information may be less constructible for experts than for nonexperts. The implication of this study for the theory of mental cuing is that it questions whether expert knowledge can provide constructible memory cues. It implies that experts should generate expert terms *less* reliably than nonexperts. Constructiblity of mental cues is important in free recall because the same mental cues present during learning must later be generated for the recall of the new information associated with those cues. This is necessary because so few overt cues are provided by the questioner in a free-recall task. Yet in this last study experts were less likely than were nonexperts to generate the same items for categories of expert information.

This failure to validate the constructibility of expert knowledge suggests that the results of the first two studies should be reconsidered. Bellezza and Buck (1988) concluded that a major factor in the superior recall of experts was the superior constructibility of the mental cues generated from the expert knowledge structures in memory. But the low reliabilities found for expert information in this last study cast doubt on this interpretation. Perhaps the superior recall performance of experts found in the first two studies was the result only of the superior associability and discriminability (Bellezza, 1986) of the mental cues and not the result of their superior constructibility. Expert terms are certainly more meaningful and imageable to the expert than to the nonexpert. Therefore, expert terms act as mental cues to which new information can be easily associated. But it is not clear that mental cues based on expert knowledge are more reliably generated.

Implications for Knowledge Elicitation

Another implication of the low reliabilities found in the third study pertains to the value of various interviewing techniques in extracting expert knowledge. The procedure used involved the generation of expert information from categories, and certainly this procedure is not identical to interview technique commonly used by knowledge engineers. However, the similarity may be enough to raise the issue of what kind of expert information is extracted when using generative, open-ended questions to interview experts. The results of the third study suggest that questions that elicit large amounts of information administered to the same expert at two different times may provide verbal protocols that overlap in content. The two protocols do not contain the same information, but neither is the information completely different. The expert may tend to repeat the same information within a single interview. Furthermore, when an expert indicates that he or she has no more to say about a topic, new information may nevertheless appear when the same questions are asked some time later. In these cases there may not be anything wrong with the information elicited, but still one has to wonder how much important expert information may be missed in these interviewing techniques.

I suggest that in the generation of information from minimal cues, as generating items to a category label or asking open-ended questions that require long responses, the information requested is underdetermined. That is, as far as the expert is concerned, it is not clear precisely what information is desired by the interviewer. As a result, the expert simply verbalizes the information that is currently salient and accessible in memory and that meets the loose constraints provided by the interviewer. Of course, what is accessible in memory on any particular occasion is dependent on a number of factors, including what the expert happens to have been previously doing or thinking. These thoughts and actions, in turn, may be partly the result of random events. The implication of this for the knowledge engineer is to provide a good deal of structure when eliciting information during an interview.

Providing structure imposes constraints regarding the information the expert is to generate. Knowledge engineers suggest that the expert be presented with specific tasks of various kinds (analysis of familiar tasks, analyzing tough cases, repertory grid analysis) relatively early in the knowledge elicitation process. Expert knowledge, by its very nature, may become organized only when the expert is reacting to a well-defined and specific problem or query. Expert knowledge is acquired in very specific sets of circumstances, so it will not be surprising if the only way it can be reliably extracted is to recreate or to describe those circumstances. In memory research, this is known as encoding specificity (Tulving, 1983).

Some Related Issues

In the knowledge-generation paradigm used here, experts seem to be less reliable in their generation of expert knowledge than are nonexperts, though, of course, the quality of the expert's knowledge may be much higher. But there are some methodological questions that must be answered before we can accept this conclusion. I would like to raise some of these questions, although I cannot fully answer them.

First, I have referred to the high-scoring participants used in these studies as clothing and football experts. However, they were really high-knowledge rather than expert participants because they were among the 10% of the most knowledgeable individuals tested. Perhaps, if "real" experts were used, such as the top 1/10 of 1% of the population, the organization and therefore the constructibility of the expert information would be higher. The problem of accounting for the lower reliability of expert information generated by experts would then disappear.

A second problem is that the time interval between the sessions in the third study might be too long. If experts and nonexperts were asked to generate the same information after an interval of only 1 day rather than 1 week, the experts might be more reliable in their responding than the nonexperts. However, this result would bring with it the unfortunate implication that expert knowledge is stable for a day but not for a week.

A third problem is that the knowledge extracted from the experts was declarative knowledge, that is, verbal knowledge rather than procedural or action knowledge. If the experts were asked to make decisions or solve problems, which is what experts typically do in our society rather than act as encyclopedic repositories of verbal information, then the reliabilities obtained should be higher. Hence, it might be that the generation task is itself the problem. People have information organized in memory for reactive tasks, and many of the techniques used in knowledge engineering are reactive tasks. So if a high-knowledge individual is given a problem, he or she will respond to it in an organized manner. Reactive and procedural knowledge may be highly organized. But generated, declarative knowledge might not be.

Generating categorical information in some area of expertise is usually unnecessary in everyday life. Hence, when it is necessary to generate declarative knowledge with minimal cues and to do so reliably from one situation to the next, a special kind of expert is needed. These are experts who are also expert pedagogues; that is, they are people who have become experts in organizing, generalizing, and declaiming their expert knowledge. They are experts who are used to talking about their area of expertise, such as theorists and teachers.

Conclusions

Expert declarative knowledge does not appear to be well organized in memory in the absence of externally imposed constraints or circumstances. Knowledge becomes structured only in response to some specific problem to be solved or set of complex information to be acted upon. One implication of this is that a sequence of unstructured or moderately structured interviews with an expert will result in new information being elicited in each interview but at a decreasing rate (Hoffman, 1987, 1989). A second implication is that the knowledge extracted from the expert will depend on the type of stimuli, problems, or circumstances he or she must react to. There is not much empirical evidence available, but we assume that when presented the same highly structured situation on two different occasions, the expert will respond in the same way. Third, because knowledge elicitation seems dependent on what information is presented to the expert, more pains should be taken by the knowledge engineer in defining the range of any domain of expertise so that representative problems from all areas of the domain can be presented to the expert. Otherwise, representation of expert knowledge may remain incomplete.

Many interesting questions remain with regard to expert knowledge, both declarative and procedural, such as how this information is represented and organized in memory, and how this information is utilized. An expert can be faced with tasks of different levels of specificity, ranging from solving a specific problem to responding to the statement, "Tell me all that you know." It seems that expert knowledge is not particularly well organized in memory in the absence of imposed multiple constraints. Perhaps expert knowledge in most domains is too vast for good memory organization. The constraints imposed on the expert are what create the organization of expert information in memory and precipitate its retrieval. Herbert Simon (1981, chap. 3) discusses the complexity of an ant's progress across the beach and makes the point that the complexity is in the external constraints imposed on the ant, not in the ant itself. The expert is like the ant only in that he or she is used to working with the many constraints that act as memory cues, and these are what organize the expert's knowledge in memory. Unless we test the expert in a similar situation, we cannot expect to obtain highly organized performance.

References

Alba, J. W., & Hasher, L. (1983). Is memory schematic? *Psychological Bulletin*, *93*, 203–231.

Anderson, J. R. (1976). *Language, memory, and thought*. Hillsdale, NJ: Erlbaum.

Anderson, J. R. (1983). *The architecture of cognition*. Cambridge, MA: MIT Press.

Anderson, J. R., & Bower, G. H. (1972). Recognition and retrieval processes in free recall. *Psychological Review*, *79*, 97–123.

Anderson, J. R., & Bower, G. H. (1973). *Human associative memory*. Washington, DC: Winston & Sons.

Ausubel, D. P., & Blake, E. B. (1958). Proactive inhibition in the forgetting of meaningful school material. *Journal of Education Research*, *52*, 145–149.

Bellezza, F. S. (1981). Mnemonic devices: Classification, characteristics, and criteria. *Review of Educational Research*, *51*, 247–275.

Bellezza, F. S. (1983). Mnemonic-device instruction with adults. In M. Pressley & J. R. Levin (Eds.), *Cognitive strategy research: Psychological foundations* (pp. 51–73). New York: Springer-Verlag.

Bellezza, F. S. (1984a). Reliability of retrieval from semantic memory: Common categories. *Bulletin of the Psychonomic Society*, *22*, 324–326.

Bellezza, F. S. (1984b). Reliability of retrieval from semantic memory: Information about people. *Bulletin of the Psychonomic Society*, *22*, 511–513.

Bellezza, F. S. (1984c). Reliability of retrieval from semantic memory: Noun meanings. *Bulletin of the Psychonomic Society*, *22*, 377–380.

Bellezza, F. S. (1986). Mental cues and verbal reports in learning. In G. H. Bower (Ed.), *The psychology of learning and motivation* (Vol. 20, pp. 237–273). New York: Academic Press.

Bellezza, F. S. (1987a). Mnemonic devices and memory schemas. In M. McDaniel & M. Pressley (Eds.), *Imagery and related mnemonic processes* (pp. 34–55). New York: Springer-Verlag.

Bellezza, F. S. (1987b). Reliability of retrieving information from knowledge structures in memory: Self information. *Bulletin of the Psychonomic Society*, *25*, 407–410.

Bellezza, F. S. (1988). Reliability of retrieving information from knowledge structures in memory: Scripts. *Bulletin of the Psychonomic Society*, *26*, 11–14.

Bellezza, F. S., & Buck, D. K. (1988). Expert knowledge as mnemonic cues. *Applied Cognitive Psychology*, *2*, 147–162.

Bower, G. H., Black, J. B., & Turner, T. J. (1979). Scripts in memory for text. *Cognitive Psychology*, *11*, 177–220.

Chi, M. T. H., & Koeske, R. D. (1983). Network representation of a child's dinosaur knowledge. *Developmental Psychology*, *19*, 29–39.

Graesser, A. C., & Nakamura, G. V. (1982). The impact of a schema on comprehension and memory. In G. H. Bower (Ed.), *The psychology of learning and motivation* (Vol. 16, pp. 60–109). New York: Academic Press.

Graesser, A. C., Woll, S. B., Kowalski, D. J., & Smith, D. A. (1980). Memory for typical and atypical actions in scripted activities. *Journal of Experimental Psychology: Human Learning and Memory*, *6*, 503–515.

Green, B. F., Wolf, A. K., Chomsky, C., & Laughery, K. (1963). Baseball: An automatic question answerer. In E. A. Feigenbaum & J. Feldman (Eds.), *Computers and thought* (pp. 207–216). New York: McGraw-Hill.

Hart, A. (1986). *Knowledge acquisition for expert systems*. New York: McGraw-Hill.

Hoffman, R. R. (1987). The problem of extracting the knowledge of experts from the perspective of experimental psychology. *The AI Magazine*, *8*, 53–64.

Hoffman, R. R. (1989, April). A brief survey of methods for extracting the knowledge of experts. In C. R. Westphal & K. McGraw (Eds.), *The SIGART newsletter* (pp. 19–27). New York: Association for Computing Machinery.

Lenat, D. B. (1984, September). Computer software for intelligent systems. *Scientific American*, pp. 204–213.

McKeithen, K. B., Reitman, J. S., Rueter, H. H., & Hirtle, S. C. (1981). Knowledge organization and skill differences in computer programmers. *Cognitive Psychology*, *13*, 307–325.

Norman, D. A. (1982). *Learning and memory*. San Francisco, CA: W. H. Freeman.

Norman, D. A., & Rumelhart, D. E. (1975). *Explorations in cognition*. San Francisco, CA: W. H. Freeman.

Olson, J. R., & Rueter, H. H. (1987). Extracting expertise from experts: Methods for knowledge acquisition. *Expert Systems*, *4*, 152–168.

Osgood, C. E. (1940). The similarity paradox in human learning: A resolution. *Psychological Review*, *56*, 132–143.

Paivio, A. (1971). *Imagery and verbal processes*. New York: Holt, Rinehart & Winston.

Rumelhart, D. E., & McClelland, J. L. (1986). *Parallel distributed processing* (Vols. 1–2). Cambridge, MA: MIT Press.

Ryle, G. (1949). *The concept of mind*. New York: Harper & Row.

Simon, H. A. (1981). *The sciences of the artificial* (2nd Ed.). Cambridge, MA: MIT Press.

Smith, E. E., Adams, N., & Schorr, D. (1978). Fact retrieval and the paradox of interference. *Cognitive Psychology*, *10*, 438–464.

Tulving, E. (1983). *Elements of episodic memory*. New York: Oxford University Press.

Underwood, B. J. (1957). Interference and forgetting. *Psychological Review*, *64*, 49–60.

13 The Role of General Ability in Cognitive Complexity: A Case Study of Expertise

Stephen J. Ceci and Ana Ruiz

Introduction

Although theories about generalization (or transfer) have abounded since the time of Aristotle's *DeAnima*, the scientific debate over it did not take shape until the beginning of the 20th century. At that time, researchers at the two main centers of "associationism" in the United States, the University of Chicago and Columbia University, argued about the nature and developmental course of transfer. At Columbia, Edward Thorndike (1905) attacked the then-popular belief in the "theory of formal disciplines," which alleged that training in one discipline enabled students to think more rationally in other disciplines. Learning Latin, for example, was thought to lead to a better understanding of English, not simply because these two languages shared many cognates but also because learning Latin was regarded as an exercise that promoted the development of logical reasoning. (Learning chess was also promoted in some quarters for the same reason.) Reasoning is reasoning, so the thinking went, and therefore learning how to reason in one context was thought to transfer to reasoning in other contexts. Thus, it was the presumed ubiquity of transfer that was responsible for its appeal. It represented a parsimonious way of accounting for the obvious fact that humans do not require explicit learning of all matters.

Thorndike (1905) challenged the view that transfer was ubiquitous by arguing instead that learning in one context only transferred to another context to the extent that the number of "identical elements" in two contexts overlapped. His model was a neural one, suggesting that the neural pathways that were excited by each element could be important because the overlap in these excited elements was responsible for transfer, when it does occur. His evidence came primarily from subjects' performance on simple perceptual discrimination tasks on which they had received prior training on one form and were then asked to solve a related but physically dissimilar form. For example, subjects who were experienced at estimating the area of a particular geometric shape were not sucessful at estimating the area of other shapes, despite similarities of parameters (Thorndike & Woodworth, 1901). Thus, Thorndike's view of the limits of transfer stood in contrast not only to the supporters of formal disciplinary training but also to those who favored a broader view of transfer that went beyond stimulus-response, or feature overlap, theory. Foremost in the latter camp was Charles Judd at the University of Chicago.

Judd (1908) argued that the degree of transfer was a function not of the number of neural paths shared by the physical features of two tasks but rather of the overlap in meaning attached to the two tasks. His argument ran along the lines of much of contemporary cognitive psychology and seemed to foreshadow Bartlett's (1932) notion of "effort after meaning" (i.e., that individuals strive to impose some sort of meaning on their perceptions). For Judd, the individual interprets and makes sense out of a task, and it is the equivalence of meaning between two tasks that gives rise to their transfer rather than the number of identical physical elements that they have in common. In support of his theory, Judd asked fifth- and sixth-grade children to throw darts at a target that was submerged under 12 inches of water until they improved at hitting it. Half of the fifth and sixth graders were also taught laws of refraction. After the children spent a comparable amount of time trying to hit the target, Judd then submerged it beneath only 4 inches of water. Now the group that had been taught the laws of refraction outperformed the group that had received only practice. This suggested to Judd that the former group had subjectively come to a shared meaning of the two tasks, and this is why they transferred expertise from one to the other, and it is also why the latter group did not.

Although others after Judd and Thorndike continued to conduct research in this area, the debate over the ubiquity of cognitive transfer abated after their deaths and was replaced by a concern over the transferability of motor skills—until quite recently.

In this chapter we shall briefly review some recent work by cognitive psychologists on the topic of transfer, with a view toward answering the following question: Do measures of so-called general intelligence, the best known of which is the IQ, predict the ease and accuracy of transfer? To address this question, we will describe a case study of two men whose IQs differ markedly attempting to solve what for them is a novel problem—one that is isomorphic with a solution they have already discovered in another realm.

Recent Research on Cognitive Transfer

Ann Brown and her colleagues (Brown, in press; Brown & Kane, 1988; Brown, Kane, & Echols, 1988) have reported a number of demonstrations of preschoolers' transfer under conditions in which they have sufficient experience in the task domain and the two tasks are similar. For example, in one series of tasks, children who learned that a certain material (e.g., grains of corn) can be transported across a river by using a straw also transferred the same solution to transporting other materials across other barriers with a straw.

Bossok and Holyoak (1989) have shown that there is considerable difficulty involved in training college students to transfer certain statistical rules (like the Poisson distribution principle), even when a great deal of support and practice is provided and, indeed, even when the transfer is "local," that is, within the same domain (albeit one that is probably not elaborately structured for these participants). Their work calls into question the presumption that transfer is a ubiquitous process that is freely invoked by those with high IQs. On the other hand, Nisbett, Fong, Lehman, and Cheng (1988) have found limited evidence that other kinds of statistical reasoning can be taught to university students and that some of them are able to transfer this knowledge to new domains immediately following training. After 2 weeks, however, much of the transfer appears lost. Thus, given the difficulty of achieving transfer *within* some domains, it should come as no surprise to find that training across knowledge domains, although it can occur, frequently does not. This has prompted one observer to make the following comment:

> The question for which we do have some empirical answers has to do with how generalizable cognitive training is from one subject area to another. As of now, the answer is not very much. (Schooler, 1989, p. 11)

According to the law of large numbers, or LLN (Nisbett et al., 1988), transfer should proceed in accordance with the rule that holds that large samples are necessary to generalize about groups that are more variable on a given attribute than groups that are less variable (Nisbett et al., 1988). Adults seem to appreciate this rule in some contexts quite well—appreciating, for example, that a small sampling of a slot machine's behavior or a rookie's batting performance is inadequate for generalizing about their performance over the long haul. Yet in other contexts, adults are quite unlikely to appreciate this rule (e.g., realizing that empathy or altruism expressed by someone they just met at a party is a poor basis for generalizing about the manifestation of such traits over the long haul of a social relationship).

Others have reported similar failures among college-aged subjects to transfer. Leshowitz (1989) gave a broad cross-section of college students "everyday" problems that contained principles taught in introductory social science courses (e.g., the need for control groups). Yet, of the hundreds of students tested, hardly any demonstrated even a semblance of methodological thinking about these problems, despite the considerable number of science and math courses they had taken.

Thus, notwithstanding the pervasive and often-stated assumption that transfer is ubiquitous, the empirical picture is far less clear. The research has prompted Rogoff (1981) to ask critically, after reviewing the cross-cultural literature on transfer and generality,

> What conclusions about generality can be drawn from successful performance on a syllogism problem? That the individual (a) will do well on the next syllogism? (b) will do well on other kinds of logic problems? (c) will be logical in many situations? or (d) is smart? (p. 271)

Regardless of whether transfer occurs frequently or only rarely, it is clear that there are large individual differences when it does occur: Some participants successfully transfer while others do not. Presently, little is known about the characteristics of those who fail or succeed.

The Issue of Generality

One of the stickiest issues facing contemporary cognitive psychology is that of *generality*. We assert that it is "sticky" because it appears to be assumed, rather than proven, that intelligent individuals are by definition those who are capable of acting intelligently in general—capable of transferring knowledge and insights across domains. It is not too great of a simplification to say that Jensen's "Type 2 intelligence," that is, the ability to draw inferences, transform symbolic input, and to correlate information from one domain with that from another (Jensen, 1980); Spearman's "g" (Spearman, 1904); Sternberg's "executive" (Sternberg, 1985); and Itzkoff's "cortical glue" (Itzkoff, 1989) are all variations on this same theme of intelligence as transfer. Consider this: In psychometric research, the often large and reliable first principal component that is extracted from a matrix of intertask correlations is interpreted as the magnitude of an individual's *general* ability to solve problems. (See Ceci, 1990, for review of this position.) The large first principal component is taken to mean that a person has a specified amount of mental power to transfer solutions across a very wide range of problems.

> An enormous amount of evidence argues that with a high "g" an individual is enabled to enter a civilized society along a number of professional, vocational pathways—medicine, law, scholarship, arts, and business. Persons with high "g" can retrain themselves to do many different tasks in one lifetime and often at a highly creative level. (Itzkoff, 1989, p. 85).

The present case study was undertaken to assess the role of one individual difference parameter in determining the degree of transfer. We asked whether an individual's IQ was related to his or her ability to succeed at a complex transfer task, one requiring multiple interaction effects and an appreciation of nonlinearity. Our aim was not simply to assess the role of IQ in a transfer task the way one might assess the role of field independence, impulsiveness, or any other individual difference measure, but to ask whether IQ, which is assumed by many to index the complexity and generality of cognitive processing, or "g," is in fact related to one's ability to solve a complex task that is homomorphic with another task one has already learned to solve in a different domain. Our goal was to provide some empirical evidence for the view that IQ is related to complexity and generality of thought, by virtue of its correlation with transfer across complex tasks. Although the hazards of generalizing from a case study do not need stressing, we felt that the following data provide a "first pass" at a thorny issue that is difficult to tackle experimentally with large samples. In the Discussion section we shall link this research to that dealing with the psychology of expertise, including the role that the elaborateness of one's knowledge base plays in developing expertise.

Background Research

In our previous research, we studied a group of men who attended harness races practically every day of their adult lives, in order to gain an understanding of the basis for their expertise at predicting post time odds (Ceci & Liker, 1986a).[1] Based on their knowledge of racing facts and their ability to correctly predict post time odds, we distinguished the men as either experts or nonexperts.

Although both experts and nonexperts were far more knowledgeable than amateurs about horse racing, and were far better at predicting post time odds than amateurs, they differed greatly among themselves. The group we called experts was rather amazing at estimating what the odds on each horse would be at post time, whereas the group called nonexperts was less dazzling. We asked the men in these groups to handicap real races as well as hypothetical races that we designed. The latter were included to separate variables that were often too correlated in actual races, to determine their importance in an expert's decision. In the hypothetical races, these variables were allowed to covary systematically.

We demonstrated that expert handicappers employed a complex, multiplicative model involving multiple interaction effects. By regressing 25 racetrack variables on experts' assessments of odds, we were able to show that simple additive models failed to account for the complexity of their decisions. Experts not only took into consideration more variables when handicapping a race, but they did not simply "add up" this information. Rather, they developed implicit algorithms that gave more or less weight to different types of information. And each type of information changed the way the experts thought about the other types.

The correlation between an expert's IQ and the *b* weight for a seven-way interactive term (which is a surrogate for their cognitive complexity and was shown to be highly correlated with their success at predicting odds) was −.07. This means that even though the greater use of complex, interactive thinking was causally related to success at predicting odds, there was no relation between such complex thinking and IQ. Thus, assessment of the experts' intelligence on a standard IQ test was irrelevant in predicting the complexity of their thinking at the racetrack (Ceci & Liker, 1986b; 1988). Within either group (experts or nonexperts), IQ was unrelated to handicapping complexity. Between groups, however, there was an invariant finding: Experts with low IQs always used more complex interactive models than did nonexperts with high IQs, and their success was due in large part to the use of these complex interactive models. IQ was useless in predicting both how complexly these experts reasoned and the number of variables they considered interactively in their judgments. (Interestingly, the success of experts at making these computations depends, in part, on their skill at doing mental arithmetic and, in particular, at subtractions that cross fifths boundaries. Yet this skill was unrelated to their scores on the mental arithmetic scale of the IQ test, too!)

It could be argued that IQ is still a valid predictor of intellectual functioning, despite the failure to find it correlated with complexity in the racetrack task. After all, these participants in the racetrack study attended the races nearly every day of their adult lives, and perhaps those with high IQs developed the complex understanding needed to estimate odds long before those with low IQs. By subjecting two of these individuals to a new task that was novel for both of them, but that depended on a similar algorithm to the one they used at the racetrack, we could assess whether the man with a high IQ would be quicker at reinventing the complex algorithm to succeed at this new task. To our knowledge, there are no data on such a question and therefore ours represent a rather crude first pass.

Method

Participants

Two men participated in this study. One man was a 46-year-old self-employed businessman with a Master's degree in mathematics education and an IQ of 121. The other man was a retired 74-year-old dockworker with an IQ of 81 and a fifth-grade education. Both men had participated in an earlier study of expertise in racetracking handicapping (Subjects 20 and 29, respectively, in the Ceci & Liker, 1986a, study). And both had been rated as experts in that study, demonstrating comparable use of a complex multiple interaction term in their decisions (see Table 6 of Ceci & Liker, 1986a).

Neither participant reported having had prior experience in the stock market before participating in this exercise: Neither man had invested in stocks in the past or claimed to have known anything about the variables that influence actual market forces, or to have played stock market games like Millionaire. To make sure that these two individuals had no substantive knowledge or beliefs about the stock market that might differentiate them at the start of this study, we asked them 24 multiple-choice questions that were designed by a market consultant to assess basic level understanding of stock market mechanisms. (See this chapter's Appendix for a sample of these.) On this quiz neither man achieved an above-chance score when answering questions about basic stock market mechanisms. While it is still possible that these men might have differed on some more subtle knowledge-based measure, it appears that neither was sufficiently knowledgeable to have had an advantage or a disadvantage because of background experience.

Procedure

Both men were presented 411 trials of a stock market game that required them to estimate which of two stocks would have the best future price-earnings (P/E) ratio (a stock's price divided by its earnings per share). One of the two stocks was always listed at the market average, and the other was a stock that was to be evaluated against the average, by estimating the probability that it would yield a higher P/E ratio than the average. To present comparative information for stocks, a commercial program developed by Value Line, Inc. for use by stock analysts was modified. The program incorporates 38 variables, but for this task only 18 were covaried in order to provide comparability with the racetrack tasks to which these men were accustomed. A single page described two stocks (one being the composite market average and the other being a novel stock) along 18 variables. The subject was charged with deciding the probability that the novel stock would return a higher P/E ratio than the market average. Table 13.1 depicts a sample page with two stocks. All stocks were identified only as numbers, to avoid any expectancies associated with certain ticker names (e.g., IBM).

The participants were informed that the information presented was fictitious and that the task was a game to see whether they could

TABLE 13.1. P/E ratio comparison of market average (#72) with another stock (#132).

Variables	#72	#132
1. Timeliness rank	1	3
2. Beta*	1.65	1.35
3. Book value/share	38.33	44.51
4. 3-month price change (%)	4.5	−2.42
5. Safety rank	2.0	1.0
6. Price stability	60	85
7. Financial strength	A+	A++
8. Market value/sales ($ millions)*	14,734.3/8685	74,267/59,041
9. 5-year dividend growth*	3.2	4.5
10. Current yield (%)	0.0	2.8
11. Current P/E ratio*	15.4	14.9
12. Current EPS*	6.01	8.72
13. Recent price	110.400	123.325
14. 12-month high/low	110.255/59.370	175.655/122.800
15. Price-book value*	2.31	2.47
16. Debt as % of capital*	5	9
17. Returned to com eq (%)	9.7	10.4
18. Return net worth (%)	9.9	19.2

* Variable was part of interaction term.

determine the rule that predicted P/E ratios. They were informed that the information to be presented was prepared in a manner so that the P/E ratio could be inferred but that it was not obvious how this was done and that their job was to try to figure out the rule. Some of the information to be presented was sufficient to predict which stocks would have the most favorable future P/E ratio, but not all information presented would be useful. Finally, even among those categories of information that were useful in determining the P/E ratio, not all were of equal importance. Each of the 18 variables listed in Table 13.1 was explained to the participants, and examples were provided until the participants indicated that they understood what the variables stood for. In addition, a glossary of the meaning of all 18 variables was available throughout the study.

The rule to establish whether a particular stock's P/E ratio would rise or fall above the market average was the same one used by these men to establish post time odds at the racetrack. In short, it was a seven-factor equation with multiple interaction effects.

In the course of presenting the 411 trials, seven variables were shown to be relevant to the P/E ratio prediction task. This was done implictly, through the use of an algorithm that was written to generate the stock market data but that was not explained to the participants. The task for the participants was to infer the nature of the algorithm and the seven-factor interactive term that determined the P/E ratios. These seven variables (denoted by asterisks in Table 13.1) were systematically weighted by the algorithm to provide deterministic outcomes; only these variables could serve as the basis for consistently accurate P/E ratio predictions. Thus, as in harness-race handicapping, occasionally irrelevant independent variables might be associated with changes in the dependent variable (P/E ratio), but over the long haul only seven variables interacted to consistently determine P/E ratios. The men were not told of the isomorphism between the manner in which these seven stock market variables could be weighted and multiplicatively combined to determine P/E ratios and the seven variables in harness racing that determined post time odds. That is, they were not expressly informed of the similarity between the rule that governed P/E ratios and the one that they routinely employed at the racetrack. The goal of presenting these trials was to see if these two individuals would realize, on the basis of the 411 feedback trials, that (a) only 7 of the 18 independent variables presented were deterministic of P/E ratios, (b) simple main effects and lower level interactive models were inadequate for determining P/E ratios, and (c) they would be able to construct more complex models akin to those they had already demonstrated in another domain (harness racing).

Although the variables used in this transfer task represent actual stock market variables that stock analysts consider important, the task was quite different from actual stock market analysis. Not only was the format for presenting the stocks different (two-stock comparisons that required a forced choice among them), but also a number of variables that analysts consider important were excluded in order to keep the number at 18. Thus, no product development, political, or historical data were provided for any stock, nor was general economic data provided that could bear on decisions. Finally, the task was quite unlike the one facing serious investors, who must take into account the joint performance of entire portfolios that were assembled to balance risk, market sensitivity, diversification, and so forth. Our goal was simply to see whether an algorithm that had been used in one setting might emerge in another in which it was instrumental.

The object of each trial was for the participants to sift through the simulated data on the 18 independent variables and predict whether the stock on that page represented a future P/E ratio superior to the market average. The participants performed this task independently and made their decisions by marking a sheet of paper given to them. At a given session (approximately 1 hour), participants were presented with, on average, 10 pages of two-stock comparisons, with a range between 4 and 12 trials. (The number of trials they were given depended on their schedules and interest levels that day. They were usually administered prior to or following the races in a clubhouse restaurant.) The 411 trials were spread out over a 7-month period. Although participants were invited to take as much time as they wished to make their selections, 95% of the one-page trials took between 5 minutes and 10 minutes. It was stressed that the stocks in this game were fictitious and not to be found in any newspaper, and that behavior could not be improved by reading financial reports or by doing any type of studying.

Following every P/E ratio prediction from a two-stock comparison, the participant was given feedback in the form of the future P/E ratios for each of the two stocks on that page. (This is similar to harness racing, where, following an expert's estimation of probable odds for horses in a race, he observes the actual odds.) A tally was kept each day so that the participants could keep track of their overall performance (i.e., the percentage of time they correctly predicted whether the novel stock would return a better P/E ratio than the market average P/E ratio). Although they were not informed about the other participant's performance by the experimenters, they often informed each other of their overall prediction rates, though they were asked not to divulge the strategies or variables they used.

Although this task was structurally similar to the one these participants confronted on a daily basis at the racetrack, there were several important differences. First, there was no actual financial risk or gain involved in this task, unlike the racetrack task. Second, participants were not provided the same level of extensive experience on this task that they had with the racetrack handicapping task. Yet within this less extensive period of experience, they were presented the actual trials much less frequently than they experienced them at the racetrack. That is, instead of 20 years of experience at the track, these men were provided only about 7 months of experience at this market task. However, to handicap the comparable number of races ($N = 411$) would take approximately 8 weeks (10 races/day × 5 racing days/week), whereas these 411 stock market trials were spaced over 7 months. Finally, the level of motivation, although seemingly high, could not be equated with that involved in harness racing for these men. The latter enterprise was a fundamental aspect of their adult lives, one that had not waned over more than 20 years of daily involvement.

Results

Figure 13.1 depicts the participants' accuracy at predicting stock increases in this fictional simulation as a function of the number of trials they observed. As can be seen, even after 411 trials, these two participants were far from adept, though they were both able to predict P/E ratios considerably better than chance, which is 50% in this two-choice task. At the end of 411 trials, both participants had acquired part of the complex seven-way interactive model that drove the simulation, to roughly similar degrees.

To analyze the implicit algorithms the men developed to guide their P/E ratio decisions, a modeling procedure based on Ceci and Liker (1986a, 1986b) was employed. This is a modification of the general linear procedure, so parameters can be estimated without fully crossing all levels of each variable with those of the others.[2] The analytic approach was to assess whether the nonadditive combination of variables (the seven-way interaction term) had a unique net effect on the participants' decisions about P/E ratios over and above the simple additive effects of the individual variables. Therefore, this interactive term was added after all of the variables were first entered into the model in order to determine whether this interaction term resulted in a reduction of the sums of squares error independently of the variables' additive effects.

The higher IQ participant used what amounted to a combination of main effects and lower order interactive effects. For example, during his final 50 trials, he combined the value of a stock's beta (an index of sensitivity to overall fluctuations in the market)[3] with its riskiness (judged against the standard of the market itself, with stocks being either above or below the market in their riskiness) and multiplied this value by two other values that

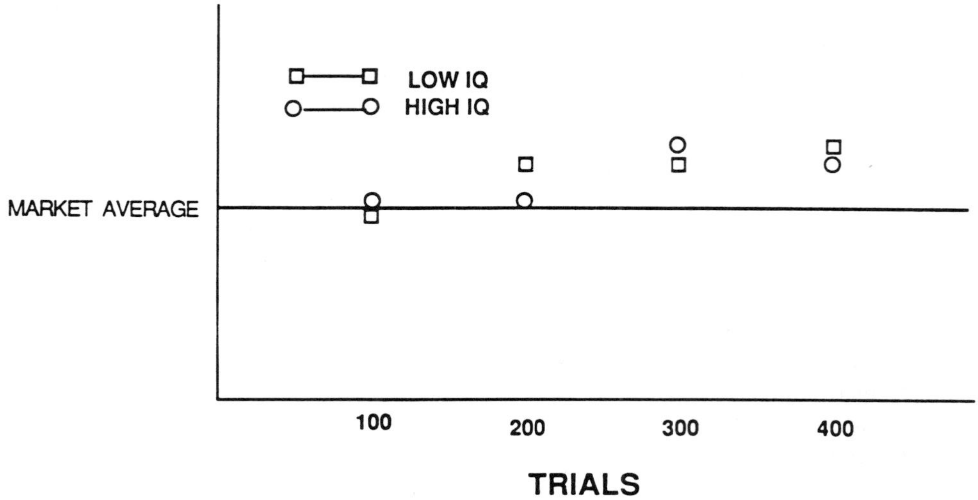

FIGURE 13.1. Mean probability of predicting whether a stock would exceed the market average.

he derived. During earlier trials, he had used other implicit models of equivalent complexity, though these resulted in somewhat inferior performance.

The lower IQ participant employed what appears to be a series of three-way interactive models. The best predictive value of these models was observed during his final 30 trials. Here he took into consideration the 5-year dividend growth of a stock, the market liquidity for that same stock (which in turn was derived by multiplying the price of the stock by the number of its outstanding shares), and its beta.

Thus, neither racetrack expert succeeded in rediscovering the same seven-way interactive model that governed their racetrack predictions, even though both were able to perform significantly above chance in predicting whether a stock's P/E ratio would exceed the market average. Their implicit algorithms were still changing at the time of the final trial, and it is almost certain that they would have continued to change with increased experience. We plan to explore this hypothesis by continuing to provide these men with additional trials. Upon completion of the study, we shall ask them to verbally describe their decision model. We will explain to them that the same model they employed at the racetrack could be used to predict P/E ratios in this task, and we will allow them additional trials to see if they can apply it in this context.

Conclusions

What should we make of these early findings, if anything? First, it is evident that the study needs extension. As already noted, perhaps these men would show even greater prediction on the stock market task if they were provided with more trials and/or if the incentive for successful P/E ratio estimation was increased (e.g., by allowing personal monetary risk). These two men may have required thousands of trials to develop the equivalent mental algorithm that they used at the racetrack, and therefore the 411 trials that were provided here might have been insufficient for attainment of the full seven-way interactive model. Also, it bears noting that we did not provide these men with any information that suggested that the same algorithm they used at the races would also be appropriate on this task. One might wonder what would happen if the men were informed of the isomorphism. We have no doubt what the answer would be, and if these men still have not discovered the full model at the end of this study we will give them this information. We expect them to rapidly and accurately apply it. It is almost certainly the case that had we informed the men of the relevance of the racetrack algorithm for the stock market task, they would have solved the latter within 200 trials (i.e., the time needed to sort out the seven variables that interacted and assign them relative weights). However, it is worth bearing in mind that the goal of this study was not to see how many trials would be necessary to develop a new mental model or to see whether these men could adapt an old model to a new context if they had been instructed to do so, but rather to see if a comparable level of cognitive complexity would spontaneously be reinvented to provide a solution in a new problem domain. That is, we wanted to see whether the cognitive complexity that was associated with expertise at the racetrack was domain-specific, or whether it was *general*. Many psychologists have speculated that expertise among those with low IQs in a specific setting would be unrelated to expertise in other settings (see Ceci & Liker, 1988). Thus, one would have thought that 411 trials would be sufficient to see whether these men would develop complex interactive models, and more important, whether the lower IQ participant would be at a disadvantage vis-à-vis the participant with the higher IQ.

It appears that neither participant gained any sudden insight into the appropriateness of his already-developed racetrack model, but each sought to develop a new model. Given the task demands and the novelty of the domain, this is perhaps a wise strategy. The alternative would be for these men to go through life trying to fit a model they developed for a highly specific situation (racing) to situations that were in need of entirely different models.

If with increased numbers of trials neither participant were to show the same complexity on this task that he routinely exhibited at the racetrack, this would suggest that cognitive complexity is domain-specific. Regardless of whether this proves true, however, from the present data it appears that IQ is unrelated to the ability to achieve such complexity when other factors like motivation and background knowledge are reasonably well controlled. While greater experience might result in the participant with a higher IQ developing a more complex model than that of the lower IQ participant, this is pure conjecture at this point. Nothing in the data distinguished the two participants—not their actual P/E ratio prediction accuracy, nor the complexity of the models they developed to achieve it.

IQ has been viewed as an excellent index of general intelligence, or "g." It is alleged to reflect the extent to which one's cognitive processes are efficient and flow into *all* cognitive endeavors (Ceci, 1990). Although cases of highly specific abilities can be found, they are usually distinguished from the case of high intelligence (i.e., high IQ) because the latter is assumed to be more general. An individual with a high IQ is supposedly someone who posseses the ability to tackle many tasks at a high level. Yet evidence is beginning to challenge this view, and the present data accord with this challenge.

According to this emerging view, IQ reflects a single type of cognizing rather than the general aptitude for complex thinking that it has been touted to be. Specifically, it appears to reflect academic-verbal performance that is highly related to the possession of a codified set of background knowledge (primarily verbal knowledge gained through schooling, and vital to performance on IQ tests, employment screening tests, and SATs). IQ and other alleged indicants of general ability are good predictors of a wide range of real-world endeavors (school, work, social satisfaction), but it appears that "prediction" and "explanation" are fundamentally disjunctive processes. In the future, psychologists might spend more energy focusing on the causal paths between IQ and real-world accomplishments, in an effort to explain the bases for these predictions. The anticipaton, based on the present findings and on other studies (Ceci & Liker, 1986a; Dörner & Kreuzig, 1983; Gardner, 1983; Schneider, Körkel, & Weinert, 1989) is that it will not be through the generalized ability to engage in complex thinking that IQ tests have their predictive power. Even low-IQ individuals may have the ability to engage in complex multicausal reasoning when care is taken to control for background knowledge and motivation (e.g., Schneider et al., 1989). This claim carries with it the following "moral": Casual impressions are apt to be misleading when it comes to judging an expert. Even an expert with a low IQ may be a good expert and may engage in reasoning that is more complex than that of a nonexpert who possesses a high IQ.

Finally, it is worth noting that the focus of this chapter has been on the likelihood of rediscovering a previously developed form of expertise; no attention was paid to the more basic question of how expertise is initially developed. To answer this, the present data are silent, but the psychology of expertise, coupled with the background findings that motivated the present study, do offer the following insights.

Although background experience is important to develop expertise, it is not simply a matter of exposure to relevant information that fosters this development. All of the men in the racetrack study had attended the races for an equivalent number of years, but not all of them developed equivalent levels of expertise. It is relatively easy to find individuals who have compiled large repositories of factual knowledge through exposure to many relevant activities, but who have no ability to integrate these facts to derive complex insights (Sternberg, 1985). Not everyone who experiences the same environment that Mozart experienced will become as insightful into musical structures as he did. Sheer facts, in the absence of an organizational plan that permits the simultaneous integration and differentiation of the facts, will not help develop expertise. This is because the expertise of the form we have studied is predicated on more than a passive factual retrieval system. Research is converging on the view that the critical factor in the development of expertise is the manner in which facts are integrated and differentiated in one's knowledge base (Chi & Ceci, 1987). Exposure to relevant information is therefore necessary but insufficient for expertise to develop.

There is no single best method for assessing the degree of differentiation/integration of one's knowledge base. It comes down to what is best for the particular problem being studied that will determine how one describes the knowledge base. For the problems that we have been struggling to answer, the best approach is one that preserves several features that we have already learned about the human information processor. First, it is not enough to have the underlying potential to develop a complex insight (e.g., the native cognitive abilities to draw inferences, deduce relations, etc.); one must also have the relevant background knowledge for these abilities to operate. Either through explicit training or through some more casual experiences, one must amass the raw materials (data) that can be integrated and differentiated. Without these materials, there can be no expertise. Even at the most basic level of cognition, the contents and structure of the knowledge base constrain the efficiency of a cognitive operation. It is illusory to think that basic cognitive efficiencies can be assessed in knowledge-free contexts (Ceci, 1990). Depending on how elaborately interconnected the data are, a cognitive operation will be more or less efficient.

In addition to possessing the underlying abilities (or hardware) in order to engage in expert decisions, and to be exposed to the relevant experiences (training or casual) to amass the needed knowledge base for these abilities to operate efficiently, one must also have the motivation to benefit from such experiences. Herein lies a source of important variance in our own studies. Some of the participants in our study of racetrack decision making report having been raised in homes that contained books, word games, and other forms of reading materials. Yet they claim to have had no interest in them while growing up, and this assertion is nearly always borne out by their low verbal intelligence scores. Of course, it can always be argued that a gene-environment correlation is responsible for this failure to take advantage of one's opportunities (i.e., those with poor genotypes for verbal learning avoid reading materials). But the point is nevertheless a valuable one for those concerned with real-world application because it underscores the need to assess one's motivation to benefit from training. This reminder is intended as a tonic for those familiar with the decision theory literature because there it has often been reported that adults, including Stanford and Ann Arbor college students, are miserable at multicausal reasoning, even with simple additive models (e.g., Brehmer, 1980; Nisbett & Ross, 1980). In fact, in his review of the past 30 years of decision theory research, Brehmer offered the following dismal conclusion:

> These results, taken as a whole, strongly suggest that people do not have the cognitive schemata needed for efficient performance in probabalistic tasks . . . even if they are given massive amounts of practice. (p. 233)

It may be that many more individuals than has been thought possess the capacity to engage in expert decision making, because the studies that led to the dismal findings were not motivationally enriched. Needlessly ungenerous estimations of individuals' cognitive complexity may inevitably follow from studies that do not supply adequate incentives for sufficient facts to be acquired, integrated, and differentiated.

Notes

1. Expertise was based on precision at predicting post-time odds rather than on more "intuitive" measures such as the number of race winners picked or the amount of money won. This was done for two reasons. First, it is not possible to gather reliable data on the actual amount of money won or lost. Individuals may over- or underestimate their winnings for a variety of reasons. For example, they may fear that the interviewer will report them to the IRS, as all payoffs that exceed 300 to 1 are supposed to be cashed at the special IRS window; they may not wish word to circulate that they had won a large payoff, for fear that others will want to borrow from them; or they may wish to impress the interviewer with their prowess. So, for these and other reasons, the use of "earnings" as a criterion was not feasible. Concerning the use of "number of winners" as a criterion of expertise, there are other problems having to do with the nature of pari-mutuel wagering. In brief, short of correctly picking winners in over 95% of races (a feat that no one has yet come even close to!), the sheer number of winners that one picks is unimportant. What really matters in pari-mutuel wagering is to avoid overvalued horses and select undervalued ones, that is, ones that the betting public has not bet commensurate with their "true" odds. Even if one were to pick 50% winners (not difficult to do, as the favorite alone wins 38% of races at the tracks studied), one could still lose money. Yet, one *could* actually win money if one picked only 5% winners, provided they were sufficiently undervalued by the public. Although there are many different models of probability that can be applied to racing data, we may take a simple model of independence between two horses in the same race to illustrate why accurately assessing true odds is what racing is all about. Suppose that the number-one horse is the heavy favorite, say 1 to 5 (20¢ on a dollar, or $2.40 return on a $2.00 wager). But suppose the expert assesses this horse's chances of winning to be more like 5 to 2 ($7.00 return for a $2.00 wager). Now suppose that another horse in that race, the number-eight horse, is being bet at 10 to 1 ($22.00 return on a $2.00 wager), even though the expert assesses this horse's "true" odds closer to 4 to 1 ($10.00 return on a $2.00 wager). According to some models, the number-one horse will win 10 out of every 17 direct matchups with the number-eight horse (i.e., at 5:2 true odds, he will win 10 out of 35 races, whereas the number-eight horse, with true odds of 4:1, will win only 7 out of 35). Yet, at 1:5 post time odds, the number-one horse is a poor wager, even though in any given race he has a better chance to win than the number-eight horse (10 to 7). A gambler will definitely lose money betting horses like the number-one horse (e.g., in the example given here, one would lose $46 every 35 matchups, assuming a constant $2.00 wager at 1:5 post-time odds with a horse whose "true" odds were 5:2). Yet, the individual who accurately assessed the mismatch between the number-eight horse's posttime odds and his "true" odds, could win a lot of money even though he selected fewer winners than his counterpart. Thus, the measure of expertise that was used was odds estimation, not winners or earnings.
2. Separate regression analyses were run for each subject in a linear model in which vectors represented the characteristics of each stock on the variables as well as on a composite seven-way interactive term. This latter term was constructed to produce P/E ratios in a manner similar to the way race horses' post-time odds were derived. The dependent variable in these analyses was the

log of the odds of the P/E ratio probabilities (see Ceci & Liker, 1986a for details). A standard stochastic error term was included. The basic model assumes that vectors combine additively and β coefficients represent the relative importance of each each in predicting P/E ratios. Of course, the computer program that generated the P/E ratios was nonadditive, because the purpose of the study was to determine whether experts who deploy this same nonadditive algorithm in another context (racing) would rediscover its relevance in a novel context.

3. Beta was derived from a least squares regression between the percent change in the long-term market average and the percent change in the short-term price of that stock, adjusted for the tendency to regress to the mean.

References

Bartlett, F. C. (1932). *Remembering: A study in experimental and social psychology*. Cambridge: Cambridge University Press.

Bossok, M., & Holyoak, K. J. (1989). Interdomain transfer between isomorphic topics in algebra and physics. *Journal of Experimental Psychology: Learning, Memory, and Cognition*, *15*, 153–166.

Brehmer, B. (1980). In one word: Not from experience. *Acta Psychologica*, *45*, 223–241.

Brown, A. L. (1989). Analogical learning and transfer: What develops? In S. Vosniadou & A. Ortony (Eds.), *Similarities and Analogical Reasoning*. Cambridge: Cambridge University Press.

Brown, A. L., & Kane, M. J. (1988). Preschool children can learn to transfer: Learning to learn and lerning from example. *Cognitive Psychology*, *20*, 493–523.

Brown, A. L., Kane, M. J., & Echols, C. H. (1986). Young children's mental models determine analogical transfer across problems with a common goal structire. *Cognitive Development*, *1*, 103–121.

Ceci, S. J. (1990). *On intelligence . . . more or less: A bio-ecological treatise on intellectual development*. Englewood Cliffs, NJ: Prentice-Hall.

Ceci, S. J., & Liker, J. (1986a). A day at the races: A study of IQ, expertise, and cognitive complexity. *Journal of Experimental Pscyhology: General*, *115*, 255–266.

Ceci, S. J., & Liker, J. (1986b). Academic and non-academic intelligence: An experimental separation. In R. J. Sternberg & R. Wagner (Eds.), *Practical intelligence: Origins of competence in the everyday world*. (pp. 119–142). New York: Cambridge University Press.

Ceci, S. J., & Liker, J. (1988). Stalking the IQ-expertise relationship: When the critics go fishing. *Journal of Experimental Pscyhology: General*, *117*, 96–100.

Chi, M. T. H., & Ceci, S. J. (1987). Content knowledge: Its restructuring with memory development. In H. W. Reese & L. Lipsett (Eds.), *Advances in Child Development and Behavior*, *20*, 91–146.

Dörner, D., & Kreuzig, H. (1983). Problemlosefahigkeit und intelligenz (Problem-solving ability and intelligence). *Psychologische Rundhaus*, *34*, 185–192.

Gardner, H. (1983). *Frames of mind: The theory of multiple intelligences*. New York: Cambridge University Press.

Itzkoff, S. W. (1989). *The making of the civilized mind*. New York: Peter Longmans.

Jensen, A. R. (1980). *Bias in mental testing*. New York: Free Press.

Judd, C. H. (1908). The relation of special training to general intelligence. *Educational Review*, *36*, (June–December), 28–42.

Leshowitz B. (1989). It's time we did something about scientific illiteracy. *American Psychologist*, *44*, 1159–1160.

Nisbett, R., Fong, G., Lehman, D., & Cheng, P. (1988). *Teaching reasoning*. Unpublished Manuscript. University of Michigan, Ann Arbor, MI.

Nisbett, R., & Ross, L. (1980). *Human inference: Strategies and shortcomings of social judgment*. Englewood Cliffs, NJ: Prentice-Hall.

Rogoff, B. (1981). Schooling and the development of cognitive skills. In H. Triandis & A. Heron (Eds.), *Handbook of cross-cultural psychology* (Vol. 4, pp. 233–294). Rockleigh, NJ: Allyn and Bacon.

Schneider, W., Körkel, J., & Weinert, F. (1989). Expert knowledge and general abilities and text processing. In W. Schneider & F. Weinert (Eds.), *Interactions among aptitudes, strategies, and knowledge in cognitive performance*. (pp. 114–136). New York: Springer-Verlag.

Schooler, C. (1989). Social structural effects and experimental situations: Mutual lessons of cognitive and social science. In K. W. Schaie & C. Schooler (Eds.), *Social structure and aging: Psychological processes*. Hillsdale, NJ: Erlbaum.

Spearman, C. (1904). General intelligence objectively determined and measured. *American Journal of Psychology*, *15*, 206–221.

Sternberg, R. J. (1985). *Beyond IQ: A triarchic framework for intelligence*. New York: Cambridge University Press.

Thorndike, E. L. (1905). *The elements of psychology*. New York: A. G. Seiler.

Thorndike, E. L., & Woodworth, R. S. (1901). The influence of improvement in one's mental function upon the efficiency of other functions. *Psychological Review*, *3*, 247–384, 553.

Appendix

The following are sample questions given to the participants to establish their lack of expertise in the stock market. (Correct answer indicated by underscore.)

1. The phrase "50½–51, 200 by 100" is:
   a) a range
   b) a special bid
   c) a size and quotation
   d) a "fill and kill" order

2. When stocks are left in a "street name" by a cash account customer, the "beneficial owner" is:
   a) the brokerage firm
   b) the Depository Trust Company
   c) a nominee
   d) the customer

3. A corporation has 1,000,000,000 shares authorized; 50,000,000 shares were issued and 1,000,000 are in the treasury. How many shares are outstanding?
   a) 1,000,000
   b) 49,000,000
   c) 50,000,000
   d) 100,000,000

4. The job of monitoring proper payment for an acceptable delivery of securities is the responsibility of:
   a) the margin dept.
   b) the cashier's dept.
   c) the purchase and sales dept.
   d) the research dept.

5. Which of the following services is not performed by a mutual fund custodian?
   a) providing investment advice
   b) paying out dividends and distributions
   c) safeguarding the fund's money and securities
   d) maintaining records for the fund

6. Prices on most OTC stocks may be found by looking:
   a) at the composite tape
   b) in the pink sheets
   c) in the Blue List
   d) in the yellow sheets

7. Bonds are normally quoted in:
   a) dollars and cents
   b) percentage par value
   c) decimal fractions
   d) dollars and eighths of dollars

8. Assuming the same issue and maturity dates, the highest yield would be earned by investing in which product?
   a) U.S. government bond
   b) Municipal bond
   c) B-rated corporate bond
   d) AAA-rated corporate bond

9. Bonds usually pay interest:
   a) monthly
   b) every 90 days
   c) every 6 months
   d) annually (only)

10. An investor purchases 100 shares of ABC Corp.'s common stock (par value $1 per share) for $5,000. If ABC Corp. goes bankrupt, the investor could not lose more than:
    a) $100
    b) $4,000
    c) $5,000
    d) Impossible to determine, as the investor's personal property may be attached.

Part V
Expert-Novice Differences and the Development of Expertise

14
Expert-Novice Differences and Knowledge Elicitation

Mícheál Foley and Anna Hart

Introduction

Some years ago it was stated that knowledge acquisition is a major bottleneck in the development of expert systems (Feigenbaum & McCorduck, 1984), and this view is still propounded. Although it is not unusual for knowledge engineers to experience great difficulty in getting experts to verbalize and formalize their knowledge, there is another fundamental problem. Often the overall objective of knowledge elicitation is to construct a computer-based system. In such a case the usefulness of the system, and therefore of the knowledge modeled in it, is assessed at least in part by the system's users. Knowledge engineering therefore involves constructing models of knowledge that can be validated by experts and that prove useful to the intended users. This means that the knowledge elicitation process should be viewed in the context of system design, and not as mining out an expert's knowledge. As Kidd says (1987), the process should take into account the different classes of users who are likely to use the system, their requirements, and the types of knowledge they bring to the problem-solving process.

The importance of the usability of a system's knowledge is stressed by Wyatt and Speigelhalter (in press) in their proposed guidelines for evaluating medical decision aids. They make a clear distinction between the validity and accuracy of the knowledge, as assessed by expert authorities, and its usefulness to the users in a practical problem-solving environment. Expert systems are often intended for use in situations that involve judgment under uncertainty and so cannot be prove correct. In such cases the usability and effectiveness of the systems are of paramount importance.

Often, systems are designed for users who are relative novices in the domain compared with the experts who supply the knowledge. Thus, the design process should take into account the differences between novices and experts. Our program of research is designed to further a general understanding of the differences between experts and novices and to study the implications for knowledge engineering.

This chapter has two aims: One is to outline what is already known about expert-novice differences, and the other is to identify key issues for knowledge engineering. There are four main sections: The first presents evidence of the importance of expert-novice differences from expert systems literature and identifies shortcomings in common methods for system development. The second summarizes the published literature on expert-novice differences. In the third section we describe an informal study that was carried out as part of our research program, and the fourth section identifies outstanding issues for knowledge engineering.

Evidence of the Need to Consider Expert-Novice Differences

Effects of the Differences

An interesting experiment that clearly shows the importance of expert-novice differences was carried out by Lamberti and Newsome (1989). They studied the relative usefulness of an expert system in the hands of experts and novices. Lamberti and Newsome were particularly concerned with the distinction between "abstract" and "concrete" representations, although they acknowledged that the terms can be "defined operationally in a number of different ways". (pp. 28, 29) The expert system was designed to assist diagnostic programmers who had to help customers who rang them about problems with system software. The expert system suggested questions to ask and provided limited explanations of why a question was asked and how a recommendation was reached. The experiment involved the solution of a sample problem in three ways: first, without the system at all; second, with a system that used abstract questions; and third, with a system that asked concrete questions. The questions were categorized as abstract or concrete by an independent set of experts. During the experiment, measures were made of the overall time taken, the time taken to answer each question, the accuracy of the answers, and the user's confidence in the system recommendation. Both experts and novices performed the task more quickly and more accurately using the system. The experts outperformed the novices whether or not they used the system, and the system was of more benefit to novice performance. Novices were quicker and more accurate in answering concrete questions than in answering abstract questions, whereas experts were quicker at answering abstract questions. Experts were actually slower than novices at answering concrete questions but were faster at answering abstract questions. Finally, the confidence the users had in the solutions was directly related to the solution time.

The distinction between abstract and concrete representations is one relevant expert-novice difference, but there may be many more. Wyatt and Emerson (1990) describe an interesting example of a system in which consideration of users' skills was critically important. The project was to build an expert system to assist in screening patients who had arrived at a casualty department (emergency room) complaining of chest pain. The system was for use by nurses in order to decide whether or not a patient was a high-risk cardiac patient and therefore needed to see a physician quickly. Wyatt and Emerson found that the diagnostic symptoms that were powerful in the hands of the physicians were not the same set as those that were usable by or useful to the nurses. It was necessary to undertake a very careful study of the symptoms and diagnoses before Wyatt and Emerson could build a suitable system. Wyatt and Emerson term their approach *pragmatic*, but there is much to be learned from their experience. Some time ago Clancey (1983) reported that knowledge bases cannot be viewed as independent of their mode of use. The way in which the knowledge base is structured limits the way in which it can be used. Knowledge about the users, about how they differ from the experts, and about the actual functional role of the system should be considered from the start of a project.

Deficiencies in Methods for System Development

It is commonly acknowledged that user involvement in system design is desirable, but there is less of a consensus as to how this can be achieved. Prototyping is commonly used as a method for system development, and prototype systems are often used to elicit information from the intended users. This can be very useful when the users are not able to explain exactly what their requirements are. It is much easier for them to offer constructive criticism of a working demonstration than to specify detailed needs in advance, especially when they are not fully aware of the capabilities of expert systems. Edmonds, Candy, Slatter, and Lunn (1990) go further and advocate that the system interface should be the first part of the system to be prototyped, that this should be used to elicit requirements, and that the design of the interface should thus drive the overall design.

Others take a more traditional view, akin to that of software engineering, in which the objective is to produce a complete and correct system specification. (Somerville, 1985). Diaper (1989) is in favor of this approach and is very skeptical about the effectiveness of prototyping. The KADS methodology (Breuker & Weilenga, 1987) aims to produce a thorough analysis of the knowledge before any implementation, although later prototyping is not precluded. The analysis comprises three stages: orientation (identifying terminology and concepts), problem identification (constructing a domain structure), and problem analysis (ensuring that the interface is suitable for the task). Each of the three stages involves a cycle of elicitation of information and subsequent analysis of it. Breuker and Weilenga suggest which knowledge acquisition and elicitation methods are appropriate at each stage, given that the stages concentrate on different levels of description. The motivation behind the more traditional approaches is that one should put a lot of effort into analysis and paper formalisms before any implementation takes place.

Some representations are referred to as *intermediate representations*, because they are used during knowledge modeling but tend to be independent of the mode of implementation. Johnson (1989) prefers to call her representations, which are based on systemic grammar networks, *mediating representations*, because she claims that they mediate "between verbal data and standardised knowledge representation schemes". (p. 181) Her aim is to "provide both a medium of communication between members of a team and a grammar of the expert's task". (p. 184)

None of the methods outlined above takes explicit account of the differences between experts and novices. KADS does acknowledge that system development is a modeling process and that the ideal system from the users' point of view may not be one that tackles all domain problems and may not behave "as an expert." It is deficient in that users are considered rather late in the process and have a major contribution after the domain has been structured. This means that mistakes may already have been made—the domain may have been structured in a way that prevents users from understanding it. Edmonds et al.'s use of interface prototypes comes closest to avoiding this problem.

Shpilberg, Graham, and Schatz (1986) describe a method for overcoming some of these difficulties. They devised a way of simulating the interaction between experts and user, and used this as a means of knowledge elicitation. Their assumption is that it is more important to model the ways in which experts and novices interact than to model the experts' knowledge. In this approach it is possible to identify some commonalities and differences between the expert and users in terms of the use of terminology and general strategy. A study of such differences can form an important part of knowledge analysis.

Whereas some expert-novice differences are domain-specific, others are more general. Neither the experts nor the novices are likely to be explicitly aware of these differences. It is the responsibility of the knowledge engineering community to try to understand both the differences and how a consideration of them can influence system design and knowledge elicitation.

Studies in Expert-Novice Differences

General

There have been several studies of expert-novice differences, although not primarily in the context of knowledge engineering. Here we present a summary of the published literature followed by a description of a preliminary study from our research. Before studying expert-novice differences, it is necessary to define *expert*. In knowledge engineering an expert is generally considered to be someone who has attained a high level of performance in the domain, as a result of years of experience. In experimental work sometimes experts of this category are used, but on occasions experts are "trained up" on artificial problems. Furthermore, the degree to which "novices" are truly novice varies considerably. The results are nonetheless interesting.

Differences in General Strategy

Early work was concerned with general strategy. People were asked to solve puzzles, and their strategies were analyzed. Newell and Simon (1972) suggested that means-end analysis is the basis for solving problems where there is an identifiable goal. A number of puzzles have been used, including the towers of Hanoi problem, and the results show that means-end analysis is used to some degree (Greeno, 1978).

Some well-known experiments were carried out to study experts and novices as they solved physics problems. The pioneers of this work were Simon and Simon (1978) and Larkin, McDermott, Simon, and Simon (1980a, 1980b). They found that experts differed from novices in four ways: They were able to reason forward and use the given facts to generate the equations they used; they solved the problems more quickly; they mentioned the equations they were going to use and then wrote them down with the values already substituted in; and some evidence showed that they generated some physical representation of the problem as opposed to a mere syntactic translation. Interestingly, when more novel problems were used, the experts reverted to a means-end method.

Sweller, Mawer, and Ward (1983) designed experiments to show the change in strategy from means-end to forward reasoning. They studied a group of subjects who were considered to be novice at the start of the experiments and then relatively expert later on. They did find a switch in strategy from a means-end approach to a forward-reasoning method.

De Groot (1965, 1966) and then Chase and Simon (1973) studied differences in the skills of chess players. De Groot (1965) analyzed the strategies and concluded that there were no differences in the breadth or the depth of search of possible moves, but that the difference lay in the players' representations of the problem.

Differences in Representation

Given the results of de Groot (1965), it was conceivable that experts are distinguished from novices merely by a superior general perceptual memory ability. This was refuted by de Groot (1966) Chase and Simon (1973) in studies of the ability of subjects to recall configurations of chess pieces on a board. Experts were far superior to novices in reconstructing valid board configurations, but not random ones. Chase and Simon concluded that experienced players perceive relations between chess pieces as "chunks" and so are able to remember a large number of pieces.

Similar results have been obtained in the tasks of playing the game of go (Reitman, 1976); remembering electronic circuit diagrams (Egan & Schwartz, 1979); and playing bridge (Engel & Bukstel, 1978). In the domain of computer programming, several experimenters have studied people's ability to recall lines of coding presented either randomly or in program order. In general, experts were good at recalling coding in program order but bad at recalling randomly ordered lines, whereas naive programmers were equally proficient at recalling lines independent of the ordering (Barfield, 1986; McKeithen, Reitman, Rueter, & Hirtle, 1981).

Further work attempted to analyze the nature of the chunks. McKeithen et al. (1981) set up an experiment requiring programmers to recall reserved words from ALGOL W, the aim being to uncover any knowledge structures. Experts tended to use programming concepts to organize a structure, whereas novices were inclined to use the literal English meaning of the words. Adelson (1981) used lines of PPL code. She found that experts' chunks contained more lines than those of the novices, that novices tended to cluster lines together on the similarity of syntax, and that there was more consistency between the experts' hierarchies than between those of the novices.

Chunking is sometimes described as "abstracting" away from the problem as it is presented. One argument is that experts form an abstract model of the problem, unlike novices (Larkin et al., 1980a). Another view is that both form representations, but the novices use concrete features and the experts use abstract ones. As indicated earlier in this chapter, the distinction between what is abstract and what is concrete depends on the domain and may be defined subjectively. Adelson tried to pursue this matter with further experiments in the domain of computer programming (Adelson, 1984) and concluded that novices do not use abstraction to the same degree as experts.

The different knowledge structures will result in differences between experts and novices when they describe various diagnostic categories. Murphy and Wright (1984) studied the categories and features used to describe psychologically disturbed children. They restricted the problem to three categories, which were chosen from lists produced by experts, but which were also usable by the novices. Subjects were asked to list the features they associated with each category. The experts produced more features than did the novices, and the experts were more consistent as a group. The experts' categories were overlapping, whereas the novices' categories were totally distinct. Furthermore, there was no evidence that the differences were due to the use of technical jargon. This means that experts are more likely to be aware of the complexity of a situation. In addition to being aware of complexities of a problem, experts react differently to apparently inconsistent information. Fiske, Kinder, and Larter (1983) showed that experts and novices react in different ways when presented with information that is inconsistent with their prior beliefs. Novices tend to be very much influenced by their initial expectations.

In summary, there is evidence that there are expert-novice differences in terms of the methods of reasoning, knowledge structures, and the ways in which information is processed. Each of these differences has implications for the expert system designer. There is scope, therefore, for further work with special regard to the implications for knowledge engineering.

Exploring the Domain of Computer Program Debugging

Previous Work in the Domain

One of Adelson's (1981) experiments was concerned with debugging skills. She was interested in the differences between "abstract" and "concrete" representations. Her fairly complicated experiment required subjects to answer questions about a program after either studying the program, solving a supposedly concrete problem associated with the program, or solving an abstract problem associated with the program. The idea was that in the first case the subjects would form "natural representations," whereas the other two problems would induce representations of the same type as the problem. Adelson's (1981) interpretation of the results, measured by subjects' abilities to answer the different types of questions, was that the natural representation formed by experts is an abstract one, whereas novices' representations are concrete. In fact, the novices were actually more accurate than the experts in answering concrete questions.

Vessey (1989) also studied expert-novice differences in the problem of debugging. She was interested in developing a theory of debugging. Vessey selected subjects who did debugging as part of their professional work. Some subjects had a long period of experience, and others had only a short period of experience. They were presented with a number of programs, each of which contained a bug. She suggested that the main factor in finding a bug was its position on a hierarchy of types of bug and that for each bug an appropriate mental model was required by the subject. Her results showed that the position of the bug in the program did not affect performance and that experts were quicker and more accurate with all of the types of bugs she used. Experts consistently used a breadth-first search. Novices also used breadth-first searches but were less able to think at a system level and sometimes resorted to depth-first searches. This implied that experts did not look straightaway for the module containing the bug, but analyzed the whole program, unearthing the bug in the process. Novices also did this but were inclined to focus too narrowly on one possibility.

Rationale for Study

Program debugging has proved to be a useful domain for studying expert-novice differences. We carried out a study of debugging as an exploratory investigation into the importance of expert-novice differences in knowledge engineering. The study was designed to provide data to formulate theories rather than to test some aspect of an existing theory. Debugging is a good example of one of the most common types of expert system application, namely faultfinding. The aim of the study was to examine the ways in which experts and novices used their knowledge to tackle problems. The areas of interest were the use of terminology and the strategies employed by participants in both easy and difficult problems. The success rate was not of prime concern, and some of the experts did not manage to solve the most complex problem.

Participants

There were three groups of participants, namely, first-year students, second-year students, and staff. There were eight students from each year group, and all were enrolled in courses involving a significant amount of programming in Pascal. The staff members were taken from the advisory group in the Computer Centre, which provides help for students' programming problems. It was assumed that the staff could be considered experts at debugging other people's programs and that students were relative novices. The extra programming expertise of the second-year students was seen as a possible indicator of a degree of debugging expertise. This then created three groups of expertise in debugging. Staff members were the experts, second-year students were intermediates, and first-year students were novices.

Procedure

Each participant was required to solve three debugging problems. A training problem (which was an error in instructions in English for washing clothes in the launderette) was conducted first to ensure that the procedure was understood. Because the training problem was composed of a simple algorithm, it enabled participants to get used to the procedure that was used, but it gave no clues as to how the debugging problems should be approached. For example, there was no possibility of the experimenter's having to talk about "procedures" or "variables," which might have given clues about the real problems. During the trial run it was possible for a participant to ask questions about the procedure, but thereafter the conversation was limited to the experimenter's asking for more precise information if required. It was explained that each program compiled, ran, and produced output, but that there was a bug in the program. The order of the last two problems was reversed for a random half of the subjects, to allow for ordering effects. The first problem was relatively easy.

The programs were divided into sections printed on cards. These sections corresponded to natural blocks of the program, for example, procedures. Only one such card was in view at any time. Two extra cards were included: one to describe what the program should do (the statement card), and one to show the erroneous

TABLE 14.1. Statement card for Problem 1.

This is a description of what the program SORT is meant to do:

This program is meant to:
(a) ask the user for an array size;
(b) set up an array of this size with random integers;
(c) print this array (before sorting);
(d) sort this array;
(e) print this array (after sorting);
(f) repeat (a) to (e) until a negative size is entered.

Files used: input (keyboard), output (screen).

TABLE 14.2. Output card for Problem 1.

```
Enter number of elements (-ve to exit): 15
Before:   5  15  30  31  33  41  51  52  59  62  78  78  84  84  88
After:    5  15  30  31  33  41  51  52  59  62  78  78  84  84  88

Enter number of elements (-ve to exit): 16
Before:   2  10  27  39  49  50  54  55  75  76  84  89  91  94  95  96
After:    2  10  27  39  49  50  54  55  75  76  84  89  91  94  95  96

Enter number of elements (-ve to exit): -1
```

TABLE 14.3a. Card 1 for Problem 1.

```
program sort (input,output);
const max=100;
type items=array[1 . . max] of integer;
var  a:items;
     ascending:boolean;
     num:integer;
```

TABLE 14.3b. Card 2 for Problem 1.

```
procedure initialize (var a:items; num:integer);
var i:integer;
begin
  for i:=1 to num do a[i]:=random (100)
end; {initialize}
```

output produced by the program (the output card). Tables 14.1, 14.2 and 14.3 show sample cards for one of the problems.

For each of the three problems, information was elicited in two tasks. First, participants were required to complete a form for each problem. These forms were composed of three columns. In each row the participant had to enter the number of the card currently being examined, any request for a new part of the program, and the reason for that request. An example of a completed card for one problem is shown in Table 14.4. From this it was possible to reconstruct the search path that the participants took through the cards, the way in which they referred to each card, and the reasons they gave for the card requests. The second task was of the nature of a "think-aloud" task. Although the study was designed to capture most of the information in writing, pilot work indicated that it was impractical to capture all of the information in this way. The sessions were tape-recorded, and the

TABLE 14.3c. Card 3 for Problem 1.

```
procedure sort_array (var a:items; num:integer;
  ascending:boolean);
var first, x:integer;
    i,j,p:integer;
begin {sort_array}
  first:=a[1];
  p:=1;
  for i:=2 to num do
    if (ascending and (a[i]<first)) or ((not ascending) and
      (a[i]>first))
    then
      begin
        first:=a[i];
        p:=i
      end;
  a[p]:=a[1];
  a[1]:=first;
  for i:=3 to num do
    begin
      x:=a[i];
      j:=i;
      while (ascending and (x<a[j-1])) or ((not
        ascending) and (x>a[j-1])) do
        begin
          a[j]:=a[j-1];
          j:=j-1
        end;
      a[j]:=x
    end
end; {sort_array}
```

TABLE 14.3d. Card 4 for Problem 1.

```
procedure print_array(var a:items; num:integer;
  prompt:string);
var i:integer;
begin
  write(prompt);
  sort_array(a,num,ascending);
  for i:=1 to num do
    write(a[i]:4);
  writeln;
end; {print_array}
```

TABLE 14.3e. Card 5 for Problem 1.

```
begin {main}
  randomize;
  ascending:=true;        {we want an ascending sort}
  num:=1;                 {makes while true first time}
  while num>0 do
  begin
    repeat
      writeln;
      write('Enter number of elements (-ve to exit): ');
      readln(num); writeln(num);
      if (num>max)
      then writeln('Sorry, the maximum number of
        elements is: ',max:3)
    until (num<max);
    if num>0 then
    begin
      initialize(a,num);
      print_array(a,num,'Before: ');
      sort_array(a,num,ascending);
      print_array(a,num,'After: ')
    end
  end
end.
```

transcripts were particularly useful for recording hypotheses or ideas that participants could not follow up immediately (e.g., if they wanted two cards but chose to look at one first) and general observations that they did not feel warranted writing down.

Each problem began with the participants' examining the statement and output cards and saying what the error was and what might be causing it. They then had to try to solve the problem by locating the bug in the program and correcting it. Participants had to ask for a particular card by writing down which part of the program they wanted to see, and why. If this request was not specific enough, the experimenter asked for a more precise request. Each time a new program card was given, the previous one was taken back. The number of the new card was entered in the left column of the form. It was possible to go back to a card that had been viewed. The task continued until the participant either resigned or felt that the problem was solved. In the former case, participants wrote down any ideas about where they thought the bug was and what they though it might be. In the latter case, they wrote down the proposed solution.

TABLE 14.4. A sample written form showing the number of the card being viewed and the participant's written request, and reason, for the next card.

Subject:	1d (novice)
Task 1:	sorting numbers
Card:	none
Request:	Card that shows the output procedure that prints the inputted line of data to the screen (I am looking for writeln statement).
Reason:	It seems as if the same line is being printed twice, after the sort procedure has already taken place.
Card:	4
Request:	The card showing the definition of 'PROMPT' where prompt is defined as 'before' or rather where there are string definitions when prompt is defined as "enter number of elements" before PRINT_ARRAY is called.
Reason:	By doing this it is possible to show where the program has gone to by the time that the output is required.
Card:	5
Request:	4
Reason:	Check my assumptions are correct.
Card:	4
Solution:	SORT_ARRAY should not be called within the PRINT_ARRAY procedure as it is not needed here as it causes the imput data to be sorted before being viewed.

The Problems

Problem 1 was fairly simple. The program procedures were called in the wrong order. There were relatively few possible explanations for the actual output. Problem 2 involved the incorrect syntax for two nested IF statements within the program. Problem 3 was the most complex. The program processed transactions and updated a master file of bank accounts. The main loop of the processing part should have finished only when all of the old master file and the transaction file had been checked. An AND had been coded as OR

TABLE 14.5. General summary of results, showing the average number of card requests and the success rates for the different groups and problems.

	Problem	Novices	Intermediates	Experts
Number of card requests per subject per example	1	3.4	5.0	3.0
	2	3.6	4.3	1.8
	3	5.4	6.6	6.3
Number of participants who believed they had solved the problem (actual number who had solved the problem)	1	8	7	4
		(6)	(7)	(4)
	2	7	5	4
		(6)	(4)	(4)
	3	6	5	3*
		(1)	(0)	(1)

* These three experts were very close to the answer, unlike the five novices and five intermediates who misjudged their success.

within a logical condition, so that processing stopped as soon as the end of either file was reached. As a result, the last account from the old master was not in the new master. The possible causes of this were not apparent in the program output.

Results and Analysis

For each problem the reasons written on the forms were examined to see whether there were any common approaches among participants. For example, in Problem 1, all reasons concerned with investigating the process of random number generation were put in one category, and those relating to a problem in the order of execution of the procedures were put in another. For each problem the resulting categories were used to classify the hypotheses that had been expressed. The initial statements made by the participants were similarly categorized. All of the requests on the forms were cross-referenced to the card to which they referred, so it was possible to see how terminology varied among participants. The search paths taken by each person were extracted from the sequence of card numbers. Table 14.5 gives a simple analysis of the number of card requests made by the three groups of each problem and their success rates. This is useful background data for the overall analysis.

Terminology used by the participants was influenced by the terms introduced on the cards. Terms used on the statement cards were probably most important in this respect, but the effects of terms in the programs themselves (e.g., procedure names) could not be discounted. It was therefore important to take into account the order in which the cards had been viewed when examining the use of terminology. As a general principle, novices were more inclined to use the terminology with which they were presented. Table 14.6 illustrates this with one of the simplest examples, although similar differences were manifest throughout the problems. The term *initialize* is the actual procedure name, which had already been seen by the 10 participants who used it. *Set up* is the terminology from the statement card for this problem (see Table 13.1), whereas *generate* is terminology introduced by the participants.

TABLE 14.6. Breakdown of terminology used to request Card 1 in Problem 1; *set up* is the terminology from the statement card, *generate* is produced by the participants, and *initialize* is the actual procedure name.

Terminology used for request	Novices	Intermediates	Experts
Generate	12.5%	0%	50%
Set up	50%	37.5%	0%
Initialize	37.5%	62.5%	50%

Not surprisingly, some novices used standard computing terminology while not being

TABLE 14.7. The percentage of each group that described each problem in terms of the symptoms, as opposed to a possible solution.

Problem	Novices	Intermediates	Experts
1	12.5%	25%	25%
2	37.5%	50%	0%
3	100%	100%	100%

TABLE 14.8. The percentage of card requests that were related to general understanding (as opposed to bug finding) for each group and problem.

Problem	Novices	Intermediates	Experts
1	18.5%	22.5%	16.7%
2	24.1%	35.3%	0%
3	37.2%	60.4%	78.6%

aware of its precise meaning. This meant that some of their statements, although apparently correct, betrayed misunderstandings when seen in context. This was observed with six novices and three intermediates, but no experts.

In general, participants followed the same pattern of describing the easier problems in terms of the assumed solution, whereas Problem 3 was described in terms of the surface features of the error. For example, one expert gave these three descriptions for the three problems: (1) "The output does not actually say what was actually done. It is more a problem with presentation as I see it. In the first set of numbers you have to look fairly closely to see there is any difference at all. These numbers are supposed to be random but they are all in increasing size already." (2) "The fact that there are 10 items all together and the fact that the positive integer count is also 10 would lead me to believe that the check for positive integers does not actually check for positive integers. It just adds one to the total whatever. The last count appears to be okay as there are three odd integers in that range. I am not sure why it says there are three odd integers in that range. I am not sure why it says there are only two not in the range one to nine." (The expert misinterpreted positive integer as even integer here.) (3) "It has left the last account off." The novices and intermediates showed a similar shift, for example: (1) "The before array is already sorted," (3) "Mr. Clews is being left off." The data for this analysis is shown in Table 14.7.

Following from this, the general strategies for Problem 3 were different from those on the other two problems. Participants tended to use "bug-finding" strategies on the first two problems, but "general understanding" enquiries on Problem 3. Table 14.8 shows the frequencies of card requests classified as bug finding or general understanding. Note in particular the way in which the experts' strategies changed for Problem 3.

An examination of the transcripts suggested that at any one time experts were considering more possible hypotheses than were the other groups. (This study was not designed to capture data for this, but later experiments in a different domain have confirmed this observation.) Experts' behavior suggested that they were more willing to drop one hypothesis in favor of another, and, unlike the novices, they did check that their solution would actually rectify the error. The data for these conjectures are incomplete, and thorough investigation would require further experimentation.

In summary, there was suggestive evidence of important differences in this domain. Differences relate to the participants' models of the problems themselves and to the strategies used. Similar differences are likely to emerge in other diagnostic tasks. We advise that knowledge engineers should take note of the differences during knowledge elicitation and construct methods to highlight such differences.

Outstanding Questions for Knowledge Elicitation

The existing literature and our study highlight some interesting issues for knowledge elicitation. Consideration of some of the issues can be incorporated into current knowledge engineering practice, whereas other matters demand further research.

One very simple issue is that the terminology used by experts and novices is likely to differ. This will mean that sometimes experts and novices have a different vocabulary, that sometimes novices may have an impoverished understanding of the terms and the assumptions behind definitions, and that sometimes novices' and experts' conceptual structures may be different. It is important that the users of a system understand the terminology used by the system, that it is "natural" to them, and that they are able correctly to interpret input and output. It is therefore important to investigate the commonalities and differences in expert and novice terminology early on in a project. It is important to note that novices may appear to understand terms and may construct apparently correct statements, but subsequent decisions or questions may betray a lack of understanding.

Second, experts and novices are likely to have different representations of problems, assumptions, and hypotheses. They are therefore likely to use different strategies. Furthermore, the strategies used by experts are likely to be essentially different on "hard" problems versus "easy" ones. In other words, a knowledge elicitation exercise on a "hard" problem will elicit different knowledge from the same technique used on an "easy" problem. Experts are likely to be more versatile and to consider a more diverse and rich set of alternatives; they may interpret evidence in different ways and keep an open mind for longer. This is very important given the almost rigid nature of many expert system interactions. The system should not encourage novices to have confidence in bad solutions, but it should increase confidence in good ones. It may have to suggest more alternatives or issue cautionary advice without causing total confusion.

In any domain the following questions are important: Do experts and novices use terminology differently? If so, how? Can a common terminology be established? Do experts and novices use different strategies? How do experts and novices differ in the number, level, and type of hypotheses that they consider during problem solving? How do experts and novices differ in their confidence in the different hypotheses and in their perceptions of the necessary and sufficient evidences which they consider? Can novices understand a system that reasons like experts? What information can be elicited from the novices, when, and how? Can an analysis of experts and novices tell us whether an expert system is feasible for a domain? How do novices respond to different forms of interaction in terms of the numbers of alternatives suggested by the system, the levels of confidence displayed by the system, and a user-driven or system-driven interaction?

Our research is addressing some of these questions, with particular regard to the numbers of hypotheses considered by experts and novices, and their confidence in those hypotheses. After we have an understanding of these issues it should be possible to devise knowledge elicitation methods that incorporate novices and that result in models of the knowledge they can use to effect.

References

Adelson, B. (1981). Problem solving and the development of abstract categories in programming languages. *Memory and Cognition*, *9*, 422–433.

Adelson, B. (1984). When novices surpass experts: The difficulty of a task may increase with expertise. *Journal of Experimental Psychology: Learning, Memory, and Cognition*, *10*, 483–495.

Barfield, W. (1986). Expert-novice differences for software: implications for problem solving and knowledge acquisition. *Behaviour and Information Technology*, *5*, 15–29.

Breuker, B., & Weilenga, B. (1987). Use of models in the interpretation of verbal data. In A. Kidd (Ed.), *Knowledge acquisition for expert systems* (pp. 17–44). New York: Plenum.

Chase, W. G., & Simon, H. A. (1973). Perception in chess. *Cognitive Psychology*, *4*, 55–81.

Clancey, W. J. (1983). The epistemology of a rule-based expert system: A framework for explanation. *Artificial Intelligence*, *20*, 215–251.

de Groot, A. D. (1965). *Though and choice in chess*. The Hague: Mouton.

de Groot, A. D. (1966). Perception and memory versus thought: Some old ideas and recent findings. In B. Kleinmuntz (Ed.), *Problem solving: Research, method, and theory* (pp. 19–50). New York: Wiley.

Diaper, D. (1989). Designing expert systems: From Dan to Beersheba. In D. Diaper (Ed.), *Knowl-*

edge Elicitation: Principles, Techniques and Applications (pp. 15–46). Chichester, England: Ellis Horwood.

Edmonds, E., Candy, C., Slatter, P., & Lunn, S. (1990). Issues in the design of expert systems for business. In D. Berry & A. Hart (Eds.), *Expert systems: Human issues*. London: Kogan Page 98–120.

Egan, D. E., & Schwartz, B. J. (1979). Chunking in recall of symbolic drawings. *Memory and Cognition*, *7*, 149–158.

Engel, R. W., & Bukstel, L. (1978). Memory processes among bridge players of differing expertise. *American Journal of Psychology*, *91*, 673–689.

Feigenbaum E., & McCorduck, P. (1984). *The fifth generation*. London: Pan Books.

Fiske, S. T., Kinder, D. R., & Larter, W. M. (1983). The novice and the expert: Knowledge-based strategies in political cognition. *Journal of Experimental Social Psychology*, *19*, 381–400.

Greeno, J. G. (1978). Natures of problem-solving abilities. In W. K. Estes (Ed.), *Handbook of learning and cognitive processes*. Hillsdale, NJ: Erlbaum.

Johnson, N. E. (1989). Mediating representations in knowledge elicitation. In D. Diaper (Ed.), *Knowledge Elicitation: Principles, Techniques and Applications* (pp. 179–194). Chichester, England: Ellis Horwood.

Kidd, A. L. (1987). *Knowledge acquisition for expert systems: A practical handbook*. London: Plenum.

Lamberti, M., & Newsome, S. L. (1989). Presenting abstract versus concrete information in expert systems: What is the impact on user performance? *International Journal of Man-Machine Studies*, *31*, 27–45.

Larkin, J., McDermott, J., Simon, D., & Simon, H. (1980a). Expert and novice performance in solving physics problems. *Science*, *208*, 1335–1342.

Larkin, J., McDermott, J., Simon, D., & Simon, H. (1980b). Models of competence in solving physics problems. *Cognitive Science*, *4*, 317–345.

McKeithen, K., Reitman, J. S., Rueter, H., & Hirtle, S. C. (1981). Knowledge organization and skill differences in computer programmers. *Cognitive Psychology*, *13*, 307–325.

Murphy, G. L., & Wright, J. C. (1984). Changes in conceptual structure with expertise: Differences between real-world experts and novices. *Journal of Experimental Psychology: Learning, Memory and Cognition*, *10*, 144–155.

Newell, A., & Simon, H. A. (1972). *Human problem solving*. Englewood Cliffs, NJ: Prentice-Hall.

Reitman, R. S. (1976). Skilled perception in go: Deducing memory structures from inter-response times. *Cognitive Psychology*, *8*, 336–356.

Shpilberg, D., Graham, L. E., & Schatz, H. (1986). EXPERTAX: An expert system for corporate tax planning. *Proceedings of the second International Conference on Expert Systems* (pp. 99–123). Oxford: Learned Information.

Simon, D. P., & Simon, H. A. (1978). Individual differences in solving physics problems. In R. S. Seigler (Ed.), *Children's thinking: What develops*. Hillsdale, NJ: Erlbaum.

Somerville, I. (1985). *Software engineering* (2nd edition) Wokingham, UK: Addison-Wesley.

Sweller, J., Mawer, R. F., & Ward, M. R. (1983). Development of expertise in mathematical problem solving. *Journal of Experimental Psychology: General*, *112*, 638–661.

Vessey, I. (1989). Toward a theory of computer program bugs: An empirical test. *International Journal of Man-Machine Studies*, *30*, 23–46.

Wyatt, J., & Emerson, P. (1990). A pragmatic approach to knowledge engineering with examples of use in a difficult domain. In D. Berry & A. Hart (Eds.), *Expert Systems: Human Issues*. London: Kogan Page 65–78.

Wyatt, J., & Speigelhalter, D. (in press). The evaluation of decision technology: 2. Laboratory Testing. *British Medical Journal*.

15 When Novices Elicit Knowledge: Question Asking in Designing, Evaluating, and Learning to Use Software

Robert Mack and Jill Burdett Robinson

Introduction

This paper describes a qualitative empirical method aimed at uncovering what computer users need to know to use a computer to accomplish tasks. The method asks users to acquire information about how to use a computer by asking questions of a more experienced user (i.e., the investigator) or a "coach" (a term we prefer). The technique is both similar to and different from qualitive think-out-loud (TOL) methods for eliciting verbalizations related to thinking and problem solving. TOL verbal protocol techniques have been widely applied in cognitive psychology (see Ericsson & Simon, 1980, 1984). Although question asking may have equally wide applicability, the focus of this paper will be on its use in studying human-computer interaction issues connected with the design and evaluation of computer systems.

We begin with a brief overview of question asking and question answering in the broader context of cognitive and social psychology. We then briefly review the few accessable studies of computer user behavior that have applied question asking to the evaluation of computer software. This includes a more in-depth discussion of a software evaluation carried out in our lab using question asking. Applications of question asking have discussed the technique mostly in an anecdotal way, without grounding the method in the larger context of possible qualitative methods, or methodological and interpretive issues (but see Kato, 1986). In a general discussion we summarize and discuss in more depth methodological issues and guidelines for practitioners of the method. We also sketch possible research directions for better understanding the methodology.

Background

Asking questions is a basic linguistic strategy people have for communicating and, in particular, acquiring knowledge about the physical or social world. Not surprisingly, question asking and question answering are basic tools of empirical inquiry in social and cognitive science for learning what someone may or may not know about some domain. More recently, question asking and answering have become topics of research in their own right in cognitive science, particularly through research of Graesser and colleagues. (For review of research in cognitive psychology and artificial intelligence see Graesser & Black, 1985.) In this paper, we focus on question-asking methods in a new domain, that of human-computer interaction (HCl), where a key focus is developing methodologies for evaluating and diagnosing problems with users' interaction with computer systems.

The HCl domain is multidisciplinary, with basic and applied goals, and its methods and professional membership draw on traditional software human factors and ergonomics, as well as cognitive psychology and computer science. Its goals are (a) to understand and improve the use of computers as tools for supporting individuals and groups in their work; (b) to apply research methods and conceptual frameworks successful in other domains of cognitive science (not only psychology but also, for example, anthropology (Suchman, 1987) and artificial intelligence research) in order to guide development of useful, usable computer tools; and (c) to discover principles of human-computer interaction. (The interested reader can find discussions of the discipline in Card & Newell, 1985; Carroll & Campbell, 1986; Landauer, 1988; Norman, 1986; Whiteside, Bennett, & Holtzblatt, 1988; and Whiteside & Wixon, 1988.)

The purpose of software evaluations in the domain of HCl is typically, on the one hand, to provide empirical user feedback aimed at improving specific systems under design and, on the other, to understand and develop general principles of human interaction with computers. Qualitative methods, such as TOL, have proved to be useful techniques for probing possible cognitive mechanisms underlying user performance and problems. (See Bainbridge, 1979, in the traditional human factors domain, and Bott, 1979; Gould, 1987; Lewis, 1982; and Mack, Lewis, & Carroll, 1983; in the human-computer interaction domain). It was in this context that Takashi Kato (1986) first proposed question asking as a potentially distinct qualitative method for evaluating user interfaces for software.

Kato (1986) wanted Japanese secretaries to think aloud as they used a word processor, in order to diagnose some suspected problems with the user interface and provide design guidance to development colleagues. Kato found the secretaries to be quite reticent, and in the interests of encouraging them to externalize their questions about how to use the word processor, Kato requested that they learn the system by asking questions rather than by using training and documentation materials normally available. Kato's variation on the TOL task turned out to be successful, in his judgment, for encouraging verbalizations that exposed problems users experienced with the word processor and that suggested possible design improvements.

TABLE 15.1. Question and answer protocol example: Kato (1986) editor problem.

(Participant types a line of text in a letter, right justifies it and presses Return. The cursor is to the left of the typed text when Return is pressed. Return splits the line at the cursor, which puts the typed text at the beginning of the next line.)

P(1) Well I want to bring back that line. How do I do this?
T(1) You entered return over here (points to the middle of the blank lin) which you want to erase. So, to get to the return mark, bring up the cursor one line and move it to the right.
T(2) Delete that mark, and the line will move up.
P(2) Press delete? And?
T(3) Press enter.

(Participant does so, restoring the situation.)

An example of Kato's (1986) question-asking data is reproduced in Table 15.1. In this protocol a user is trying to understand how to "correct" the unexpected result of hitting the Return key. Unlike the same key on a typewriter, the Return key on many word processors (certainly the one in Kato's study) inserts a new blank line in the text of a document file rather than simply positioning the typing point at the beginning of the next line. The user asks how to "bring back that line." The coach explains what happened and how to delete the blank line. Notice the follow-up question about how to complete the delete operation. The example illustrates the procedural character of the question, closely tied to a user's immediate task, and illustrates how the answer to the question enables the user to proceed successfully, if not to understand something novel about the system. The investigator in turn has good grounds for inferring that this user did not understand the function of the Return key for starting new lines or that lines are represented by characters that can be entered or deleted (see also Mack, Lewis, & Carroll, 1983).

In reflecting on the success of this approach for uncovering gaps in users' knowledge and skill regarding how to use computers to accomplish tasks, Kato (1986) proposed that question asking might actually be a useful alternative to conventional TOL methods. (For a seminal and comprehensive analysis and justification of the technique, see Ericsson & Simon, 1984; for application to human-computer interaction, see Mack, Lewis, & Carroll, 1983, and Lewis, 1982.) Kato proposed that question asking is a more natural way for computer users to express what they do not know or the problems they experience. Asking questions is a typical way in which real users resolve problems they are having, where the questions may be posed to colleagues or hot-line experts. Of course, questions arise in thinking aloud also, as we have observed (e.g., Mack, Lewis, & Carroll, 1983), but often they are only one category of verbalization among many others, some of which may owe more to the demands of the TOL task (e.g., "Tell (the investigator) what you are thinking") than the user's perceived needs for information (e.g., "How would I (the user) do that?"). Moreover, unlike TOL interactions, question asking as Kato and others have used it involves a quid pro quo in that users' verbalizations create opportunities for users to learn something about the system they are using. In our view, *question asking is more task-oriented than thinking out loud*.

TABLE 15.2. Question and answer protocol example: Knox, Bailey, and Lynch's (1989) software oscilloscope evaluation.

Subject: Umn . . . so you push the buttons . . . and each time you push it, its going to (motions with hand down the channel 1 LED column) . . . like if we were on 500 millivolts, does it go down? Is that it, if you push the button once (points to the channel 1 button), does it go down?

Tutor: I'm not going to answer yes or no just now.

Let me give you something to do and we will see how you approach it.

What I would like you to do now is this.

The square wave is channel 1 and sawtooth is channel 3. I would like you to position the sawtooth above the square wave.

(. . .)

Subject: OK . . . (Turns the POSITION knob clockwise. The square wave moves up on the display.)

Tutor: Ok. Let's hold. And what happened?

Subject: Ok. The moved the other one up.

Tutor: And what did you want to do?

Subject: I wanted to . . . to move the lower one. (Points to the sawtooth.)

(. . .)

Tutor: Let me call your attention to some things then.

I want you to focus on this grey area here. (Tutor points to the graphic area surronding the buttons and the knobs.) Does that seem to be . . . communicate anything to you? . . . . This grey area?

Subject: Umm . . . I guess (moves his finger over the grey graphic area) . . . these (points to the knobs) have to be . . . unity . . . in conjunction with that (points to the row of buttons) is what I . . . what my thinking is.

Tutor: How do you mean?

Subject: Ok. These are all used in conjunction with each other that's the only thing I can . . . .

A handful of other studies have used question asking in one form or another to evaluate software and contribute to design, including a study of our own, which we will discuss in some detail below. An example of question asking in the service of interface evaluation is a recent evaluation carried out by Knox, Bailey, and Lynch (1989). In contrast to Kato's qualitative application of question asking Knox et al. categorized and compared types of questions asked in the course of using different techniques for operating a computer interface that implemented oscilloscope function. They also obtained quantitative performance data and analyzed both types of data to develop a converging picture of users' relative facility with different interface techniques. An example is excerpted in Table 15.2. The specifics are less important here than the contrast in how question asking and answering were carried out. It is clear from the short excerpt that the investigator is probing users with questions of his own, in some cases asking users to carry out tasks to test their understanding or focus their attention.

This application of question asking is also distinguished from Kato's (1986) study (and others we know about) by Knox et al.'s (1989) attempt to classify and count questions. The authors do not provide much detail, but presumably their intent was to provide some basis for comparing multiple techniques for implementing oscilloscope operations using both quantitative and qualitative data. In any case, the general rationale for such comparative judgments is interesting: Granting individual differences in question-asking propensity, on balance more intuitive systems ought to induce fewer questions than less intuitive systems.

More frequent questions for some class of function or technique should indicate usability problems. The Knox et al. evaluation may exemplify this general rationale, but it remains to be seen if reliable, systematic classifying and counting for *comparative* purposes can be scaled up to more complex software user interfaces (and questions pertaining to those interfaces).

A second example of question asking is found in an unpublished study by Robert Campbell (1990; Campbell & DiBello, 1989). Campbell's goal was to refine a system for on-line help for a PC text editor. Conventional on-line help systems are notoriously unhelpful because they tend to organize information about system usage in formal categories that are not related to how users may think about tasks or problems. (See Campbell, 1990, for a discussion.) The text editor Campbell was trying to improve, for example, organized help information by alphabetical enumeration of the editor's commands. In contrast, Campbell wanted to organize and describe help topics in task-oriented categories. Based on qualitative think-aloud and quantitative performance evaluation of a prototype task-oriented help system, Campbell decided to obtain data for a second iteration on the system by having users frame questions directly to the investigator.

In Table 15.3, for example, a user requests help for selecting text, in the context of moving a paragraph. The initial help system presented separate information about either moving text segments or selecting text, but did not integrate these two operations. The user had to distinguish these two operations and search for help in two separate places, and then integrate the two operations him- or herself. The answer to the user's question clearly integrates the two operations. Question asking proved effective in refining specific design elements of the the first iteration of the help system. Campbell (1990) engaged in more free-form interaction than Kato (1986), using questions as springboards into not only what users thought they needed to know but also how they might want it organized. Campbell was also able to develop a better overall "model" of his users' notion of task-oriented help for other topics that went

TABLE 15.3. Question and answer protocol example: Campbell (1990) editor help protocol.

U:	How do I move a few chunks of text?
T:	To move a paragraph, mark the top line with alt and 1.
U:	Is there a way to go a paragraph at a time?
T:	No. There isn't. Ok. Now you mark the bottom line, and then you move the cursor to where you want it to go and press alt m for move and then you press alt u to get rid of the mark. It looks like you need a new line after than paragraph. It's below . . . it makes a new line below the cursor.

beyond specific questions raised in question-asking protocols.

Questions that users ask (implicitly or explicitly) in the context of soliciting help or advice from help systems, manuals, consultants, or experts would seem to qualify as question-asking data aimed at understanding users' knowledge and skill, and contributing to the design of user assistance. Alty and Coombs (1981), Pollack (1985), McKendree and Carroll (1986), and Aaronson and Carroll (1987) all have examined advice-giving dialogues between expert consultants and novices on programming problems, or the use of applications such as text editors. These studies have been attempts to improve advice giving and possibly develop requirements for intelligent (automated) advice-giving systems.

We believe, however, that there is a distinction to be made between advice giving (or coaching, as we have applied it) and these consulting interactions. The latter emphasize the interaction between the expert and the novice, whereas the former emphasizes the interaction between the novice and the system (as in Kato's, 1986, evaluation). Advice-giving situations also involve larger scale problems and issues that are less closely tied to immediate use of the computer, as well as specific actions. Moreover, the interactions between advisee and advisor are more complex, in the nature of tutorials. They require establishing more context by the consultant with respect to what the user is doing and the problem he or she needs to solve.

These applications of question asking (by computer users) suggest the potential value of the technique. However, each study had different goals (e.g., qualitative evaluation of a single system versus quantitative comparisons) and involved methodological decisions that are not well articulated (e.g., whether to probe user questions and stage answers or not). Our goal in the remainder of this chapter is to motivate a broader exposition of methodological issues relating to question asking by users. We begin by discussing in some detail an exploratory study of question asking of our own.

Case Study of Question Asking

Our own work was motivated less by reticent users (recall that this was Kato's, 1986, motivation) than by a desire to obtain more focused verbal protocols from computer users. Earlier work by the first author (Mack et al., 1983) and others (e.g., Bott, 1979) demonstrated to our satisfaction that verbal protocol methods in widespread use in cognitive psychology could be usefully applied in the HCL domain to study people's interaction with computers. However, it was clear that verbal protocols provide diverse and voluminous data that require considerable review, selection and interpretation. In contrast, the questions that users asked about specific problems or information needs seemed an especially useful way to focus users' verbalizations about their experience using computers.

We decided to apply this "coaching" method to an evaluation of a workstation with a graphical, direct manipulation style of user interaction. In addition, we were interested in a qualitative assessment of the claims made for the naturalness and intuitiveness of this type of interface for novice users of computers (see, e.g., Smith, Irby, Kimball, Verplank, & Harslem, 1982). The workstation was an Apple Lisa system,[1] now defunct but quite innovative for its price range when it appeared. (Several discussions of the Lisa workstation exist; Carroll & Mazur, 1986; see Erhardt, 1983; Seybold & Seybold, 1984; and Smith et al., 1982; and Williams, 1983.)

TABLE 15.4. Subtasks users performed (subtasks in parentheses were not explicit in tasks but attempted by one or more users).

Getting started (first task)
Open blank document
Type memo (see Text entry)
Print
Store
Document management
Open blank document
Retrieve old document
Store
Print
(Name)
(Delete)
Text entry
End lines (word wrap)
(Immediate typo, by backspace)
Center
Underline
Text revision
Insert (space; character; line)
Delete (character; word; line; paragraph)
Move (paragraph)
Replace (string)
Text formatting
Margins
Tabs

Approach

We asked four relatively computer-naive secretaries to use the system for 4 hours to complete seven standard letter-composition and editing tasks adapted from other studies (Mack et al., 1983; Carroll, Mack, Lewis, Grischkowsky, & Robertson, 1985). Each task consisted of simpler subtasks, and these are listed in Table 15.4. Examples include subtasks such as, Open the hard disk directory (the so-called Profile), Open a document (file), Delete a word, Print a document, and so on. We provided access to documentation and even some quick reference cards modeled on the earlier study by Carroll, et al. (1985). The key instruction to users, as it turned out, was the invitation to simply ask the experimenter questions about what to do or how to do things, with the expectation that questions would result in answers. Table 15.5 summarizes guidelines for the user's question asking and the coach's question answering. A

TABLE 15.5. Mack and Burdett Q/A rules.

Users instructed to:
1. Ask relatively specific, procedural questions.
2. Try to answer own question first, but not engage in extensive problem solving.
3. Focus on getting tasks done, as you would in a real work setting.

Coach instructed to:
1. Reply with specific procedural answers, to the underlying form of the question (interpret if needed).
2. Do not tutor, do not explain at length.
3. Keep users moving on task, keep initiative with them.

TABLE 15.6. Total number of question and answer (Q/A) dialogue types.

Type of Sequence	Participant 1	2	3	4	
Q/A	42	21	6	31	(100)
NQ/A	5	3	7	3	(18)
Q/NA	8	7	0	3	(18)
Q/A(+)	25	13	5	13	(56)
Total Q/A sequences*	80	44	18	50	*a*
Total Q/A exchanges**	(140)	(65)	(25)	(74)	*b*

* Total individual Q&A sequences.
** Total individual Q&A exchanges (including multiple exchanges in prompt Q/A+ sequences).

key methodological decision on our part was to keep users focused on accomplishing tasks, that is, to keep users *task-oriented*. In this context, question asking was only part of their experience, a means to accomplishing tasks we gave them.

Results and Discussion

Overview of User Experience and Performance

Users were successful at some tasks and had problems with others. They asked questions about some tasks and in the context of some problems, and not others. The following are some high-level results of this overall experience of use. On average, users completed about five of the seven text-editing tasks, specifically, typing two new letters and making revisions in three old letters. The mean time to complete these tasks was 2 hours and 50 minutes. Users spent on average 5 minutes reviewing the quick reference cards we prepared for them (times ranged from 3.5 to 8.75 minutes). It took users on average 27.7 minutes to get to their first blank typing page (i.e., inside the word processor) and about 58 minutes on average to complete their first task (typing a simple one-page memo).

Table 15.4 summarizes the categories of subtasks given to participants as part of the seven tasks we asked them to complete. The mean number of subtasks carried out over the seven main tasks was 83 (range from 33 to 107). The seven tasks had 70 subtasks built into them as task scenario instructions. Specifically, users were given hard-copy instructions for each of the seven tasks, indicating what letter content to compose or revisions to make in an existing letter. The additional subtasks involved correcting mistakes or meeting additional goals users set for themselves (indicated in parentheses in Table 15.4).

With respect to question asking, the first observation to make is that users generally were able to adhere to the ground rules (see Table 15.5) we set for asking questions. Users were generally able to ask sensible questions (an exception is discussed below), and they did so in the context of trying to translate general instructions into system-specific actions, and to solve problems they experienced in carrying out actions. Users did not always rely on the coach but tried to carry out this translation of goals into actions on their own as well. We conclude that users were *active learners*, taking initiative and trying to solve problems on their own, a characteristic observed in other contexts (Carroll & Mack, 1983, 1984; see also Carroll & Rosson, 1987).

Table 15.6 provides some relevant descriptive statistics. About 26 percent of the subtasks involved questions. The mean number of question-and-answer (Q/A) sequences was 76 (ranging from 25 to 140). In addition, we counted the problems that users experienced, indicated in Table 15.7. About 26% of the subtasks involved problems. We will not analyze these in depth here; however, it is worth noting that the mean number of prob-

TABLE 15.7. Summary of number of problems for which no questions were asked; problems for which one or more questions were asked; and subtasks without problems, for which one or more questions were asked (PROB means *problem*; Q, *question asked*; NQ, *no question asked*; NPROB, *no problem*).

Task Scenario	Participant 1	2	3	4	
1 Prob/NQ	0	0	3	1	
Prob/Q	5	8	2	6	
NProb/Q	3	0	1	1	
2 Prob/NQ	2	1	6	0	
Prob/Q	7	0	1	1	
NProb/Q	3	0	1	1	
3 Prob/NQ	0	4	2	2	
Prob/Q	3	2	1	1	
NProb/Q	4	5	0	2	
4 Prob/NQ		2	2	1	
Prob/Q		5	0	3	
NProb/Q		3	0	2	
5 Prob/NQ		3	0	0	
Prob/Q		1	1	0	
NProb/Q		1	0	0	
6 Prob/NQ		1	0	2	
Prob/Q		3	1	2	
NProb/Q		1	0	0	
7 Prob/NQ			3		
Prob/Q			1		
NProb/Q			0		
Tot Prob/NQ	02	11	16	06	35
Tot Prob/Q	15	19	07	13	54
Tot NProb/Q	10	09	02	05	28

lems was 21 (ranging from 17 to 30). About half of these problems, 46% were associated with questions.

Users did have difficulty posing specific questions for the first task when they were quite unfamiliar with how to do formerly familiar tasks (letter typing and revision) with the new computer system. Some questions in particular were aimed at clarifying the task scenarios, not at how to do the tasks in terms of the computer. This suggests a *start-up phase* in which the Q/A dialogues may be somewhat more open-ended in the beginning of a user session, on the part of both user and coach. We also provided a brief description of basic concepts and operations needed to get started on a quick reference card, to which we referred in the beginning of the study.

Users' questions were also not always very specific or technical. Users often referred to questioned elements or relevant interface features elliptically and anaphorically. All of the question-asking protocols we have discussed provide examples, and we will discuss more from our own protocols below. Recall the example in Table 15.1 in which the user characterizes the task of joining two halves of a line split by inadvertently hitting the Return key as, "I want to bring back that line," where what is needed is to delete the line-break character entered when the Return key was hit. Or, consider the example in Table 15.2 in which the user refers anaphorically to the change in level of a signal in a display (". . . does it go down?"). Users in our study were likely taking advantage of the shared visual context between them and the coach, which is generally appropriate in normal conversation. We allowed this in the interests of not seeming to "correct" their language or increasing the burden of verbalizing questions, and because we generally felt confident inferring what users were referring to. In cases where the question was not clear (or the coach wanted clarification for other reasons), the coach asked clarifying questions. With this overview of users' experience and performance, we now turn to a more detailed analysis of questions and answers.

Analysis of Questions and Answers

We organize the remainder of our discussion of our specific study around the following questions:

1. What types of question-and-answer dialogues occurred?
2. What were questions about?
3. Did users ask the right questions?
4. Were answers to questions effective?
5. What did we learn from questions?
6. How are questions interpreted?

We will discuss broader methodological and interpretive questions related to these questions later in our General Discussion. Let us explain our motivations for using these questions by briefly discussing in a qualitive way the

TABLE 15.8. Question and answer Q/A(+) protocol example: how to change the document title.

Context: User has finished typing a new document. Document is displayed in open window with a default title "untitled".

Q12: How would you change the title?
Ans: (no time to answer)
Q13: Instead of "untitled" I want to call it something else.
Ans: Select the name of the icon of the document, and type what you want.

Act: Points to title in document window title area.
Act: Presses mouse button to select
Out: Nothing (cannot change title in document window)

TABLE 15.9. Question and answer (Q/A) protocol example: verifying text insertion.

Context: User has inserted some text within a string. Original text has reflowed to accommodate new text. She is not sure if new text actually was inserted.

Q55: Can I type over these to make a correction? I've added a word but it still wouldn't be right would it? I know I've added a word (but) will it just move the rest of the sentence over?
Ans: Yes, it will.

question-and-answer sequences summarized in Tables 15.8, 15.9, and 15.10.

In Table 15.8, a user wants to change the name of a document she has been composing from the default "Untitled" to something else specified in the task instructions. The user has a document opened up in a window, which includes the usual window management controls for moving, sizing, and closing, and a title field containing a default title, "Untitled." Users need to change the title not of the open document but the title of the icon that represents the closed document. The goal of naming the document, upon completing its content, is given to the user in written task instructions and in the form of a high-level goal, not the system-specific actions needed. After reading this instruction, the user asks quite directly, "How would you change the title?" and asks an unsolicited follow-up question, possibly to clarify her first question. The coach replies in a seemingly straightforward way that the user needs to select the name of the icon for the document and type the title desired. At this point the user "makes a mistake": She tries to select and type over the default name "Untitled" in the open document window, which will not work, rather than closing the document to get the icon and typing over the name "Untitled" displayed with the icon.

The user's question is a straightforward "how-to" question at some level of generality, that is, without referring to specific system features. The answer was also at some level of generality beyond listing the specific physical actions needed. In fact, the answer turned out to be too general. The icon needed implies that the current document window is closed and that the document is represented in the form of an icon. The user had dealt with document icons, opening them into windows and closing them back into icons. Nonetheless, the answer was insufficient and led to an erroneous interpretation by the user. The user's follow-up question and the coach's answer led to the "correct" procedure.

The first observation to make is that the question was not ultimately answered in a one-shot answer. The user's question was correct in that it was about the relevant subtask and identified the action needed to accomplish the subtask. The initial answer was not effective, because it assumed the user knew about a prerequisite action that she evidently did not know about. The answer led to a sequence of questions and answers. Note also the questions raised *for the designer* about the design of the interface. The distinction between open windows and icons did not seem readily available to the user, at least in this context, and the confusion raises a design issue: Why can't the title of the document file be changed in the context of an open document window? Why does the user have to worry about this distinction?

In contrast, in Table 15.9 a user queries the outcome of an operation, expressing disbelief that inserted text reflows existing text. The coach confirms her characterization without elaborating on it. This problem, like the earlier problem with the Return key in the

Kato (1986) protocol, reinforces observations we have made elsewhere about the unintuitiveness of many word-processing operations. (See Mack, 1983, 1989, for an extended discussion of this and related problems; see also Mack et al., 1983; and Douglas & Moran, 1983.) We have attributed these comprehension problems to prior experience with conventional typewriting.

What Types of Question-and-Answer Dialogues Did We Find?

Turning to our first question, return to Table 15.6, which summarizes four types of question-and-answer dialogues we observed. About one half of the Q/A sequences were one-shot questions answered by the coach, with no follow-up by the user or the coach. In contrast, about one third of the Q/A sequences, which we will designate as Q/A(+), involved follow-up, where a user and coach alternated questions and answers. These extended sequences, illustrated in Table 15.8 were driven by a need to clarify ambiguous questions or complex answers. For example, in some cases it was not immediately obvious how to provide a simple one-shot answer to a question, or an intended one-shot answer turned out not to be sufficient to resolve a question.

Table 15.10 provides another example. Here a user wants to underline a word in a partially underlined phrase and asks a straightforward "how-to" question. The user had positioned the pointer under the first letter of the to-be-underlined word but had not yet selected it. The coach answered that the user simply needed to select the Underline action from the Typestyle options (pull-down) menu. This action will work only if the user has selected the "object," or word in this case, to be underlined. The user did not do this, and the subsequent dialogue involves conveying the need for this prerequisite action.

This entire episode, classified as a Q/A(+) sequence, involves eight embedded questions related to a single subtask (underlining a word). In this case, it was simply not sufficient for the coach to identify the relevant action (Underline) and the menu where it could be found. The user asked follow-up questions that filled in aspects of the procedure the coach assumed the user would know. In these cases the coach provided an answer that was evidently too long or too complex to remember, and the user needed something closer to prompting to unpack the answer.

One-shot Q/A sequences and longer sequences constitute most of the questions and answers. However, in about 9% of the Q/A sequences, users asked questions that the coach did not or could not answer before the user seemingly went on to do what he or she wanted to do, and in another 9% of the cases, the coach simply intervened and answered a question that was not asked.

TABLE 15.10. Question and answer (Q/A+) protocol example: how to underline a word.

Q123: How do I underline?

Ans: Choose the action underline from the typestyle options menu.

Act: Opens typestyle options pull-down menu. Moves pointer to highlight underline option. Selects. No word is selected so nothing is underlined. There is no feedback message or indication.

Q124: Would it do it if I just point at Typestyle options menu?

Ans: Do it.

Q125: Can I ignore this right here (referring to the insertion point at the beginning of the to-be-underlined word). I want to underline (the word) "thrid."

Ans: Move the mouse (pointer) to the beginning of (the word) "third" and hold down the button.

Q126: Do I move under it?

Ans: Move it to the immediate left, same line. Hold down the mouse button and move to the end of the word. Then let up on the button.

Act: Selects and highlights the word "third" by dragging (i.e., moving pointer over the string while holding down the mouse button).

Q127:

Ans: Now select the action underline from the typestyle options menu

Q128: Will it . . . (does not finish).

Ans:

Act: Opens typestyle options pull-down, and the underline option to underline the target word "third."

What Are Questions About?

The analysis of Q/A sequences by form provides a first-pass characterization of these interactions, but we are most interested in what they are *about*, that is, their content. At one level, questions are about actions needed to accomplish subtasks built into the task scenarios users were asked to carry out. That is, virtually all of the questions could be classified as about the subtasks summarized in Table 15.4 (e.g., naming a document, or underlining a word). However, it became useful to interpret questions at another level of description, pertaining to a more fine-grained specification of elements composing computer operations needed to accomplish subtasks. (Kato, 1986, proposed a similar scheme for analyzing the content of questions.) We developed these interpretations for the purpose of composing answers to questions and for developing more general, retrospective interpretations of what the question might be telling us about the user's knowledge or lack of knowledge.

The procedural elements include the purpose or goal of the procedure (e.g., delete a word); specific user actions that accomplish the goal (e.g., select the word, select the command *cut*); and one or more outcomes (or system actions, resulting in word deletions, text reflowing, and so on). These elements may seem straightforward, and they suggest representations of procedural knowledge proposed in cognitive psychology. (See Evans, 1988, for a recent survey related to knowledge engineering methods; see also Norman, 1986.) In fact, there are additional action and outcome elements characteristic of computer operations that are not always immediately obvious to users, at least to novices. Actions can involve additional elements that may vary in relatedness to the ultimate goal. The full sequence for deleting a word, for example, involves prerequisite actions (select the word, select an edit pull-down, select the "cut" option from the pull-down). Similarly, outcomes have multiple aspects that are not always directly related to the goal of an action. For example, deleting a word not only excises a string but also reflows the paragraph containing the word, and inserting a word not only enters new text but also reflows surrounding text. Table 15.9 provides an example in the context of text insertion. The user asks a question about the outcome of inserting text, which we interpret as expressing puzzlement over the automatic reflowing of text that accompanies insertion of text. Reflowing happened when the user inserted (simply by typing), but the user could not believe it.

Figure 15.1 schematizes the propositional content of several key types of questions in a way that emphasizes these procedural elements (i.e., goals, actions, and outcomes) and other distinctions relating to the context in which questions about these elements are salient to users. By *propositional content* we simply mean that these specific procedural interpretations are abstractions from surface forms of questions that may be quite elliptical and ambiguous when taken out of context. The analysis scheme expresses system distinctions that we would not claim novice users would express directly in questions.

Notice first the distinction between questions that anticipate an action (or actions) and questions about actions that have been carried out (questions about outcomes). Those that precede seem to request *verifying* a planned action or expectation relating a goal (or subgoal), the latter possibly expressed in terms of a hypothesized action or in terms of how to bring about an outcome. Alternatively, questions may request the coach to *specify* an action that would accomplish that goal (or outcome). In some cases, the user's request or specific plan he or she wants to verify may actually pertain to some problem incurred by a previous mistaken action. In contrast, questions that follow an action seem to request *verifying* outcomes or the user's understanding of what happened. In many cases, the outcome may involve detecting the result of a mistake or some discrepancy between what the user expected and what seemed to happen. Most questions before or after are "how" questions, rather than "why" questions, although we have indicated examples of both. It is not clear why more "why" questions did not occur. It may be

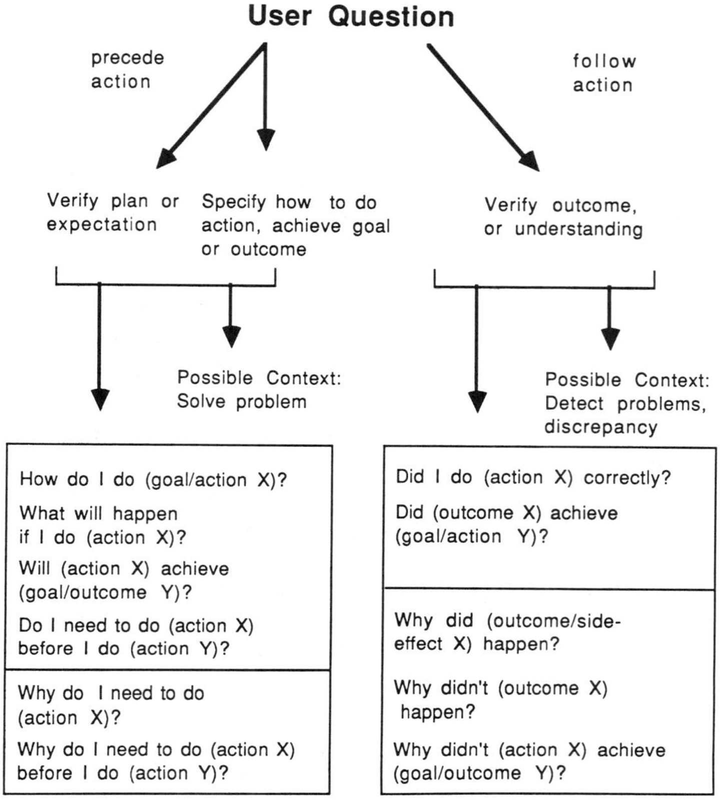

FIGURE 15.1. Content of user questions indicating propositional content and category of user need.

that the instructions to users were successful in constraining them to ask procedural "how" questions, and not questions that would likely have required more explanation. This scheme provides a heuristic framework for classifying user questions. It was also implicit in the process by which the coach understood and provided answers to questions.

Did Users Ask the Right Questions?

Ultimately, users are asking the right questions if the answers they get are effective. However, the effectiveness of the question in soliciting a useful answer depends also on the coach's ability to interpret the question. In general, users' questions were useful in creating an opportunity for coach and user to engage in a dialogue that helped the user. Users were able to pose questions in terms of the task scenario they were given and in terms of what they understood about possible workstation operations that enabled those tasks to be accomplished. This understanding was partial, but sufficient to pose questions.

As we indicated before, the surface form of users' questions was typically nontechnical if not ambiguous with respect to the "correct" description of what the user needed to know—expressed, for example, in terms of the procedural elements of the operation in question. The coach either had to request clarification, or, as was more often the case, infer an underlying "ideal" question. Table 15.11 can reinforce the point quickly with another example. Here, a user poses a valid query about how to change the margins of a document but does so in a very ambiguous way. The user's question (Question 24 in the table) is not a clear question about how to change document margins. The coach might have

TABLE 15.11. Question and answer protocol example: how to change the margins of a document.

Context: User has a document open in a window. User wants to change the margins for the entire document.

Q24: How would I change one file to go into another to change margins?

Ans: Select the whole document and then select the option "show margin & tab ruler."

Act: Opens "file and print" pull-down menu (mistake)
Act: Opens "format" menu
Act: Selects "show margin & tab ruler" option.

asked the user to clarify the question, but in fact there was sufficient context that the coach was confident the question pertained to changing margins. Context in this case referred to instructions in the task scenario (which described a goal but not the needed operations specific to the computer) and to the coach's understanding of the user's immediately preceding actions.

Were Answers to Questions Effective?

How reliably did the coach interpret the user's problem and generate an answer that solved it? The simplest indicator of an effective answer is that a user can take effective action with respect to his or her goal. One quarter of the answers were manifestly not successful, because they resulted in an exchange of follow-up questions by users and clarifications by the coach (recall Table 15.6). We have already discussed examples of answers that failed at least initially, for example, the user in Table 15.8, who was told how to change a document name but continued with an incorrect action (from a system perspective). The user posed a valid question, but the coach's answer assumed that the user knew more than was the case about needed prerequisite actions. We should note that in all such cases, the failure was not in inferring what the user needed to know or do but in responding with the most comprehensible level of action. Where users did not understand the answer, their next action would be incorrect, leading to clarification by the coach or follow-up questions by the user. In either case, the user and coach worked until the user could proceed. In this sense, answers were always ultimately "correct" with respect to leading to correct actions.

The correctness of the next action, however, remains a limited indicator. It is not applicable, of course, to questions that request verification that some user's action produced the correct result and did not involve some overt action by the user (the example in Table 15.8). Furthermore, we did not assess whether users really understood an answer in whole or in part even if they correctly executed the next action. We do not know if users learned anything that might help them if they should need to do the same or similar thing again. These are issues that could be tackled systematically via follow-up probing (as Knox et al., 1989 demonstrated) and/or additional scenarios that provide repeated opportunities to carry out operations.

We should note that initial failures can still be informative to the investigator with respect to developing interpretations. They are a form of probing that may elicit more information from the user. For example, in both examples above, the failure indicated that there were prerequisite actions that the coach assumed the users understood but that they did not understand, based on their activity following the coach's answer. Failures may also be informative for users to the extent that they result in additional elaboration by the coach.

What Did We Learn From Question Asking?

In general, the utility of question asking depends on the purpose of the evaluation. Our goal was exploratory. We wanted to develop a global, qualitative characterization of the intuitiveness for users of the Lisa workstation environment. There are other possibilities, of course, and these have implications for how evaluations using question asking are conducted. We will discuss these more general issues in the final section. Here, we comment briefly on the utility of our specific evaluation.

Our working hypothesis was that aspects of the workstation that were not intuitive to users as they tried to accomplish tasks should lead to both questions and problems, both of which should indicate possible areas of improvement in the interface or requirements for help and instruction. We made two general observations. (We remind the reader of our earlier disclaimer that the Lisa workstation was not necessarily intended to be used the way we asked users to use it, that is, without the tutorial and without encouragement to use documentation, instead relying on coaching. We believe we did a better job of assisting users than the documentation does, but we did not assess this.) First, the questions users ask raise questions, in turn, for investigators (designers) about the software design and about the presence or absence of relevant user skill and knowledge. Questions can suggest specific implications for specific design elements of the software. For example, the problem with changing the document title (Table 15.8), suggests that the system should allow users to replace document titles no matter where they appear, unless there are strong implementation reasons for not doing so. The problem with underlining (Table 15.10), might suggest a problem specific to the underlining operation, for example, that it might be more intuitive to allow an underline "mode" in the context of which the typestyle of existing text could be changed.

In other cases, questions suggested more general issues regarding the fit (or lack of fit) between the user's knowledge and expectations, and the design of the computer system. Several questions, for example, collectively reinforced generalizations we have made elsewhere (Mack et al., 1983) about users' expectations and interaction problems in the context of using more conventional, text-based, and command- and menu-based interfaces. The Lisa workstation exploited several metaphors in the interface design that invited comparision to noncomputer tasks and methods, in particular: (a) the so-called desktop metaphor for representing computer objects like files or file collections, and (b) object-action order of carrying out operations. We were especially interested in the extent to which these metaphors actually aided users in figuring out how to use the system. (See Carroll, Mack, & Kellogg, 1988, for a general discussion of software metaphors in this connection.) In fact, these metaphors, as implemented in the workstation and under the conditions of use we imposed, did not necessarily help users easily accomplish tasks that were hypothesized to exploit the metaphors.

One example involved a method of starting a new document from a document template that took the form of an icon of a "stationery pad." Instead of "creating a new file," users are supposed to select the Stationery icon and then the menu action *Tear off*. This method seems aimed at exploiting an analogy to starting new paper notes by tearing off stationery sheets. In fact, users made mistakes starting new documents, which indicated that this implementation did not seem readily inferable from the more familiar paper method.

Problems with object-action dialogue provide a second example. The general hypothesis here is that object-oriented interfaces, in which users specify an object (e.g., word, file, icon) and then an action to operate on it (e.g., open, print), are easier to understand because the interface style exploits an analogy to how objects in the physical world are manipulated. We were somewhat surprised to find cases for all users in which they seemed to select actions first (e.g., underline) before selecting an object (e.g., word or phrase). The fact that all users tried to carry out actions before specifying objects at one time or another suggests a problem more general than the specific operations that were problematic. For example, users may not understand the object-action metaphor. Consequently, we may want to make this interaction model more salient to users or allow both orders for operations.

Finally, we note that questions were not the only input into the process of developing design interpretations. We are focusing on questions in this paper, but in practice we found that the problems users experienced were equally interesting and often interpretable to some level even when they were not associated with questions or other TOL comments. (Recall in Table 15.6 that about one half of the problems experienced did not stimulate questions and were solved by users on their own.) Put another way, all of the users' behavior was informative and relevant to developing interpretations. Overt behavior (especially problems) and questions complemented each other. Questions helped the investigator home in on the possible cause of a mstaken action or pointed to some gap in knowledge that might have resulted in a mistake.

How Are Questions Interpreted?

Our experience in interpreting questions was quite similar to our experience in making sense of TOL protocols, with the additional task of answering users' questions. That is, interpretations were aimed not only at forming retrospective generalizations but also at developing effective answers to solve users' problems as they arose during immediate task performance. In both cases, Ericsson and Simon's (1980, 1984) comprehensive and seminal analysis of the appropriate use and interpretation of verbal protocols appears to apply to question asking as well. As Ericsson and Simon observed, individual verbalizations typically make sense only in context. In our experience there are multiple sources of contextual information:

1. Known scenario requirements.
2. User actions (especially problems).
3. Cumulative understanding of user by coach.

First, of course, our users were given specific tasks to perform, in our case in the form of word-processing scenarios. We knew roughly what sort of activities a user ought to be carrying out at any given time. Task instructions place some bounds on the questions a user is likely to ask or the problems he or she is likely to experience. Second, users' actions before and after a question, along with shared visual context, can tell us something about the likely source or motivation for their questions.

It was also possible to build up intuitions over time about what the user probably knew. We noted earlier, for example, that some coach answers actually referred to subgoals, not specific actions, based on the assumption that users would be able to unpack the subgoal into familiar actions. In some cases, the coach made the wrong inference and the answers were not understood. However, in many cases, users were able to do this in a way that justified the coach's strategy of answering with subgoals, not simply a list of elementary actions. For example, the coach did not feel compelled to describe how to manipulate the mouse on the physical desktop to move the pointer, beyond the very first task scenario, but simply referred to using the mouse to select something. Avoiding overspecification of procedures may seem obvious for frequent low-level interactions, but finding the right level for more complex operations is not so obvious.

General Discussion

The preceding discussed question asking in more depth than other case studies the authors know about, but it likely raises as many questions about the method as it answered. In this final section we want to discuss broader methodological and interpretive issues relating to the use of question asking, motivated by our experience with the technique but also transcending it. Specifically, we revisit three issues:

1. Why question asking should be used.
2. Methodological issues in conducting question asking (and answering).
3. Understanding questions and question interpretation.

What follows is a framework of methodological and research issues that we believe can guide practitioners of the method and focus research aimed at understanding question-asking behavior itself.

Why Question Asking Should Be Used

First, we set question asking in the broader context of empirical software evaluation methods, in particular think-out-lond (TOL) methods. In the HCI domain, we want to learn useful things about users' experience with computer systems, in order to improve the systems, guide design, and achieve some more general understanding of the mechanisms that underlie computer use as a cognitive skill. There are multiple empirical methods for obtaining information about users' interactions with computers. These include quantitative measurements of task performance, aimed at establishing benchmarks to compare systems or tracking possible improvement in iterative versions of a particular system. They include qualitative verbal protocol methods, such as TOL methods, which are aimed at externalizing users' thoughts and problem-solving strategies underlying performance. Qualitative information is useful, of course, for diagnosing user problems and contributing insight into possible design solutions. In this respect, question asking and thinking out loud have similar goals.

We believe Ericsson and Simon's (1980, 1984) analysis of verbal protocol methods and data is applicable for characterizing the type of knowledge and mental processing that questions are likely to uncover. In Ericsson and Simon's analysis, verbal protocols are hypothesized to encode what people are currently attending to in conscious attention ("heeded" information) as part of the mental processing needed to carry out tasks (solve problems). Moreover, the expression ("encoding") of information can be more or less effortful, and more or less directly tied to a person's problem solving. Much of human cognition is not readily accessible to consciousness or is difficult to verbalize. To the extent that this is the case, requiring people to verbalize can be intrusive and can interfere with task performance, resulting in selective and fragmentary protocols. This, in turn, increases the burden of analyzing and interpreting verbalizations.

A reasonable working hypothesis is that users' questions also express the content of conscious attention, with the following additional task demand that users focus on, and form questions expressing gaps in their knowledge about how to translate goals into specific operations. The questions are, How cognitively demanding is this task demand? and How valid are these questions as indicators of what users really need to know? Recall in this context the original rationale for question asking, namely, that it is potentially a more natural accompaniment to thinking and problem solving in a way that conventional thinking aloud is not. People do spontaneously ask questions that are accurate enough expressions of what they need to know to do something, especially when those questions actually help them solve problems. In contrast, people do not necessarily verbalize out loud at large, at least as adults, or for the benefit of an observer who does not reinforce questions by answering them. Certainly the manifest content and typical goal of asking questions seems situated in the realm of conscious problem solving, where people set goals; develop plans to achieve them, typically with limited resources (information); and resolve problems that arise. This is the realm of bounded rationality, in contrast to knowledge that may be automatized and not readily accessible to consciousness (Simon, 1969; Newell & Card, 1985).

Figure 15.2 expands this characterization in the context of learning and using computers. The distinctions shown are prominent in an information-processing paradigm of cognitive psychology, and we have adapted Norman's (1986) qualitative description of knowledge levels tailored for human-computer interaction. At the top is general, largely declarative knowledge about a task domain, that is, about relevant goals and tasks, appropriate methods, and objects constituitive of the domain. At the other end are more procedural kinds of knowledge, which relate to actually doing things within the domain, or what we can call routinized cognitive skills underlying overt behavior in the task domain.

In the HCI domain, we are especially interested in how these levels of knowledge are acquired for a new domain. Experience (Carroll & Mack, 1985; Douglas & Moran, 1983) suggests that a key acquisition process involves trying to map what one knows of a familiar domain that seems similar to the comparable level of knowledge or skill needed to function in the new domain: for example, trying to learn word-processing tasks and operations by analogy to typewriting tasks and methods. This particular example has been rather exhaustively analyzed, which has led to

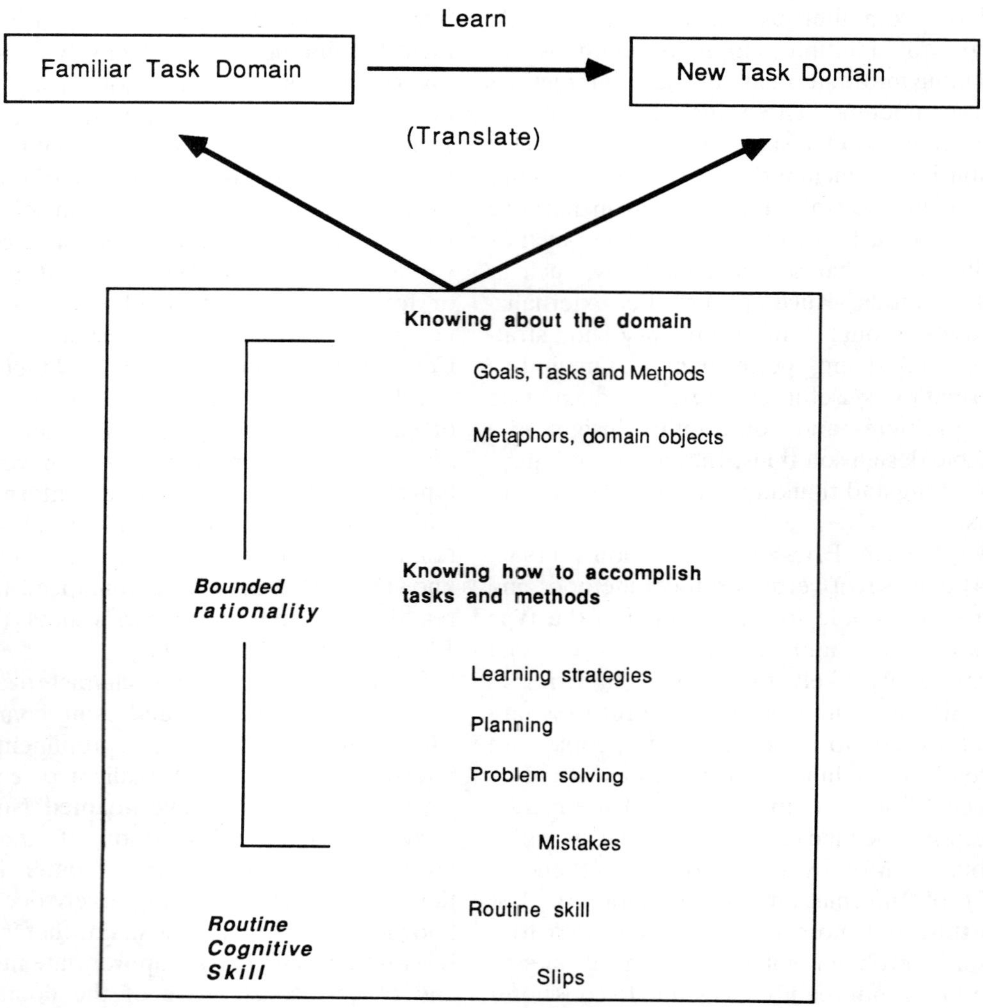

FIGURE 15.2. Levels of knowledge and cognitive skill underlying the understanding and use of computer systems.

broader characterizations of the possible role of metaphors in learning computer systems. (See Carroll et al., 1988, for latest analysis.) In this context, general domain knowledge influences the diverse expectations that users may have about the structure and operation of a new system, supporting useful inferences in some cases and creating problematic inferences in other cases. The latter can result in *mismatches* between goals and appropriate methods (from the designer's perspective) and misinterpretations of novel features of the system. (For examples in text editing see Carroll & Mack, 1983, 1984.) These cognitive errors or mistakes likely diminish with learning, as skills become more routine, at which point mistakes are likely to have a simpler origin as "slips" whose consequences may or may not require problem solving to recover.

In the context of these distinctions, we are most confident that question asking pertains most directly to the middle of the knowledge spectrum, that is, to procedural knowledge needed for conscious planning and execution of specific actions to accomplish specific goals. However, it is clear that this type of knowledge needs to be set in the broader context of other knowledge not tied to procedures or to ver-

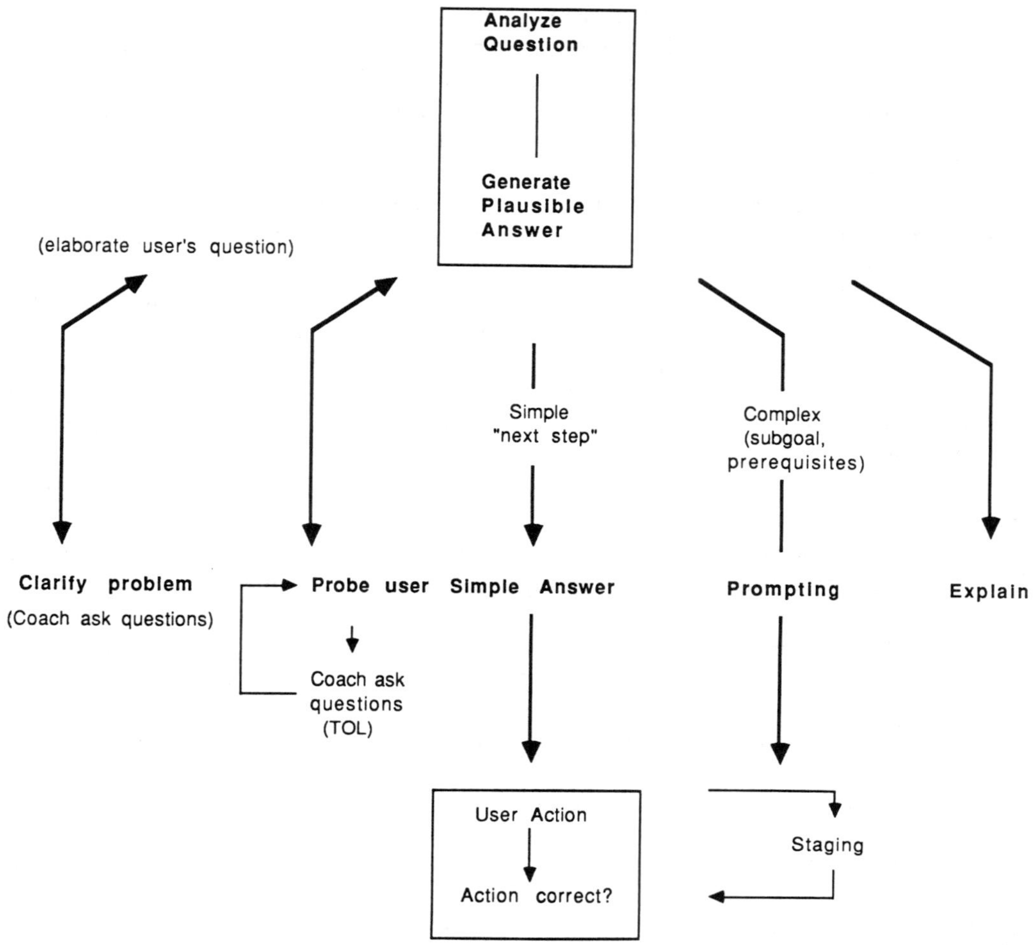

FIGURE 15.3. Methodological issues when coaches ("experts") generate answers to the questions users ask.

balizable or conscious mental processes. Access to, or evidence for, these types of knowledge likely involves more interpretation, additional probing, and/or behavioral evidence based on converging types of empirical evaluation (e.g., performance analysis of user problems).

Methodological Issues in Conducting Question Asking (and Answering)

Granted that question asking is selected as an appropriate empirical method for our goals, questions arise about how to carry out Q/A interactions, beyond the general guidelines that were summarized in Table 15.5. Recall that our guiding principle was to keep users task-oriented. We did not want to make them dependent on us or distract them from accomplishing tasks by extensive tutoring or explanations. We wanted question asking to be concomitant to an active learning process that involved users' taking initiative themselves to solve problems, with questions reflecting genuine gaps in understanding that could not be filled by some level of exploration on the user's part. This is a fine balancing of investigator intentions, and other guidelines can be motivated by articulating these other possibilities. We have represented these choices in more depth in Figure 15.3. This is not a process model of how a coaching interaction is generated, but a heuristic way of indicating distinctions and choices for the investigator/coach.

Notice first in Figure 15.3 the choices relating to the depth of (a) *clarifying* users' questions (and their underlying problems) by asking users questions, and (b) *prompting* users through a sequence of steps instead of simply providing one-shot answers to questions. Clarification is motivated in part by the ground rules in Table 15.5. The coach may want to ensure that the user has exhausted his or her own initiative in solving problems or answering questions first (e.g., "Do you have any idea how to do (goal/action) *X*?"). Or the coach may want to ensure that the user has framed a question that expresses as much as possible what the user understands (e.g., "Can you be a little more specific about what (goal/action *X*) you are referring to?"). Prompting sequences are motivated by those cases of questions that were not understood or were too complex to be followed without clarifying follow-up. They may occur when simple answers turn out to be inadequate. However, investigators may also anticipate the inadequacies of simple answers and may therefore stage answers in a way that anticipates prompting users through a sequence of steps. Knox et al. (1989) deliberately staged answers to questions in a prompted sequence, progressively disclosing finer grained answers until coach and user concluded that both understood what they needed to understand.

Returning to Figure 15.3 again, note next the choices the investigator faces about how deeply to (a) *probe* users' assumptions and thoughts (beyond clarifying the user's questions) by, for example, the asking questions of the user, more in the style of prompted thinking aloud, and (b) *explain* more general concepts and principles underlying users' questions (and problems), again beyond simply answering with procedural aspects of their interaction. Probing moves question asking toward a mixed dialogue that includes follow-up questions by the coach to probe the user's thinking and expectations more broadly, along the lines of TOL protocols. Explanations could be driven by "why" questions that really require placing some action (or outcome) in a rationale or context that does not simply pertain to the procedural details of an operation queried.

The distinctions in Figure 15.3 help us to compare and contrast various studies relating to question asking. Our study, for example, was quite Spartan in clarifying and prompting, or probing and explaining, compared to that by Campbell (1990) and Kato (at least based on sample protocols in Kato's 1986 report). Campbell also exemplifies more explanatory answers that, for example, try to relate one question to another or one aspect of a user's task to another. The Knox et al. (1989) study, in contrast, exemplifies a mixed dialogue of question asking by the investigator to probe users' problems, where in some cases variations on a user task were staged in the interest of testing understanding or focusing attention of some aspect of the interface. These are all interesting variations on the basic technique of inviting and responding to questions people have about what they are doing in some task. The variations entail possible trade-offs in what the investigator might learn. For example, the possible benefits of probing and clarifying for the investigator's interpretation of users' questions needs to be set against possibly distracting users from the task at hand and possibly influencing how users think about the system. Prompting may ensure more accurate coaching but may also compromise users' initiative, leading them to expect the coach to tell them what to do.

The framework in Figure 15.3 may also let us draw contrasts to studies that involve more radical variations on the basic method. Miyake and Norman (1979; see also O'Malley, Draper, & Riley, 1984) and Robertson and Swartz (1988), for example, asked users to verbalize questions that arose in using a text-editing application, but did not necessarily answer those questions. Robertson and Swartz were interested in characterizing the possible role of questions in what they called *self-learning*, that is, questions as a strategy users might have for organizing their own problem solving. We believe answering questions is important to the success of question asking, but it would be interesting to know how comparable question asking would be without answering questions. Examining their illustrative protocols suggests types of questions quite similar to those that were distinguished in Figure 15.1.

We note, finally, that question asking can be turned around completely in a methodology with similar intent, but with different task demands and trade-offs. Graesser and colleagues (Graesser & Black, 1985) have pioneered the use of asking systematic "why" and "how" questions of people engaged in a variety of cognitive tasks, including reading stories; understanding procedures; and, more recently, using computers. That is, the investigator does not focus on questions users ask or on answering questions users might have (expressed overtly or not). Instead, Graesser, Lang, and Horgan (1988) ask people questions to infer evidence for how people mentally represent procedures implicit in tasks underlying these different domains. (See Graesser & Black, 1985, for a wide-ranging discussion of these studies and many others; see also Graesser et al., 1988; and Graesser & Murray, 1989.)

Several methodological considerations will motivate the variations in question-and-answer dialogues sketched in Figure 15.3. One is the *purpose of the evaluation*, as we suggested earlier. Most uses to date have been exploratory, in the spirit of TOL protocols, where investigators want to develop qualitative interpretations of users' interactions with systems. Here, questions are a source of data. Mixing answers to questions with TOL probes might provide wider variety of data on users' thinking, beyond what users ask about. An interest in understanding or developing requirements for more effective help or intelligent on-line assistance might lead an investigator to focus on generating and assessing the effectiveness of different kinds of explanations and prompting when answering questions. For example, we might be especially interested in explanations that go beyond conveying narrow procedural aspects of users' problems and questions. (For recent analyses of explanatory and advice-giving dialogues, see Zissos & Witten, 1985; and Cawsey, 1989). Stricter adherence to Q/A rules over tasks, users, and possibly multiple systems might be needed if an investigator wanted to systematically and reliably classify and count questions, perhaps in the interest of comparing interfaces in terms of measures based on the frequency and category of user questions (exemplified by Knox et al.'s, 1989, comparative evaluation). Of course, purpose and style of interaction can be thought of as roughly independent decisions. However, in practice, comparative evaluations may be harder to carry out with a mixed style of interaction because it may be harder to ensure that question asking and answering are carried out in comparable ways across systems, techniques, or tasks.

The *amount of context available to the investigator* may also influence departures from simple Q/A dialogues. We found that in many cases, it did not seem necessary to ask users to clarify their questions, because it seemed clear from the context of what the user had done or was supposed to be working on. (They were given specific tasks in our study.) Investigators who lack such context, or who desire more clarity, might decide to probe users to clarify what they are asking (or thinking). Alternatively, we might be interested in knowing more detail about what aspects of the interface seemed salient to the user, and we might want to press the user to frame a more specific question in order to reveal more about the user's general expectations or understanding of how the system operated.

Understanding Questions and Question Interpretation

We have discussed methodological choices and variations in question asking and answering, including the rationale for these variations. Question asking, however, is a complex and interesting cognitive task apart from its utility as a method for evaluating software. The complexity of asking questions and learning from them (via, e.g., a coach or tutor) makes us most comfortable in recommending the use of question asking in an exploratory and diagnostic capacity, and in the context of other evaluation methods. This recommendation could equally apply, of course, to the use of

TOL protocols, although there is currently more experience to draw on in assessing the applicability of TOL to research problems (Ericsson & Simon, 1980, 1984). Research on question asking, as well as on the cognitive mechanisms underlying human interaction with computers, has the potential for providing firmer grounding for question asking as a methodology.

A useful start would be to develop our working hypothesis that users understand procedures in terms that map to some degree to real system structure and function. This step alone provides a useful heuristic framework for describing, and comparing and contrasting, users' questions, and the gaps in knowledge that seem to motivate them. It is tempting to try to formalize this characterization of procedural knowledge along the lines of a GOMS-like cognitive task analysis that decomposes user tasks into goals, methods, operators, and rules for selecting among alternative methods (Card, Moran, & Newell, 1983), or still further as production systems (Kieras & Polson, 1985). Here, the goals are to analyze hypothesized cognitive representations underlying behavior and to predict error-free performance and some aspects of learning (given additonal processing assumptions).

These schemes have had some success in simple and highly controlled task situations (e.g., Card et al., 1983; Polson, 1988), but their general utility remains to be demonstrated in applied settings with ecologically valid tasks that involve a wider range of types of knowledge. Although questions and problems are a frequent and important part of users' real experience with computers, these models have not been very successful, in this author's opinion, in modeling such behavior. Production system schemes tend to be quite procedural, and it is not at all clear how they would cover the full range of knowledge levels that we know are relevant to understanding users and their interaction with systems (especially problems). These criticisms about the scope, theoretical interest, and practicality for applied problems (like software development) of formal cognitive models in HCI (if not in other arenas) are well known (See Carroll & Campbell, 1986; Carroll et al., 1988; and Karat, 1988) although not universally accepted (see Newell & Card, 1985; Olson & Olson, 1990; Reitman-Olson, 1988).

On the question-asking side we would be interested in understanding how questions are generated and what the pragmatics are of deciding to ask questions. We want to account for the distinctions in Figure 15.1, and the decisions that users make about when to ask questions overtly and when to generate (or explore) solutions to problems on their own. Research on question generation in cognitive psychology and artificial intelligence is emerging that should be applicable to tackling these questions. (See Graesser & Black, 1985, for useful collection of papers.) We also want to characterize more than the procedural content of questions.

If formal, deductive cognitive models are premature, we still see heuristic, qualitative utility in procedural representations. As we noted earlier, questions that *users* have can help stimulate and resolve questions that *designers* have about how users might interact with the systems that are being designed. Questions can serve as an "intuition pump" for designers, helping them to get beyond the manifest content of questions and to create more integrative interpretations that have general diagnostic value for understanding the knowledge needed to effectively use computer systems. Formal representations or models may also aid or support this analysis and interpretation process. Further discussion of "qualitative modeling" of user behavior in a design evaluation context can be found in Kieras (1988), Karat (1988), Karat and Bennett (1991). However, design implications (or even implications for modeling cognitive processes) derive from an interpretative process, and we see the modeling medium as the HCI/design professional himself or herself.

Conclusions: Prospects for Coaching

We believe that Kato (1986) is correct in identifying question asking as a distinctive and useful method complementary to more familiar think-out-loud (TOL) methods for acquiring qualitative information. His claim that asking questions is more natural and task-oriented than general TOL instructions also seems legitimate. The techniques are not exclusive of course, and variations of question asking are possible that use questions as springboards into more extensive dialogues between users and coaches about what the user experiences.

It should also be clear that, like TOL verbalizations, question asking is itself a complex behavior entailing methodological decisions and trade-offs that we have identified. As an empirical method, we are most confident in recommending question asking in an exploratory and diagnostic role, and in the context of other evaluation methods (i.e., not as stand-alone method). Question asking as we have applied it seems more focused on procedural knowledge than TOL verbalizations, and this may or may not be appropriate to an investigator's goals. The analyses in this paper provide distinctions that we believe are useful both for guiding practitioners in the use of the method in a usability engineering context and for guiding research on question asking as a form of behavior in its own right.

We have restricted our discussion to computer usage and novice computer users because that is the domain in which we have worked. It is worth considering the potential broader application of the method to knowledge engineering enterprise. Hoffman's (1987) analysis of knowledge extraction methods suggests that question asking shares attributes of techniques that limit an informant's (or expert's) access to relevant information, encouraging him or her to externalize the need for that information (possibly in the form of questions). Anna Hart (this volume) provided an illustration at the conference (see Acknowledgments). She recounted a suggestion from a colleague about how to better motivate experts to describe their methods for classifying geological objects: The colleague suggested that instead of asking experts to describe how they classified each of a set of objects presented to them, she might have the expert guess the identity of concealed objects by asking questions he or she felt were necessary to identify the item. Asking an expert directly how he or she makes a judgment or arrives at some conclusion may not engage that expert's knowledge and problem-solving skills in as task-oriented a way as arranging a situation in which the expert needs information and expresses that need in questions of his or her own choosing.

We would like to close by offering an additional perspective, not exclusive to the HCI arena but certainly more salient for it. The cognitive science of human-computer interaction has a strong interventionist possibility, where HCI professionals contribute to the design of computer software via empirical evaluation and psychological analysis of users' interaction with computers. This systematic empirical and iterative design process in a software engineering context has come to be called *usability engineering* to emphasize the focus on the users (versus the software technology) and on engineering (versus theory). (For a general treatment of usability engineering, see Whiteside et al., 1988.) To the extent that our understanding of users' performance, problems, and mental processes can substantively drive aspects of design and design modification in an engineering context, the "correctness" of interpretations of qualitative data can be inferred from demonstrated iterative improvement in the system. In some cases, specific improvements might be traced to specific types of feedback (e.g., Campbell's, 1990, refinement of task-oriented help topics), but more likely the relationship between our "theory" of the user and the implications embodied in the design is not direct or deductive. In this usability engineering context, we "test" our "theory" of the user in a more holistic way in the context of assessing the quality of the user's interaction with the system. What is true for usability engineering seems to be emerging for knowledge engineering: Representations of experts' knowledge are created from interpretations of empirical data, emerging over time as a function of iterative refinement, in the context of continual use of these representations in knowledge-based systems that are used to solve problems in the domain.

Notes

1. Apple and Lisa are registered trademarks of Apple Computer, Inc. Regarding the Lisa computer workstation, a disclaimer is in order: The Lisa workstation was not necessarily intended to be learned without instruction. Indeed, the system was accompanied by documentation, and an on-line tutorial on basic elements of user. We know of no software system, Lisa included, that claims to be so intuitive for users that they can simply walk up and use it without training or effort. This is a worthy goal, however, and we were interested in the extent to which the Lisa approximated this possibility through its graphical direct manipulation interface style.

Acknowledgments. We thank John Gould, Jakob Nielsen, and Robert Campbell for useful comments on earlier presentations of the research reported here, and we thank Robert Hoffman for extensive comments on the current chapter. This chapter is based on a presentation, "Coaching Computer Users as a Technique for Software Evaluation and Design," given at the Seventh Annual Adelphi Conference on Applied Experimental Psychology: Expert Systems and the Psychology of Expertise, May 5, 1988, Adelphi University, Long Island, New York.

References

Aaronson, A., & Carroll, J. (1987). The answer is in the question: A protocol study of intelligent help. *Behavior and Information Technology*, *6*, 393–402.

Alty, J., & Coombs, M. (1981). Communicating with university computer users: A case study. In M. Coombs & J. Alty (Eds.), *Computing skills and the user interface* (pp. 7–71). London: Academic.

Bainbridge, L. (1979). Verbal reports as evidence of the process operator's knowledge. *International Journal of Man-Machine Studies*, *11*, 411–436.

Bott, R. (1979, March). *A study of complex learning: Theory and methodologies*. (CHIP Report 82). University of California, San Diego, La Jolla, California.

Campbell, R. (1990). *Online assistance: Conceptual issues*. (Research Report RC 15407). New York: IBM, Thomas J. Watson Research Center.

Campbell, R., & DiBello, L. (1989). *An empirical evaluation of a system for on-line assistance*. Unpublished manuscript.

Card, S., Moran, T., & Newell, A. (1983). *The psychology of human-computer interaction*. Hillsdale, NJ: Erlbaum.

Card, S., & Newell, A. (1985). The prospects for psychological science in human-computer. *Human-Computer Interaction*, *1*, 209–242.

Carroll, J., & Campbell, R. (1986). Softening up hard science: Reply to Newell and Card. *Human-Computer Interaction*, *2*, 227–249.

Carroll, J., & Mack, R. (1983). Actively learning to use a word processor. In W. Cooper (Ed.), *Cognitive aspects of skilled typewriting* (pp. 259–282). New York: Springer-Verlag.

Carroll, J., & Mack, R. (1984). Learning to use a word processor: By doing, thinking and knowing. In M. Schneider & J. Thomas (Eds.), *Human factors in computer systems* (pp. 13–51). Norwood, NJ: Ablex.

Carroll, J., & Mack, R. (1985). Metaphor, computing systems, and active learning. *International Journal of Man-Machine Studies*, *91*, 39–57.

Carroll, J., Mack, R., & Kellogg, W. (1988). In M. Helander (Ed.), *Handbook of human-computer interaction* (pp. 67–85). New York: Springer-Verlag.

Carroll, J., Mack, R., Lewis, C., Grischkowsky, N., & Robertson, S. (1985). Exploring exploring a word processor. *Human-Computer Interaction*, *1*, 283–307.

Carroll, J., & Mazur, S. (1986, November). Lisa Learning. *Computer Magazine*. pp. 35–49.

Carroll, J., & Rosson, M. B. (1987). The paradox of the active user. In R. Hartson (Ed.), *Advances in human-computer interaction* (pp. 80–111). Norwood, NJ: Ablex.

Cawsey, A. (1989). Explanatory dialogues. *Interacting With Computers*, *1*, 73–92.

Douglas, S., & Moran, T. (1983). Learning text-editing semantics by analogy. *Proceedings of CHI'83 Conference on Human Factors in Computer Systems* (pp. 207–211). Boston, Massachusetts. New York: Association for Computing Machinery.

Erhardt, J. (1983, January). Apple's Lisa: A personal office system. *The Seybold Report on Professional Computing*. Media, PA. Seybold Publications, Inc.

Ericsson, K., & Simon, H. (1980). Verbal reports as data. *Psychological Review*, *87*, 215–251.

Ericsson, K., & Simon, H. (1984). *Verbal protocol analysis*. Cambridge, MA: MIT Press/Bradford Books.

Evans, J. St. B. T. (1988). The knowledge elicitation problem: A psychological perspective. *Behavior and Information Technology*, *7*(2), 111–130.

Gould, J. (1988). How to design usable systems: Towards a practical GOMS model methodology for user interface design. In M. Helander (Ed.), *Handbook of human-computer interaction* (pp. 757–790). New York: Elsevier Science Publishers.

Graesser, A., & Black, J. (Eds.). (1985). *The psychology of questions*. Hillsdale, NJ: Erlbaum.

Graesser, A., Lang, K., & Horgan, D. (1988). A taxonomy for question generation. *Questioning Exchange*, *2*, 3–15.

Graesser, A., & Murray, K. (1989). A question-answering methodology for exploring a user's acquisition and knowledge of a computer environment In S. Robertson, W. Zachary, & J. Black (Eds.), *Cognition, computing, and interaction* 237–267. Norwood, NJ: Ablex.

Hoffman, R. (1987). The problem of extracting the knowledge of experts from the perspective of experimental psychology. *AI Magazine*, *8*, 53–67.

Karat, J. (1988). Approximate modeling as an aid to software design. *Human Factors Society Bulletin*, *31*, 1–3.

Karat, J., & Bennett, J. (1991). Working within the design process: Taking design seriously. In J. Carroll (Ed.), *Designing interaction: Psychology at the human-computer interface*. Cambridge, MA: Cambridge University Press.

Kato, T. (1986). What "question asking protocols" can say about the user interface. *International Journal of Man-Machine Studies*, *25*, 659–673.

Kieras, D. (1988). Towards a practical GOMS model methodology for user interface design. In M. Helander (Ed.), *Handbook of human-computer interaction* (pp. 135–158). New York: Elsevier Science Publishers.

Kieras, D., & Polson, P. (1985). An approach to the formal analysis of user complexity. *International Journal of Man-Machine Studies*, *22*, 365–394.

Knox, S. T., Bailey, W. A., & Lynch, E. F. (1989). Directed dialogue protocols: Verbal data for user interface design. In *Proceedings of CHI, 1989* (pp. 283–288). Austin, TX, April 30–May 4. New York: Association for Computing Machinery.

Landauer, T. (1988). Research methods in human-computer interaction. In M. Helander (Ed.), *Handbook of human-computer interaction* (pp. 905–928). New York: Elsevier Science Publishers.

Lewis, C. (1982). *Using the "thinking aloud" method in cognitive interface design*. (IBM Research Report RC 9265), Yorktown Heights, NY: IBM Thomas J. Watson Research Center.

Mack, R. (1983). *Understanding text-editing: Evidence from predictions and descriptions given by computer-naive people*. (Research Report RC 10333). Yorktown Heights, NY: IBM, Thomas J. Watson Research Center.

Mack, R. (1989). Understanding and learning text-editing skills: Observations on the role of new user expectations. In S. Robertson, W. Zachery, & J. Black (Eds.), *Cognition, computers and interaction*. (pp. 304–337). Hillsdale, NJ: Erlbaum.

Mack, R., Lewis, C., & Carroll, J. (1983). Learning to use a word processors: Problems and prospects. *ACM Transactions on Office Information Systems*, *1*(0), 254–271.

McKendree, J., & Carroll, J. (1986). Advising roles of a computer consultant. In *Proceedings CHI'86 Human Factors in Computer Systems* (pp. 35–40). Boston, MA: April 13–17. New York: Association for Computing Machinery.

Miyake, N., & Norman, D. (1979). To ask a question, one must know enough to know what is not known. *Journal of Verbal Learning and Verbal Behavior*, *18*, 357–364.

Norman, D. (1986). Cognitive engineering. In D. Norman & S. Draper (Eds.), *User-centered system design* (pp. 31–61). Hillsdale, NJ: Erlbaum.

Olson, J., & Olson, G. (1990). The growth of cognitive modeling in human-computer interaction since GOMS. *Human-Computer Interaction, 5*, 221–266.

O'Malley, C., Draper, S., & Riley, M. (1984). *Constructive interaction: A method for studying user-computer-user interaction*. In *Proceedings INTERACT 84: Human Factors in Computer Systems* (pp. 269–274). London, UK: August 23–27. New York: Elsevier Science Publishers.

Pollack, M. (1985). Information sought and information provided: an empirical study of user/expert dialogues. In *Proceedings CHI'85: Human Factors in Computer Systems* (pp. 155–159). San Francisco, CA: April 14–18. New York: Association for Computing Machinery.

Polson, P. (1988). A quantitative theory of human-computer interaction. In J. Carroll (Ed.), *Interfacing thought: Cognitive aspects of human-computer interaction* (pp. 184–235). Cambridge, MA: MIT Press.

Reitman-Olson, J. (1988). Cognitive analysis of people's use of software. In J. Carroll (Ed.), *Interfacing thought: Cognitive aspects of human-computer interaction* (pp. 260–293). Cambridge, MA: MIT Press.

Robertson, S., & Swartz, M. (1988). Why do we ask ourselves questions? Question generation during skill acquisition. *Questioning Exchange*, *2*, 47–51.

Seybold, J., & Seybold, A. (1984). Apple's Macintosh and Lisa 2. *Seybold Report on Professional Computing*, *2*(6), 6.

Shneiderman, B. (1980). *Software psychology: Human factors in computer and information systems*. Cambridge, MA: Winthrop.

Simon, H. (1969). *The sciences of the artificial*. Cambridge, MA: MIT Press.

Smith, D., Irby, C., Kimball, R., Verplank, B., & Harslem, E. (1982, April). Designing the STAR interface. *Byte*, pp. 242–282.

Suchman, L. (1987). *Plans and situated actions: The problem of human-machine communication*. New York, Cambridge University Press.

Whiteside, J., Bennett, J., & Holtzblatt, K. (1988). Usability engineering: Our experience and evolution. In M. Helander (Ed.), *Handbook of human-computer interaction* (pp. 791–818). New York: Elsevier Science Publishers.

Whiteside, J., & Wixon, D. (1988). Discussion: Improving human-computer interaction: A quest for cognitive science. In J. Carroll (Ed.), *Interfacing thought: Cognitive aspects of human-computer interaction* (pp. 260–293). Cambridge, MA: The MIT Press.

Williams, G. (1983, February). The LISA computer system: Apple designs a new kind of machine. *Byte*, pp. 33–50.

Zissos, A., & Witten, I. (1985). User modeling for a computer coach: A case study. *International Journal of Man-Machine Studies*, *23*, 729–750.

16
The Programmer's Burden: Developing Expertise in Programming

Robert L. Campbell, Norman R. Brown, and Lia A. DiBello

Our goal in this chapter is to describe and illustrate a developmental approach to programming knowledge. We argue that the prevailing practice of making binary novice-expert comparisons is of little value in the psychology of programming. We argue instead for a constructivist developmental approach, which uses Piagetian structural interviews and longitudinal tape diaries to gather information, and which builds multiple-level sequence models to describe the development of programming skill. With clinical interviews, we have elicited information about the learning histories of expert programmers. With tape diaries, we have followed the course of learning by individual programmers for up to 2 months. From these sources, we propose a developmental sequence of types of problems that programmers try to solve when learning the Smalltalk language.

We believe that our approach can lead to more detailed developmental descriptions of programming knowledge, to improvements in programming education, and to the construction of expert systems to aid in learning to program. This chapter presents the case for a developmental approach to expertise in programming and, in a schematic way, charts the evolution of our current research. We establish the importance of expertise in programming as a research area and criticize the current reliance on a binary novice-expert paradigm in psychological research in programming. We describe the basic assumptions of a constructivist developmental approach to expertise in programming.

Because there were no prior developmental analyses of programming skill for us to test, we chose an exploratory research strategy. In Study 1, we gave structured interviews to expert professional programmers in a range of specialties. We found that expert programmers are often able to describe major qualitative changes that occurred in their development.

We found these results encouraging but realized that our investigations needed a sharper focus. It became clear to us that programming is not a unitary developmental domain. We decided to focus on object-oriented programming (specifically the Smalltalk language) as our domain. We found object-oriented programming interesting in part because of widespread reports that experienced procedural programmers "become novices again" when learning Smalltalk. We also wanted to try a prospective, longitudinal research technique that might uncover details of programmers' learning paths likely to be missed in retrospective interviews.

In Study 2, we asked professional programmers who were beginning to learn Smalltalk to keep "tape diaries" of all of their work sessions for periods of up to 2 months. From the tape diaries, we were able to document in detail the programming tasks that the programmers worked on and the difficulties they reported. We found that our professional programmers used learning strategies that required them to take up the "programmer's burden" of understanding what happens in each line of Smalltalk code. Such strategies run counter to the incentives for code reuse built into the Smalltalk environment and produce frustration for the programmer.

The Study 2 data also gave us examples of problems that programmers worked on at different points in their progress. Using these data (along with some Study 1 material covering more advanced work), we were able to propose a sequence of seven developmental levels of expertise in Smalltalk. The refinement and testing of this model will be a major goal of our future developmental research.

We conclude that developmental models and research methods are eminently applicable to expertise in programming. We discuss some implications for the study of expertise in general: how developmental methods might supplement the usual approaches to knowledge elicitation, and how the tape diary method might be useful for nondevelopmental approaches to knowledge elicitation as well.

Importance of Programming Expertise

Computer programming is a rich domain for the psychology of expertise. It takes 5 or more years of work to become an expert programmer (diSessa, 1989; Kurland, Mawby, & Cahir, 1984). Currently, researchers cannot describe all of the skills and knowledge that expert programmers have, nor can they explain how programmers come to know what they know.

Yet an improved understanding of programming expertise and its acquisition could have major implications for the way that programmers are trained. There are tremendous individual differences in ability among programmers, even experienced programmers (Curtis, 1988). There are correspondingly large differences in programmer productivity. Differences in expertise must be in large part responsible for these productivity differences.

Programming education is weakly developed at the professional level. Classroom education does not deal with the problems that professional programmers confront, nor does it try to impart the strategies that expert programmers use. Programming courses have beginning and intermediate levels; advanced courses are extremely rare. Professional programmers learn on their own or through an apprenticeship process that is hardly ever deliberately managed (Brooks, 1989). A better understanding of programming expertise and its acquisition could lead to major improvements in programming education.

What we are proposing to do in studying expertise in programming is a form of knowledge elicitation. But our concerns are somewhat different from those that prevail in the expert systems field. Expert system builders typically concentrate on the end state, that is, on the knowledge held by bona fide experts. We are interested not only in the knowledge of expert programmers but also of those at various levels below true experts, and in the path that they travel as they acquire expert knowledge. Our concern with developmental paths toward expertise also differentiates our work from cognitive research on expertise (see the section headed Contrasts With Cognitive Science Approaches).

We think it is premature to judge which aspects of programming knowledge can or should be presented in the form of rule-based systems. The kind of research that we describe is necessary to lay the groudwork for future expert systems, because it can reveal what programmers do well, what they do not do well, and when. Such information is needed to design intelligent programming aids; to simulate the performance of expect programmers; and, for that matter, to evaluate the feasibility of building expert systems at all (cf. Klein, this volume; Klein, Calderwood, & MacGregor, 1989). Much of what we learn about programming is aimed as much at the design of teaching and the allocation of programming projects as at the design of systems.

The Novice-Expert Paradigm

Our talk of processes and of multiple levels is characteristic of developmental psychology. It is not characteristic of expert systems work, nor of human-computer interaction research. Important questions about expertise cannot be answered, however, unless we adopt a developmental approach.

In human-computer interaction, *expert* and *novice* are central categories. Design recommendations constantly invoke them (e.g., menus for novices, command lines for experts). So do empirical studies. But the meanings assigned to these terms are unsatisfactory. There is a well-ingrained novice-expert paradigm, in which the following are true:

1. Novice-expert is a binary distinction.
2. Static comparisons are made between expert knowledge and novice knowledge.
3. Experts are regarded as people with a certain amount of experience, not people who meet specific criteria of knowledge and skill.

In his review of work on expertise in programming, Mayer (1988) explicitly affirms this paradigm: "In reviewing each study . . . I use the term 'novice' to refer to the least experienced group of users, the term 'expert' to refer to the most experienced group of users, and groups with intermediate levels of expertise generally are ignored" (p. 570). In the work reviewed by Mayer, *expert* can mean anything from having passed several programming courses to having 10 years of experience as a professional programmer.

Mayer (1988) expects this binary distinction to bear a lot of descriptive weight. He proposes that novice programmers differ from experts in four general areas: syntactic knowledge, semantic knowledge, schematic knowledge, and strategies. In the arena of syntactic knowledge, experts "recognize syntactically correct code more rapidly and accurately" and novices "recognize syntactically correct code more slowly and inaccurately" (p. 574). In the arena of strategies, experts "decompose the program into finer parts and explore more alternative solutions," among other things, and novices "focus on specific solutions and low level program modules in debugging" (p. 574).

We will return to Mayer's (1988) claim later (see the section headed Utility of Developmental Levels). For the moment, we contend that a binary distinction between novice and expert is woefully inadequate in face of the enormous variations in skill, knowledge, and productivity that exist among programmers. Laboratory studies have found performance differences of two orders of magnitude (on measures such as time to find bugs) between professional programmers doing simple tasks (Curtis, 1988). The difference in productivity between a student who has passed one computer science course and a "superprogrammer" at the top of the industry has to be much, much larger (Brooks, 1989). Trying to span this vast territory with two categories of skill makes it inevitable that one researcher's "expert" will be another researcher's "novice." Researchers who make these terms their entire working vocabulary will be unable to offer coherent generalizations.

In general, we believe that the novice-expert paradigm has put a crimp on the psychology of programming. Some investigations have merely endeavored to show that expert programmers are like experts in other fields: If chess experts are better than novices at remembering meaningful (but not scrambled) configurations of chess pieces, then expert programmers will be better at recalling meaningful (but not scrambled) ALGOL programs (McKeithen, Reitman, Reuter, & Hirtle, 1981). Other investigations have adhered more closely to the specifics of programming: It has been found that experienced programmers are better than novices at forming high-level representations of the programs they read (e.g., Adelson, 1984; Jeffries, Turner, Polson, & Atwood, 1981); that they are better at finding bugs, particularly those that depend on such a conceptual understanding (e.g., Gould, 1975); and that they are more likely to use standard programming plans and to compose those plans according to rules of programming discourse (Soloway & Ehrlich, 1984). Some think-aloud protocol studies have disclosed expert programmers' problem-solving strategies in detail (e.g., Guindon, Krasner, & Curtis, 1987; Rosson & Gold, 1989). Very little has been done, however, to situate such results in the landscape of programming, in the range of proficiency that exists, in the range of programming specialties that exists, in the range of skills that every proficient programmer needs to acquire. And researchers will have trouble finding their bearings as long as *novice* and *expert* are used, without nuance, as convenient labels for experimental groups differing in amount of experience.

To paraphrase Karmiloff-Smith (1981), we can look for novice-expert differences, or we can study the development of expertise. Looking for novice-expert differences in programming has not served researchers well. Maybe studying the development of expertise in programming will serve better.

A Developmental Approach

We have chosen to adopt a developmental approach to expertise in programming. We draw our general assumptions from constructivist approaches to psychological development. In consequence, we seek to describe the development of programming skill as a sequence of levels of expertise, and we have adopted developmental research methods, specifically, longitudinal methods that follow learning over time.

Developmental Assumptions

There is a wide range of approaches to developmental psychology. (For a contemporary survey, see Collins, 1982.) We draw from *constructivist* approaches, such as those of Piaget (1975/1985), Vygotsky (1934/1987), Bickhard (Campbell & Bickhard, 1986), and Feldman (1980). On the constructivist view, development is not the passive absorption of structures from the environment, nor the accretion of atoms of knowledge. Constructivist approaches maintain that learners actively construct their knowledge and that what they are capable of constructing depends on what they already know. In Campbell and Bickhard's (1986) framework, there are two main types of constructive processes: learning (the active construction of new knowledge by variation and selection) and reflective abstraction (coming to know properties of one's prior knowledge and skills).

Constructivist approaches maintain that development is qualitative change. The course of development in any domain is structured into sequences of steps; there may also be universal transitions that are "bigger" and "more important," normally called stages. (For a discussion of different conceptions of stages, see Campbell & Bickhard, 1986.) Critical to the description of development in any domain will be a sequence model for that domain.

Finally, our brand of constructivism contends that development is domain-specific. The detailed sequence of development will differ from one domain to another. When learners go through comparable changes (e.g., stages) in different domains, they will not necessarily be in the same stage in social role-taking skills as they are in addition and subtraction skills. Where Piaget was concerned with development by children in "universal" domains—those that everyone masters without having to be specifically taught—those of us who study expertise are concerned with nonuniversal domains, which are not learned by everyone and have to be taught (Feldman, 1980).

Contrasts With Cognitive Science Approaches

Our developmental approach can be fruitfully contrasted with the well-known studies of expertise that have been conducted by cognitive scientists (Chase & Simon, 1973; Chi, Glaser, & Rees, 1982; Larkin, 1981). Developmental and cognitive science approaches agree that expertise is domain-specific and that the appropriate description of expertise in chess may be quite different from the appropriate description of expertise in programming.

Whereas developmental approaches are committed to developmental sequence models with multiple levels, cognitive science approaches have perpetuated the novice-expert paradigm. Cognitive science studies of expertise—whether concerned with the recall of configurations of chess pieces (Chase & Simon, 1973) or solution strategies and categorization of textbook Newtonian mechanics problems (Larkin, 1981; Chi et al., 1982)—

make binary contrasts between experts and novices, and describe their differences. No doubt these researchers acknowledge the existence of some kind of continuum between novices and experts, but they consider a focus on the endpoints of that continuum to be methodologically adequate.

Cognitive science models also lack the commitment to constructive developmental processes characteristic of developmental models. The information-processing models favored by many cognitive science researchers are currently incapable of accounting for the observed differences between novices and experts in areas like Newtonian physics and mental calculation, as leading researchers in this tradition have recently acknowledged (Neches, Langley, & Klahr, 1987, pp. 32–34). Some critics contend that this is an inadequacy in principle (Bickhard & Terveen, 1991; Campbell, Carroll, & DiBello, 1989; Dreyfus & Dreyfus, 1986).[1] For all of these reasons, we consider a developmental approach to be preferable to the currently popular cognitive science approaches to expertise.

Exploratory Interviews

The prescription to take a constructivist approach to development is an extremely broad one. In an area like number development in children, developmental analyses are already there to be refined and tested. By contrast, when we began this research, no serious developmental analysis of programming had been proposed. We decided, therefore, to begin with exploratory research. An additional constraint is that we were not expert programmers ourselves. We did not presume, as psychologists sometimes do, to know the secret processes in programming that remain hidden from programmers themselves. Like anthropologists beginning to study an alien culture, we wanted to take seriously what programmers say about their working methods, their experience, and the skills they are trying to master. Ecological studies of the social context of programming, such as those of Curtis, Krasner, and Iscoe (1988) and Scribner (1989), tend to take this approach, but cognitive studies of programming generally have not. As soon as we have a reasonable account of programming in terms close to the programmers' own, we assumed, then we may begin to discover things that the programmers did not notice or could not tell us about.

We began with exploratory methods based on what programmers could tell us in their own words. We used a form of structured interview with probes and elaborations, modeled after Piaget's (1929/1960) *méthode clinique*. Interviews of this type are familiar in expert systems work (Hoffman, 1987, 1989). The use to which we put them is a little different. Although we sought descriptions of everyday working methods (familiar tasks), we also tried to get experts to describe their development as programmers. And we were not trying to render their decisions and discriminations into rules. We wanted information that would help us shape our hypotheses about the development of programming skill and that would serve as a "sanity check" on our belief that such development could be described as a sequence of qualitatively different developmental levels. These were the major goals of Study 1.

Study 1: Structured Interviews With Expert Programmers

Method

We talked to seven professional programmers in the Computer Science Department at IBM Research, designated here as P1 through P7. All had at least 5 years of professional experience and were considered to be experts by their peers. Their backgrounds were diverse: C, Pascal, LISP, Prolog, Assembler. Three of them (P1, P2, and P3) had extensive experience in the object-oriented language Smalltalk and were experts in procedural language programming before taking up Smalltalk. We recruited these programmers informally, based on personal recommendations. We told them that we wanted to interview them about their working methods and about their development as programmers.

TABLE 16.1. Core questions for the structured interviews of study 1.

Background
Describe your background: years programming professionally, education, number of languages, and when each language was learned.
Current working methods
When you are planning a long-term project, how do you go about conceiving the project? What does it "look like" in your mind at the beginning?
How does the way you think about the project change as you work on it?
As you are working on the project, how do you know when there's a problem? What tells you?
When you know there's a problem, but you aren't sure exactly what is causing it, how do you go about finding it?
Development as a programmer
Compare the way you do things now with how you did them when you were a beginning programmer.
When you moved to a new language, how much of what you knew about other languages helped you to learn the new one? Did you have to become a beginner again with the new environment?
Telling experts from those not so advanced
Can you tell by reading another programmer's code how expert that person is?

The interviews were structured. We asked all of the programmers certain core questions, but we let them provide extended examples or develop interesting points as they arose. Our core questions, in the order we asked them, are presented in Table 16.1. Each interview took 60 to 90 minutes. All of the interviews were recorded on audiotape and transcribed for later interpretation.

Results

Our purpose in conducting these interviews was frankly exploratory. We were interested in the presence or absence of general patterns in what programmers told us. Fine-grained analysis of the transcripts struck us as premature, though it could always be done on these data later in the light of hypotheses that might require it. In our presentation here, we omit the programmers' answers to questions about their current working methods. All gave us extensive descriptions that differed in the degree to which they claimed to use top-down approaches. Their answers are interesting from a psychology of programming standpoint, but working methods are one of the better explored areas in that discipline, and we are not ready at this time to present a developmental analysis of them. Instead we will focus on programmers' views of their own and others' development, and on their ability to recognize the work of other experts.

Qualitative Change

Two of our programmers (P5 and P7) essentially described their current working methods and did not describe their early experiences as being seriously different. Those who did believe their working methods had changed identified specific shifts in their understanding of programming. P6 identified some difficulties that she had encountered in moving from Assembler to Prolog programming. P4, who had considerable experience teaching C and Pascal, gave a series of examples of qualitative changes:

> You learn in stages. In the beginning you start writing very small programs and everything is a global variable. And then you learn about local variables and passing arguments in routines. And then everything becomes arguments and local variables. You start passing these huge variable lists around. And then all of a sudden you get to bigger things and it becomes a mess to do that, so you start going back to global variables.
>
> It's similar with structures: The beginner doesn't understand structure at all and tries to make a global thing, which is a disaster. The intermediate person tends to break things up too much and always wants . . . well, any piece of code that gets replicated becomes a subroutine. This isn't always right because there's a lot of overhead involved in that. An advanced person understands the balance between when he should be breaking things up and when he shouldn't. When he should be using arguments and when he should be using global variables. He understands better how to package up data in appropriate data structures.
>
> The novice doesn't understand data structures at all. The intermediate understands somewhat and probably thinks the solution to everything is a linked list or big array of something. The person who is advanced understands the advantages and disadvantages of each kind and what is appropriate. He has an underlying feeling of what's going on in the

machine and what will be fast and what will be slow, what makes the most sense to solve the problem. Because you can solve each problem all different ways. Picking the right things can make it hard or easy. That appreciation helps too.

P4 labeled three distinct stages (novice, intermediate, expert) for several subareas of C programming. (In passages omitted here, he extended the analysis to general program structure, on the one hand, and to specific strategies for writing loops, on the other.) P1, P2, and P3 also referred to specific developmental milestones that they passed. Their sharpest recollections pertained to learning the object-oriented language Smalltalk. In fact, we will reserve P2's account for our presentation of developmental levels of Smalltalk learning. P3 began to learn Smalltalk after being a professional programmer for over 15 years. He found it very difficult and identified a sharp transition in learning the language:

There's this tremendously steep learning curve in learning Smalltalk. Everybody that I've ever talked to who's a Smalltalk programmer talks about climbing the cliff, or climbing the mountain. . . . When you grab the extra rock and it breaks off in your hand, and you're about to fall down the cliff, you can go ask somebody. . . . But there's still this cliff, and even in the best of circumstances it seems to be something that people remember, not for the rest of their lives, but at least for several years afterwards. . . .

It's real easy to tell if somebody's on one side of the cliff or on the other side. . . . They are over it. Their whole way of thinking about it is, "I can work. I understand what I'm doing here." They can go find things. And once they're there, I think it's hard to determine how proficient they are, short of seeing their code, or some long interviewing.

But it seems to be pretty easy to tell if somebody's into Smalltalk, but not over the mountain. You can tell—it's not quite visible wounds, but you know.

Here, P3 asserted that there is a major transition in his and others' Smalltalk learning, a threshold at which they begin to have confidence in their ability to navigate in the environment and apply basic programming strategies. The programmer's feelings about Smalltalk help to indicate where he or she is in the climb.

Our final example of qualitative change comes from P1, who describes how basic concepts of Smalltalk have *changed their meaning* for him. Specifically, his understanding of a fundamental doctrine, that everything is done by sending a message to an object, changed over time.

I can actually point to two times. Once at Tektronix and then when I was teaching Smalltalk here last year. When I felt everything fell into place. . . . In both those cases I reviewed what I know about Smalltalk and [thought], "Oh, so that's like that and this is like this. . . ."

[The first case:] Like sending a message, that this template is just the message template and you're just sending to the metaclass of this thing. During one of these sessions I rethought things. I thought, "Oh yeah, so that's a message, *everything* is a message." It would be things like that that I would relearn. It would be like the things I didn't fully understand the first time because I didn't have all the knowledge. . . . So in the first pass-over, you kind of gloss over some things and . . . when you do the second pass . . . there is just a chance to piece things together. . . . It's a bootstrap problem, you have to kind of go slow and iterate.

[The second case:] I discovered that creating a new subclass was that the thing in the [Class Hierarchy Browser] window was just sending a new message, and I knew all along that everything was an object. I heard that a million times. But here it drove that home. I saw that *everything* is an object. Everything is sending a message to an object, this too; this thing I (had) just considered system magic.

From these sorts of examples, we concluded that programmers did experience qualitative change in their development and that they could give us useful information about development levels and transitions in acquiring expertise in programming. Assuming that programmers move through multiple developmental levels in acquiring expertise, we wanted to know whether experts could recognize where other programmers were in the course of development.

Recognizing Expert Work

A major interest of ours was experts' ability to tell whether a program that they read was the work of another expert. Program critiquing is an important skill, and it is interesting to know what kinds of discriminations expert programmers can make when they critique programs. For our own future research efforts, we wanted to know whether some experts could judge other programmers' work by level (using criteria that we, as nonexperts, might lack skill in applying). Hence, in our interviews, we asked our experts whether they could tell from reading code, without knowing the other programmer's credentials or past projects, whether he or she was an expert.

Except for P7, who had little to say about the subject, all of our programmers claimed that they could draw extensive inferences about expertise from reading another programmer's code. P2 gave several examples of "novice" traits that could show up in a programmer's Smalltalk code, including use of a "brute force" approach as opposed to an "elegant" one, and reduplicating existing classes in the class hierarchy instead of creating appropriate subclasses; P1 also gave examples of amateurish Smalltalk design. P3 spoke of stylistic matters that arise for him when reading code:

> I might not understand why he's done it the way he did it, or it might be obvious, oh gee, he's carefully thought through some orthogonal issues here regarding the data structure or something, but. . . . It's a good analogy to . . . writers writing poetry [who] might each have their own quite different view of what poetry is, but each can read the other one, and say, "OK, this guy's a real poet, not a hack."

P3 spoke at some length about the value of reading well-written code and appreciating its style. His comments support the view that aesthetic criteria, frequently mentioned by programmers (e.g., Lammers, 1986) but usually ignored by psychologists (Leventhal, 1988), become important at higher levels of expertise in programming.

Of course, one expert may produce a solution quite different from another's, even to the point of bizarreness. In that case, an ambiguity arises: Did the other programmer know what he or she was doing, or blunder into an unusual approach that happened to work? P4 suggested resolving this ambiguity by asking the other programmer to defend the design:

> The next thing is you start talking about design. You get a feel for how they would put something together and while the experts might all have their own ways of doing it, typically they can agree on . . . why the other guy's structure has its pros and cons. Whereas a novice may lay something out and if you point out the holes in that system, or the deficiencies, they are less likely to be able to defend them. I can look at something somebody else did, and he can look at mine and we may have done them radically different ways; both are good and he can chew mine apart and I can tear his apart, and he can tell me why and we both *see* the validity of each other's arguments.

Programmers P2, P5, and P6 also mentioned the value of asking the other programmer to defend the design.

Expert programmers may not, of course, be as good at detecting expert work as they claim they are. Researchers need to test such assertions by asking experts to rate programs produced by people at different skill levels. And much remains to be learned about the criteria that experts apply when making these discriminations. (For an initial exploration of program critiquing by experts, see Mastaglio & Rieman, 1989.) But the remarks made by six of the seven programmers we interviewed indicate that the judgment of expertise by expert programmers is a promising area of investigation.

Conclusion

Programming Expertise in General

Our interviews support the developmental position that expert programmers, when they describe themselves as developing at all, experience their development as qualitative. They are not just becoming faster and more facile with an unchanging set of component skills; they are not just aggregrating atoms of programming knowledge. Across the different programming domains, there is a progression from "cookbook" to "intuitive" understanding, of the sort described by Dreyfus and Dreyfus (1986).

This is generic information about expertise, containing little that is specific to programming. The development of expertise in many a domain, from driving to igneous geology to bass clarinet playing, could be described as a progression from "cookbook" to "intuitive." Cognitive science accounts (e.g., Chi et al., 1982) endeavor to explain such progressions in terms of the acquisition of "chunks," possibly through the operation of a self-modifying production system. Accounts opposed to cognitive science (e.g., Dreyfus & Dreyfus, 1986) invoke the development of "holistic situation recognition" through an unspecified learning mechanism (see Cooke, this volume). It is not clear whether a generic account of expertise can discriminate between the two approaches, despite the difference in their theoretical commitments. Nor will a generic account of programming expertise help anyone teach programming skills. To be useful, an account of expertise in programming should not emphasize properties that experts everywhere are expected to have; it should emphasize the specific skills that are acquired (call them chunks, holistic situations, or what you may) in learning C or Smalltalk, and the manner in which they are acquired.

The interviews point toward valuable specifics by providing examples of developmental milestones and major transitions in different domains of programming. Changes in the use of data structures in C (cited by P4) and the understanding of navigation in the Smalltalk environment (mentioned by P3) are candidate markers for developmental levels in those domains. It seemed to us that to extend our investigation of expertise in programming, we would have to focus on a specific programming domain, with the aim of identifying developmental milestones and learning strategies specific to that domain.

Utility of Interviews

We also had some questions about the utility of structured interviews in continuing our work. Though appropriate for the global questions of Study 1, the interviews were somewhat of a blunt instrument. Our interviews were both introspective and retrospective: Programmers had to reflect on the way they work now and then compare it to the way they worked in the past. Because the interviews were retrospective, they were not an ideal way to follow the course of development. Programmers often had to recall events from several years' distance, which led to a loss of detail about projects, problems, and experiences.

Moreover, programmers are not used to talking about their own development. P5, who claimed that he had always used the same approach to design, told us later that he had not truly understood the purpose of the interview at the time. By contrast, P1 and P4 had taught programming and had had the opportunity to witness the development of other programmers. P6 had made a transition to logic programming (Prolog). P1, P2, and P3 had made a recent transition to object-oriented programming and had more to say about their Smalltalk learning than about their prior development as procedural programmers; perhaps they had gained a dizzyingly new perspective from climbing the Smalltalk mountain. More generally, we wonder if confrontation with an alternative to one's standard programming practices, either through teaching or through struggling with a new paradigm, is necessary to stimulate reflection about one's development as a programmer.

In addition, it may be asking too much to expect expert programmers to be able to define all of the developmental levels in their domain, even if they are sympathetic to such a conception and, once given a model, can readily rate the level at which another programmer is performing. Feldman and his colleagues, in their collaborations with experts in a number of fields, have not been able to extract developmental levels directly by interviewing experts (Feldman, 1988).

There are a number of ways to solve the problem of developmental detail. Interviews could be given more structure that focuses on developmental questions. More technical programming detail could be elicited by enlisting the services of an expert programmer as co-interviewer. The critical decision method (Klein et al., 1989) could be used to focus on programming projects that were pivotal in the development of experienced programmers.

Even with all of these improvements, however, interviews are still retrospective. We wanted to be able to follow the course of development in detail as it occurs. To do this we needed to adopt longitudinal methods, which follow change in an individual over time. Our structured interviews, then, left us with some valuable leads about the development of programmers, but also left us with a concern for sharpening our focus: for restricting ourselves to a particular domain of programming and tracking the detailed course of development in that domain.

Focusing on a Specific Domain of Expertise

Domains of Programming

When we considered the diversity of programming activity covered in the Study 1 interviews, it became clear to us that researchers need to classify the domains of programming. We use the plural because programming is clearly not a unitary domain. One can be an expert C programmer and flounder hopelessly in Smalltalk. One can be an expert builder of electronic mail systems and not know what to do in avionics. As these examples suggest, we propose to carve programming up into programming language families and application families.[2] Thus, programming languages within a family, such as Pascal and C among procedural languages, may be closely enough related that an expert in one can quickly become an expert in another. But the transition from procedural to object-oriented languages can force the programmer to become a novice again (as P1, P2, and P3 made clear in their interviews).

Although an expert C programmer may become a novice again in Smalltalk, we suspect that such a programmer will not learn Smalltalk at the same rate, or in the same way, that a nonprogrammer would. We hypothesize that becoming proficient in *some* area of programming makes a permanent difference. Proficient programmers know general programming concepts (e.g., stacks, registers, and buffers) and general programming skills (e.g., how to build a sort routine) that presumably will carry over to a new area and will guide their learning trials. They may have different goals, in programming and in learning about programming, than nonprogrammers would have had a chance to develop.

To continue our work, then, we needed to zero in on a specific language family and a specific application family. We selected Smalltalk (an object-oriented language) and visual interface programming (the application for which Smalltalk is most widely used). We also decided to focus on learners already proficient in some area of programming. Most Smalltalk learners, especially in industrial settings, are already proficient programmers. And Smalltalk is radically different from the languages that they already know.

Smalltalk Programming

Smalltalk (Digitalk, 1988; Goldberg & Robson, 1983) is a leading object-oriented language.[3] The object-oriented programming paradigm is radically different from standard procedural programming (BASIC, C, Pascal, FORTRAN); functional programming (LISP); or logic programming (Prolog). For introductions to object-oriented programming, see Robson (1981) and Pascoe (1986); Rosson and Alpert (1990) outline some of the psychological issues that it raises. In procedural languages, computation is accomplished by active procedures manipulating passive data structures. In object-oriented programming, computation is accomplished by communication among *objects*. An object is defined as a local bundle of procedures and data. All processing in the system is the result of messages being sent to objects. Objects have internal *methods* that implement their responses to the messages they receive. For instance, strings of letters are objects that have a variety of built-in behavior, such as printing themselves. Sending the message `printString` to the string `'This is a string'` will cause it to execute an internal method and print itself.[4]

Smalltalk is a pure object-oriented language; even adding numbers is accomplished by sending messages to objects. It has a number of features that are radically different from those of conventional programming languages:

1. Communicating objects may be natural for some operations (e.g., asking for a checking account balance by sending an accountBalance message to a CheckingAccount object) but are clearly not for others (e.g., adding 1 + 1 by sending the message + with the parameter 1 to the object 1).

2. Smalltalk objects are instances of classes, and classes are organized into a hierarchy. Classes can inherit methods from superordinate classes in the hierarchy: For instance, Array inherits methods from FixedSizeCollection, OrderedCollection, Collection, and Object. The same message name can trigger different methods for different classes, a property called polymorphism; new means one thing when sent to Array and a different thing when sent to Integer. Programmers have to develop special skills for searching through the Smalltalk environment to find useful methods.

3. Smalltalk is not just a programming language; it is an environment stocked with a built-in program libary consisting of classes and methods. "Writing a program" is not producing a unitary piece of code; it is making numerous local changes and additions to the program library. For instance, writing a program to shuffle arrays randomly might consist of adding a new subclass Random to the class Integer and adding new methods to the class Array. "Reading a program" means finding such local fragments and understanding their interactions.

4. Smalltalk is often used to prototype visual user interfaces, but the classes and methods that make up visual interface objects, like windows and menus, and that control their behavior, interact in complex ways. Classes Model, Pane, and Dispatcher, their methods, instance variables, and manifold interactions take months to learn, according to our interviews with P2 and P3 in Study 1.

5. Smalltalk promotes object-oriented design. Proponents of object-oriented design claim that designing applications in terms of communicating objects is more faithful to the designer's initial ideas, and hence more natural, than designing in terms of data structures and generalized procedures to operate on them (e.g., Rosson & Alpert, 1990). Whether the naturalness claim is true or not, it is clear that object-oriented design is very different from the functional design that procedural programmers are accustomed to.

Smalltalk, then, may offer a rich source of learning difficulties for programmers already proficient in procedural languages, requiring them to become novices again for a time.

Longitudinal Methods

To track programmers' difficulties in learning Smalltalk, we turned to longitudinal methods, which follow change in an individual over time. Many developmental questions cannot be settled empirically without tracking development longitudinally (de Ribaupierre, 1989; Hoppe-Graff, 1989; Wohlwill, 1973). Longitudinal studies have been used by developmentalists to follow the course of learning and stage transition (e.g., Inhelder et al., 1976). Longitudinal methods have even made an incursion into the study of expertise (Staszewski, 1988; Walton, Adams, Goldsmith, & Feldman, 1988).

When carried out in the field, longitudinal studies have the additional advantage of allowing researchers to follow the natural course of learning in its natural setting, not just for short stretches in the laboratory. They make it possible to observe the sequence of problems that programmers deal with. Longitudinal studies can be used heuristically, to provide information about problems and strategies that can be used to construct developmental level models. They can also be used to test such models (e.g., by determining whether individuals actually pass through all of the levels that the model specifies). Although developmentalists often recommend longitudinal studies for model-testing purposes only (de Ribaupierre, 1989), we chose to employ longitudinal methods for exploratory purposes. Our specific technique was to collect tape diaries of programmers' work sessions.

Study 2: Tape Diaries of Programmers Learning Smalltalk

Our investigation of Smalltalk focused on initial learning by professional programmers. Object-oriented languages are usually learned by programmers already experienced in procedural languages like C, Pascal, and FORTRAN, or functional languages like LISP. We wanted to know how Smalltalk learning is affected by prior experience with more conventional languages. We also wanted to identify developmental markers for the early levels of Smalltalk learning.

Method

In this study, we asked programmers to keep *tape diaries* for us. We put portable audiotape recorders in the programmers' offices and asked them to dictate one diary entry for each work session with Smalltalk. We did not ask them to do concurrent thinking out loud; in our experience, thinking out loud while working on Smalltalk problems is too difficult, and the evidence obtained from it too dilute, to be useful in an exploratory investigation.

Instead, we asked programmers to summarize each work session with Smalltalk. We asked them to give the date and time that they began working, who else they were working with, the tasks that they wanted to accomplish, and the things that they wanted to learn about Smalltalk during the session. We asked them to describe what problems they were having with Smalltalk, both generally and in the current task, what they found hard to understand, what they now understood for the first time, and the strategies they used to solve problems or to understand Smalltalk. They were encouraged to talk about Smalltalk in general, including their likes and dislikes about the language, rather than confining their discussions to the current programming problem.

We collected tape diaries from four people: three programmers who were already expert in procedural or functional languages (P8, P9, and P10), and one nonexpert programmer (P**). All were learning Smalltalk/V286 (Digitalk, 1988) and were working on the language at a steady pace during our study. All four were employees of the IBM T. J. Watson Research Center. The diaries began with the learners' first serious Smalltalk session and ended between 2 weeks and 2 months later. We were largely successful in getting programmers to make a diary entry after each work session (some also made comments during the session). One programmer sometimes took notes of sessions and recorded his comments later rather than doing so immediately.

The diaries were transcribed for later interpretation. We have marked all diary entries with the number of days since the programmer began to participate. Although not as precise as the number of hours spent programming, the number of days roughly indicates how experienced each programmer was with Smalltalk.

Results

Here we focus on the learning strategies used by the three expert procedural programmers in our study. We were also able to document typical programming problems, which will be used below to illustrate developmental levels (see also Campbell, 1989). The professionals' approaches to Smalltalk seemed markedly similar to us, differing noticeably from the approach used by the single nonexpert programmer and the approaches we ourselves, as nonprofessional programmers, had used while learning Smalltalk.

The Programmer's Burden

The Smalltalk programming environment encourages the use of its built-in "program library," the objects and methods that come with the system. Programmers are expected to reuse existing methods, relying on a general understanding of how a method works, supplemented by trial and error, rather than studying each line of code in the method. The minimal use of comments in Smalltalk (typically one comment at the top of a method) also favors such an approach. Experienced programmers, however, are reluctant to use methods without studying them first:

[P10, Day 24]: Being a programmer, I have this burden of trying to understand all the bits and bytes. That's my disadvantage, that I was dealing with languages that are, compared to Smalltalk, lower level languages. I hope somebody you randomly catch on the street, who doesn't know programming, would actually have less of a headache trying to understand. I guess most people would just, in that case, skip getting into details trying to understand. . . . I guess I would have to get used to the different kind of work that Smalltalk demands. . . . You don't have to understand all the stuff that you are using.

P2, in our Study 1 interviews, had made the same observation:

I think that programmers, coming at least from my background, have this thing where they don't want to use somebody's code. They think that they can write it better or do it better or they want to understand how the guts of it work, before they blindly use it. And that is something, to some extent, in an object-oriented environment, you should abandon.

Our tape diaries exhibit a number of strategies that professional programmers adopt when taking up the programmer's burden. These include focusing on the details of Smalltalk syntax, learning the class hierarchy by reading code, and conducting small "experiments" to assess the properties of the Smalltalk environment.

Mastering Syntax and Precedence

Early in the course of learning (typically the first week), all three of our programmers strove to master the details of Smalltalk syntax, including the quirky order of precedence in which different kinds of messages are sent to objects:

[P9, Day 1]: I've been looking at the precedence order of unary, binary, and keyword messages. That all seems pretty straightforward. . . . I feel that I need to learn more about the syntax so that I can understand this code better.

[P10, Day 2]: Smalltalk evaluates an arithmetic expression strictly from left to right. Just as you accept this weirdness, there are exceptions to the rule. Then you have the fact that unary messages are evaluated first, then binary ones, and then these other keyword things. So, if you have an expression, you have to kind of hop around to see any unary ones, and calculate them, and then binary ones.

This short-lived, intense concentration on syntax was also observed by Scholtz and Wiedenbeck (1989) in a study of experienced C and Pascal programmers beginning to work in the Icon language.

While learning the syntax, the professional programmers often attempted detailed mental simulations of what happens when each line of code is executed:

[P10, Day 4]: Whenever I learn a new programming language, I try to mimic the behavior of the computer calculation. So things like precedence and the order of calculation are very important for me to understand. In Smalltalk, I still have this confusion: "What happens when?"

Our programmers' concern about details extended beyond syntax, to seeking an in-depth understanding of each method they used.

Learning the Hierarchy by Reading Code

Professional programmers sought to understand Smalltalk methods that they encountered by reading the code for the method, then searching for the additional methods used in that method's code, reading their code, and so forth. For instance, a programmer who wants to break up a string of letters into separate words may discover the instance method asArrayOfSubstrings for class String. On reading the code for asArrayOfSubstrings, the programmer discovers that the messages on:, atEnd, peek, and position are being sent to instances of class ReadStream. To understand these messages, the programmer has to look up the four methods. Because of inheritance, the methods may belong to superclasses of ReadStream and not to ReadStream itself. Moreover, the methods may refer in their code to unfamiliar methods of other classes. Because methods belong to classes, and classes participate in a hierarchy with inheritance and polymorphism, trying to gain an in-depth understanding of a method leads to long searches through the hierarchy to the read scattered bits of code.

[P9, Day 7]: I find I make a lot of depth-first regressions through the code. I'll be looking at a method and I try to look at things that are new that I haven't seen. . . . Sometimes that takes me through several layers deep. . . . The main frustration in trying to do that is just locating where the methods are. . . . You can guess fairly easily what the class of an object is that is receiving a message. But when you go there in the Class Hierarchy Browser, and the method isn't there, you have to look through the superclass clain in order to find it, and that can be pretty frustrating.

I guess the reason I'm using this depth-first approach is exactly because I feel that I know most of the syntax. So at this point . . . it is a matter of learning the whole hierarchy, and what's there, and what the classes do, and then the details of what the methods are, and the protocols for the classes. . . . It makes more sense to start with something concrete, so that . . . it sort of builds a more natural understanding of what the things are doing.

Such depth-first regressions, however, can be counterproductive in trying to learn Smalltalk. Trying to find all of the instance variables of a class, which can also lead to extensive searches, P10 (Day 21) ruefully concluded, "You browse until you get bruised." He later (Day 45) decided as follows:

I have to get used to the fact that I don't have to know all [about] this stuff that I'm using. There is obviously a very delicate balance on how much one should browse through classes. Recently, I caught a complete listing of Smalltalk, and it's a huge document. So that was an illustrative way for me to see how much would I have to look through if I wanted to know all of it. In languages that I've learned before, like C or Pascal or whatever, you would really try to learn all of the commands and keywords as quite probably you would need most of them.

The programmers in our sample were not content with depth-first search, however. They also engaged in active hypothesis testing.

Experimentation

Professional programmers carried out little "experiments" in code to test their understanding of Smalltalk. They were always trying to "break the clock" (P10, Day 24). In the diaries, we found experiments to test hypotheses about the meaning of #, which is used to indicate arrays and symbols, and ∧, which is used to return an object explicitly. Such experiments are conspicuously absent from the diary of P**, the nonprofessional programmer in our sample, who worried about crashing the system with unwise experiments and concentrated on developing theories about the operation of Smalltalk instead.

[P9, Day 1]: I've been trying to figure out what exactly # implies. So I was trying earlier to embed a method inside an array that started with #, and it didn't work. So I'm not quite sure what that implies, or whether there is some way to do it, what I was trying to. Maybe there isn't.

In fact, Smalltalk does not permit embedding a method inside an array. P10 carried out a more extensive set of experiments while trying to understand how ∧ is used and how pointers are defined in Smalltalk.

[Day 7]: Here is one confusion. . . . Array a and b . . . have the same elements at the same places, and a == b returns true. Again, I don't understand—this is a small thing—what this ∧ is doing. I took it away, and I ran [the code] without it, and it still returns true. I mean, I understand why it's true—these two objects have the same elements at the same places. They are equal, so the true is returned.

But here comes the confusion. . . . Now we have three . . . [temporary] variables, a, b, and c, and the first one gets assigned an array a := #(1 2 3 4). . . . Then, the second one, b, is equal to a, and then c is equal to b. Finally, this == sign [in a == b] is checked, and it returns true.

Thinking C way, that sounds OK. So we have an array and a is actually the name of that array, that is, a pointer to the physical location where the array is stored. Then b is equal to a—we have another memory cell b that contains the address of this dimensioned array. So it points to the same place where a points to. And then c points to b, and b points to a, and everything's nice. [The expression] c == a returns true because we had one object all the time and all these pointers pointed to the same place. Good.

But here are two small checks that I ran. Before the last line, I put again a := #(2 5 7). Now, if my C way of thinking works, I completely change the content of array a, but I didn't change the values of pointers b and c. So this == should still return true because we have again just one object involved, and all these pointers point to the same thing. Instead, it returns false, so I don't know what to think about it.

So I ran another small experiment. Again I added this line—it is next to the last one—but [now] a is equal to the same thing, that is a := #(1 2 3 4), and then I evaluated it, and now it returns *true*. In other words, well, if I thought of b and c as some kind of pointers to [the] array, they are just not that.

P10 concluded that Smalltalk pointers do not behave like pointers in C. In his pointer experiment, analogies to the C language play a prominent role.

Analogical Understanding of Smalltalk Concepts

In the early phases of Smalltalk learning, the programmers in our sample often made analogies to languages that they knew.

[P8, Day 1]: My feeling is that [Smalltalk] is not that different from languages that I'm familiar with; there are similarities to bits and pieces of other things. Some aspects of the object hierarchy . . . seem similar to me to the scoping of Pascal variables. . . . Smalltalk seems similar to some other interpreted languages like LISP or APL, where if I type in the name of something in LISP or APL, I am returned its value. . . . Messages are primarily names of functions. And what one does is send the name of a function to an object.

[P9, Day 7]: What was more interesting to me was that the methods were there to use [class OrderedCollection] in a way that I was used to using other data structures. I think that that's the way I've been approaching a lot of Smalltalk. As I don't think there are any big innovations in data structures. As far as I can see, they use the normal sorts of arrays and lists and so forth. So the interesting thing becomes to me, "How do I use those same data structures in Smalltalk? How do I understand the methods in terms of the data structures that I am used to?"

[P10, Day 2]: Arrays are quite like lists in LISP.

Some of these analogies are useful; others are thoroughly misleading. Thinking of Smalltalk arrays as data structures is reasonable, as Smalltalk distinguishes data types (without calling them that). Assimilating arrays to lists in LISP, though, is an error. Lists in LISP are variable in size; arrays in Smalltalk are of fixed size.

After discovering, in the excerpt quoted above, that his C and LISP anaiogies had given him an incorrect model of pointers in Smalltalk, P10 decided as follows:

It is good that the author of the book [Smalltalk/V286 manual] did not impose too much of this analogy-type thinking upon me. I prefer if you tell me, "Now this is a new language, and let's think about it from scratch as something totally new," and give me the weirdness of it, and the behavior of it, I'll accept it as is.

Others share this sentiment. In the structured interviews, P1 declared that his general knowledge of programming had helped him learn Smalltalk but that his knowledge of specific languages had been an impediment.

Can experienced programmers learn Smalltalk by just accepting "the weirdness" of it? Attempts to assimilate Smalltalk to familiar languages seem inevitable, and, when wrong, can serve as positive occasions for learning. And although the explicit verbal hypotheses in our tape diaries come from the first week of learning (similar analogies are reported by Scholtz & Wiedenbeck, 1989), analogies to other languages continue to influence Smalltalk programmers' style for a long time (according to P1, P2, and P3 in the structured interviews).

Conclusion

Learning Strategies

Using tape diaries, we were able to document the problems that Smalltalk learners work on, session by session, as they begin to master the language. Programmers were also able to describe their attitudes toward Smalltalk and their strategies for learning it. The professional programmers in our sample, when undertaking to learn the Smalltalk environment, adopted the programmer's burden. They were unwilling to use existing code without first understanding it in depth. They sought to understand the syntax in depth, to support mental simulation of the execution of each line of code. They conducted depth-first searches through the class hierarchy to find each method or instance variable mentioned in the method code that they were trying to understand. They wrote and ran experimental pieces of code to test their understanding of the system. They frequently made analogies to other languages that they knew.

The information that the tape diaries disclosed about learning strategies has important implications for Smalltalk pedagogy. Object-oriented environments like Smalltalk are supposed to promote code reuse. By taking up the programmer's burden, professional programmers are working against the grain of Smalltalk, and they may end up frustrated when the burden proves too large. Ways need to be found to induce programmers to drop the burden, or to turn it to the learner's advantage by directing inquiries more fruitfully.

Utility of Tape Diaries

Tape diaries have the advantage over interviews of being prospective and longitudinal. We could follow learning session by session for an extended period of time. Details emerged that would be hard to recover from retrospective interviews.

Nonetheless, tape diaries have limitations. They are not concurrent thinking out loud; they are largely retrospective descriptions of a work session, albeit immediately after the fact. The most serious limitation is that they are tiring for the programmer. On top of the work of learning Smalltalk, there is the extra effort of reflecting on the work and dictating a 10- to 15-minute summary each day. At the beginning of the study, tape diaries may have been rewarding for the programmers because they provided an occasion for self-discovery; as the novelty of keeping diaries wears off, the commitment to keep them may come to be seen primarily as a service to the experimenter.

Repeated visits and encouragement from the experimenter were necessary to keep programmers talking. Indeed, in a 4-week field study of a help system that involved fewer visits from the experimenter, users' comments became telegraphic as time went on (Campbell, Roemer, & Mack, 1989). Thinking out loud is not a normal part of programmers' work. And although is not conversation, it is a form of socialized speech, normally done (in the laboratory) in the presence of another person. Models (e.g., Ericsson & Simon, 1984) that treat thinking out loud as a purely cognitive phenomenon, explicable in terms of working memory, are missing something important. Because of the work involved in keeping tape diaries, none of the programmers in Study 2 would continue beyond 2 months. In consequence, the diaries covered only the beginning of Smalltalk learning.

Despite the limitations of tape diaries, they are a rich source of information. Documenting learning strategies by no means exhausts them. The diaries document the sequence of problems that programmers work on, session by session. They also record programmers' feelings about what they know and what they do not know, and their contentment or frustration with Smalltalk. A sequence of the types of problems programmers work on can be treated as a developmental sequence. In the next section, we use data from the tape diaries, supplemented by material from the Study 1 interviews, to generate a sequence of seven developmental levels for Smalltalk learning.

Developmental Levels for Smalltalk Learning

Developmental Sequence Models

What kind of developmental sequence is appropriate for describing Smalltalk learning, given our current understanding? A developmental approach to expertise uses a sequence with multiple levels to describe the path of acquisition. A sequence of developmental levels is a sequence of qualitatively distinct abilities, in which learners must move through the levels in order. We can distinguish two kinds of developmental sequences in specific domains of expertise: sequences with issue-based levels and sequences with proficiency-based levels.

A sequence of *issue-based levels* is an ordering of the major kinds of problems learners work on, that is, of the stumbling blocks they encounter. Within a complex domain of expertise, learners tackle different problems at different points in their development. Some problems cannot be tackled until others have been solved. Problems that occupy much of the learner's attention for a long time later cease to be a pressing concern. For instance, Smalltalk programmers actively work for a long time on strategies for building windows. This is an especially difficult problem when the windows behave differently from the the standard windows in the Smalltalk interface. Eventually, window building is mastered, the strategies become recipes, and window building is replaced by other active problems, such as doing proper object-oriented design.

A sequence of *proficiency-based levels* is an ordering of understandings or skills or approaches to problems within the domain. Proficiency-based levels are those we have in mind pretheoretically when we suppose that beginners, intermediates, proficient performers, experts, and grandmasters can be identified in any domain. We use patterns of performance on standard tasks to "diagnose" the proficiency-based level at which the learner is functioning. For example, preferring the longer row in Piaget's (1941/1952) number conservation task might be taken as evidence that a child is preoperational. A certain pattern of performance on a hypothetical window-building task (e.g., building standard Smalltalk windows quickly and easily, while being unable to build some kinds of windows with non-standard behavior) might indicate that the learner is intermediate on that skill.

Proficiency-based levels are meant to apply to a wide range of subskills in the domain: For instance, if a learner is intermediate in window building, we can ask whether he or she is intermediate in program comprehension or critiquing. And if we have an account of window building at different levels that specifies performance patterns on standard window-building tasks, we can identify performance patterns for all levels from beginner through grandmaster. By contrast, issue-based levels single out those subskills that the learner is currently working on. The question is not whether the learner is intermediate in window-building skills; the question is whether the learner is currently trying to master window-building skills, and perhaps which specific skills are being worked on. With issue-based levels, it does not make sense to ask whether the learner is at the same level in understanding syntax as in window-building skills; these are different problems, to be worked on at different points in development.

Work by Feldman and his colleagues has produced proficiency-based sequence models for specific domains of expertise: six levels of map-drawing skill (Feldman, 1980) and eight levels of juggling (Walton et al., 1988). Although the usual terminology of beginner, intermediate, and expert is employed, the number of steps in these models varies with the domain, as do the markers used to define the levels. Thus, the levels of juggling expertise are defined in terms of number of balls juggled, height of toss, timing, and so on. (Table 16.2 displays the eight levels of juggling, with partial descriptions of accomplishments at each level.) Moreover, each of the individual subskills (Walton et al. call them *elements*) can be individually scored at one of the eight levels. For instance, there is a height of toss criterion for each of the eight levels. By contrast, an issue-based level sequence might identify a level in which jugglers work on height of toss or on juggling four balls instead of three.

Under either conception of level, learners can function at different levels: Under issue-based levels, they can be working on problems at different levels; under proficiency-based levels, they can function at different levels for different subskills. Properly speaking, we should identify the most advanced level at which a learner functions, or the predominant level (Feldman, 1980).

Seven Levels of Smalltalk Learning

Our data about Smalltalk learning specify a series of types of problems that program-

TABLE 16.2. Proficiency-based levels of performance in juggling (abridged from Walton et al., 1988; used by permission).

1. **Raw Beginner.** No sense of timing or pattern whatsoever. Cannot throw 1 ball from hand to hand with any consistency. NO PATTERN
2. **Beginner.** Can throw 2 balls from hand to hand with some consistency of accuracy and timing. Dominant hand much better than non-dominant hand. Rigid posture with hands held too high and/or too far in front of body. Cannot maintain exchanges without moving entire body. EXCHANGE
3. **Novice I.** Can put 2 exchanges back to back (1 cycle). Cannot maintain more than 10 cycles. Accuracy and timing sufficient to juggle, but are inconsistent, and pattern falls apart. Non-dominant hand has improved but there is still imbalance between hands. Body control sufficient but variable: there is still rigidity and chasing of pattern. JUGGLING
4. **Novice II.** Can maintain between 10 and 20 cycles. Accuracy, timing, and plane consistent but deteriorate over juggling span. Pattern width generally is still too narrow. Imbalance between hands is not as noticeable. Posture more relaxed and not much running after balls. Is comfortable with juggling. Can begin to do variations with 3 balls and some partnership juggling.
5. **Intermediate I.** Accuracy timing pattern width and plane fairly consistent over juggling span. Non-dominant hand almost as proficient as dominant hand. Can put two 3 ball variations together. Can juggle items of slightly different weights and sizes. Beginning partnership juggling with more than 3 balls. Can maintain more than 20 cycles. Body is relaxed, only chases patterns on new variations. VARIATIONS
6. **Intermediate II.** Does variations with 3 balls smoothly. Can juggle 5 balls. Can vary patterns in idiosyncratic ways. Hands of essentially equal adeptness. All basic elements are secure enough to be varied at will without loss of pattern and only begin to deteriorate as pattern gets more complex. STABILITY
7. **Craftsman.** Levels 1 through 6 are rock steady; all basic elements are completely automatized. Objects of different types can be juggled together with no difficulty. Difficult patterns can be maintained alone and in concert with other jugglers. AUTOMATIZATION
8. **Expert.** Can perform the most difficult routines. Juggles at highest degree of technical proficiency with style and creativity. Recognized by other jugglers as being in top echelon. Can juggle 7 balls, 5 clubs. HIGHEST SKILL and HIGHEST STYLE

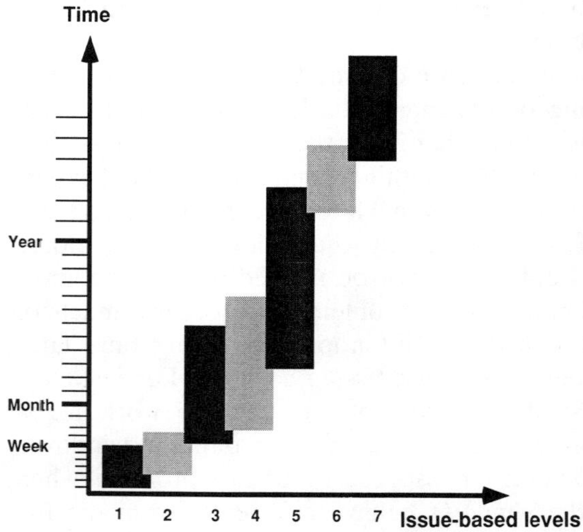

FIGURE 16.1. Issue-based developmental levels for Smalltalk. The bar chart conveys the overlapping of the levels, whose beginnings are ordered but whose endings are not, and the nonlinear time scale indicates that the higher levels take longer than the lower levels.

mers try to solve. Study 2 provides a detailed description of the problems that programmers try to solve in early learning; the P1, P2, and P3 interviews of Study 1 give a less detailed account of the problems tackled by more advanced programmers. These two data sets provide a basis for a sequence of issue-based levels. We propose that Smalltalk learning involves movement through a sequence of seven levels. Each developmental level is defined in terms of the key Smalltalk issues that the programmer has to work on (except for the highest, where the issues remain to be clarified).

1. Interacting with the visual interface.
2. Syntax rules and order of precedence.
3. Locating classes and methods.
4. Class versus instance distinction.
5. Model–Pane–Dispatcher.
6. Object-oriented design.
7. (Grandmaster level).

Because these are issue-based levels, they can overlap (we convey this relationship graphically in Figure 16.1). Our levels are sequenced in terms of the first attempts to deal with specific Smalltalk issues, not the completion of work on those issues. Thus, Level 4 issues appear while programmers are still contending with Level 3 issues; Level 5 issues appear while Level 4 issues are still active, and so forth. These issue-based levels form a "soft sequence," in which only the beginnings of the levels are constrained to appear in order. Proficiency-based levels form a "hard sequence," in which, in a given subskill area, one level must end before the next level can begin. We offer brief descriptions, with examples, of each level in the sequence. An extended treatment, with multiple examples of problems for each level, is available in Campbell (1989) and a briefer treatment in Campbell (1990).

Levels 1–4: Getting Started

Levels 1–4 are the beginning of learning Smalltalk. Our primary evidence for these levels comes from the tape diaries of Study 2. Level 1, learning the visual interface of Smalltalk, is normally taken care of in a day or two. The Smalltalk/V interface is a somewhat old-fashioned windowed environment. A typical problem at Level 1 is learning how to resize a window gracefully.

At Level 2, mastery of Smalltalk syntax (e.g., types of messages and order of precedence) is the issue. The problems that P9 and P10 tried to solve (see Study 2) by conducting "experiments"—the meaning of # and the meaning of := and ==—are typical of Level 2. Conscientious programmers can master the slender syntax within 2 weeks.

Level 3 involves locating appropriate classes and methods in the class hierarchy. Because of the large number of built-in classes and methods, and the additional complexities of inheritance and polymorphism, developing useful search skills begins a week or two into the learning process and can take several months. A typical problem at Level 3 is P8's search (day 22) for a method to add an element to an array. Level 3 may also include the first uses of the Smalltalk debugger.

Level 4 centers on the distinction between class and instance variables, and class and instance methods. A typical example is P8's attempt (day 23) to write a new method as an instance method when it must be a class method, accompanied by his comment that he did not understand which methods "belong in the instance bucket or in the class bucket." Completion of Levels 3 and 4 (resulting in a firm understanding of basic distinctions in the Smalltalk environment and solid skills for navigating in it) may be what gives rise to the feeling of having successfully climbed the Smalltalk mountain (mentioned by P3 in Study 1).

Our tape diaries leave off at 2 months, and our interviews do not pick up until later. In consequence, there is a period between 2 months and 18 months in which search and debugging skills and much of visual interface programming are being mastered, for which we currently lack data. Although Level 3 and 4 problems extend into this period and Level 5 problems begin during it, further investigations of it may add new levels to the sequence.

Level 5: Model–Pane–Dispatcher

Level 5 involves mastery of Model, Pane, and Dispatcher, the complex of classes that control the behavior of windows and menus. Every window or piece of a window in Smalltalk/V must have a model that dictates its input behavior, a pane that regulates its appearance on the screen, and a dispatcher that controls its output behavior. Communication between model, pane, and dispatcher often becomes highly complex.

> [P2]: The model tells the pane in an indirect way that it has changed and then the pane ends up asking the model, anyway, what the contents of its pane should now be. There are other examples of how the communication is just not straightforward. It is not transparent at all; it's kind of in this black box and once you take off the covers and try to look inside, it becomes almost like worms in there—interweaved.

We believe that it takes 6 months or more to master these window classes. When asked in the Study 1 interviews how long it had taken him, P2 replied, "How long have I been programming in Smalltalk?!" He had been using Smalltalk for 2 years.

According to P2, there were three distinct sublevels within his understanding of Model–Pane–Dispatcher. His initial use of it was to borrow existing methods to build windows and menus that behaved in standard Smalltalk ways.

I don't think I would call it a gradual evolution because there was sort of a point where I was able to use it and build interfaces that I wanted to build without understanding its insides that well.

When a major project required him to produce windows and menus with nonstandard behavior, a new level of understanding was required.

There were times, where [as] I said, I wanted to understand its insides. One particularly good example is that the model informs the pane to update itself. I went through [the interaction] many times, I understood it for the time being, when I had to understand it. And then the next time I had to deal with it—I had forgotten because it was so ornate—it didn't stick. So I went through that several times and I am at the point now where I know how that particular mechanism works quite well. So now I have it down. But it really was sort of different stages of understanding.

P2 acknowledged that there are other Model–Pane–Dispatcher interactions that he would still have to learn:

That is absolutely true and if I had to look at it now, I would have to reunderstand it all over again, to look at the sort of bizarre interactions among the three parts to see how they work out.

Campbell (1989; 1990) gives extended examples of problems at the first and second sublevels of Level 5, for instance, a recipe for building a complex window that may include nonstandard menu interactions, and an example of the breakdown of a standard debugging strategy. Level 5 completes the acquisition of programming skills in the strict sense. Although an accomplished Level 5 programmer may be able to build virtually anything buildable in Smalltalk, he or she may still lack a good sense of object-oriented design.

Levels 6 and 7: Object-Oriented Design

What distinguishes Level 6 from Level 5 is not a further increment of programming skill, but whether the programmer thinks in terms of aesthetic. The Smalltalk environment can be put in the service of design ideas better expressed in conventional programming languages:

[P2]: You can really be thinking about things in the way that you just thought about them before you started doing Smalltalk, in really functional terms, and sort of translating that into objects, rather than doing this sort of classic object-oriented design.

Level 6 programmers design in terms of objects and their communication, not in terms of function. Programmers P1 and P3, whom we interviewed in Study 1, operate primarily at Level 6. Programmer P2 points out that there is proficient Smalltalk programming and

There is really naturally [doing object-oriented design]. Even though there is [a tendency among] proponents of object-oriented design to think that [thinking in terms of communicating objects] is just natural for everybody. By naturally, I mean I think of systems as the object classes that are involved or the kinds of objects that are involved—it is just second nature.

Object-oriented designers treat objects anthropomorphically, as agents. They may ask, "Who does this?" when inquiring about an object.

I am still thinking about functionality a lot. I am not anthropomorphizing the objects all that much. In some cases, in the snippet [from a study] that I saw, [a leading Smalltalk expert] is really thinking about "this guy" that works with this other "guy" and these "guys" we are talking about are things like recipes and shopping lists. That is just not something he is cognizant of. It is second nature, the way he thinks about this stuff.

P2 functioned primarily at Level 5 when this interview was done, and his description of the next higher level tends to conflate Levels 6 and 7 (a confusion any developmental stage theorist would happily expect!). More recent interviews with Level 6 programmers (Campbell, 1989; 1990) suggest that they sharply distinguish design from programming issues and that they do their designing in terms of objects, but quite self-consciously. And they consider "grandmasters" to be more expert than they are.

P2 mentions a leading Smalltalk expert who thought out loud while solving a design problem in a study by Rosson and Gold (1989). He is considered a "grandmaster" by other Smalltalk programmers, and is probably at Level 7. According to P1's interview, Level 7 programmers have memorized all of the standard classes and methods in the Smalltalk environment—an estimable feat, but not the most informative from a developmental standpoint. More interestingly, P1 claimed that Level 7 programmers do object-oriented design without any self-consciousness; it is "second nature" for them.

Utility of Developmental Levels

When professional programmers undertake to learn Smalltalk, they appear to take on problems in a specific order, which can be described by a sequence of seven issue-based developmental levels. Using data from the interviews of Study 1 and the tape diaries of Study 2, we have been able to document typical problems faced by programmers at Levels 1 through 6; Campbell (1989) has fleshed out the set of illustrative problems with additional interview material covering Levels 5 and 6.

The level model needs elaboration and empirical testing. By producing such a model, however, we have provided prima facie evidence for the applicability of developmental models and methods to the psychology of programming. The course of learning to program in Smalltalk has a rich structure that could never be recovered from investigations done within the novice-expert paradigm.

In fact, now that we have identified seven developmental levels, we can revisit Mayer's (1988) claims about novice and expert programmers (see the section headed The Novice-Expert Paradigm, above). Mayer asserted that expert programmers could be distinguished from novices because they have greater command of syntax and better debugging strategies. The problem is that, in Smalltalk, "expert" syntactic knowledge and "expert" debugging strategies do not reflect the same level of understanding. A professional programmer will be adept at recognizing syntactically correct Smalltalk code within a month (end of Level 2), yet may still be refining his or her debugging skills well into Level 5 (more than a year later). The enhanced descriptive power of developmental level models of programming skill is hard to challenge. And research in this vein has only just begun.

Conclusion: Developmental Research on Expertise

Our goal in this work has been to show how a developmental analysis of programming expertise might proceed. Using structured interviews, we have been able to elicit a considerable amount of information about the learning histories of expert programmers and their judgments of other programmers' expertise. Using tape diaries, we have been able to follow the course of learning by individual programmers in detail for up to 2 months. From these sources, we were able to propose an issue-based model of developmental levels of expertise in Smalltalk programming.

As preliminary work, what we have reported is mainly of value for the research directions that it indicates. These break down into three categories: (a) further descriptive research on programming, (b) applications of developmental research to programming education, and (c) implications for knowledge elicitation in general.

Developmental Description

To have a satisfactory description of the acquisition of expertise in Smalltalk, we need to extend our descriptive research. This work could start with the collection of tape diaries to document progress between 2 and 18 months of experience with Smalltalk. The tape diary data would drive revisions of the issue-based level model and help us to inventory the Smalltalk programming strategies (search through the hierarchy, tracing communication between objects, debugging) that develop during this time.

For exploratory purposes, we have chosen to follow programmers working on tasks of their own selection. Our goal has been to discover the series of learning problems that Smalltalk learners tackle, and the set of programming and design skills that they come to acquire. In knowledge elicitation terms (Hoffman, 1987), we have sought to understand the familiar tasks that programmers carry out before monitoring their performance on contrived tasks (such as tasks with imposed limitations on time or resources).

We consider it important to give priority to tasks that programmers select for themselves because the entire field of psychology of programming has been harmed by a premature focus on contrived tasks. Instead of trying to understand how programmers solve problems and learn in their normal work context, far too many studies, for reasons of convenience, have scrutinized first-year computer science students comprehending different styles of 50-line programs. Computer science students are not working programmers, and real programs, which are more likely to be 5,000 or 50,000 lines long, pose entirely different problems of comprehension and maintenance than do 50-line laboratory toys (Curtis, 1986; Curtis et al., 1988; Sheil, 1981).

However, after the necessary exploratory work has been completed, we could turn to cross-sectional investigations that focus on specialized skills, like debugging, aesthetic judgment, and initial design strategies, and that compare programmers at different developmental levels. These studies could observe programmers undertaking contrived tasks selected to be reasonably representative of the material they normally work on. Tracking subskills would enable us to build a sequence of proficiency-based levels and appropriate diagnostic tasks for following such a sequence.

Beyond these cross-sectional investigations, studies that track programmers' progress through a major project, or their transition between levels, by repeatedly presenting similar tasks could be the next step. The outcome we ultimately aim for is tested developmental sequence models, both issue- and proficiency-based, an inventory of the important programming strategies for Smalltalk, and some account of the constructive process by which programmers move from one level to the next. Should this research strategy prove fruitful for Smalltalk programming, it can and should be extended to other programming language and application families.

Our sketch has slighted the social aspects of learning to program, which are clearly important both early on (there are distinct Smalltalk cultures, for instance at Tektronix, in which groups of programmers learn Smalltalk together and in which common guidelines for programming style have been proposed; Rochat, 1986) and later when programmers work in teams on most large projects (Curtis et al., 1988). Work underway by Scribner (1989; Scribner & Sachs, 1989) should help us to frame appropriate research questions here.

Programming Education

Although the developmental description of expertise in programming has a long way to go, it is capable of widespread applications in the near term. The primary applications are in programming education. We use this term in a broad sense, to include standard classroom presentations, apprenticeship on the job, and enhancements to the programming environment, including online assistance and tutoring systems. Issue and proficiency-based developmental level models can provide clear targets for what programmers should be learning.

The developmental level model should help us to frame reachable performance targets—not too big a jump from what programmers at a given level already know. We hope they will prove useful in managing apprenticeship: helping programmers and their managers choose projects that will provide the right amount of challenge to help programmers grow, rather than boring them or overburdening them. They should prove useful in improving tutoring systems, replacing the single expert model or ideal student model (Anderson, 1987) used in current systems with a model pitched a level above the programmer's current level. Knowledge of developmental levels may come in useful for other expert systems applications in programming, in which expert-level advice has to be assimilated by users at levels significantly below expert themselves (cf. Hart, this volume).

Developmental research on learning strategies (e.g., our Study 2) also provides information that teachers and designers of tutoring systems need to know. For instance, can professional programmers be induced to drop the programmer's burden while learning Smalltalk? Can teaching strategies take advantage of the programmer's burden while preventing excessive searches of the environment and other sources of frustration?

The relationship between developmental research on programming and its educational applications is reciprocal. The success or failure of teaching strategies based on developmental models constitute a partial test of the models and may provide grounds for revising them. However, the implications of developmental research in general, and of specific methods like tape diaries, extend beyond programming to other fields of expertise.

Knowledge Elicitation in General

We believe that the developmental approach to expertise in programming has some implications for the way that other domains of expertise are studied. Expertise cannot be understood merely as an end state. Experts get where they are by following complex developmental paths.

There are many areas of expertise in which it is not enough to extract rules and criteria from those who are already experts; it is necessary to know how, and through what stages, they got to be that way, and how others might do the same. Typically these will be domains, like programming, which many people need to master, typically over a long period of time, and expert systems can assist practitioners but will not seriously reduce the need for learning. In Feldman's (1980) system, these are "cultural" or "discipline-based" domains, which must be learned by large numbers of people, and in which guideposts or scaffolding to assist in learning are widely needed. In such domains, developmental research will be useful.

By contrast, there are domains of the sort Feldman calls idiosyncratic or unique. Few people have followed the learning paths in these domains; it is hard to separate the essential from the accidental in those learning paths; and there may be little incentive for others to become experts, by retracing these steps or by other means. Skill at repairing obsolete equipment, at interpreting data obtained from obsolete apparatus, at reading the handwriting of Thomas Aquinas, at identifying the approximate date and soloists for a Sun Ra Arkestra recording, are all examples of idiosyncratic expertise. In such areas, when the expertise is needed at all, it is clearly desirable to reconstruct the knowledge of existing experts and, where possible, to render it into rules. When training new experts is undesirable, developmental investigations of old experts will usually be undesirable as well.

Finally, our tape diary technique may be useful even when developmental investigations are not called for. Tape diaries are a particularly useful technique for tracing learning histories. But they can also be used in investigations that focus only on end-state experts. Charting an expert's work over a period of days via tape diary may convey details about familiar tasks, and the expert's manner of encountering and solving them, that interviews, whether structured or unstructured, tend to miss. Tape diaries are kept in the expert's normal working environment, and, although retrospective, stick closer in time to the expert's experiences than do interviews. Except when they interfere with the expert's work, tape diaries may be an excellent source of information about experts' familiar tasks and customary working methods.

Notes

1. Another school of thought likens the acquisition of expertise to historical changes in scientific theories. Thus, learning Newtonian mechanics is assimilated to the historical transition between Scholastic and Newtonian physics (diSessa, 1983; McCloskey, 1983). This conception can be challenged philosophically and developmentally (Campbell, Carroll, & DiBello, 1989). But it

seems outright irrelevant to a recently evolved technological field like programming: What is the appropriate historical analogy? What do programmers recapitulate in their development?

2. Strictly speaking, avionics and Smalltalk programming are fields, not domains. *Domain* is a psychological concept. We define a domain as an area of representation for which the learner has common learning and problem-solving heuristics. This definition implies that domains need not correspond to externally defined fields, and that the identification of domains is an empirical question.
3. There are several versions of Smalltalk, which differ in many details. Our examples are exclusively drawn from Smalltalk/V286 (Digitalk, 1988).
4. We follow standard computer science practice by citing Smalltalk code in a special type font.

Acknowledgments. Portions of this chapter were presented at the ONR Workshop on Models of Complex Human Learning, Cornell University, and at the Software Psychology Society colloquium, George Washington University. All three authors were at the IBM T. J. Watson Research Center when these studies were conducted. Lia DiBello collected six of the seven general interviews. Norman Brown conducted the tape diary study. Thanks to Linda Celtruda and Phyllis Aycock for transcribing the interviews and diaries, to Robert Hoffman for many excellent editorial suggestions, and to Bob Cooper and Stephen Payne for their comments on earlier versions.

References

Adelson, B. (1984). When novices surpass experts: The difficulty of a task may increase with expertise. *Journal of Experimental Psychology: Learning, Memory, and Cognition*, *10*, 483–495.

Anderson, J. R. (1987). Production systems, learning, and tutoring. In D. Klahr, P. Langley, & R. Neches (Eds.), *Production system models of learning and development* (pp. 437–458). Cambridge, MA: MIT Press.

Bickhard, M. H., & Terveen, L. (1991). *The impasse of artificial intelligence and cognitive science*. Unpublished manuscript, Lehigh University, Bethlehem, PA.

Brooks, R. (1989). Personal communication, Austin, TX, October 3, 1989.

Campbell, R. L. (1989). *Developmental levels and scenarios for Smalltalk programming*. (IBM Research Report RC15305). IBM T. J. Watson Research Center, Yorktown Heights, NY.

Campbell, R. L. (1990). Developmental scenario analysis of Smalltalk programming. In J. C. Chew & J. C. Whiteside (Eds.), *Proceedings of the CHI'90 Conference on Human Factors in Computing Systems* (pp. 269–276). New York: Association for Computing Machinery.

Campbell, R. L., & Bickhard, M. H. (1986). *Knowing levels and developmental stages*. Basel: Karger.

Campbell, R. L., Carroll, J. M., & DiBello, L. A. (1989, June). *Human-computer interaction: The case for a developmental approach to expertise*. Paper presented at the Jean Piaget Society meeting, Philadelphia.

Campbell, R. L., Roemer, J. M., & Mack, R. L. (1989, May). *Extending the scope of field research in HCI*. Panel presentation at the CHI'89 Conference on Human Factors in Computing Systems, Austin, TX.

Chase, W. G., & Simon, H. A. (1973). The mind's eye in chess. In W. G. Chase (Ed.), *Visual information processing*. New York: Academic Press.

Chi, M. T. H., Glaser, R., & Rees, E. (1982). Expertise in problem solving. In R. J. Sternberg (Ed.), *Advances in the psychology of human intelligence* (Vol. 1, pp. 7–75). Hillsdale, NJ: Erlbaum.

Collins, W. A. (Ed.). (1982). *The concept of development*. Hillsdale, NJ: Erlbaum.

Curtis, B. (1986). By the way, did anyone study any real programmers? In E. Soloway & S. Iyengar (Eds.), *Empirical studies of programmers* (pp. 256–262). Norwood, NJ: Ablex.

Curtis, B. (1988). The impact of individual differences in programmers. In G. C. van der Veer, T. R. G. Green, J.-M. Hoc, & D. M. Murray (Eds.), *Working with computers: Theory versus outcome* (pp. 279–294). London: Academic Press.

Curtis, B., Krasner, H., & Iscoe, N. (1988). A field study of the software design process for large systems. *Communications of the ACM*, *31*, 1268–1287.

de Ribaupierre, A. (1989). Epilogue: On the use of longitudinal research in developmental psychology. In A. de Ribaupierre (Ed.), *Transition mechanisms in child development: The longitudinal perspective* (pp. 297–317). Cambridge: Cambridge University Press.

Digitalk. (1988). *Smalltalk/V286 tutorial and programming handbook*. Los Angeles: Digitalk.

diSessa, A. (1983). Phenomenology and the evolution of intuition. In D. Gentner & A. L. Stevens (Ed.), *Mental models* (pp. 15–33). Hillsdale, NJ: Erlbaum.

diSessa, A. (1989, June). *Local sciences: Viewing the design of human-computer systems as cognitive science*. Paper presented at the Kittle House Workshop on Cognitive Theory and Design in Human-Computer Interaction, Chappaqua, NY.

Dreyfus, H. L., & Dreyfus, S. E. (1986). *Mind over machine: The power of human intuition and expertise in the era of the computer*. New York: The Free Press.

Ericsson, K. A., & Simon, H. A. (1984). *Protocol analysis: Verbal reports as data*. Cambridge, MA: MIT Press.

Feldman, D. H. (1980). *Beyond universals in cognitive development*. Norwood, NJ: Ablex.

Feldman, D. H. (1988). Personal communication, Medford, MA, December 6, 1988.

Goldberg, A., & Robson, D. (1983). *Smalltalk-80: The language and its implementation*. Reading, MA: Addison-Wesley.

Gould, J. D. (1975). Some psychological evidence on how people debug computer programs. *International Journal of Man-Machine Studies*, *7*, 151–182.

Guindon, R., Krasner, H., & Curtis, B. (1987). Breakdowns and processes during the early activities of software design by professionals. In G. M. Olson, S. Sheppard, & E. Soloway (Eds.), *Empirical studies of programmers: Second workshop* (pp. 65–82). Norwood, NJ: Ablex.

Hoffman, R. R. (1987). The problem of extracting the knowledge of experts from the perspective of experimental psychology. *AI Magazine*, *8*(2), 53–67.

Hoffman, R. R. (1989, April). A brief survey of methods for extracting knowledge from experts. In C. R. Westphal & K. McGraw (Eds.), *The SIGART Newsletter* (pp. 19–27). New York: Association for Computing Machinery.

Hoppe-Graff, S. (1989). The study of transitions in development: Potentials of the longitudinal approach. In A. de Ribaupierre (Ed.), *Transition mechanisms in child development: The longitudinal perspective* (pp. 1–30). Cambridge: Cambridge University Press.

Inhelder, B., Ackermann-Valladão, E., Blanchet, A., Karmiloff-Smith, A., Kilcher-Hagedorn, H., Montangero, J., & Robert, M. (1976). Des structures cognitives aux procédures de découverte [From cognitive structures to discovery procedures]. *Archives de Psychologie*, *44*, 57–72.

Jeffries, R., Turner, A., Polson, P., & Atwood, M. (1981). The processes involved in designing software. In J. R. Anderson (Ed.), *Cognitive skills and their acquisition* (pp. 255–283). Hillsdale, NJ: Erlbaum.

Karmiloff-Smith, A. (1981). Getting developmental differences or studying child development? *Cognition*, *10*, 151–158.

Klein, G. A., Calderwood. R., & MacGregor, D. (1989). Critical decision method for eliciting knowledge. *IEEE Transactions on Systems, Man, and Cybernetics*, *19*, 462–472.

Kurland, D. M., Mawby, R., & Cahir, N. (1984). The development of programming expertise in adults and children. In D. M. Kurland (Ed.), *Symposium: Developmental studies of computer programming skills*. (Technical Report No. 29). Bank Street College of Education, New York City.

Lammers, S. (1986). *Programmers at work: Interviews*. Redmond, WA: Microsoft Press.

Larkin, J. H. (1981). Enriching formal knowledge: A model of learning to solve textbook physics problems. In J. R. Anderson (Ed.), *Cognitive skills and their acquisition* (pp. 311–334). Hillsdale, NJ: Erlbaum.

Leventhal, L. (1988). Experience of programming beauty: Some patterns of programming esthetics. *International Journal of Man-Machine Studies*, *28*, 525–550.

Mastaglio, T., & Rieman, J. (1989). *How experts model the expertise of novice programmers: A protocol analysis of LISP code evaluation*. Unpublished manuscript, Department of Computer Science, University of Colorado, Boulder.

Mayer, R. E. (1988). From novice to expert. In M. Helander (Ed.), *Handbook of human-computer interaction* (pp. 569–580). Amsterdam: North-Holland.

McCloskey, M. (1983). Naive theories of motion. In D. Gentner & A. L. Stevens (Eds.), *Mental models* (pp. 299–324). Hillsdale, NJ: Erlbaum.

McKeithen, K. B., Reitman, J. S., Reuter, H. H., & Hirtle, S. C. (1981). Knowledge organization and skill differences in computer programmers. *Cognitive Psychology*, *13*, 307–325.

Neches, R., Langley, P., & Klahr, D. (1987). Learning, development, and production systems. In D. Klahr, P. Langley, & R. Neches (Eds.), *Production system models of learning and development* (pp. 1–53). Cambridge, MA: MIT Press.

Pascoe, G. A. (1986). Elements of object-oriented programming. *Byte*, *11*(8), 139–144.

Piaget, J. (1952). *The child's conception of number*. New York: Humanities Press. (Originally published 1941.)

Piaget, J. (1960). *The child's conception of the world*. Totowa, NJ: Littlefield, Adams. (Originally published 1929.)

Piaget, J. (1985). *The equilibration of cognitive structures: The central problem of intellectual development*. Chicago: University of Chicago Press. (Originally published 1975.)

Robson, D. (1981). Object-oriented software systems. *Byte*, *6*(8), 74–86.

Rochat, R. (1986). *In search of good Smalltalk programming style*. (Technical Report No. CR–86–19). Computer Research Laboratory, Tektronix Inc., Beaverton, OR.

Rosson, M. B., & Alpert, S. R. (1990). The cognitive consequences of object-oriented design. *Human-Computer Interaction*, *5*, 345–379.

Rosson, M. B., & Gold, E. (1989). Problem-solution mapping in object-oriented design. In N. Meyrowitz (Ed.), *OOPSLA'89 Conference Proceedings* (pp. 7–10). New York: Association for Computing Machinery.

Scholtz, J., & Wiedenbeck, S. (1989). *Learning second and subsequent programming languages: A problem of transfer*. (Report Series #80). Department of Computer Science and Engineering, University of Nebraska, Lincoln, NE.

Scribner, S. (1989, March). *Learning and working in an activity theory perspective*. Paper presented at the American Educational Research Association meeting, San Francisco, CA.

Scribner, S., & Sachs, P. (1989). *A study of on-the-job training*. Final report submitted to National Center on Education and Employment, Washington, DC.

Sheil, B. A. (1981). The psychological study of programming. *ACM Computing Surveys*, *13*, 101–120.

Soloway, E., & Ehrlich, K. (1984). Empirical studies of programming knowledge. *IEEE Transactions on Software Engineering*, *SE–10*, 595–609.

Staszewski, J. J. (1988). Skilled memory and expert mental calculation. In M. T. H. Chi, R. Glaser, & M. J. Farr (Eds.), *The nature of expertise* (pp. 71–128). Hillsdale, NJ: Erlbaum.

Vygotsky, L. S. (1987). Thinking and speech. In L. S. Vygotsky, *Problems of general psychology*. New York: Plenum (Originally published 1934).

Walton, R. E., Adams, M. L., Goldsmith, L. T., & Feldman, D. H. (1988, June). *Affect and development in a nonuniversal domain*. Paper presented at the Jean Piaget Society meeting, Philadelphia, PA.

Wohlwill, J. F. (1973). *The study of behavioral development*. New York: Academic Press.

Part VI
Overview

17 The Psychology of Expertise and Knowledge Acquisition: Comments on the Chapters in This Volume

Stephen Regoczei

Introduction

The following commentaries on the chapters in this volume start with a few humble premises. The realization is inescapable that the enterprise of psychology is vast, has a long history, and is deeply divided into often non-communicating subfields. The enterprise of artificial intelligence (AI) is much younger than psychology and yet is also divided into areas that barely communicate with each other. The divisions in AI often parallel those in psychology and often seem to be motivated by similar theoretical and methodological concerns. In addition, the enterprise of expert system construction, and in particular knowledge acquisition (KA), is but one tiny corner of AI, yet the activities focused on expertise are expanding beyond the narrow confines of computer science. It is as if a new field concerning itself with expertise, knowledge, and expert system construction—a field as yet unnamed—is trying to be born. It is an event exciting to witness, but it leaves the poor commentator with no fixed or generally agreed-upon platform—no firm Newtonian coordinate system from which to survey and summarize multidisciplinary research.

And yet, some framework has to be used. So I propose to make use of the basic model of the KA process that was described in the second chapter of this volume, and I would like to refer the reader to the figures, especially Figure 2.3. I will be referring to that basic model as I discuss the contributions to this volume, in order. For each chapter, I try to mention its key concerns and contributions, and I also point to lingering issues and problems.

Cognitive Theory and Expertise

The chapter by Cooke (chap. 3) directly addresses the issue of the nature of expertise. Much of the material in chapter 3 would be new to the typical implementor or researcher in the expert systems field. This seems paradoxical. Surely, if one is eager to build expert systems, first one would try to find out about expertise. Hence, a chapter outlining current work in the psychology of expertise should not be especially informative to the AI practitioner. Yet, because of the historical development of the AI field, or because psychology of the past had little to offer to those studying knowledge, or because psychology traditionally focused on process rather than content, the cultural gulf between AI and psychology is deep. Although one often has to read between the lines to see it, psychology is often construed as being partly, if not largely, irrelevant to AI (cf. Newell, 1983). For instance,

> It is by no means evident just what the intelligence of people and the intelligence of computers have in common beyond their ability to perform certain tasks. (Simon & Kaplan, 1989, p. 4)

Indeed, I have heard it expressed by some members of one particular school of thought in

AI that the bulk of psychology is mere "folk knowledge."

The misunderstanding by psychologists of the engineering nature of AI work is a bit more difficult to demonstrate. However, Cooke's chapter provides a good example that highlights some of the common misconceptions about expert systems and KA. The reader is asked to "consider trying to list the steps required to tie a shoe" (p. 30). The point Cooke is trying to make is the following: Certain knowledge, such as tying of shoelaces, cannot be verbalized. Cooke calls such knowledge *procedural*, an unfortunate term. Perhaps referring to it as *athlete knowledge* as opposed to *coach knowledge* would be more descriptive. The essential point about this example is not that the hypothesized knowledge behind the activity is procedural but that no explicit *conceptual* model for it exists in the mind of the informant. Specifiability and describability for KA depends on how well the knowledge has been conceptualized and formalized (Sowa, 1983). Tying shoelaces is unformalized, but, for example, the biochemistry of digestion is well worked out.

To get a better perspective on the problem, let us contrast the question, How do you tie your shoe? with the question, How do you digest your food? If anything is procedural, to use Cooke's terminology, digestion is. Informants should be mystified and tongue-tied. Yet some people—in fact, a handful of experts—can describe very accurately the mechanical and biochemical processes that take place in digestion. Now, one may raise possible objections to placing shoelace tying and digesting parallel to each other. One may argue that we digest with our body, and we do not think about it. Well, who would claim to be certain that shoelace tying is not done by the body and that there is not very much thinking being done about it? Just like digestion. This is a puzzle for psychology, namely, the mind-body problem, compounded by the conscious-unconscious dichotomy. Any solution to it could be of considerable help in AI.

Shoelace tying is a bad example of a typical expert system for the following reason: It involves material handling rather than information handling. From the AI perspective, tying shoelaces is not an expert systems issue but a robotics issue, because it is the material-handling problem that is the significant and difficult component. Robotics concentrates on material-handling problems, and expert systems concentrate on information processing. Building the robot might not be easy, but as soon as it is well built there will be inside the robot a knowledge-based system that could be programmed by imitation to tie shoelaces. Sensors attached to the fingers of a person would record the movements digitally to be repeated by the robot. This is a special case of KA as defined in chapter 2 of this volume. It is debatable whether it should be considered a "true" KA problem in the sense that it is *solely* a KA problem.

But the most important misconstrual comes to the surface when the informant is asked to "list the steps required" to tie the shoelaces. Are there steps there? We can analyze the process post hoc into numerous, ontologically different sequences or hierarchies. But the process itself is not performed in a stepwise, algorithmic way. It is continuous. It can be *approximated* by a discrete, step-by-step, or digital process, but that is not how people do it. So how could they be expected to be able to describe it as such?

Algorithm construction is a nontrivial activity. We teach it in universities. Generations of students who sweat blood trying to acquire programming skills will testify that it is nontrivial. But above all, algorithm and rule construction is not the job of the informant; it is the job of the analyst. That is why we say that the construction of the externalized knowledge is the result of a collaborative effort between analyst and informant. Analysts are not simply note takers but also participants in a creative activity. They do not simply get the knowledge that is already there "as is." Rather, they add conceptual structure to it and create something new.

Cooke's chapter broaches at several places the issue of the verbalizability of knowledge. Discussions of this problem commonly fail to draw the distinction between the following statements: (a) X is verbalizable, and (b) it is possible to verbalize about X. The distinction is that under the first instance we expect X to be fully expressed or fully described, whereas in the latter case some verbal behavior can be engaged in that enables an astute observer to generate conclusions about X. The difficulty of verbalization under the first statement should not be taken to imply the *impossibility* to engage in verbal behavior under the second. I would like to emphasize that knowledge is constructed, or reconstructed, by the analyst on the basis of clues provided by the informant. Impossibility of a full description does not preclude the ability to supply lots of other clues.

To put the matter more clearly, Hoffman (personal communication) pointed out that it is helpful to distinguish: (a) knowledge that is verbalizable only with effort or the use of special tasks, and (b) knowledge that is not verbalizable in principle. We know that the first kind of knowledge exists, but there is actually no empirical demonstration as yet that the second kind also exists. So KA analysts should not start despairing as yet.

To summarize, Cooke's chapter directly addresses the question of how to bridge the gulf between the psychology of expertise and expert system implementation. The answer should be; By carefully examining the concerns of both fields of activity.

The chapter by Schumacher and Czerwinski (chap. 4) performs a useful service by bringing attention to the importance of mental models in KA, at least the mental models of the expert informant. But it shies away from mentioning other mental modellike embodied representations such as the schemas, rules, images, and so forth—all of the things that make the cognitive approach so ontologically fruitful in KA work (Gardner, 1985). At the same time, one finds taxonomic difficulties with other, related entities such as abstract models and conceptual models. Reading this chapter, one gets the feeling that its authors are uncomfortable with ethereal entities such as "concepts" and "abstractions," and that is why they choose to declare themselves exclusively interested in the mental models of *physical* systems.

Chapter 4 does not raise the basic question: Mental models as opposed to *what*? This question is crucial in understanding what a change of direction the mental models approach occasioned in AI. Mental models are to be contrasted with propositions, assertions, beliefs and other languagelike representational entities. Mental models are thinglike and can be manipulated as something thinglike. The focus on mental models broke the stranglehold that language-based proposition metaphors had on expert system work (Johnson-Laird, 1983; Gardner, 1985). Mental models are *constructs*. Schumacher and Czerwinski start their chapter with the well-known story of the six blind men. They proceed to compare the elephant with mental models, the point being that there are many ways to look at elephants, none of which is *the* real way, and likewise there are different ways of looking at the concept of mental models. The parallel is not entirely apt, however. The fable of the six blind men illustrates perspectivism. But mental models illustrate constructivism first, and perspectivism only secondarily. Mental models are constructs, both in the sense of what is in people's heads and as tools of analysis. Who constructs them and how? That is where perspectivism comes in.

The process of forming, attributing, testing, and revising mental models is a crucial phase in KA projects. Mental models are constructs in terms of which we—or KA analysts—construe what is in the expert's head. Mental models of mental models are constructs in terms of which we understand how the analyst reconceptualizes the knowledge that is supposed to be in the expert's head.

Having said this, we regretfully have to conclude that in terms of the basic model for KA, chapter 4 dodges some truly difficult issues. Restricting attention to physical systems may be unwise because it excludes so much. The central task in the KA process is the conceptualization by the analyst of the expert's knowledge. Or, to use the terminology of Schumacher and Czerwinski, the analyst forms mental models of the informant's mental models. "Mental models of mental models" is the pivotal entity in KA. If Schumacher and Czerwinski restrict themselves to *physical* systems—that is, they are only willing to look at models of physical systems—then they are excluding from their study considerations of how the analyst forms mental models of another person's thoughts on the basis of verbal input. This is a fatal omission in the study of KA. We need to know how the analyst conceptualizes, if for no other reason than to be able to perform quality control on the process.

It is crucial to the design of the KA process that we study the analyst's formation of a mental model of the expert's knowledge. How does the analyst go about reconceptualizing the expert's knowledge? How does the analyst go about forming mental models of the expert's mental models? Obviously, we cannot concentrate exclusively on the mental models that are in the head of the expert. And by the way, the mental models of the *users* of the expert system are also important (Hart & Foley, this volume).

To further illustrate how restricting our attention to physical systems excludes too much, we note that even a computer system is not merely a physical entity. It is a hybrid or composite entity, consisting of physical and ideational components, commonly referred to as hardware and software, respectively. A strict focus on the physical might exclude the understanding of software, for example.

Finally, chapter 4 gives an interesting glimpse of the inadequacy of techniques such as multiple scaling in KA. The reason is that such clumsy mathematical techniques destroy the structure among the various fragments and concepts that have been elicited. It is precisely this fine structure, not preserved by statistical techniques, that constitutes the knowledge we are after. The meaningful preservation of fine structure in the modeling of mental models has perhaps reached a pinnacle in the work on "conceptual graphs", described by Sowa in Chapter 5 of this volume.

Knowledge Elicitation Methods

Is KA for expert systems something like the cramming of knowledge into students by a teacher? To raise this issue, one can hardly do better than to quote the opening paragraph in Gordon's chapter (chap. 6):

> KA for expert systems is a subclass of any instructional situation whereby knowledge must be externalized from a human expert and transferred to one or more "systems." These systems have historically been other people, as in education and training. (p. 99)

This would be quite helpful, if we actually knew more about education and training, and how exactly the externalization of the knowledge is to be carried out. The transfer, needless to say, is not without its difficulties. My suspicion is that rather than using school instruction to help in KA, we can expect research results in KA to be applied in teaching. This is because teachers usually do not quite know *how* they do what they do. The teacher finds out about a subject, usually from books. He or she prepares the notes for the lecture, walks into class, and delivers the lecture. This is a very common activity; hundreds of thousands of teachers do it regularly. Some are even experts at it. Yet if we go to a teacher and say, "You are an expert. Tell us how you do it. Tell us how you go through the KA and knowledge delivery cycle. How do you prepare for lectures?" I suspect that the answers of these expert teachers would not always be illuminating. They cannot quite conceptualize how they do it, without additional help.

I do not see this as a cause for despair. On the contrary, to my way of thinking, this is the most exciting result of work with expert and knowledge-based systems: It makes visible what was invisible before. We did not know that we did not know how to prepare for a lecture. This "mental experiment" helps us realize that the knowledge of the teacher is athlete knowledge, not coach knowledge.

An even more significant example is the understanding of natural language text, such as English prose. Many people are experts at understanding text of various kinds: literary critics, lawyers, and in fact any scholar who reads a great deal. Now, let us suppose that we try to do KA with one of these experts, and we ask, "Please tell us how you go about understanding language." Or we give him a piece of text and make the request, "Don't tell us what it means, just tell us how you go about finding its meaning." The expert might respond by becoming tongue-tied and incoherent. The knowledge/skill/expertise of understanding text is not verbalizable, not because it is some mysterious procedural knowledge, but simply because we do not yet have a conceptual model of the understanding process. And this is a sobering realization. Expert systems force us to confront the limits of our own conceptualized and conceptualizable knowledge.

The use of the terms *declarative* and *procedural* by Gordon is unfortunate for two reasons. First, it can be confused with two ways of representing knowledge: declarative versus procedural (Newell, 1982). If someone remains within the bounds of cognitive psychology, this may cause no problems. But in an AI-cognitive science mixed company, the AI people can get impatient and annoyed.

But that is not the main reason why the use of the two terms is misleading. The main reason is that this dichotomy misdiagnoses the problem. According to chapter 6, declarative knowledge is articulate and verbalizable, and procedural knowledge is not because it is expressed only through doing—it is knowing *how* to do something (i.e., it is athlete knowledge, not coach knowledge). But action or procedure is not the cause of the difficulty. (See the digestion example above.) What matters is having a conceptual model of the knowledge. Once we have this, articulation becomes possible, or at least more likely. Without a conceptual model, our explanations are incoherent and contentless. To conclude, the basing of elicitation techniques on the procedural-declarative dichotomy, a dichotomy I believe is fundamentally mistaken, certainly will hinder acceptance by the AI community of some very useful components of cognitive theory described by Gordon.

Regrettably, even poor Gilbert Ryle (1949) gets dragged into this. Something that we just do versus something that we think about—that is Ryle's distinction. That sounds very much like automaticity versus conscious, premeditated, monitored action, and not procedural versus declarative. Unfortunately, Reber (1985) also gives the link to Ryle and even bolsters it by the shoelace example.

In terms of the basic model for KA, there is not much in chapter 6 that would help us beyond elicitation. The production of the knowledge design document is not addressed. The analysis essentially ends with the permanent record of the interview. Stopping so short of a way into the examination of the KA process may not offer sufficient help for knowledge base construction. It would also probably not help us very much with preparing lectures.

There is a very important reason why we should not stop so short into the KA process. At least the conceptualization process by the KA analyst has to be addressed. As described in chapter 2 in connection with the basic model for KA, it is the job of the analyst to conceptualize; we should not expect the expert informant to do so. Digestion is very procedural, yet it can be described. Description of the parts of a sailboat is declarative—but very difficult if not impossible for someone who does not know sailing terminology. The most procedural of things, such as golf swings and tennis forehands, have been conceptualized in great detail and are described by coaches over and over again. Why does Gordon think that task knowledge, the closest standard term I can find to her term *procedural*, is inextricably linked with nonverbalizability? To neutralize these concerns, we have to emphasize the constructive power and responsibilities of the KA analyst or knowledge engineer.

In terms of the baseline model for the KA process, the Ford and Adams-Webber chapter (chap. 7) makes two important points: (a) that the expert thinks in terms of personal constructs, and (b) that the analyst conceptualizes the expert's knowledge in terms of personal constructs. Like Sowa's conceptual graphs, this view directly addresses the conceptualization of the world by the expert. But it also emphasizes that the KA analyst, in turn, conceptualizes the expert's knowledge and that the resulting model is also a construct. At this point, some readers would tend to replace the phrase *personal construct* with *mental model* (cf. Schumacher & Czerwinski, this volume).

But this may be too hasty a substitution. The bipolar nature of personal constructs makes possible their systematic handling through the repertory grid. The significance of the repertory grid method is increasing not only in applied fields such as market research, but also in KA.

But are bipolar personal constructs the right sort of constructs to *always* pay attention to in KA? Let me first consider the qualifier *personal*. Much of the knowledge that we want to encapsulate in expert systems is public knowledge, which is a social construct created and used by a group. Medical knowledge is an example. The emphasis on private, individual, personal knowledge may be a misdirecting of our attention. As for *constructs*, I am very much in favor of taking constructs seriously. But I feel uneasy about the strictly bipolar nature of Kelly's (1955) personal constructs. This may be more a result of the dominant cultural thinking (at least in Kelly's time) in terms of polar opposites—yes-no, good-bad, communist-capitalist—rather than in terms of how people in general actually think. In our more fluid, more mentalistic, and more perspectivist age, the bipolar nature of constructs seems to me to be a dispensable feature.

Ah, but if we accept the above suggestion, what will happen to the repertory grid? The repertory grid as a KA tool suffers from a great shortcoming: It is not accompanied by *verbal* protocols. The various constructs and features are mere words and numbers to both informant and analyst. I would like to know about the *concepts* behind the words and numbers—not only the concepts that show up in the grid as constructs but also the concepts that somehow cannot quite make it into the grid because they just do not quite seem to fit. I suspect that such misfit concepts would show up in a protocol or would come to light in the course of a "post-grid introspection." And I would further venture to guess that they would turn out to be *crucial* concepts. Thus, I will hold on to my skepticism about exclusive reliance on any one KA technique, especially one based solely on filling out preprinted forms with numbers or ratings scales. I guess I am permanently imprinted with the horror of filling out income tax returns. Not only are they tedious, but also they represent only the bones of my life, not its flesh and meaning.

Finally, it is important to note that personal construct theory emphasizes domain knowledge rather than task knowledge. What the expert thinks about, how he or she construes the world according to personal constructs, dominates the consideration of what he or she is going to do with the knowledge. *What there is* dominates the consideration of *how to act*. This runs counter to the more procedure-oriented approaches in AI (and task analysis work on human factors psychology). Earlier work in AI emphasized the processes and procedures, such as search and deduction using rules. Problem solving was considered in terms of actions and processes—not in terms of the *what*, but in terms of the *how*. Thus, much of KA was oriented toward task knowledge rather than domain knowledge. With an emphasis on the constructs that the expert uses, we can hope that the balance will be restored and that more attention will be paid to the landscape or backdrop against which the expert's decisions play themselves out. And if some analysts prefer to use the terminology of mental models to refer to personal constructs, that may even bring about new insights and connections.

The chapter by Prerau, Adler, and Gunderson (chap. 8) is the only one in the volume that covers the entire expert system development life cycle, from knowledge elicitation to working software. For that reason alone, it should be carefully considered, although not too much detail is given on the actual management of the project (for that, see Prerau, 1990). The expert informants were interviewed by KA analysts, but we do not know if a knowledge design document was produced. We also do not know how the knowledge that would have been contained in the design document was actually reworked so as to be implementable. Nevertheless, we know that because KEE was the expert system shell used, the knowledge and expertise ended up in the form of frames, rules, and LISP functions.

The COMPASS expert system analyzes hardware fault messages and suggests maintenance actions. Is this a good (typical) example of an expert system? From the point of view of expertise and the psychology of expertise, this is a very restricted, artificial domain. As for it being an expert system, a jaundiced-eyed software engineer would say that he did nothing more than fix the interface of a system that was badly designed in the first place and had an incomprehensible reporting system. This is software maintenance, or fixing after the fact, he might continue. The rejoinder to such comments is that in order to build a good human-computer interface, we have to model (or at least be aware of) a great deal of the thinking and knowledge of the typical (i.e., expert) user within the computer system. This is definitely expert systems work. Looking at it, we realize that expert system work today is actually much broader than would be suggested by early MYCIN-like examples.

Turning to the psychology component, unfortunately, the terminology of chapter 8 concerning the different kinds of knowledge is rather confusing. I had trouble with some of Prerau et al.'s distinctions. For example, they want to contrast "expertise knowledge" with "experiential knowledge." Although I kept saying to myself, "But expertise knowledge *is* also experiential," I managed, I think, to detect the main issue. The authors are trying get at the important distinction between book learning, on the one hand, and knowledge picked up on the job, on the other. The trouble is that either type of learning can be superficial, and we would like to ignore superficial kinds of knowledge. We would prefer to concentrate on the deeper, but still experiential, expertise knowledge. This knowledge is deeply experienced and is not merely known but also understood.

This distinction reminds me of the one the Japanese draw between a *jitsu* and a *do*. A *jitsu* (as in *jujitsu*) is a mere technique that one may repeat mechanically without any feeling or understanding. A *do* (as in *judo*) may involve the same kind of mechanics, but is imbued with the spiritual value and is deeply felt.

There is much here that is ill-understood and requires a broader set of concepts and terms to discuss. Clearly, there is a need for a better taxonomy and nomenclature to describe the various types of knowledge and understanding, and work such as that in Prerau et al.'s chapter points the way.

A method of widening the so-called KA bottleneck is to use software to help in the KA process. Unfortunately, most such software is aimed at circumventing or bypassing the KA analyst. The hypothesis is that the KA analyst is the troublemaker, and if we could only get rid of the analyst by replacing him or her with a piece of software, then expert's could directly enter all of their knowledge into the computer. This picture seems to ignore the essential role that the KA analyst plays in conceptualizing and formalizing the expert's knowledge such that it can be captured in the knowledge design document and further reformulated for implementing a knowledge base. I am not claiming that software is useless in KA. On the contrary, it is essential for recording of the knowledge. But if software is to replace the services provided by the KA analyst, then it would have to contain a conceptualizing module that assists in the cooperative construction of recordable knowledge.

McGraw advocates the use of software to aid in the KA process, but she very correctly focuses on *helping* the analyst rather than trying to replace him or her. The software she describes, HyperKAT, is likened to software that assists software engineering. The conceptualization and design is still conducted by the analyst, but the recording of the results, updating of the permanent record of the interviews, and the generation and verification of the knowledge design document is assisted by the software. In terms of the model presented in chapter 2, HyperKAT helps the analyst to maintain both the permanent record of the interviews and the knowledge design document.

There are two opposing schools of thought on software: One is to replace people and hence do things automatically without assistance from people. The other is to use software to enhance human capabilities: Such software has to stimulate the user and enable him or her to perform better, perhaps by making some of the mundane clerical or bookkeeping jobs easier. HyperKAT subscribes to the second school of thought. Perhaps only one final word of advice is in order to those who are ready to use software *now*. Something simple and immediate, such as HyperCard with some customization or a word-processing package, is better than keeping records by hand and for the time being is better than waiting for some highly elaborate software to come along at some time in the future.

There is a recent shift of concerns in AI well summarized by Stefik (1986):

> The most widely understood goal of artificial intelligence is to understand and build autonomous, intelligent, thinking machines. A perhaps larger opportunity and complementary goal is to understand and build an interactive knowledge medium. (p. 34)

I see Klein's chapter (chap. 10) as a manifestation of this shift. Although the possibility of using expert systems to deliver knowledge and expertise to the users is left open, Klein by no means restricts himself to such vehicles. He is more interested in what Stefik calls an interactive knowledge medium, which does not replace intelligent beings but enhances their performance. He wants to build up a knowledge base for general purposes, not only to encapsulate it within an expert system. The knowledge preserved within the corporation can be transmitted for future use in decision making, technology transfer, and training.

Klein wants to turn knowledge into a corporate asset. This parallels developments in business information systems, where corporations want to turn their data base and information into a corporate asset. Klein's dream—and the dream of others—will be realized if we can put a dollar figure on the knowledge design document, enter that dollar figure on the assets side of the balance sheet, get the entry accepted by the auditors, and get a bank loan on it using it as collateral. Then it is "real" (i.e., it is in the domain of discourse of business). This is a radical proposal—to treat knowledge as a corporate asset—but because we live in what is called the Age of Information, we had better develop some new ways of taking inventory.

Unfortunately, Klein's chapter does not emphasize the role of the KA analyst as an active participant in constructing publicly examinable, archivable knowledge. In his chapter we note the following statement:

> The term *knowledge engineering* was coined to describe the activities of computer programmers *eliciting rules* from experts and coding them in LISP or Prolog in a knowledge base. (p. 173)

And further down, we note the following:

> It is hard to elicit knowledge rules from domain experts. . . . There were no generally accepted techniques for *eliciting implementable rules*. . . . [italics added] (p. 173)

Is there any reason for us to expect that such techniques *would* exist? It is not the job of the expert to provide implementable rules. In light of the baseline for KA (Fig. 2.3), we can note that rules are not elicited from experts. The rules are constructed by the analyst on the basis of the information provided by the expert, in whatever form the expert happens to provide it. It is the KA analyst who conceptualizes the knowledge. We also note that programmers may be just about the worst people to elicit (or use!) knowledge. Their tendency is to get too bitsy-bytesy too soon and lose sight of the larger picture.

Just in case it seems that I am too harsh on the computer community, there is an indictment by Klein that deserves careful attention:

> After a hundred years of claiming to be a science, after several decades of cognitive simulation and information processing and information science, experimental psychology had not addressed the problem of how to efficiently elicit or organize knowledge in general, let alone the knowledge of experts. (p. 173)

I really cannot add any more to what Klein says, except to strongly urge the reader to look at his chapter very carefully. The reason for doing so is that those who do not understand history are doomed to relive it.

Is *engineering* a good metaphor for whatever it is that Klein wants to do to the knowledge? In chapter 2, Hirst and I expressed reservations about the term, but I find it hard to offer a better alternative. Klein wants to do *knowl-*

edge archiving; perhaps that is what he should call the process. If the knowledge is to be archived as stories of the oral tradition, then perhaps we need knowledge anthropologists. The main use for the knowledge, however, is knowledge reuse, or knowledge recycling. The images multiply. Perhaps the phrase *knowledge engineering* is not so bad, after all.

In spite of the technological change, one thing has to be emphasized: There is nothing new about capturing world experience and world knowledge. Traditionally, in so-called primitive societies, cultural knowledge was encapsulated in fairy tales and myths. In our business schools, knowledge has been propagated through the case study method. In earlier times, prospective civil servants were trained on Thucydides' histories and the great political writings of Rome. For the man-on-the-street, proverbs gave quick and easy access to instant wisdom. So we must be cautious in thinking that knowledge has never been archived before, whether or not expert systems bring about an invisible revolution for archiving knowledge.

Klein gives some superb examples of the use of "bottled" knowledge inside or outside expert systems. Klein's accounts of knowledge reuse, or recycling, in the form of stories are excellent. However, it might be a good idea to enrich them by explicit representations of concepts such as "concept maps" (Novak & Gowin, 1984) or conceptual graphs (Sowa, this volume). This is a new version of the transmission of common culture and yet another step in the migration from orality to literacy.

Psychological Research on Expertise

The chapter by Sternberg and Frensch (chap. 11) makes (at least) two important points: first, that expertise is partly attributional—so there is more so-called expertise than intrinsic, cognitional components—and second, that being an expert has costs as well as benefits, and inflexibility can be one of the handicaps.

In terms of the basic model of the KA process, the Sternberg and Frensch chapter addresses the cognitive architecture of the expert and what it means to be an expert in term of this cognitive architecture. But it also considers the social interaction between the expert and the external world. That expertise is attributional means that the performance of the expert matches the expectations of the social group (e.g., being a witch doctor). From our present perspective, the witch doctor's expertise may look flawed. Our urge to play the omniscient, detached, privileged, "objective" observer would stop us from considering the witch doctor as a genuine expert. But chapter 11 cautions us to assume a more humble attitude as KA analysts, as opposed to trying to set up ourselves as the judges of what constitutes "true" knowledge. Sternberg and Frensch advance an interesting architecture for expert systems. This architecture is more complex—and more interesting—than an inference engine based on a von Neuman machine and a deductive paradigm. Unfortunately, there is no explicit guidance on how the analyst should conceptualize the expert's knowledge taking into account this new design possibility. But given the cognitive architecture as a framework, we could hazard some guesses. If the architecture of the expert system is conventional, then we must conclude on the basis of what the chapter presents that much of the expert's expertise or knowledge cannot be captured. If the architecture is the new triarchic/global/local scheme suggested by Sternberg and Frensch, then we have some interesting possibilities. The analyst can put together the knowledge design document to match the hypothesized or suggested model of the architecture of the cognition of the expert, and presumably get an expert system with enhanced capabilities and greater psychological authenticity.

To follow the lines of reasoning laid out by Sternberg and Frensch, we could start wondering about the KA analyst as an expert-at-being-an-analyst. Must the analyst become inflexible, as chapter 11 suggests? Would the analyst become *less* useful on new projects? It is interesting that in the literature on expert systems, the analyst is usually something of an innocent novice who has never encountered the informant's subject matter before. But if the analyst is an expert-at-being-an-analyst, then presumably the theory in the Sternberg and Frensch chapter would apply. Such analysts should be jaded and inflexible as experts. Perhaps, therefore, novices make the best analysts. Would the seasoned, experienced analyst start interfering with and correcting the expert (cf. Prerau, this volume)?

Sternberg and Frensch say nothing about knowledge elicitation, yet their theory inevitably raises questions. Is the automatized knowledge in their local processor verbalizable? Does it need to be verbalized if it is a compiled copy/version of what is in the global domain? Nothing is said about embodiment versus external representation. This is not surprising, because the authors' main concern is expertise and not expert systems implementation. But from this perspective of our concerns, however, we can note that the automatized local module is embodied. Is it possible to represent the "content" of this module externally—as, for example, in the form of a neural net—leaving aside any considerations of verbalizability? Clearly, there is much here that requires further examination.

In other chapters, we saw stories of the tongue-tied expert who cannot verbalize his so-called procedural knowledge. What does the analyst do to stimulate the informant to divulge his knowledge, or at least divulge some clues about his knowledge? He could ask questions, request that the expert talk, present some material, and ask for response, or he could feed back some rephrased material from earlier sessions and ask for corrections and comments. But suppose the expert's tongue still will not loosen? Anna Hart told an amusing story in her conference presentation about a geological expert who just would not talk. When presented with rock samples, he would point and identify the samples. But he could not say how he did it. Hart was called in to help, and she suggested to turn the expert around, with his back to the rock samples. The expert's communication channel was therefore greatly restricted. Now he had to ask questions about the samples, and the questions he asked gave clues about the criteria he was using to classify (Hart, personal communication). This is an ingenious solution, but this is perhaps something of a special case. It would be desirable to have a general version of this technique.

This is where Bellezza's (chap. 12) theory of mental cuing may come to our assistance. Cuing could be looked upon as being similar to the "leading questions" a lawyer addresses to an eye witness: What the lawyer wants to hear is shaped by the question. Cuing is relevant to KA because leading questions are used by the analyst, and sometimes *have* to be used in order to get the job done. The analyst has to lead the informant and has to place a framework upon the informant and upon the interview because the KA work of the analyst is goal-oriented. Intuitively, it seems plausible that what the expert remembers to say (i.e., what occurs to the expert to say) is influenced by the cues, such as questions or tasks set in front of the expert. How to cue the expert most effectively? With a Rorschach test? With a discussion of the weather? With a discussion of his childhood experiences? What would prompt or stimulate the expert best without introducing any bias? We simply do not know; the matter has not been studied. If there is going to be help, it may very well come from studies of mental cuing.

The way Bellezza construes it, knowledge structures are highly organized sets of information in memory that contain conceptual components that are meaningful and imageable. Concepts, words, and images put forward by the analyst can provide memory cues to which new information can be associated. From a general perspective, one can look upon all of natural language understanding as a process of cuing: There is no knowledge "in" the text. What there is, is a set of cues that stimulate the reader to recreate old knowledge or to add new knowledge. Some sets of words and some pieces of text are more powerful in eliciting an informative response than others. If we only knew which! So Bellezza's topic goes straight to the heart of the informant-analyst elicitation process. What turns the expert on? This could very well be the key question about *any* series of KA interviews.

In terms of the basic model (Fig. 2.3), the cues passed out by the analyst place a framework upon the expert. The cues guide the expert's thinking, and they influence which words, concepts, and images get activated. As Bellezza states, because activated concepts and words are imageable and meaningful to the expert, they easily become associated to new verbal information. An obvious application of memory cuing is to use it to structure further interviews. The cues could be fed back (hopefully, in a nondirective or nonintrusive way). Appropriate cues may be extracted from the transcripts of earlier interviews (e.g., "When I was in the air force . . ." or "When I was working in France . . .") and reused in subsequent sessions with the expert. This "question probe" method is one possible knowledge elicitation task (Hoffman, Shanteau, Burton, & Shadbolt, 1991; Shadbolt, Hoffman, Burton, & Klein, 1991).

The Ceci and Ruiz chapter (chap. 13) has a very inspiring conclusion:

> Casual impressions are apt to be misleading when it comes to judging an expert. Even an expert with a low IQ may be a good expert, and more complex than a nonexpert who possesses a high IQ. (p. 227)

This is a clear warning to those who would tend to equate IQ with expertise.

The best way to comment on the Ceci and Ruiz chapter is to contrast it with that of Sternberg and Frensch. Ceci would like to know, If an agent is good at one task, then is he good at another similar task? And will he be able to learn other tasks better (transference)? Sternberg and Frensch provide the answer: If an agent is good at a task, if he is an expert at it, then he may actually be *worse* at doing or learning other tasks. In fact, the expert can become inflexible, and this is one of the costs of being an expert. No doubt this insight may throw some light on some of the more puzzling results that Ceci and Ruiz obtained.

I would like to emphasize an important methodological issue: How an individual will conceptualize, how he will build mental models or conceptual models, cannot be deduced from, or studied through, aggregate, statistical results. Each informant has to be studied in meaningful detail. Only after this KA-oriented empirical work has been done, as for example, as described in the Foley and Hart chapter, only after that can we start meaningful clustering and classifying.

Ceci and Ruiz's two experts failed to make a connection, that is, failed to detect an isomorphism between horse racing and the stock market. Similar failures of seeing isomorphisms are discussed in general terms by Simon (1969), using tic-tac-toe as an example. The point is that the construction of abstract isomorphisms or homomorphisms is sometimes a very difficult, or at least a nonobvious, activity. When a person is confronted with the similarity, he or she might say, "Oh, I never thought of it that way!" This problem of recontextualization, or reframing, or making connections, or seeing parallelisms, or forming new gestalts out of the same old material, can be a very difficult task.

The language in which problems are phrased can give crucial clues as to what domain of knowledge, what domain of discourse, what representation of knowledge, and what connection of the knowledge to other things are to be admissible or will be activated. This is perhaps a reason why the two experts did not make a connection between horse racing and stock market forecasting (see also Bellezza, this volume).

There may be yet another reason, and Ceci and Ruiz allude to this under their discussion of motivation. Problems presented in a symbolic form, such as the stock market data, would be very sensitive as to how the problem is being presented. I would venture to say that even the way the problem is typeset on the page could make a difference. The different presentations seem to activate different processes for processing the problems. Problems presented ecologically would call for a very different approach than problems presented symbolically (i.e., phrased in terms of symbols).

We teach this fragmentation technique in both math and computer science. We show the students how to act and think like little automata and to empathize with the machine. We encourage them to play at being computers. But this is coping with the domain of discourse of the fragmented, analytical world. It does not always carry over to holistic, ecologically situated everyday life. A person acting or behaving in a fragmented, analytical way in everyday life might be characterized as strange or weird. So if a problem is phrased in everyday terms, the analytical techniques may not be useful. They may reside in a different knowledge domain. Problems presented in an ecologically situated way, versus those in verbal or symbolic form, may even activate different processing mechanisms. Knowledge and expertise seem categorized and compartmentalized. These compartments, called domains of discourse, are cued by the language we use, and similar clues. Perhaps this is another reason why the two experts failed to make a connection between horse racing and the stock market.

To continue on this theme, I would like to know how the experts *themselves* conceptualized the task. This kind of investigation would be in line with the KA-paradigm approach. Ceci and Ruiz conceptualize performance in terms of "a coupled, multiplicative model" (not in terms of simple additive models) using "implicit algorithms" and "a seven-factor equation with multiple interaction effects". (p. 225)

Ceci and Ruiz interpret the results in their own terms, not in the terms of the experts. Chapter 13 therefore illustrates the difference between the older, traditional paradigm on the research of expertise and the newer, KA-oriented paradigm exemplified by the model presented in chapter 2. The key distinguishing feature is the analyst's conceptualizing, if possible, the expert's knowledge in the same terms that the expert uses to conceptualize it. Any deviation from this—or any reorganization, reengineering, or redesign of the knowledge—is done for an explicit purpose, and an audit trail is kept of the changes.

Ceci and Ruiz's work falls within the traditional paradigm of psychological research. Certain additions to their methodology would be necessary to bring it in line with the KA paradigm. One would extend the work by interviewing the experts to get verbal expressions of the mental models that (they think) they were using. On the basis of these interviews and transcripts, the analyst would try to conceptualize the agents' thinking as they actually carried out the tasks. None of this was done. Rather, numbers were crunched within the conceptual framework of the experimenter, not in the conceptual framework of the experts. (The gulf between the two is bridged by clever phrases such as *implicit*, *in effect*, and *in reality*.)

Expert-Novice Differences and the Development of Expertise

The Foley and Hart chapter (chap. 14) consists of two major sections: The first is a new approach to expert systems development that takes into account the user of the system, not just the expert who supplies the knowledge. The second section is an interesting piece of empirical work that illustrates some of the difficulties of empirical investigation within the KA paradigm.

There is nothing modest about what Anna Hart is trying to do both in her book (1986) and in the Foley and Hart chapter. She is actually expecting expert systems to be useful to users. Why is this such a revolutionary position? More precisely, if the purpose of creating systems is not the usefulness of the system, then what is all this research concerning expert systems aiming to do? It seems that expert systems often benefit the *builders* of the system and not any potential user community. This in itself is an embarrassing situation, one to which Hoffman alluded in his introduction to this volume (chapter 1). But let us continue. If the mental world of the user is to be looked at seriously, as opposed to the knowledge of the expert, which could be considered as a surrogate for "objective knowledge," then we are forced to take cognizance of an entirely new requirement for psychological authenticity. The system actually has to fit the psychology of the user. As yet, we do not know how to always make systems do that. We certainly have no widely accepted theory for it. And we are not quite sure at present how to evaluate whether we have been of service to the user. Companies like IBM have been acknowledged masters at managing user expectations, and hence they get good marks for user satisfaction and service. Perhaps we should follow their example and engineer not only systems but also users. Human engineering seems only a minor extension of knowledge engineering.

The opening statement by Foley and Hart contains the traces of a basic misapprehension:

> Some years ago it was stated that knowledge acquisition is a major bottleneck in the development of expert systems . . . It is not unusual for knowledge engineers to experience great difficulty in getting experts to verbalize and formalize their knowledge . . . (p. 233)

This shows a common misconception about KA: Experts do not formalize their knowledge; analysts do. It is very difficult to get any coherent discussion going as long as everyone does not understand this basic point.

Then, of course, elicitation is not the major bottleneck; KA is. KA includes not only elicitation but also the explication of the knowledge, the forming of conceptual models by the analyst, and the formalization of a partial model of the expert's knowledge (see Fig. 2.2). In other words, the main task is the producing of a usable knowledge design document in the terms of the basic model of KA.

I also feel uncomfortable with the expert versus novice dichotomy: It seems to be misfocusing our attention. The dichotomy sets up those-who-know against those-who-do-not-know. This split is not very flattering to the latter group. Also, one would think that it would be rather difficult to acquire knowledge from the latter group when their defining characteristic is that they do not have the knowledge. Of course, they have some knowledge, and this knowledge may be crucial to the whole enterprise of building usable systems.

To avoid these problems, a much better dichotomy is one profession versus another (i.e., Profession A versus Profession B). A good example mentioned by Foley and Hart is nurses versus doctors. These are both medical professions, but the knowledge of nurses is not merely an abridged version of the knowledge of doctors. Nurses and doctors deal with different aspects of the medical process, and they are both essential. A doctor would not do the nurse's job and vice versa. Here is a case where an expert system encapsulating the expertise of one profession to make it usable by the other could be engineered to take into account both domains of discourse.

In collecting information about the world views of experts and novices, terminology is a major problem. The Foley and Hart chapter correctly points out that the terminology used by experts and novices is different, and that this terminology should be matched up. But the study of terminology is itself very difficult. An excellent example is the painstaking research required in Hoffman's (1990) study of terrain analysis, describing the terminology used in aerial photo interpretation. The trouble is that careful study of terminology is possible for an expert, but not for a novice. The reason is, to paraphrase Tolstoy, that all experts in a given field are alike, but each in his or her own way. Foley and Hart's suggestion may be doomed unless we can categorize users and establish a common vocabulary for the categories. This is why expert systems of Profession A used by Profession B are more feasible and useful than experts producing systems for novice.

The problem with chapter 14 is that programming and debugging are poor kinds of domain knowledge to test this very interesting proposal on. The world of programming is a microworld, not ecologically situated, but presented in terms of symbols. Furthermore, student advisors may not have much insight when it comes to programming or even debugging—not very prime specimens. (Editor's note: See the bibliography on expertise in programming, presented at the end of this volume.)

Perhaps I read too much into chapter 14, but what I think I saw in it was this: the necessity to mesh together and harmonize expert and user world views. Awareness of this, I predict, will be crucial for building usable expert systems. And the chapter does promote the idea of usable systems. The world model of the expert and the world model of the user should both be acquired and then meshed together, melded in the implementation stage. But this chapter gives no indication of precisely how this is to be done. The world, or domain knowledge, of the expert and novice are different—different vocabulary, different concepts, different mental models, different strategies. The main task is how to specify the knowledge domains and how to integrate the world models of experts and users. The empirical work in the paper does not give us much guidance on this crucial point. Foley and Hart establish that novices, intermediates, and experts are different, and we believe it, but they do not tell us what to *do* about this difference when it comes to building the desired usable expert system.

Chapter 14 falls within the KA paradigm on the basis of Foley and Hart's interviewing technique. We cannot quite call their technique experimental, since the authors clearly state that

> The study was conceived as an exploratory exercise providing data to formulate theories rather than an experiment to test some aspect of an existing theory. (p. 238)

There is no need to be apologetic about this not being an "experiment." Such *empirical* "exploratory exercises" are precisely the main tools in the KA paradigm. What is being collected is empirical information about conceptualizations. Foley and Hart are doing the right thing, but because the methodology of the KA approach is not clearly spelled out, it is interesting that they feel constrained to volunteer an explanation.

To summarize a theme of chapter 14, the necessity of a new architecture for expert systems is becoming a pressing problem, and new structures for KA will probably be one solution. But what the Foley and Hart chapter does not say is how to structure the knowledge base. If I may hazard an answer, we should look at the domain of discourse of the expert and the domain of discourse of the user, both in terms of the concepts and the terminology being used. Then we should look for commonalities. Next, we should ask the expert to rephrase his knowledge using the concepts available to the user, and try to control the terminology, keeping track of discrepancies. It may be easier to teach new terminology to users than to try to teach totally new concepts. Then, we can try to build teaching modules to enrich the user's vocabulary before using the system. This would mean that the system must have a model of the conceptual models of the user and must be able to update it as the thinking of the user changes. Although this sounds complicated, if the chapter's main thesis is correct, and if we want "useful" expert systems, it may very well turn out to be the case that a teaching module will be a necessary part of a "usable" expert system. The conventional wisdom in KA holds that the expert is to articulate his or her knowledge in statements, presumably in well-formed English sentences. Yet, empirical observations under actual field conditions indicate that informants are often reluctant to do this. Is it not time to absorb some obvious lessons from this fact?

Questions, like declarative statements, carry information. More precisely, the declarative component of a question, as opposed to the interrogative import, can be analyzed for the concepts "underlying" the text, as can be any other natural language fragment. The advantage of using questions is that the discussion is focused on the knowledge needs of the informant. The analyst can conceptualize and record the knowledge, as if the text had been composed of statements.

Anna Hart turned the tables on the geology expert: She forced the expert to ask the questions. Her trick helps us to see that questions contain knowledge also; it is not only answers through which we can elicit knowledge.

The chapter by Mack and Robinson (chap. 15) elaborates on the techniques of acquiring knowledge from questions that are asked. Mack and Robinson's technique may be especially useful with novices. It would be interesting to see it extended and developed for dealing with more experienced informants. Knowledge selection can be as difficult as knowledge acquisition. At times, too much information is available. The questions can help the analyst to concentrate on what is essential.

In a way, chapter 15 addresses the issue raised by Foley and Hart. If we want to design the expert system to address the needs of its end users, let the prospective users help guide the process of designing the knowledge base, through techniques such as question asking. The central issue then is not just what the expert knows but also what the user needs to know to make good use of a system. Thus, the Mack and Robinson chapter goes beyond the expert system development framework, which the basic model of the KA process reflects, and looks at the problem of using the ecologically situated final expert system from the point of view of the user.

Mack and Robinson's chapter also raises the question of whether there is such a thing as programming expertise. The answer might be no. Please note that I am not disputing the existence of excellent programmers whose performance far surpasses that of the average programmer. I am also not questioning the necessity of knowing the basics of programming. Chess masters and duffers alike have to know how a knight may be moved. I am merely questioning the existence of a publicly examinable, teachable, transmittable, and acquirable body of knowledge that we could call programming expertise. Excellent programmers exist, but they may be born, as well as made.

I may be wrong in my suspicions, and there are many workers in the software engineering field who hope that my reservations are unfounded. Be that as it may, the chapter by Campbell, Brown, and DiBello (chap. 16) elegantly sidesteps my concerns by denying the importance of the novice-expert comparison. Instead, they take the position—not unlike the one that dominates Japanese manufacturing design—that anything good can be made a little better. Therefore, they study excellent performance in programming using a constructivist-developmental approach. They do not try to acquire the expert's entire stock of knowledge all at once, hoping to explain what makes his or her performance so good. Instead, through interviews they elicit information about the learning histories of expert programmers. It is worth emphasizing the significance of this incremental approach. To try to take hold of *all* of the knowledge of an expert may be unrealistically ambitious. If, however, we can approach the problem in an incremental, developmental way, then it may very well turn out that some good, useful expert system software might be produced before the funding for a development project runs out. Question: How does one move a mountain? Answer: One shovelful at a time.

Conclusions

With the clarity of hindsight, we can see that expert systems or knowledge-based systems have never been all that comfortable within AI. The culture of AI is perhaps too ephemeral, if not ethereal. It is forever dealing in conundrums such as Turing tests, halting problems, and nonmonotonic logics, and it is forever generating new metaphors such as neural nets, adaptive systems and genetic algorithms. Such conundrums and metaphors are the lifeblood of theoreticians, who are forever pushing the field into new realms. Expert systems, on the other hand, with all their "brittleness" problems, all the convolutions involved in forming conceptual models of the user's mental models, all the details involved in intelligent tutoring and natural language interfaces, all the scaling and validation issues involved in automated knowledge acquisition—all of this may make expert systems a legitimate subfield of AI only on a temporary basis.

One broad, but definite conclusion is that expert systems are moving out of AI, whether AI wants them to go or not. We are in the midst of a major realignment, perhaps. Expert systems are moving into engineering and numerous technical fields, management information systems (MIS), geographic information systems (GIS), and countless other applications. With the clarity of hindsight, people in the MIS area have actually been concerned with expertise (and hence, expert systems) since the early 1970s (McCleod, 1986). Researchers in the area of judgment and decision making are realizing that their work on judgment "hit rates," "linear decision models," and other topics in the 1960s was modeling . . . guess what . . . expertise (Hoffman et al.,

1991). People are coming from all directions and are changing the shape of the field called expert systems. We even see heralds of a new age, as in Klein's (this volume) pronouncement, Whether or not expert systems ever work by truly simulating expertise, the technology of KA can be used in a grand new enterprise—the recording, preservation, and dissemination of knowledge.

In the early years of KA and expert systems, knowledge elicitation proceeded largely by a reliance on unstructured interviews. This knowledge elicitation technique seems to come naturally to those whose primary job is to do what they learned: to write code. From this confluence of trends and training came the so-called knowledge acquisition bottleneck. Enter experimental psychologists. Experimental psychologists in their learning laboratory, and applied psychologists in their task analyses, have been studying the acquisition of knowledge and skill for many years. It is only natural that their methods and ideas should be applied and adapted for helping in KA.

But by turning to experimental psychology, KA has fished in only one of the possible streams. Alternative empirical, clinical, and worldly approaches may help provide methods and insights. I am thinking here of such things as social psychology methods used in research on eyewitness testimony, counseling sessions, journalistic interviewing techniques, lawyers talking with clients, architects getting specifications, fire claim adjusters, and so on.

For example, consider a knowledge elicitation session to be like a counseling session, in which the counselor helps the client articulate his or her ideas, feelings, and difficulties. This hearkens back to the common pronouncement in the expert systems literature that experts can become confused, and even bewildered and tongue-tied. Certainly, social-psychological factors are involved (and are uncontrolled) in most knowledge elicitation sessions, but when dealing with experts—who can span the spectrum of human differences—knowledge elicitation can be made effective only if the interviewer is something of an analyst. An expert may, for example, be totally amazed that anyone cares, or may feel threatened, or may simply be inarticulate.

Beyond this particular counseling example, KA may benefit by looking at methods and results from such fields as ethnography, ethnomethodology, and anthropological linguistics. Enrichment of the "culture" of expert systems may be the result. The pragmatics of human communication is the common substantive thread. And a translation process is the common method: What the client or informant says and does must be translated by the expert (lawyer, accountant, physician, etc.) into the vocabulary of his expertise.

But where would this leave experimental psychology, and especially applied cognitive psychology? Where will the current changes and potential new avenues leave those who are interested in knowledge elicitation? With regard to expert systems work that is going on out in the trenches, the position of experimental psychology is insured. For knowledge elicitation, psychological expertise and research are both essential. One can easily envision, and argue for, a role for experimental psychologists in teams of expert system developers. The psychologist would concentrate on knowledge elicitation and the keeping of knowledge base records and data. He or she might also have input into the choice of representational formats (e.g., reasoning in this domain seems to rely a lot on framelike categorizations). The knowledge base programmers would concentrate on implementation, but then would collaborate with the psychologist with regard to interface design and system evaluation methodology.

Where will current trends and possibilities leave the academic cognitive scientist who is interested in simulations based on evidence from basic psychological research? The position of such experimental psychologists in the "expert systems game" is also well insured. Indeed, empirical research may be the new paradigm for expert systems, just as it may be a new paradigm for AI as a whole (Buchanan, 1987; McCarthy, 1984; Mitchell & Welty, 1988).

References

Buchanan, B. G. (1987). *Artificial intelligence as an experimental science*. (Report No. KSL–87–03), Knowledge Systems Laboratory, Stanford University, Stanford, CA.

Gardner, H. (1985). *The mind's new science: A history of the cognitive revolution*. New York: Basic Books.

Hart, A. (1986). *Knowledge acquisition for expert systems*. New York: McGraw-Hill.

Hart, A. (personal communication, May 1987). Conference on Expert Systems and the Psychology of Expertise, Adelphi University, Garden City NY.

Hoffman, R. R. (personal communication, May 1987). Conference on Expert Systems and the Psychology of Expertise, Adelphi University, Garden City NY.

Hoffman, R. R. (1990). What's a hill? Computing the meanings of topographic and physiographic terms. In U. Schmitz, R. Schutz, & A. Kunz (Eds.), *Linguistic approaches to artificial intelligence* (pp. 97–128). Frankfurt, Germany: Verlag Peter Lang.

Hoffman, R. R., Shanteau, J., Burton, A. M., & Shadbolt, N. R. (1991). *The cognition of experts*. Unpublished manuscript, Department of Psychology, Adelphi University, Garden City, NY.

Johnson-Laird, P. N. (1983). *Mental models*. Cambridge, MA: Harvard University Press.

Kelly, G. A. (1955). *The psychology of personal constructs*. New York: Norton.

McCarthy, J. (1984, Fall). President's message. *The AI Magazine*, *5*, 7–8.

McCleod, R. (1986). *Management information systems*. Menlo Park, CA: Science Research Associates.

Mitchell, J., & Welty, C. (1988). Experimentation in computer science: An empirical view. *International Journal of Man-Machine Studies*, *29*, 613–624.

Newell, A. (1982). The knowledge level. *Artificial Intelligence*, *18*, 87–127.

Newell, A. (1983). Some intellectual issues in the history of artificial intelligence. In F. Machlup & U. Mansfield (Eds.), *The study of information: Interdisciplinary messages* (pp. 187–227). New York: Wiley.

Novak, J. D., & Gowin, D. B. (1984). *Learning how to learn*. Cambridge: Cambridge University Press.

Prerau, D. S. (1990). *Developing and managing expert systems: Proven techniques for business and industry*. Reading, MA: Addison-Wesley.

Reber, A. S. (1985). Declarative versus procedural knowledge. In *The Penguin dictionary of psychology* (p. 385). New York: Penguin Books.

Ryle, G. (1949). *The concept of mind*. London: Hutchinson.

Shadbolt, N., Hoffman, R. R., Burton, A. M., & Klein, G. A. (1991). *Eliciting knowledge from experts: A methodological analysis*. Unpublished Manuscript, Department of Psychology, University of Nottingham, Nottingham, England.

Simon, H. A. (1969). *The sciences of the artificial*. Cambridge, MA: MIT press.

Simon, H. A., & Kaplan, C.A. (1989). Foundations of cognitive science. In M. I. Posner (Ed.), *Foundations of cognitive science* (pp. 1–47). Cambridge, MA: MIT Press.

Sowa, J. (1983). *Conceptual structures: Information processing in mind and machine*. Reading, MA: Addison-Wesley.

Stefik, M. (1986, Spring). The knowledge medium. *The AI Magazine*, *7*, 34–36.

Appendix A
Bibliography: Psychological Theory and Reviews

Akin, O. (1980). *Models of architectural knowledge*. London: Pion.

Anderson, J. R. (1983). *The architecture of cognition*. Cambridge, MA: Harvard University Press.

Anderson, J. R. (1985). *Cognitive psychology and its implications*. New York: Freeman.

Anderson, J. R. (1987). Skill acquisition: Compilation of weak-method problem solutions. *Psychological Review*, *94*, 192–210.

Anderson, N. H. (1987). A cognitive theory of judgment and decision. In B. Brehmer (Ed.), *Proceedings of the Conference on Subjective Probability, Utility, and Decision Making* (pp. 23–46). Amsterdam: North-Holland.

Anzai, Y., & Simon, H. A. (1979). The theory of learning by doing. *Psychological Review*, *86*, 124–140.

Arkes, H. R., & Hammond, K. R. (Eds.). (1986). *Judgment and decision making: An interdisciplinary reader*. Cambridge, UK: Cambridge University Press.

Ashton, R. H. (1982). *Human information processing in accounting*. Sarasota, FL: American Accounting Association.

Ashton, R. H. (1983). Researching in auditing decision making: Rationale, evidence, and implications. *Canadian Certified General Accountant's Monographs* (No. 6). Vancouver, BC: Canadian General Accountants Association.

Beach, L. R., Barnes, V. E., & Christensen-Szalanski, J. J. J. (1986). Beyond heuristics and biases: A contingency model of judgmental forecasting. *Journal of Forecasting*, *5*, 143–158.

Benner, P. (1984). *From novice to experts: Excellence and power in clinical nursing practice*. Menlo Park, CA: Addison-Wesley.

Bereiter, C., & Scardamalia, M. (1985). Cognitive coping strategies and the problem of inert knowledge. In S. F. Chipman, J. W. Segal, & R. Glaser (Eds.), *Thinking and learning* (Vol. 2) (pp. 65–80). Hillsdale, NJ: Erlbaum.

Berry, D. C. (1987). The problem of implicit knowledge. *Expert Systems*, *4*, 144–151.

Berry, D. C., & Broadbent, D. E. (1986). Expert systems and the man-machine interface. *Expert Systems*, *3*, 228–231.

Berry, D. C., & Broadbent, D. E. (1987). Expert systems and the man-machine interface: 2. The user interface. *Expert Systems*, *4*(1), 18–28.

Beyth-Marom, R., & Dekel, S. (1989). *An elementary approach to thinking under uncertainty*. Hillsdale, NJ: Erlbaum.

Bloom, B. S., & Broder, L. J. (1950). *Problem-solving processes of college students*. Chicago: University of Chicago Press.

Bloomfield, B. P. (1988). Expert systems and human knowledge: A view from the sociology of science. *AI & Society*, *2*(1), 17–29.

Bobrow, D. G., & Norman, D. A. (1975). Some principles of memory schemata. In D. G. Bobrow & A. Collins (Eds.), *Representation and understanding: Studies of cognitive science* (Vol. 1, pp. 1–34). New York: Academic Press.

Bowden, R. J. (1989). Feedback forecasting games: An overview. *Journal of Forecasting*, *8*, 117–128.

Boyd, R. (1979). The role of metaphor in conceptual change. In A. Ortony (Ed.), *Metaphor and thought* (pp. 356–408). Cambridge: Cambridge University Press.

Bransford, J. D., Franks, J. J., Vye, N. J., & Sherwood, R. D. (1989). New approaches to instruction: Because wisdom can't be told. In S. Vosniadou & A. Ortony (Eds.), *Similarity and analogical reasoning* (pp. 470–497). Cambridge: Cambridge University Press.

Bransford, J. D., & Stein, B. S. (1984). *The IDEAL problem solver*. San Francisco: Freeman.

Broadbent, D. E. (1989). Lasting representations and temporary processes. In H. L. Roediger & F. I. M. Craik (Eds.), *Varities of memory and consciousness: Essays in honor of Endel Tulving* (pp. 211–233). Hillsdale, NJ: Erlbaum.

Brooks, L. R. (in press). Concept formation and particularizing learning. In G. Hanson & R. Davis (Eds.), *Information, language, and cognition: Vancouver studies in cognitive science* (Vol. 1). Vancouver, BC: University of Vancouver Press.

Brown, A. L. (1978). Knowing when, where, and how to remember: A problem of metacognition. In R. Glaser (Ed.), *Advances in instructional psychology* (Vol. 1, pp. 77–165). Hillsdale, NJ: Erlbaum.

Brown, A. L. (1989). Analogical learning and transfer: What develops? In S. Vosniadou & A. Ortony (Eds.), *Similarity and analogical reasoning* (pp. 369–412). London: Cambridge University Press.

Brown, J. S., & deKleer, J. (1981). Mental models of physical mechanisms and their acquisition. In J. R. Anderson (Ed.), *Cognitive skills and their acquisition* (pp. 285–308). Hillsdale, NJ: Erlbaum.

Bruner, J. S., Goodnow, J. J., & Austin, A. A. (1956). *A study of thinking*. New York: Wiley.

Bushke, H. (1976). Learning is organized by chunking. *Journal of Verbal Learning and Verbal Behavior*, *15*, 313–324.

Card, S., Moran, T., & Newell, A. (1983). *The psychology of human-computer interaction*. Hillsdale, NJ: Erlbaum.

Carey, S. (1985). Are children fundamentally different kinds of thinkers and learners than adults? In S. F. Chipman, J. W. Segal, & R. Glaser (Eds.), *Thinking and learning skills: Vol. 2, Research and open questions*. Hillsdale, NJ: Erlbaum.

Carroll, J. M., & Thomas, J. C. (1982). Metaphor and the cognitive representation of computing systems. *IEEE Transactions on Systems, Man, and Cybernetics*, *12*, 107–116.

Cazden, C. B. (1976). Implications for instructional research. In D. Klahr (Ed.), *Cognition and instruction* (pp. 317–323). Hillsdale, NJ: Erlbaum.

Chandrasekaran, B., & Josephson, J. R. (1987). Modularity of domain knowledge. *International Journal of Expert Systems*, *1*, 1–16.

Chapman, L. J., & Chapman, J. (1971, November). Test results are not what you think they are. *Psychology Today*, *18–22*, 106–110.

Chase, W. G., & Ericsson, K. A. (1981). Skilled memory. In J. R. Anderson (Ed.), *Cognitive skills and their acquisition* (pp. 141–189). Hillsdale, NJ: Erlbaum.

Chatalie, P., Dubois, D., & Prade, H. (1987). An approach to approximate reasoning based on Dempster rule of combination. *International Journal of Expert Systems*, *1*, 68–86.

Cheng, P. W. (1985). Restructuring versus automaticity: Alternative accounts of skill acquistion. *Psychological Review*, *92*, 414–423.

Chi, M. T. H., & Glaser, R. (1985). Problem-solving ability. In R. J. Sternberg (Ed.), *Human abilities: An information processing approach* (pp. 227–250). New York: Freeman.

Chi, M. T. H., Glaser, R., & Farr, M. J. (Eds.). (1989). *The nature of expertise*. Hillsdale, NJ: Erlbaum.

Chi, M. T. H., Glaser, R., & Rees, E. (1982). Expertise in problem solving. In R. Sternberg (Ed.), *Advances in the psychology of human intelligence* (Vol. 1, pp. 17–76). Hillsdale, NJ: Erlbaum.

Christensen-Szalanski, J. J., & Beach, L. R. (1984). The citation bias: Fad and fashion in the judgment and decision literature. *American Psychologist*, *39*, 75–78.

Clarkson, G. P. E. (1962). *Portfolio selection: A simulation of trust investment*. Englewood Cliffs, NJ: Prentice-Hall.

Cleaves, D. A. (1987). Cognitive biases and corrective techniques: Proposals for improving elicitation procedures for knowledge based systems. *International Journal of Man-Machine Studies*, *27*, 155–166.

Clement, J. (1983). A conceptual model discussed by Galileo and used intuitively by physics students. In D. Gentner & A. Stevens (Eds.), *Mental models* (pp. 325–339). Hillsdale, NJ: Erlbaum.

Cohen, L. J. (1981). Can human irrationality be experimentally demonstrated? *Behavioral and Brain Sciences*, *4*, 317–331.

Collett, P. (1979). The repertory grid in psychological research. In G. P. Ginsburg (Ed.), *Emerging strategies in social psychological research*. Chichester, England: Wiley.

Collins, A., & Gentner, D. (1986). How people construct mental models. In D. Holland & N. Quinn (Eds.), *Cultural models in language and thought* (pp. 243–265). Cambridge, UK: Cambridge University Press.

Coltheart, V., & Walsh, P. (1988). Expert knowledge and semantic memory. In M. M. Gruneberg, P. E. Morris, & R. N. Sykes (Eds.),

Practical aspects of memory: Current research and issues (Vol. 1, pp. 459–465). Chichester, England: Wiley.

Craik, F. I. M., & Lockhart, R. S. (1972). Levels of processing: A framework for memory research. *Journal of Verbal Learning and Verbal Behavior, 11*, 671–684.

Craik, K. (1967). *The nature of explanation*. Cambridge: Cambridge University Press.

Dahlstrom, D. (1989). Worlds of knowing and nonmonotonic reasoning. *IEEE Transactions on Systems, Man, & Cybernetics, 19*, 626–623.

Dawes, R. M. (1975). The mind, the model, and the task. In H. L. Castellon & F. Restle (Eds.), *Proceedings of the Seventh Annual Indiana Theoretical and Cognitive Psychology Conference* (pp. 119–129). Bloomington, IN: University of Indiana Press.

Dawes, R. M. (1982). The robust beauty of improper linear models in decision making. In D. Kahneman, P. Slovic, & A. Tversky (Eds.), *Judgment under uncertainty: Heuristics and biases* (pp. 391–407). New York: Cambridge University Press.

Dawes, R. M., & Corrigan, B. (1974). Linear models in decision making. *Psychological Bulletin, 81*, 95–106.

Deffenbacher, K. A. (1988). Eyewitness research: The next ten years. In M. M. Gruneberg, P. E. Morris, & R. N. Sykes (Eds.), *Practical aspects of memory, Vol. 2: Current research and issues* (pp. 20–26). Chichester, England: Wiley.

de Groot, A. D. (1965). *Thought and choice in chess*. The Hague: Mouton.

de Groot, A. D. (1966). Perception and memory versus thought: Some old ideas and recent findings. In B. Kleinmuntz (Ed.), *Problem solving: Research, method, and theory* (pp. 19–50). New York: Wiley.

deKleer, J., & Brown, J. S. (1981). Mental models of physical mechanisms and their acquisition. In J. R. Anderson (Ed.), *Cognitive skills and their acquisition* (pp. 285–308). Hillsdale, NJ: Erlbaum.

deKleer, J., & Brown, J. S. (1983). Assumptions and ambiguities in mechanistic mental models. In D. Gentner & A. L. Stevens (Eds.), *Mental models* (pp. 155–190). Hillsdale, NJ: Erlbaum.

Dewey, J. (1933). *How we think*. Boston: Heath.

diSessa, A. (1983). Phenomenology and the evolution of intuition. In D. Gentner & A. L. Stevens (Eds.), *Mental models* (pp. 15–33). Hillsdale, NJ: Erlbaum.

Dixon, N. (1981). *Preconscious processing*. Chichester, England: Wiley.

Eberts, R., & Simon, N. (1984). Cognitive requirements and expert systems. In G. Salvendy (Ed.), *Human-computer interaction*. Amsterdam: Elsevier.

Edwards, W. (1968). Conservatism in human information processing. In B. Kleinmuntz (Ed.), *Formal representation of human judgment* (pp. 17–52). New York: Wiley.

Edwards, W., & von Winterfeldt, D. (1986). On cognitive illusions and their implications. *Southern California Law Review, 59*, 401–451.

Einhorn, H. J. (1974). Expert judgment: Some necessary conditions and an example. *Journal of Applied Psychology, 59*, 562–571.

Einhorn, H. J. (1982). Learning from experience and suboptimal rules in decision making. In D. Kahneman, P. Slovic, & A. Tversky (Eds.), *Judgment under uncertainty: Heuristics and biases* (pp. 268–283). New York: Cambridge University Press.

Einhorn, H. J., & Hogarth, R. M. (1981). Behavioral decision theory: Processes of judgment and choice. *Annual Review of Psychology, 32*, 53–88.

Eliot, L. B. (1987, August). Investigating the nature of expertise: Analogical thinking, expert systems, and thinkback. *Expert Systems, 4*(3), 190–195.

Elstein, A. S., Shulman, L. S., & Sprafka, S. A. (1978). *Medical problem solving*. Cambridge, MA: Harvard University Press.

Enkawa, T., & Salvendy, G. (1989). Underlying dimensions of human problem solving and learning: Implications for personnel selection, training, task design, and expert system. *International Journal of Man-Machine Studies, 30*, 235–254.

Ennals, R. (1988). Can skills be transferable? In G. Göranzon & I. Josefson (Eds.), *Knowledge, skill, and artificial intelligence* (pp. 67–75). London: Springer-Verlag.

Epstein, W. (1967). *Varities of perceptual learning*. New York: McGraw-Hill.

Evans, D. A. (1989). Issues of cognitive science in medicine. In D. A. Evans & V. L. Patel (Eds.), *Cognitive science in medicine: Biomedical modeling* (pp. 1–19). Cambridge, MA: Bradford Books/MIT Press.

Evans, J. St. B. T. (1987). Human biases and computer decision making: A discussion of Jacob et al. *Behavior and Information Technology, 6*, 483–487.

Evans, J. St. B. T. (1989). *Biases in human reasoning: Causes and consequences*. Hillsdale, NJ: Erlbaum.

Feigenbaum, E. A. (1989). What hath Simon wrought? In D. Klahr & K. Kotovsky (Eds.), *Complex information processing: The impact of Herbert A. Simon* (pp. 165–182). Hillsdale, NJ: Erlbaum.

Finin, T. (1988). Refault reasoning and stereotypes in user modeling. *International Journal of Expert Systems*, *1*, 131–158.

Fischoff, B. (1986). Decision making in complex systems. In E. Hollnagel, G. Mancini, & D. D. Woods (Eds.), *Intelligent decision support in process environments* (pp. 61–86). Berlin: Springer-Verlag.

Fishbein, M., & Ajzen, I. (1972). Attitudes and opinions. *Annual Review of Psychology*, *23*, 487–544.

Fiske, S. T., & Linville, P. W. (1980). What does the schema concept buy us? *Personality and Social Psychology Bulletin*, *6*, 543–600.

Fitts, P. M. (1964). Perceptual-motor skill learning. In A. W. Melton (Ed.), *Categories of human learning* (pp. 243–285). New York: Academic Press.

Forbus, K., & Gentner, D. (1986). Learning physical domains: Towards a theoretical framework. In R. M. Michalski, J. Carbonell, & T. Mitchell (Eds.), *Machine learning: An artificial intelligence approach* (Vol. III, pp. 311–348). Los Altos, CA: Morgan Kaufmann.

Ford, K. M., Adams-Webber, J. R., Stahl, H. R., & Bringmann, M. W. (1990). Constructivist approaches to automated knowledge acquisition. In K. L. McGraw & C. R. Westphal (Eds.), *Readings in knowledge acquisition* (pp. 34–54). New York: Ellis Horwood.

Fox, J. (1986). Knowledge, decision making, and uncertainty. In W. A. Gale (Ed.), *Artificial intelligence and statistics* (pp. 57–76). Reading, MA: Addison-Wesley.

Fransella, F., & Bannister, D. (1971). *Inquiring man: The theory of personal constructs*. Harmondsworth, England: Penguin.

Gagne, R. M. (1962). The acquisition of knowledge. *Psychological Review*, *69*, 92–104.

Gentner, D. (1983). Structure mapping: A theoretical framework for analogy. *Cognitive Science*, 7(2), 155–170.

Gentner, D. (1989). The mechanisms of analogical learning. In S. Vosniadou & A. Ortony (Eds.), *Similarity and analogical reasoning* (pp. 199–241). London: Cambridge University Press.

Gentner, D., & Gentner, D. R. (1983). Flowing waters or teeming crowds: Mental models of electricity. In D. Gentner & A. L. Stevens (Eds.), *Mental models* (pp. 99–129). Hillsdale, NJ: Erlbaum.

Gentner, D., & Jeziorski, M. (1989). Historical shifts in the use of analogy in science. In B. Gholson, W. R. Shadish, R. A. Beimeyer, & A. Houts (Eds.), *The psychology of science: Contributions to metascience* (pp. 296–325). Cambridge: Cambridge University Press.

Gentner, D., & Landers, R. (1985, November). Analogical reminding: A good match is hard to find. *Proceedings of the International Conference on Systems, Man, and Cybernetics* (pp. 607–613). New York: IEEE.

Gentner, D., & Stevens, A. (Eds.). (1983). *Mental models*. Hillsdale, NJ: Erlbaum.

Gibson, E. J. (1969). *Principles of perceptual learning and development*. New York: Appleton Century Crofts.

Gill, S. (1988). Intelligence and social action: Education and training. In G. Göranzon & I. Josefson (Eds.), *Knowledge, skill, and artificial intelligence* (pp. 77–91). London: Springer-Verlag.

Glaser, R. (1976). Cognitive psychology and instructional design. In D. Klahr (Ed.), *Cognition and instruction* (pp. 303–315). Hillsdale, NJ: Erlbaum.

Glaser, R. (1984). Education and thinking: The role of knowledge. *American Psychologist*, *39*, 92–104.

Glaser, R. (1987). Thoughts on expertise. In C. Schooler & W. Schaie (Eds.), *Cognitive functioning and social structure over the life course* (pp. 81–94). Norwood, NJ: Ablex.

Glaser, R. (1989). Expertise and learning: How do we think about instructional processes now that we have discovered knowledge structures? In D. Klahr & K. Kotovsky (Eds.), *Complex information processing: The impact of Herbert A. Simon* (pp. 269–282). Hillsdale, NJ: Erlbaum.

Glaser, R., & Chi, M. T. H. (1988). Overview. In M. T. H. Chi, R. Glaser, & M. J. Farr (Eds.), *The nature of expertise* (pp. xv–xxviii). Hillsdale, NJ: Erlbaum.

Goldberg, L. R. (1970). Man versus model of man: A rationale, plus some evidence for a method of improving clinical judgment. *Psychological Bulletin*, *73*, 422–432.

Golde, R. A. (1970). *Can you be sure of your experts*? New York: Award Books.

Gomez, L. M., & Dumais, S. T. (1986). Putting cognition to work: Examples from computer system design. In T. J. Knapp & L. C. Robertson (Eds.), *Approaches to cognition: Contrasts and controversies* (pp. 267–290). Hillsdale, NJ: Erlbaum.

Gorman, M. E., & Carlson, W. B. (1990). Interpreting invention as a cognitive process: The case of A. G. Bell, T. Edison, and the telephone. *Science, Technology, and Human Values, 15*, 131–164.

Greeno, J. G. (1978). The nature of problem solving abilities. In W. K. Estes (Ed.), *Handbook of learning and cognitive processes, Volume 5* (pp. 239–270). Hillsdale, NJ: Erlbaum.

Greeno, J. G. (1983). Conceptual entities. In D. Gentner & A. Stevens (Eds.), *Mental models* (pp. 227–251). Hillsdale, NJ: Erlbaum.

Gregg, L. W. (1976). Methods and models for task analysis in instructional design. In D. Klahr (Ed.), *Cognition and instruction* (pp. 109–115). Hillsdale, NJ: Erlbaum.

Groner, R., Groner, M., & Bischof, W. F. (1983). The role of heuristics in models of decision. In R. W. Scholz (Ed.), *Decision making under uncertainty* (pp. 87–108). Amsterdam: North-Holland.

Gullers, P. (1988). Automation-skill-apprenticeship. In B. Göranzon & I. Josefson (Eds.), *Knowledge, skill, and artificial intelligence* (pp. 31–38). London: Springer-Verlag.

Halasz, F., & Moran, T. P. (1982). Analogy considered harmful. *Human Factors in Computing Systems Proceedings*. Washington, DC: National Bureau of Standards.

Halpern, D. F. (1984). *Thought and knowledge: An introduction to critical thinking*. Hillsdale, NJ: Erlbaum.

Hayes-Roth, B. (1977). Evolution of cognitive structures and processes. *Psychological Review, 84*, 260–278.

Hayes-Roth, F. (1978). Learning by example. In A. M. Lesgold et al. (Eds.), *Cognitive psychology and instruction* (pp. 27–38). New York: Plenum.

Hayes-Roth, F., Klahr, P., & Mostow, D. J. (1981). Advice-taking and knowledge refinement: An iterative view of skill acquisition. In J. R. Anderson (Ed.), *Cognitive skills and their acquisition* (pp. 231–254). Hillsdale, NJ: Erlbaum.

Henle, M. (1962). On the relation between logic and thinking. *Psychological Review, 69*, 366–378.

Hesse, M. B. (1981). The function of analogies in science. In R. D. Tweney, M. E. Doherty, & C. R. Mynatt (Eds.), *On scientific thinking* (pp. 345–348). New York: Columbia University Press.

Hilton, J. (1988). Skill, education, and social value: Some thoughts on the metonymy of skill and skill transfer. In B. Göranzon & I. Josefson (Eds.), *Knowledge, skill, and artificial intelligence* (pp. 93–101). London: Springer-Verlag.

Hink, R. F., & Woods, D. L. (1987, Fall). How humans process uncertain knowledge. *The AI Magazine, 8*(3), 41–53.

Hinsley, D. A., Hayes, J. R., & Simon, H. A. (1977). From words to equations: Meaning and representation in algebra and word problems. In M. A. Just & P. A. Carpenter (Eds.), *Cognitive processes in comprehension* (pp. 62–68). Hillsdale, NJ: Erlbaum.

Hoffman, P. J., Slovic, P., & Rorer, L. G. (1968). An analysis of variance model for the assessment of configural cue utilization in clinical judgment. *Psychological Bulletin, 69*, 338–349.

Hoffman, R. R. (1980). Metaphor in science. In R. P. Honeck & R. R. Hoffman (Eds.), *Cognition and figurative language* (pp. 393–423). Hillsdale, NJ: Erlbaum.

Hoffman, R. R., & Deffenbacher, K. C. (in press). A brief survey of the history of applied cognitive psychology. *Applied Cognitive Psychology*.

Hogarth, R. (1987). *Judgement and choice*. New York: Wiley.

Holding, D. H. (1985). *The psychology chess skill*. Hillsdale, NJ: Erlbaum.

Holland, J., Holyoak, K. J., Nisbett, R., & Thagard, P. (1986). *Induction: Processes of inference learning, and discovery*. Cambridge, MA: MIT Press.

Holyoak, K. J. (1984). Mental models in problem solving. In J. R. Anderson & S. M. Kosslyn (Eds.), *Tutorials in learning and memory: Essays in honor of Gordon Bower* (pp. 193–218). San Francisco: Freeman.

Horn, J. L. (1968). Organization of abilities and the development of intelligence. *Psychological Review, 75*, 242–259.

Jacob, V. S., Gaultney, L. D., & Salvendy, G. (1986). Strategies and biases in human decision making and their implications for expert systems. *Behavior and Information Technology, 5*, 119–140.

Janik, A. (1988). Tacit knowledge, working life, and scientific method. In B. Göranzon & I. Josefson (Eds.), *Knowledge, skill, and artificial intelligence* (pp. 53–63). London: Springer-Verlag.

Janik, A. (1990). Tacit knowledge, rule-following, and learning. In B. Göranzon & M. Floriz (Eds.), *Artificial intelligence, culture, and language: On education and work* (pp. 45–55). London: Springer-Verlag.

Jenkins, J. J. (1978). Four points to remember: A tetrahedral model of memory experiments. In L. Cermak & F. Craik (Eds.), *Levels of processing and human memory*. Hillsdale, NJ: Erlbaum.

Johnson, J. (1988). Expertise and decision under uncertainty: Performance and process. In M. T. H. Chi, R. Glaser, & M. J. Farr (Eds.), *The nature of expertise* (pp. 209–228). Hillsdale, NJ: Erlbaum.

Johnson, P. E. (1983). What kind of expert should a system be? *Journal of Medicine and Philosophy*, *8*, 77–97.

Johnson-Laird, P. N. (1980). Mental models in cognitive science. *Cognitive Science*, *4*, 71–115.

Johnson-Laird, P. N. (1983). *Mental models*. Cambridge, MA: Harvard University Press.

Josefson, I. (1987). Knowledge and experience. *Applied Artificial Intelligence*, *1*(2), 173–180.

Jungerman, H. (1983). Two camps of rationality. In R. W. Scholz (Ed.), *Decision making under uncertainty* (pp. 63–86). Amsterdam: North-Holland.

Kahneman, D., Slovic, P., & Tversky, A. (Eds.). (1982). *Judgment under uncertainty: Heuristics and biases*. New York: Cambridge University Press.

Kahneman, D., & Tversky, A. (1973). On the psychology of prediction. *Psychological Review*, *80*, 237–251.

Kahneman, D., & Tversky, A. (1982). On the study of statistical intuitions. In D. Kahneman, P. Slovic, & A. Tversky (Eds.), *Judgment under uncertainty: Heuristics and biases* (pp. 493–508). New York: Cambridge University Press.

Kahneman, D. & Tversky, A. (1984). Choices, values, and frames. *American Psychologist*, *39*, 341–350.

Keil, F. C. (1984). Transition mechanisms in cognitive development and the structure of knowledge. In R. J. Sternberg (Ed.), *Mechanisms of cognitive development* (pp. 81–89). San Francisno: Freeman.

Kidd, A. L. (1985). Human factors in the design and use of expert systems. In A. Monk (Ed.), *Fundamentals of human-computer interactions* (pp. 237–247). New York: Academic Press.

Kidd, A. L., & Cooper, M. B. (1985). Man-machine interface issues in the construction and use of an expert system. *International Journal of Man-Machine Studies*, *22*, 91–102.

Kintsch, W. (1988). The role of knowledge in discourse comprehension: A construction-integration model. *Psychological Review*, *95*, 163–182.

Klein, G. A. (1987). Applications of analogical reasoning. *Metaphor and Symbolic Activity*, *2*, 201–218.

Kolodner, J. L. (1983). Towards an understanding of the role of experience in the evolution from novice to expert. *International Journal of Man-Machine Studies*, *19*, 497–518.

Kolodner, J. L. (1984). Towards an understanding of the role of experience in the evolution from novice to expert. In M. J. Coombs (Ed.), *Developments in expert systems* (pp. 95–116). London: Academic Press.

Kolodner, J. L., & Riesbeck, C. (1986). *Experience, memory, and reasoning*. Hillsdale, NJ: Erlbaum.

Kolodner, J. L., & Simpson, R. L. (1984). Experience and problem solving: A framework. In *Proceedings of the Sixth Annual Conference of the Cognitive Science Society* (pp. 239–243). Boulder, CO.

Kowalik, J. (1986). *Knowledge-based problem solving*. Englewood Cliffs, NJ: Prentice-Hall.

LaBerge, D. (1976). Perceptual learning and attention. In W. K. Estes (Ed.), *Handbook of learning and cognitive processes* (Vol. 4, pp. 237–273). Hillsdale, NJ: Erlbaum.

LaFrance, M. (1989). The quality of expertise: Understanding the differences between experts and novices. In C. R. Westphal & K. L. McGraw (Eds.), *Special Issue of ACM SIGART Newsletter on Knowledge Acquisition* (No. 108, pp. 6–14). New York: Special Interest Group on Artificial Intelligence, Association for Computing Machinery.

LaFrance, M. (1990). The special structure of expertise. In K. L. McGraw & C. R. Westphal (Eds.), *Readings in knowledge acquisition* (pp. 55–70). New York: Ellis Horwood.

Lancaster, J. S. (1990). Cognitively based knowledge acquisition: Capturing categorical, temporal, and causal knowledge. In K. L. McGraw & C. R. Westphal (Eds.), *Readings in knowledge acquisition* (pp. 184–199). New York: Ellis Horwood.

Larkin, J. H. (1979). Processing information for effective problem solving. *Engineering Education*, 285–288.

Lee, W. (1971). *Decision theory and human behavior*. New York: Wiley.

Lenat, D. B., & Feigenbaum, E. A. (1987, August). On the thresholds of knowledge. In *Proceedings of the International Joint Conference on Artificial Intelligence* (pp. 1173–1182). Milano, Italy: International Joint Conferences on Artificial Intelligence.

Lesgold, A. M. (1984). Acquiring expertise. In J. R. Anderson & S. M. Kosslyn (Eds.), *Tutorials in*

Learning and memory: Essays in honor of Gordon Bower (pp. 31–60). San Francisco: Freeman.

Levi, K. (1989). Expert systems should be more accurate than human experts: Evaluation procedures from human judgement and decision-making. *IEEE Transactions on Systems, Man, and Cybernetics, 19*, 647–657.

Logan, G. D. (1985). Skills and automaticity: Relations, implications, and future directions. *Canadian Journal of Psychology, 39*, 367–386.

Logan, G. D. (1988). Automaticity, resources, and memory: Theoretical controversies and practical implications. *Human Factors, 30*, 583–598.

Logan, G. D. (1988). Toward an instance theory of automatization. *Psychological Review, 95*, 492–527.

Lonergan, B. J. F. (1953). *Insight*. New York: Harper & Row.

Luchins, A. S. (1942). Mechanization in problem solving. *Psychological Monographs, 54*(Whole No. 248).

Mancaster-Ramer, A., & Lindsay, R. K. (1988). Linguistic knowledge as expertise. *International Journal of Expert Systems, 1*, 329–343.

Mandler, G. (1962). From association to structure. *Psychological Review, 69*, 415–427.

Mandler, G. (1967). Organization and memory. In K. W. Spence & J. T. Spence (Eds.), *The psychology of learning and motivation* (Vol. 1, pp. 327–372). New York: Academic Press.

Mayer, R. F. (1988). From novice to expert. In M. Holander (Ed.), *Handbook of human-computer interaction* (pp. 569–580). Amsterdam: North-Holland.

McClelland, D. C. (1973). Testing for competence rather than "intelligence." *American Psychologist, 28*, 1–14.

McClosky, M., & Santee, J. (1981). Are semantic memory and episodic memory distinct systems? *Journal of Experimental Psychology: Human Learning and Memory, 7*, 66–71.

McGuire, W. J. (1960). A syllogistic analysis of cognitive relationships. In C. I. Hovland & M. J. Rosenberg (Eds.), *Attitude, organization, and change* (pp. 65–111). New Haven: Yale University Press.

McNeill, D., & Levy, E. Conceptual representations in language activity and gesture. In R. Jarvella & W. Klein (Eds.), *Speech, place, and action* (pp. 271–296). New York: Wiley.

Means, B., & Gott, S. (1988). Cognitive task analysis as a basis for tutor development: Articulating abstract knowledge representations. In J. Psotka, L. D. Massey, & S. A. Mutter (Eds.), *Intelligent tutoring systems: Lessons learned* (pp. 35–58). Hillsdale, NJ: Erlbaum.

Meehl, P. E. (1954). *Clinical versus statistical prediction: A theoretical analysis and a review of evidence*. Minneapolis, MN: University of Minnesota Press.

Neves, D. M., & Anderson J. R. (1981). Knowledge compilation: Mechanisms for the automatization of cognitive skills. In J. A. Anderson (Ed.), *Cognitive skills and their acquisition* (pp. 57–84). Hillsdale, NJ: Erlbaum.

Newell, A. (1980). Reasoning, problem solving, and decision processes: The problem space as a fundamental category. In R. S. Nickerson (Ed.), *Attention and performance VIII* (pp. 693–717). Hillsdale, NJ: Erlbaum.

Newell, A. (1982). The knowledge level. *Artificial Intelligence, 18*, 87–127.

Newell, A. (1985). Duncker on thinking: An inquiry into progress in cognition. In S. Koch & D. E. Leary (Eds.), *A century of psychology as a science* (pp. 392–419). New York: McGraw-Hill.

Newell, A., Shaw, J. C., & Simon, H. A. (1958). Elements of a theory of human problem solving. *Psychological Review, 65*, 151–166.

Newell, A., & Simon, H. A. (1972). *Human problem-solving*. Englewood Cliffs, NJ: Prentice-Hall.

Nickerson, R. S., Perkins, D. N., & Smith, E. E. (1985). *The teaching of thinking*. Hillsdale, NJ: Erlbaum.

Norman, D. A. (1983). Some observations on mental models. In D. Gentner & A. L. Stevens (Eds.), *Mental models* (pp. 7–14). Hillsdale, NJ: Erlbaum.

Norman, D. A. (1986). New views of information processing: Implications for intelligent decision support systems. In E. Hollnagel, G. Mancini, & D. D. Woods (Eds.), *Intelligent decision support in process environments*. Berlin: Springer-Verlag.

Norman, D. A. (1987). Cognitive engineering-cognitive science. In J. M. Carroll (Ed.), *Interfacing thought: Cognitive aspects of human-computer interaction* (pp. 323–336). Cambridge, MA: The MIT Press.

Olson, D. R. (1976). Notes on a cognitive theory of instruction. In D. Klahr (Ed.), *Cognition and instruction* (pp. 117–120). Hillsdale, NJ: Erlbaum.

Oppenheimer, J. R. (1956). Analogy in science. *American Psychologist, 11*, 127–135.

Paul, I. M. (1967). The concept of a schema in memory theory. In R. R. Holt (Ed.), *Motives and thought: Essays in honor of David Rapoport* (pp.

219–258). New York: International Universities Press.

Payne, J. W. (1982). Contingent decision behavior. *Psychological Bulletin*, *94*, 382–401.

Perfetti, C. A., & Lesgold, A. M. (1977). Discourse comprehension and sources of individual differences. In M. Just & P. Carpenter (Eds.), *Cognitive processes in comprehension* (pp. 141–183). Hillsdale, NJ: Erlbaum.

Perfetti, C. A., & Lesgold, A. M. (1979). Coding and comprehension in skilled reading. In L. B. Resnick & P. Weaver (Eds.), *Theory and practice of early reading* (Vol. 1, pp. 57–84). Hillsdale, NJ: Erlbaum.

Pitz, G. E., & Sachs, N. J. (1984). Judgment and decision making: Theory and applications. *Annual Review of Psychology*, *35*, 139–163.

Polanyi, M. (1966). *The tacit dimension*. Garden City, NY: Doubleday.

Polya, G. (1957). *How to solve it: An aspect of mathematical method*. Princeton, NJ: Princeton University Press.

Porter, D. B. (1987). Classroom teaching, implicit learning, and the deleterious effects of inappropriate explication. *Proceedings of the Human Factors Society 31st Annual Meeting*, 289–292.

Posner, M. I. (1988). Introduction: What is it to be an expert? In M. T. H. Chi, R. Glaser, & M. J. Farr (Eds.), *The nature of expertise* (pp. xxix–xxxvi). Hillsdale, NJ: Erlbaum.

Rabinowitz, M. (1985). Cognitive structure and process in highly competent performance. In F. D. Horowitz & M. O'Brien (Eds.), *The gifted and talented: Developmental perspectives* (pp. 75–98). Washington, DC: American Psychological Association.

Rabinowitz, M. (1990). Prerequisite knowledge for learning and problem solving. In C. Hadley, J. Houtz, & A. Baratta (Eds.), *Cognition, literacy, and curriculum* (pp. 47–58). Norwood, NJ: Ablex.

Rabinowitz, M., & Chi, M. T. H. (1987). An interactive model of strategic processing. In S. J. Ceci (Ed.), *Handbook of cognitive, social and neuropsychological aspects of learning disabilities*, *2*, (pp. 83–102). Hillsdale, NJ: Erlbaum.

Raiffa, H. (1968). *Decision analysis: Introductory lectures on choice under uncertainty*. Reading, MA: Addison-Wesley.

Rasmussen, J. (1986). *Information processing and human-machine interaction: An approach to cognitive engineering*. New York: Elsevier Science Publishing.

Reder, L. M. (1987). Beyond associations: Strategic components in memory retrieval. In D. S. Gorfein & R. R. Hoffman (Eds.), *Memory and learning: The Ebbinghaus Centennial Conference* (pp. 203–220). Hillsdale, NJ: Erlbaum.

Reder, L. M., & Anderson, J. R. (1980). A partial resolution of the paradox of interference: The role of integrating knowledge. *Cognitive Psychology*, *12*, 447–472.

Reilly, B. A., & Doherty, M. E. (1989). A note on the assessment of self-insight in judgment research. *Organizational Behavior and Human Decision Processes*, *44*, 122–131.

Reitman, W. (1964). Heuristic decision procedures, open constraints, and the structure of ill-defined problems. In M. W. Shelley & G. L. Bryan (Eds.), *Human judgments and optimality*. New York: Wiley.

Resnick, L. B. (1981). Instructional psychology. *Annual Review of Psychology*, *32*, 659–704.

Rosenbloom, P. S., & Newell, A. (1983). The chunking of goal hierarchies: A generalized model of practice. *Proceedings of the International Machine Learning Workshop*, Manticello, IL.

Rouse, W. B., & Morris, N. M. (1986). On looking into the black box: Prospects and limits in the search for mental models. *Psychological Bulletin*, *100*, 349–363.

Rumelhart, D. E. (1980). Schemata: The building blocks of cognition. In R. J. Spiro, B. C. Bruce, & W. F. Brewer (Eds.), *Theoretical issues in reading comprehension* (pp. 33–58). Hillsdale, NJ: Erlbaum.

Rumelhart, D. E., & Norman, D. A. (1981). Analogical processes in learning. In J. R. Anderson (Ed.), *Cognitive skills and their acquisition* (pp. 335–359). Hillsdale, NJ: Erlbaum.

Schank, R., & Abelson, R. P. (1977). *Scripts, plans, goals, and understanding*. Hillsdale, NJ: Erlbaum.

Schmidt, R. A. (1982). The schema concept. In J. A. S. Kelso (Ed.), *Human motor behavior: An introduction* (pp. 219–235). Hillsdale, NJ: Erlbaum.

Schneider, A., & Shiffrin, R. (1985). Categorization (restructuring) and automatization: Two separable factors. *Psychological Review*, *92*, 424–428.

Schneider, W. (1985). Training high-performance skills: Fallacies and guidelines. *Human Factors*, *27*, 285–300.

Schneider, W. & Shiffrin, R. M. (1977). Controlled and automatic human information processing: I. Detection, search, and attention. *Psychological Review*, *84*, 1–60.

Schön, D. (1979). Generative metaphor: A perspective on problem-setting in social policy. In A. Ortony (Ed.), *Metaphor and thought*. New York: Cambridge University Press.

Schön, D. (1983). *The reflective practitioner*. New York: Basic Books.

Schumacher, R. M. (1987). Acquisition of mental models. In J. Flach (Ed.), *Proceedings of the Fourth Annual Mid-Central Human Factors/Ergonomics Conference* (pp. 142–148). Santa Monica, CA: Human Factors Society.

Schumacher, R. M., & Gentner, D. (1988). Transfer of training as analogical mapping. *IEEE Transactions on Systems, Man, and Cybernetics*, *18*, 592–600.

Shanteau, J. (1978). When does a response error become a judgmental bias? Commentary on "Judged frequency of lethal events." *Journal of Experimental Psychology: Human Learning and Memory*, *4*, 579–581.

Shanteau, J. (1984). Some unasked questions about the psychology of decision makers. In M. E. El-Hawary (Ed.), *Proceedings of the IEEE Conference on Systems, Man, and Cybernetics* (pp. 23–45). New York: IEEE.

Shanteau, J. (1988). Psychological characteristics and strategies of expert decision makers. *Acta Psychologica*, *68*, 203–315.

Shanteau, J., & Nagy, G. F. (1984). Information integration in person perception: Theory and application. In M. Cook (Ed.), *Progress in person perception* (pp. 48–86). London: Methuen.

Shaw, M. E. (1981). *Think again*. Englewood Cliffs, NJ: Prentice-Hall.

Shaw, R., & Wilson, B. E. (1976). Abstract conceptual knowledge: How we know what we know. In D. Klahr (Ed.), *Cognition and instruction* (pp. 197–221). Hillsdale, NJ: Erlbaum.

Shiffrin, R. M., & Dumais, S. T. (1981). The development of automatism. In J. R. Anderson (Ed.), *Cognitive skills and their acquisition* (pp. 111–140). Hillsdale, NJ: Erlbaum.

Shiffrin, R. M., & Schneider, W. (1977). Controlled and automatic human information processes: II. Perceptual learning, automatic attending, and a general theory. *Psychological Review*, *84*, 127–190.

Simon, H. A. (1957). *Models of man: Social and rational*. New York: Wiley.

Simon, H. A. (1973). Does scientific discovery have a logic? *Philosophy of Science*, *40*, 471–480.

Simon, H. A. (1973). The structure of ill-structured problems. *Artificial Intelligence*, *4*, 181–201.

Simon, H. A. (1974). How big is a chunk. *Science*, *183*, 482–488.

Simon, H. A. (1978). Information processing theory of human problem solving. In W. K. Estes (Ed.), *Handbook of Learning and Cognitive Processes*, *5*, 271–295.

Simon, H. A. (1988). Rationality as process and as a product of thought. In D. E. Bell, H. Raiffa, & A. Tversky (Eds.), *Decision making: Descriptive, normative, and prescriptive interactions* (pp. 58–77). New York: Cambridge University Press.

Simon, H. A. (1989). The scientist as problem solver. In D. Klahr & K. Kotovsky (Eds.), *Complex information processing: The impact of Herbert A. Simon* (pp. 375–398). Hillsdale, NJ: Erlbaum.

Simon, H. A., & Barenfeld, M. (1969). Information processing analysis of perceptual processing in problem solving. *Psychological Review*, *76*, 473–483.

Slovic, P. (1982). Toward understanding and improving decisions. In W. C. Howell & E. A. Fleishman (Eds.), *Human performance and productivity* (Vol. 1, pp. 157–183). Hillsdale, NJ: Erlbaum.

Slovic, P., Fischoff, B., & Lichtenstein, S. (1977). Behavioral decision theory. *Annual Review of Psychology*, *28*, 1–39.

Sprio, R. J., Feltovich, P. J., Coulson, R. L., & Anderson, D. K. (1989). Multiple analogies for complex concepts: Antidotes for analogy-induced misconception in advanced knowledge acquisition. In S. Vosniadou & A. Ortony (Eds.), *Similarity and analogical reasoning* (pp. 498–531). Cambridge: Cambridge University Press.

Sternberg, R. J. (1977). *Intelligence, information processing, and analogical reasoning: The componential analysis of human abilities*. Hillsdale, NJ: Erlbaum.

Sternberg, R. J. (1984). Facets of human intelligence. In J. R. Anderson & S. M. Kosslyn (Eds.), *Tutorials in learning and memory: Essays in honor of Gordon Bower* (pp. 137–165). San Francisco: Freeman.

Sternberg, R. J. (1985). *Human abilities: An information processing approach*. San Francisco: Freeman.

Strauss, S., & Stavy, R. (1981). U-shaped behavioral growth: Implications for theories of development. In W. W. Hartup (Ed.), *Review of child development research* (pp. 547–599). New York: Academic Press.

Tolcott, M. A., Marvin, F. F., & Lehner, P. E. (1989). Expert decision making in evolving situations. *IEEE Transactions on Systems, Man, and Cybernetics*, *19*, 606–615.

Tversky, A., & Kahneman, D. (1971). The belief in the "law of small numbers." *Psychological Bulletin*, *76*, 105–110.

Tversky, A., & Kahneman, D. (1983). Extensional versus intuitive reasoning: The conjunction fallacy in probability judgment. *Psychological Review*, *90*, 293–315.

Tweney, R. D., Doherty, M. E., & Mynatt, C. R. (Eds.). (1981). *On scientific thinking*. New York: Columbia University Press.

VanLehn, K. (1983). Human skill acquisition: Theory, model, and psychological validation. In *Proceedings of AAAI–83* (pp. 420–423). Los Altos, CA: Kaufmann.

VanLehn, K. (1989). *Cognitive procedures: The acquisition and mental representation of basic mathematical skills*. Cambridge, MA: MIT Press.

Vekker, L. M. (1974). *Mental processes* (Vol. 1). Leningrad: Leningrad University Press. (In Russian).

Vekker, L. M. (1976). *Mental processes: Thinking and the intellect* (Vol. 2). Leningrad: Leningrad University Press. (In Russian).

Vekker, L. M. (1981). *Mental processes: The agent, the experience, action, and consciousness* (Vol. 3). Leningrad: Leningrad University Press. (In Russian).

Vosniadou, S., & Ortony, A. (Eds.). (1989). *Similarity and analogical reasoning*. Cambridge: Cambridge University Press.

Wagner, R. K., & Sternberg, R. J. (1986). Tacit knowledge and intelligence in the everyday world. In R. J. Sternberg & R. K. Wagner (Eds.), *Practical intelligence: Nature and origins of competence in the everyday world* (pp. 51–83). Cambridge: Cambridge University Press.

Walker, E., & Stevens, A. (1986). Human and machine knowledge in intelligent systems. In E. Hollnagel, G. Mancini, & D. D. Woods (Eds.), *Intelligent decision support in process environments* (pp. 421–433). Berlin: Springer-Verlag.

Wallsten, T. S. (1983). The theoretical status of judgmental heuristics. In R. W. Scholz (Ed.), *Decision making under uncertainty* (pp. 21–37). Amsterdam: North-Holland.

Wason, P. C., & Evans, J. St. B. T. (1975). Dual processes in reasoning. *Cognition*, *3*, 141–154.

Wason, P. C., & Johnson-Laird, P. N. (1972). *Psychology of reasoning: Structure and Content*. Cambridge, MA: Harvard University Press.

Wertheimer, M. (1945). *Productive thinking*. New York: Harper & Row.

Wickelgren, W. (1974). *How to solve problems*. San Francisco: Freeman.

Wickens, C. D. (1984). Processing resources in attention. In R. Parasuraman & D. R. Davies (Eds.), *Varieties of attention* (pp. 63–102). New York: Academic Press.

Wright, G. (1984). *Behavioral decision theory: An introduction*. Beverly Hills, CA: Sage.

Wright, G., & Ayton, P. (1986). Subjective confidence in forecasts: A response to Fischoff and MacGregor. *Journal of Forecasting*, *5*, 117–124.

Zimmer, A. C. (1984). A model for the interpretation of verbal predictions. In M. J. Coombs (Ed.), *Developments in expert systems* (pp. 235–248). New York: Academic Press.

Appendix B
Bibliography: Empirical and Experimental Investigations of Expertise

Adelson, B. (1984). When novices surpass experts: The difficulty of a task may increase with expertise. *Journal of Experimental Psychology: Learning, Memory, and Cognition*, *10*, 483–495.

Aikin, O. (1980). *Models of architectural knowledge*. London: Pion.

Anderson, J. R. (1981). Effects of prior knowledge on memory for new information. *Memory & Cognition*, *9*, 237–246.

Anderson, J. R. (1982). Acquisition of a cognitive skill. *Psychological Review*, *89*, 369–406.

Andreassen, P. B., & Kraus, S. J. (1990). Judgmental extrapolation and the salience of change. *Journal of Forecasting*, *9*, 347–372.

Arkes, H. R., & Freedman, M. R. (1984). A demonstration of the costs and benefits of expertise in recognition memory. *Memory & Cognition*, *12*, 84–89.

Ashton, R. H. (1982). *Human information processing in accounting*. Sarasota, FL: American Accounting Association.

Ashton, R. H. (1983). Research in auditing decision making: rationale, evidence, and implications. *Canadian Certified General Accountants Association Monographs* (No. 6). Vancouver, BC: Canadian General Accountants Association.

Bailey, W. A., & Kay, D. J. (1987). Structural analysis of verbal data. In J. M. Carroll & P. P. Tanner (Eds.), *Human factors in computing systems and graphics interface* (pp. 297–301). New York: Association for Computing Machinery.

Bassok, M., & Holyoak, K. J. (1989). Interdomain transfer between isomorphic topics in algebra and physics. *Journal of Experimental Psychology: Learning, Memory, and Cognition*, *15*, 153–166.

Beach, K. D. (1988). The role of external mnemomic symbols in acquiring an occupation. In M. M. Gruneberg, P. E. Morris, & R. N. Sykes (Eds.), *Practical aspects of memory: Current research and issues* (Vol. 1, pp. 342–346). Chichester, England: Wiley.

Beck, M. B., & Halfon, E. (1991). Uncertainty, identifiability, and the propagation of prediction errors: A case study of Lake Ontario. *Journal of Forecasting*, *10*, 135–162.

Begg, I., & Denny, D. (1969). Empirical reconciliation of atmosphere and coversion interpretations of syllogistic reasoning errors. *Journal of Experimental psychology*, *81*, 351–354.

Bellezza, F. S., & Buck, D. K. (1988). Expert knowledge as mnemonic cues. *Applied Cognitive Psychology*, *2*, 147–162.

Benner, P. (1984). *From novice to expert: Excellence and power in clinical nursing practice*. Menlo Park, CA: Addison-Wesley.

Berry, D. C., & Broadbent, D. E. (1984). On the relationship between task performance and associated verbalizable knowledge. *Quarterly Journal of Experimental Psychology*, *36A*, 209–231.

Berry, D. C., & Broadbent, D. E. (1988). Interactive tasks and the implicit-explicit distinction. *British Journal of Psychology*, *79*, 251–272.

Beyth-Marom, R. (1982). How probable is probable? A numerical translation of verbal probability expressions. *Journal of Forecasting*, *1*, 257–269.

Beyth-Marom, R., & Arkes, H. R. (1983). Being accurate but not necessarily Bayesian: Comments on Christensen-Szalanski and Beach. *Organizational Behavior and Human Performance*, *31*, 255–257.

Bhaskar, R., & Simon, H. A. (1977). Problem solving in semantically rich domains. *Cognitive Science*, *1*, 193–215.

Blesser, B., & Ozonoff, D. (1972). A model for the radiologic process. *Radiology*, *103*, 515–521.

Bradley, J. V. (1981). Overconfidence in ignorant experts. *Bulletin of the Psychonomic Society*, *17*, 82–84.

Brehemer, B. (1980). In one word: Not from experience. *Acta Psychologica*, *45*, 223–241.

Broadbent, D. E., & Aston, B. (1978). Human control of a simulated economic system. *Ergonomics*, 1035–1043.

Broadbent, D. E., FitzGerald, P., & Broadbent, M. H. P. (1986). Implicit and explicit knowledge in the control of complex systems. *British Journal of Psychology*, *77*, 33–50.

Brooks, L. (1978). Nonanalytic concept formation and memory for instances. In E. Rosch & B. Lloyd (Eds.), *Cognition and categorization* (pp. 169–211). Hillsdale, NJ: Erlbaum.

Brown, J. S., & Burton, R. R. (1978). Diagnostic models for procedural bugs in basic mathematical skills. *Cognitive Science*, *2*, 155–192.

Bushke, H. (1976). Learning is organized by chunking. *Journal of Verbal Learning and Verbal Behavior*, *15*, 313–324.

Canfield, T. H. (1941). Sex determination of day-old chicks. *Poultry Science*, *20*, 327–328.

Carlson, R. S., Sullivan, M. A., & Schneider, W. (1989). Practice and working memory effects in building procedural skill. *Journal of Experimental Psychology: Learning, Memory, and Cognition*, *15*, 517–526.

Carroll, J. S., & Payne, J. W. (1977). Judgments about crime and the criminal: A model and a method for investigating parole decisions. In D. B. Sales (Ed.), *Perspectives in law and psychology: The criminal justice system* (pp. 191–240). New York: Plenum.

Carroll, J. S., & Siegler, R. S. (1977). Strategies for the use of base-rate information. *Organizational Behavior and Human Performance*, *19*, 392–402.

Ceci, S. J., & Liker, J. (1986). A day at the races: A study of IQ, expertise, and cognitive complexity. *Journal of Experimental Psychology: General*, *115*, 255–266.

Ceci, S. J., & Liker, J. (1986). Academic and non-academic intelligence: An experimental separation. In R. J. Sternberg & R. Wagner (Eds.), *Practical intelligence: Originals of competence in the everyday world*. (pp. 119–142), New York: Cambridge University Press.

Ceci, S. J., & Liker, J. (1988). Stalking the IQ-expertise relationship: When the critics go fishing. *Journal of Experimental Psychology: General*, *117*, 96–100.

Chan, S. (1982). Expert judgments made under uncertainty: Some evidence and suggestions. *Social Science Quarterly*, *63*, 428–444.

Chandrasekaran, B. (1990, Winter). Design problem solving: A task analysis. *The AI Magazine*, 59–71.

Charness, N. (1976). Memory for chess positions: Resistance to interference. *Journal of Experimental Psychology: Human Learning and Memory*, *2*, 641–653.

Charness, N. (1979). Components of skill in bridge. *Canadian Journal of Psychology*, *33*, 1–16.

Charness, N. (1981). Search in chess: Age and skill differences. *Journal of Experimental Psychology: Human Perception and Performance*, *7*, 467–476.

Charness, N. (1989). Expertise in chess and bridge. In D. Klahr & K. Kotovsky (Eds.), *Complex information processing: The impact of Herbert A. Simon* (pp. 183–208). Hillsdale, NJ: Erlbaum.

Charness, N., & Campbell, J. I. D. (1988). Acquiring skill at mental calculation in adulthood: A task decomposition. *Journal of Experimental Psychology: General*, *117*, 115–129.

Chase, W. G. (1983). Spatial representations of taxi drivers. In D. R. Rogers & J. H. Slobada (Eds.), *Acquisition of symbolic skills* (pp. 391–405). New York: Plenum.

Chase, W. G., & Ericsson, K. A. (1981). Skilled memory. In J. R. Anderson (Ed.), *Cognitive skills and their acquisition*. Hillsdale, NJ: Erlbaum.

Chase, W. G., & Simon, H. A. (1973). The mind's eye in chess. In W. G. Chase (Ed.), *Visual information processing*. New York: Academic Press.

Chase, W. G., & Simon, H. A. (1973). Perception in chess. *Cognitive Psychology*, *4*, 55–81.

Chen, Z., & Daehler, M. W. (1989). Positive and negative transfer in analogical problem solving by 6-year-old children. *Cognitive Development*, *4*, 327–344.

Cheng, D. W., Holyoak, K. J., Nisbett, R. E., & Oliver, L. M. (1986). Pragmatic versus syntactic approaches to training deductive reasoning. *Cognitive Psychology*, *18*, 293–328.

Chi, M. T. H. (1978). Knowledge structures and memory development. In R. Siegler (Ed.), *Children's thinking: What develops*? (pp. 73–96). Hillsdale, NJ: Erlbaum.

Chi, M. T. H., Bassok, M., Lewis, M. W., Reimann, P., & Glaser, R. (1989). Self-explanations: How students study and use examples in learning to solve problems. *Cognitive Science*, *13*, 145–182.

Chi, M. T. H., Feltovich, P. J., & Glaser, R. (1981). Categorization and representation of physics problems by experts and novices. *Cognitive Science*, *5*, 121–152.

Chi, M. T. H., Glaser, R., & Rees, E. (1982). Expertise in problem solving. In R. J. Sternberg (Ed.), *Advances in the psychology of human intelligence* (Vol. 1, pp. 7–75). Hillsdale, NJ: Erlbaum.

Chi, M. T. H., & Koeske, R. D. (1983). Network representation of a child's dinosaur knowledge. *Development Psychology*, *19*, 29–39.

Chiesi, H. L., Spilich, G. J., & Voss, J. F. (1979). Acquisition of domain-related information in relation to high and low domain information. *Journal of Verbal Learning and Verbal Behavior*, *18*, 257–273.

Christensen, E. E., Murray, R. C., Holland, K., Reynolds, J., Landay, M., & Moore, J. G. (1981). The effect of search time on perception. *Radiology*, *138*, 361–365.

Christensen-Szalanski, J. J., Diehr, P. H., Bushyhead, J. B., & Wood, R. W. (1982). Two studies of good clinical judgment. *Medical Decision Making*, *3*, 275–284.

Clancey, W. J. (1988). Acquiring, representing, and evaluating a competence mode of diagnostic strategy. In M. T. H. Chi, R. Glaser, & M. J. Farr (Eds.), *The nature of expertise* (pp. 343–418). Hillsdale, NJ: Erlbaum.

Claparède, E. (1917). La psychologie de l'intelligence. *Scientia*, *22*, 353–368.

Claparède, E. (1934). Le genese de l'Hypothèse: Étude experimentelle. *Archiv de Psychologie*, *24*, 1–154.

Clark, J. M., & Paivio, A. (1989). Observational and theoretical terms in psychology: A cognitive perspective on scientific language. *American Psychologist*, *44*, 500–512.

Clarkson, G. P. E. (1962). *Portfolio selection: A simulation of trust investment*. Englewood Cliffs, NJ: Prentice-Hall.

Clemen, R. T. (1985). Extraneous expert information. *Journal of Forecasting*, *4*, 329–348.

Clement, J. (1981). Analogy generation in scientific problem solving. In *Proceedings of the Third Annual Conference of the Cognitive Science Society* (pp. 137–140). Berkeley, CA: Cognitive Science Society.

Clement, J. (1983). A conceptual model discussed by Galileo and used intuitively by physics students. In D. Gentner & A. L. Stevens (Eds.), *Mental models* (pp. 325–340). Hillsdale, NJ: Erlbaum.

Clement, J. (1988). Observed methods for generating analogies in scientific problem solving. *Cognitive Science*, *12*, 563–586.

Connelly, D. P., & Johnson, P. E. (1980). The medical problem-solving process. *Human Pathology*, *11*, 412–419.

Cooke, N. M., Durso, F. T., & Schvaneveldt, R. W. (1986). Recall and measures of memory organization. *Journal of Experimental Psychology: Learning, Memory, and Cognition*, *12*, 538–549.

Cooper, G., & Sweller, J. (1987). Effects of schema acquisition and rule automatization on mathematical problem solving transfer. *Journal of Educational Psychology*, *79*, 347–362.

Cox, J. R., & Griggs, R. A. (1982). The effects of experience on performance in Wason's selection task. *Memory & Cognition*, *10*, 496–502.

Cross, M. (1988). The changing nature of the engineering craft apprenticeship system in the United Kingdom. In B. Göranzon & I. Josefson (Eds.), *Knowledge, skill, and artificial intelligence* (pp. 151–158). London: Springer-Verlag.

Daan, H., & Murphy, A. H. (1982). Subjective probability forecasting in the Netherlands: Some operational and experimental results. *Meteorological Research*, *35*, 99–112.

Dawes, R. W. (1971). A case study of graduate admissions: Application of three principles from human decision making. *American Psychologist*, *26*, 180–188.

Dawson, V. L., Zeitz, C. M., & Wright, J. C. (1989). Expert-novice differences in person perception: Evidence of experts' sensitivity to the organization of behavior. *Social Cognition*, *7*, 1–30.

Deakin, J. M., & Allard, F. (1991). Skilled memory in expert figure skaters. *Memory & Cognition*, *19*, 79–86.

Dee-Lucas, D., & Larkin, J. H. (1986). Novice strategies for processing scientific texts. *Discourse Processes*, *9*, 329–354.

deGroot, A. D. (1965). *Thought and choice in chess*. The Hague: Mouton.

deJong, T., & Ferguson-Hessler, M. G. M. (1986). Cognitive structures of good and poor novice problem solvers in physics. *Journal of Educational Psychology*, *78*, 279–288.

De Keyser, V. (1987). How can computer-based visual displays aid operators? *International Journal of Man-Machine Studies*, *27*, 471–478.

deKleer, J., & Brown, J. S. (1983). Assumptions and ambiguities in mechanistic mental models. In D. Gentner & A. Stevens (Eds.), *Mental models* (pp. 155–190). Hillsdale, NJ: Erlbaum.

Detmer, D. E., Fryback, D. G., & Gassner, K. (1978). Heuristics and biases in medical decision making. *Journal of Medical Education*, *53*, 682–683.

Diamond, R., & Carey, S. (1986). Development changes in the representation of faces. *Journal of Experimental Child Psychology*, *23*, 1–22.

diSessa, A. A. (1983). Phenomenology and the evolution of intuition. In D. Gentner & A. L. Stevens (Eds.), *Mental models* (pp. 15–43). Hillsdale, NJ: Erlbaum.

Dougherty, J. W. D., & Keller, C. M. (1982). Taskonomy: A practical approach to knowledge structures. *American Ethnologist*, *9*, 763–774.

Dulany, D. E., Carlson, R. A., & Dewey, G. I. (1984). A case of syntactical learning and judgment: How conscious and how abstract? *Journal of Experimental Psychology: General*, *113*, 541–555.

Dunbar, K., & Klahr, D. (1989). Developmental differences in scientific discovery processes. In D. Klahr & K. Kotovsky (Eds.), *Complex information processing: The impact of Herbert A. Simon* (pp. 109–133). Hillsdale, NJ: Erlbaum.

Duncker, K. (1945). On problem solving (L. S. Lees, Trans.). *Psychological Monographs*, *58*(Whole No. 270), 1–113.

Ebbesen, E., & Konečni, V. (1975). Decision making and information integration in the courts: The setting of bail. *Journal of Personality and Social Psychology*, *32*, 805–821.

Edmundson, R. H. (1990). Decomposition: A strategy for judgmental forecasting. *Journal of Forecasting*, *9*, 305–314.

Egan, D. E., & Schwartz, B. J. (1977). Chunking in the recall of symbolic drawings. *Memory & Cognition*, *7*, 149–158.

Eich, J. E. (1980). The cue-dependent nature of state dependent retrieval. *Memory & Cognition*, *8*, 157–173.

Einhorn, H. J. (1972). Expert measurement and mechanical combination. *Organization Behavior and Human Performance*, *7*, 86–106.

Einhorn, H. J. (1974). Expert judgment: Some necessary conditions and an example. *Journal of Applied Psychology*, *59*, 562–571.

Elstein, A. S., Dod, J., & Holzman, G. B. (1989). Estrogen replacement decisions of third-year residents: Clinical intuition and decision analysis. In D. A. Evans & V. L. Patel (Eds.), *Cognitive science in medicine: Biomedical modeling* (pp. 21–49). Cambridge, MA: Bradford Books/MIT Press.

Engel, R. W., & Bukstel, L. (1978). Memory processes among bridge players of differing expertise. *American Journal of Psychology*, *91*, 673–689.

Ericsson, K. A., & Polson, P. G. (1988). A cognitive analysis of exceptional memory for restaurant orders. In M. T. H. Chi, R. Glaser, & M. J. Farr (Eds.), *The nature of expertise* (pp. 23–70). Hillsdale, NJ: Erlbaum.

Ericsson, K. A., & Sjaszewski, J. J. (1989). Skilled memory and expertise: Mechanisms of exceptional performance. In D. Klahr & K. Kotovsky (Eds.), *Complex information processing: The impact of Herbert A. Simon* (pp. 235–268). Hillsdale, NJ: Erlbaum.

Ettenson, R. T., Shanteau, J., & Krogstad, J. (1987). Expert judgment: Is more information better? *Psychological Reports*, *60*, 227–238.

Evans, D. A., & Gadd, C. S. (1989). Managing coherence and context in medical problem solving discourse. In D. A. Evans & V. L. Patel (Eds.), *Cognitive science in medicine: Biomedical modeling* (pp. 211–255).

Evans, J. St. B. T. (1987). Human biases and computer decision-making: A discussion of Jacob et al. *Behaviour and Information Technology*, *6*(4), 483–487.

Feltovich, P. J., Spiro, R. J., & Coulson, R. L. (1989). The nature of conceptual understanding in medicine: The deep structure of complex ideas and the development of misconceptions. In D. A. Evans & V. L. Patel (Eds.), *Cognitive science in medicine: Biomedical modeling* (pp. 113–172). Cambridge, MA: Bradford Books/MIT press.

Fischoff, B. (1975). Hindsight/Foresight: The effect of outcome knowledge on judgment under uncertainty. *Journal of Experimental Psychology: Human Perception and Performance*, *1*, 288–299.

Fischoff, B., Slovic, P., & Lichtenstein, S. (1977). Knowing with certainty: The appropriateness of extreme confidence. *Journal of Experimental Psychology: Human Perception and Performance*, *3*, 552–564.

Fischoff, B., Slovic, P., & Lichtenstein, S. (1988). Knowing what you want: Measuring labile values. In D. E. Bell, H. Raiffa, & A. Tversky (Eds.), *Decision making: Descriptive, normative, and prescriptive interactions* (pp. 398–421). New York: Cambridge University Press.

Fishburn, P. (1974). Lexicographic order, utilities, and decision rules: A survey. *Management Science*, *20*, 1442–1471.

Fiske, S. T., Kinder, D. R., Larter, W. M. (1983). The novice and the expert: Knowledge-based strategies in political cognition. *Journal of Experimental Social Psychology*, *19*, 381–400.

Flores, B. E., & White, E. M. (1989). Subjective versus objective combining of forecasts: An experiment. *Journal of Forecasting*, *8*, 331–342.

Fong, G. T., Krantz, D. H., & Nisbett, R. E. (1988). The effects of statistical training on thinking about everyday problems. In D. E. Bell, H. Raiffa, & A. Tversky (Eds.), *Decision making: Descriptive, normative, and presciptive interactions* (pp. 299–340). New York: Cambridge University Press.

Forbus, K. D. (1983). Qualitative reasoning about space and motion. In D. Gentner & A. L. Stevens (Eds.), *Mental models* (pp. 53–74). Hillsdale, NJ: Erlbaum.

Foss, J. E., Wright, W. R., & Coles, R. H. (1975). Testing the accuracy of field textures. *Proceedings of the American Soil Science Society*, *39*, 800–802.

Fox, J., Myers, C. D., Greaves, M. F., & Peagram, S. (1987). A systematic study of knowledge base refinement in the diagnosis of lukemia. In A. Kidd (Ed.), *Knowledge acquisition for expert systems: A practical handbook* (pp. 73–90). New York: Plenum.

Frey, P. W., & Adesman, P. (1976). Recall memory for visually presented chess positions. *Memory & Cognition*, *4*, 541–547.

Gaeth, G. J., & Shanteau, J. (1984). Reducing the influence of irrelevant information on experienced decision makers. *Organized Behavior and Human Performance*, *33*, 263–282.

Gagne, R., & Smith, E. (1962). A study of the effects of verbalization on problem solving. *Journal of Experimental Psychology*, *63*, 12–18.

Gardiner, P. C., & Edwards, W. (1975). Public values: Multiattribute utility measurement in social decision making. In M. Kaplan & S. Schwartz (Eds.), *Human judgment and decision processes*. New York: Academic Press.

Garland, L. H. (1960). The problem of observer error. *Bulletin of the New York Academy of Medicine*, *36*, 569–584.

Gentner, D. (1988). Expertise in typewriting. In M. T. H. Chi, R. Glasser, & M. J. Farr (Eds.), *The nature of expertise* (pp. 199–241). Hillsdale, NJ: Erlbaum.

Gentner, D., & Collins, A. (1981). Studies of inference from lack of knowledge. *Memory & Cognition*, *9*(4), 434–443.

Gentner, D., & Gentner, D. R. (1983). Flowing waters or teeming crowds: Mental models of electricity. In D. Gentner & A. Stevens (Eds.), *Mental models* (pp. 99–130). Hillsdale, NJ: Erlbaum.

Gentner, D., & Toupin, C. (1986). Systematicity and surface similarity in the development of analogy. *Cognitive Science*, *10*, 277–300.

Gick, M. L., & Holyoak, K. J. (1980). Analogical problem solving. *Cognitive Psychology*, *12*, 306–355.

Gick, M. L., & Holyoak, K. J. (1983). Schema induction and analogical transfer. *Cognitive Psychology*, *15*, 1–38.

Godden, D. R., & Baddeley, A. D. (1975). Context-dependent memory in two natural environments: On land and under water. *British Journal of Psychology*, *66*, 325–333.

Goldberg, L. R. (1959). The effectiveness of clinicians' judgments: The diagnosis of organic brain disease from the Bender-Gestalt Test. *Journal of Consulting Psychology*, *23*, 25–33.

Goldberg, L. R. (1968). Simple or complex processes? Some research on clinical judgments. *American Psychologist*, *23*, 483–496.

Goldberg, L. R. (1970). Man versus model of man: A rationale, plus some evidence for a method of improving clinical judgment. *Psychological Bulletin*, *73*, 422–432.

Goldin, S. E. (1978). Effects of orienting tasks on recognition of chess positions. *American Journal of Psychology*, *91*, 659–671.

Goldin, S. E. (1978). Memory for the ordinary: Typicality effects in chess memory. *Journal of Experimental Psychology: Human Learning and Memory*, *4*, 605–616.

Gorman, M. E. (1989). Error, falsification, and scientific inference: An experimental investigation. *Quarterly Journal of Experimental Psychology*, *41A*, 385–412.

Gottlieb, J. E., & Pauker, C. G. (1988). Whether or not to administer amphotericin to an immunosuppressed patient with hematologic malignancy and undiagnosed fever. In D. E. Bell, H. Raiffa, & A. Tversky (Eds.), *Decision making: Descriptive, normative, and prescriptive interactions* (pp. 569–587). New York: Cambridge University Press.

Greeno, J. G. (1983). Conceptual entities. In

D. Gentner & A. L. Stevens (Eds.), *Mental models* (pp. 227–252). Hillsdale, NJ: Erlbaum.

Groen, G. J. & Patel, V. L. (1985). Medical problem solving: Some questionable assumptions. *Medical Education*, *19*, 95–100.

Groen, G. J., & Patel, V. L. (1988). The relationship between comprehension and reasoning in medical expertise. In M. T. H. Chi, R. Glaser, & M. J. Farr (Eds.), *The nature of expertise* (pp. 287–310). Hillsdale, NJ: Erlbaum.

Halpern, D. F., & Irwin, F. W. (1973). Selection of hypotheses as affected by their preference values. *Journal of Experimental Psychology*, *101*, 105–108.

Hammond, K. R. (1966). Clinical inference in nursing: A psychologist's viewpoint, *Nursing Research*, *15*, 27–38.

Hammond, K. R., Frederick, E., Robillard, N., & Victor, D. (1989). Application of cognitive theory to the student-teacher dialogue. In D. A. Evans & V. L. Patel (Eds.), *Cognitive science in medicine: Biomedical modeling* (pp. 173–210). Cambridge, MA: Bradford Books/MIT Press.

Hammond, K. R., Kelly, K. J., Schneider, R. J., & Vancini, M. (1966). Clinical inference in nursing: Information units used. *Nursing Research*, *15*, 236–243.

Hartley, J., & Homa, D. (1981). Abstraction of stylistic concepts. *Journal of Experimental Psychology: Human Learning and Memory*, *7*, 33–41.

Harvey, N. (1990). Effects of difficulty on judgmental probability forecasting of control response efficacy. *Journal of Forecasting*, *9*, 373–388.

Hatano, G., & Osawa, K. (1983). Digit memory of grand experts in abacus-derived mental calculation. *Cognition*, *15*, 95–110.

Hawkins, D. (1983). An analysis of expert thinking. *International Journal of Man-Machine Studies*, *18*(1), 1–47.

Hayes, J. R. (1965). Problem typology and the solution process. *Journal of Verbal Learning and Verbal Behavior*, *4*, 371–379.

Hayes, J. R. (1989). Writing research: The analysis of a very complex task. In D. Klahr & K. Kotovsky (Eds.), *Complex information processing: The impact of Herbert A. Simon* (pp. 209–234). Hillsdale, NJ: Erlbaum.

Hayes, N. A., & Broadbent, D. E. (1988). Two modes of learning interactive tasks. *Cognition*, *28*, 249–276.

Hayes-Roth, B., & Hayes-Roth, F. (1977). Concept learning and the recognition and classification of exemplars. *Journal of Verbal Learning and Verbal Behavior*, *16*, 119–136.

Hegarty, M., Just M. A., & Morrison, I. R. (1988). Mental models of mechanical systems: Individual differences in qualitative and quantitative reasoning. *Cognitive Psychology*, *20*, 191–236.

Hink, R. F., & Woods, D. L. (1987). How humans process uncertain knowledge. *AI Magazine*, *8*, 41–53.

Hoffman, R. R., & Conway, J. (1990). Psychological factors in remote sensing: A review of some recent research. *Geocarto International*, *4*, 3–21.

Hogarth, R. (1974). Process tracing in clinical judgment. *Behavioral Science*, *19*, 298–313.

Holding, D. H., & Pfau, H. D. (1985). Thinking ahead in chess. *American Journal of Psychology*, *98*, 271–282.

Holding, D. H., & Reynolds, R. I. (1982). Recall or evaluation of chess positions as determinants of chess skill. *Memory & Cognition*, *10*, 237–242.

Holt, D. L. (1987). Auditors' base rates revisited. *Accounting, Organizations, and Society*, *12*, 571–578.

Holyoak, K. J., & Koh, K. (1987). Surface and structural similarity in analogical transfer. *Memory & Cognition*, *15*, 332–340.

Hughes, H. D. (1917). An interesting corn seed experiment. *The Iowa Argiculturalist*, *17*, 424–425.

Hutchins, E. (1983). Understanding micronesian navigation. In D. Gentner & A. L. Stevens (Eds.), *Mental models* (pp. 191–226). Hillsdale, NJ: Erlbaum.

Isenberg, D. J. (1988). How senior managers think. In D. E. Bell, H. Raiffa, & A. Tversky (Eds.), *Decision making: Descriptive, normative, and prescriptive interactions* (pp. 525–539). New York: Cambridge University Press.

Jacob, V. S., Gaultney, L. D., & Salvendy, G. (1986). Strategies and biases in human decision making and their implications for expert systems. *Behaviour and Information Technology*, *5*(2), 119–140.

Jacoby, L. L., & Dallas, M. (1981). On the relation between autobiographical memory and perceptual learning. *Journal of Experimental Psychology: General*, *110*, 306–346.

Jagacinski, R. J. (1978). Describing the human operator's internal model of a dynamic system. *Human Factors*, *20*, 425–483.

Jenkins, J. J. (1985). Acoustic information for objects, places, and events. In W. H. Warren & R. E. Shaw (Eds.), *Persistence and Change:*

Proceedings of the First International Conference on Event Perception (pp. 115–138). Hillsdale, NJ: Erlbaum.

Johnson, J. J. (1988). Expertise and decision making under uncertainty: Performance and process. In M. T. H. Chi, R. Glaser, & M. J. Farr (Eds.), *The nature of expertise* (pp. 209–228). Hillsdale, NJ: Erlbaum.

Johnson, P. E., Duran, A. S., Hassebrock, F., Moller, J., Prietula, M., Feltovich, P., & Swanson, D. B. (1981). Expertise and error in diagnostic reasoning. *Cognitive Science*, *5*, 235–283.

Johnson, P. E., Hassebrock, F., Duran, A. S., & Moller, J. H. (1982). Multimethod study of clinical judgment. *Organizational Behavior and Human Performance*, *30*, 201–230.

Johnson, P. E., Nachtsheim, C., & Zualkerman, I. (1987). Consultant expertise. *Expert systems*, *4*(3), 180–188.

Johnson, P. E., Zualkerman, I., & Garber, S. (1987). Specification of expertise. *International Journal of Man-Machine Studies*, *26*, 161–181.

Johnson-Laird, P. N., & Wason, P. C. (1970). A theoretical analysis of insight into a reasoning task. *Cognitive Psychology*, *1*, 134–148.

Josefson, I. (1988). The nurse as engineer: The theory of knowledge in research in the care sector. In B. Göranzon & I. Josefson (Eds.), *Knowledge, skill, and artificial intelligence* (pp. 9–30). London: Springer-Verlag.

Kahneman, D., & Tversky, A. (1982). On the study of statistical intuitions. *Cognition*, *11*, 123–141.

Kassirer, J. P., Kupiers, B. J., & Gorry, G. A. (1982). Towards a theory of clinical expertise. *The American Journal of Medicine*, *73*, 251–259.

Katz, R. W., & Murphy, A. H. (1990). Quality/value relationships for imperfect weather forecasts in a prototype multistage decision-making mode. *Journal of Forecasting*, *9*, 75–86.

Kessel, C. J., & Wickens, C. D. (1982). The transfer of failure-detection skills between monitoring and controlling dynamic systems. *Human Factors*, *24*, 49–60.

Kieras, D. E., & Bovair, S. (1984). The role of a mental model in learning to operate a device. *Cognitive Science*, *8*, 255–274.

Klein, G. A. (1986). Validity of analogical predictions. *Technological Forecasting and Social Change*, *30*, 139–148.

Klein, G. A., Calderwood, R., & Clinton-Cirocco, A. (1986). Rapid decision making on the fire ground. *Proceedings of the Human Factors Society 30th Annual Meeting* (pp. 576–580). Santa Monica, CA: Human Factors Society.

Klein, G. A., & Weitzenfeld, J. (1982). The use of analogues in comparability analysis. *Applied Ergonomics*, *13*, 99–104.

Kleinmuntz, B. (Ed.). (1968). The processing of clinical information by man and machine. In B. Kleinmuntz (Ed.), *Formal representations of human judgment* (pp. 149–186). New York: Wiley.

Klemp, G. O., & McClelland, D. G. (1986). What characterizes intelligent functioning among senior managers? In R. J. Sternberg & R. K. Wagner (Eds.), *Practical intelligence: Nature and origins of competence in the everyday world* (pp. 31–50). Cambridge: Cambridge University Press.

Kolers, P. A. (1975). Memorial consequences of automatized encoding. *Journal of Experimental Psychology: Human Learning and Memory*, *1*, 689–701.

Kolodner, J. L. (1983). Maintaining organization in a dynamic long-term memory. *Cognitive Science*, *7*, 243–280.

Kornell, J. (1987). Formal thought and narrative thought in knowledge acquisiton. *International Journal of Man-Machine Studies*, *26*, 203–212.

Kotovsky, K., & Fallside, D. (1989). Representation and transfer in problem solving. In D. Klahr & K. Kotovsky (Eds.) *Complex information processing: The impact of Herbert A. Simon* (pp. 69–108). Hillsdale, NJ: Erlbaum.

Krogstad, J. L., Ettenson, R. T., & Shanteau, J. (1984). Context and experience in auditors' materiality judgments. *Auditing*, *4*, 54–73.

Kulkanni, D., & Simon, H. A. (1988). The process of scientific discovery: The strategy of experimentation. *Cognitive Science*, *12*, 139–175.

Kundel, H. L., & LaFollette, P. S. (1972). Visual search patterns and experience with radiological images. *Radiology*, *103*, 523–528.

Kundel, H. L., & Nodine, C. F. (1975). Interpreting chest radiographs without visual search. *Radiology*, *116*, 527–532.

Kundel, H. L., & Nodine, C. F. (1978). Studies of eye movements and visual search in radiology. In J. W. Senders, D. F. Fisher, & R. A. Monty (Eds.), *Eye movements and the higher psychological functions* (pp. 317–328). Hillsdale, NJ: Erlbaum.

Kundel, H. L., & Wright, D. J. (1969). The influence of prior knowledge on visual search strategies during the viewing of chest radiographs. *Radiology*, *93*, 315–320.

Kupiers, B., & Kassirer, J. P. (1984). Causal reasoning in medicine: Analysis of a protocol. *Cognitive Science*, *8*, 363–385.

Kupiers, B., Moskowitz, A. J., & Kassirer, J. P. (1988). Critical decisions under uncertainty. *Cognitive Science*, *12*, 177–210.

Kusterer, K. C. (1978). *Know-how on the job: The important working knowledge of "unskilled" workers*. Boulder, CO: Westview.

LaFrance, M. (1990). The special structure of expertise. In K. L. McGraw & C. R. Westphal (Eds.), *Readings in knowledge acquisition* (pp. 55–70). New York: Ellis Horwood.

Lambert, D. M., & Newsome, S. L. (1989). Presenting abstract versus concrete information in expert systems: What is the impact on user performance? *International Journal of Man-Machine Studies*, *31*, 27–45.

Lane, D. M., & Robertson, L. (1979). The generality of the levels of processing hypothesis: An application to memory for chess positions. *Memory & Cognition*, *7*, 253–256.

Larkin, J. H. (1981). Enriching formal knowledge: A model for learning to solve textbook physics problems. In J. R. Anderson (Ed.), *Cognitive skills and their acquisition* (pp. 311–334). Hillsdale, NJ: Erlbaum.

Larkin, J. H. (1983). The role of problem representation in physics. In D. Gentner & A. L. Stevens (Eds.), *Mental models* (pp. 75–98). Hillsdale, NJ: Erlbaum.

Larkin, J. N., McDermott, J., Simon, D., & Simon, H. (1980). Expert and novice performance in solving physics problems. *Science*, *208*, 1335–1342.

Larkin, J. H., McDermott, J., Simon, D., & Simon, H. (1980). Models of competence in solving physics problems. *Cognitive Science*, *4*, 317–345.

Lawrence, J. A. (1988). Expertise on the bench: Modeling magistrates' judicial decisions. In M. T. H. Chi, R. Glaser, & J. M. Farr (Eds.), *The nature of expertise* (pp. 229–259). Hillsdale, NJ: Erlbaum.

Lee, T. J. (1987). Expert judgments of political riskiness. *Journal of Forecasting*, *6*, 51–66.

Lees, F. P. (1974). Research on the process operator. In E. Edwards & F. P. Lees (Eds.), *The human operator in process control* (pp. 386–455).

Lesgold, A. Rubinson, H., Feltovich, P. Glaser, R. Klopfer, D., & Wang, Y. (1988). Expertise in a complex skill: Diagnosing x-ray pictures. In M. T. H. Chi, R. Glaser, & M. J. Farr (Eds.), *The nature of expertise* (pp. 311–342). Hillsdale, NJ: Erlbaum.

Lewicki, P. (1985). Nonconscious biasing effects of single instances on subsequent judgments. *Journal of Personality and Social Psychology*, *48*, 563–574.

Lewicki, P. (1986). Processing information about covariance that cannot be articulated. *Journal of Experimental Psychology: Learning, Memory, and Cognition*, *12*, 135–146.

Lewicki, P., Czyzewska, M., & Hoffman, H. (1987). Unconscious acquisition of complex procedural knowledge. *Journal of Experimental Psychology: Learning, Memory, and Cognition*, *13*, 523–530.

Lewicki, P., Hill, T., & Bizot, E. (1988). Acquisition of procedural knowledge about a pattern of stimuli that cannot be articulated. *Cognitive Psychology*, *20*, 24–37.

Lewis, C. (1981). Skill in algebra. In J. R. Anderson (Ed.), *Cognitive skills and their acquisition* (pp. 85–110). Hillsdale, NJ: Erlbaum.

Lewis, M. W., & Anderson, J. R. (1985). Discrimination of operator schemata in problem solving: Learning from examples. *Cognitive Psychology*, *17*, 26–65.

Lichtenstein, S., & Fischoff, B. (1977). Do those who know more also know more about how much they know? The calibration of probability judgments. *Organizational Behavior and Human Performance*, *20*, 159–183.

Lichtenstein, S., & Fischoff, B. (1980). Training for calibration. *Organizational Behavior and Human Performance*, *26*, 149–171.

Lichtenstein, S., & Newman, J. R. (1967). Empirical scaling of common verbal phrases associated with numerical probabilities. *Psychonomic Science*, *9*, 563–564.

Lichtenstein, S., Slovic, P., Fischoff, B. Layman, M., & Combs, B. (1978). Judged frequency of lethal events. *Journal of Experimental Psychology: Human Learning and Memory*, *4*, 551–578.

Littman, D. C. (1987). Modeling human expertise in knowledge engineering: Some preliminary observations. *International Journal of Man-Machine Studies*, *26*, 81–92.

Luchins, A. S. (1942). Mechanization in problem solving. *Psychological Monographs*, *54*(6, Whole No. 248).

Lunn, J. H. (1948). Chicken sexing. *American Scientist*, *36*, 280–281.

Lusted, L. B. (1960). Logical analysis and Roentgen diagnosis. *Radiology*, *74*, 178–193.

Manis, M., Dovalina, I., Avis, N. E., & Cardoze, S. (1980). Base rates can affect individual predictions. *Journal of Personality and Social Psychology*, *38*, 231–248.

Marcel, A. T. (1983). Conscious and unconscious perception: An approach to the relations between phenomenal experience and perceptual processes. *Cognitive Psychology*, *15*, 238–300.

Mathews, B. P., & Diamandopoulos, A. (1990). Judgmental revision of sales forecasts: Effectiveness of forecast selection. *Journal of Forecasting*, *9*, 407–416.

McCloskey, M. (1983). Naive theories of motion. In D. Gentner & A. L. Stevens (Eds.), *Mental models* (pp. 299–324). Hillsdale, NJ: Erlbaum.

McDermott, J., & Larkin, J. H. (1978). Representing textbook physics problems. In *Proceedings of the Second National Conference of the Canadian Society for Computational Studies of Intelligence* (pp. 156–164). Toronto, Ontario: University of Toronto Press.

McNeill, B. J., Pauker, S. G., & Tversky, A. (1988). On the framing of medical decisions. In D. E. Bell, H. Raiffa, & A. Tversky (Eds.), *Decision making: Descriptive, normative, and prescriptive interactions* (pp. 562–568). New York: Cambridge University Press.

Means, M. L., & Voss, J. F. (1985). Star Wars: A developmental study of expert and novice knowledge structures. *Journal of Memory and Language*, *24*, 746–757.

Medin, D. L., Dewey, G. I., & Murphy, T. D. (1983). Relation between instance and category learning: Evidence that abstraction is not automatic. *Journal of Experimental Psychology: Learning, Memory and Cognition*, *9*, 607–625.

Meehl, P. E. (1954). *Clinical versus statistical prediction: A theoretical analysis and a review of evidence*. Minneapolis, MN: University of Minnesota Press.

Meyer, D. E., & Schvaneveldt, R. W. (1976). Meaning, memory structure, and mental processes. *Science*, *192*, 27–33.

Mitroff, I. I. (1974). *The subjective side of science: A philosophical inquiry into the psychology of the Apollo moon scientists*. Amsterdam: Elsevier.

Miyake, N., & Norman, D. A. (1979). To ask a question, one must know enough to know what is not known. *Journal of Verbal Learning and Verbal Behavior*, *18*, 357–364.

Morris, P. E. (1988). Expertise and everyday memory. In M. M. Gruneberg, P. E. Norris, & R. N. Sykes (Eds.), *Practical aspects of memory: Current research and issues* (Vol. 1, pp. 459–465). Chichester, England: John Wiley.

Mullin, T. (1989). Experts' estimation of uncertain quantities and its implications for knowledge acquisition. *IEEE Transactions on Systems, Man, and Cybernetics*, *19*, 616–625.

Murphy, G. L., & Wright, J. C. (1984). Changes in conceptual structure with experts: Differences between real-world experts and novices. *Journal of Experimental psychology: Learning, Memory, and Cognition*, *10*, 144–155.

Murphy, H. A., & Winkler, R. L. (1974). Credible interval temperature forecasting: Some experimental results. *Monthly Weather Review*, *102*, 784–794.

Myles-Worsley, M., Johnston, W., & Simons, M. (1988). The influence of expertise on X-ray image processing. *Journal of Experimental Psychology: Learning, Memory, and Cognition*, *4*, 553–557.

Mynatt, C. R., Doherty, M. E., & Tweney, R. D. (1978). Consequences of confirmation and disconfirmation in a simulated research environment. *Quarterly Journal of Experimental Psychology*, *30*, 395–406.

Norman, G. R., Brooks, L. R., & Allen, S. W. (1989). Recall by expert medical practitioners and novices as a record of processing attention. *Journal of Experimental Psychology: Learning, Memory, and Cognition*, *15*, 1166–1174.

Norman, G. R., Rosenthal, D., Brooks, L. R., Allen, S. W., & Muzzin, L. J. (1989). The development of expertise in dermatology. *Archives of Dermatology*, *125*, 1063–1068.

Northcraft, G. B., & Neale, M. A. (1987). Experts, amateurs, and real estate: An anchoring-and-adjustment perspective on property pricing decisions. *Organizational Behavior and Human Performance*, *39*, 84–97.

Novick, L. R. (1988). Analogical transfer, problem similarity, and expertise. *Journal of Experimental Psychology: Learning, Memory, and Cognition*, *14*, 510–520.

Novick, L. R. (1990). Representational transfer in problem solving. *Psychological Science*, *1*, 128–132.

O'Connor, M., & Lawrence, M. (1989). An examination of the accuracy of judgmental confidence intervals in time series forecasting. *Journal of Forecasting*, *8*, 141–156.

Olson, C. L. (1976). Some apparent violations of the representativeness heuristic in human judgment. *Journal of Experimental Psychology: Human Perception and Performance*, *2*, 599–608.

Oskamp, S. (1962). The relationship of clinical experience and training methods to several criteria of clinical prediction. *Psychological Monographs*, *76*(Whole No. 547).

Oskamp, S. (1965). Over-confidence in case study judgments. *Journal of Consulting Psychology*, *29*, 261–265.

Patel, V. L., Evans, D. A., & Groen, G. J. (1989).

Biomedical knowledge and clinical reasoning. In D. A. Evans & V. L. Patel (Eds.), *Cognitive science in medicine: Biomedical modeling* (pp. 53–112). Cambridge, MA: Bradford Books/MIT Press.

Patel, V. L., Evans, D. A., & Kaufman, D. R. (1989). A cognitive framework for doctor-patient interaction. In D. A. Evans & V. L. Patel (Eds.), *Cognitive science in medicine: Biomedical modeling* (pp. 257–312). Cambridge, MA: Bradford Books/MIT Press.

Patel, V. L., & Groen, G. J. (1986). Knowledge-based solution strategies in medical reasoning. *Cognitive Science*, *10*, 91–116.

Pauker, S. G., Gorry, G. A., Kassirer, J. P., & Schwartz, M. D. (1976). Towards the simulation of clinical cognition. *The American Journal of Medicine*, *60*, 981–998.

Payne, J. W. (1976). Task complexity and contingent processing in decision making: An information search and protocol analysis. *Organizational Behavior and Human Performance*, *16*, 366–387.

Payne, J. W., Braunstein, M. L., & Carroll, J. S. (1978). Exploring predecisional behavior: An alternative approach to decision making. *Organizational Behavior and Human Performance*, *22*, 17–44.

Pennington, N. (1987). Stimulus structures and mental representations in explicit comprehension of computer programs. *Cognitive Psychology*, *19*, 295–341.

Perby, M. L. (1988). Computerization and skill in local weather forecasting. In B. Göranzon & I. Josefson (Eds.), *Knowledge, skill, and artificial intelligence* (pp. 39–52). London: Springer-Verlag.

Perfetti, C. A., & Lesgold, A. M. (1979). Coding and comprehension in skilled reading. In L. B. Resnick & P. Weaver (Eds.), *Theory and practice of early reading* (Vol. 1, pp. 57–84). Hillsdale, NJ: Erlbaum.

Phelps, R. H., & Shanteau, J. (1978). Livestock judges: How much information can an expert use? *Organizational Behavior and Human Performance*, *21*, 209–219.

Porter, D. B. (1987). Classroom teaching, implicit learning, and the deleterious effects of inappropriate explanation. In *Proceedings of the Human Factors Society 31st Annual Meeting* (pp. 289–292). Santa Monica, CA: Human Factors Society.

Posner, M. I. (1988). Introduction: What is it to be an expert? In M. T. H. Chi, R. Glaser, & M. J. Farr (Eds.), *The nature of expertise* (pp. xxix–xxxvi). Hillsdale, NJ: Erlbaum.

Premkumar, G. (1989). A cognitive study of the decision-making process in a business context: Implications for design of expert systems. *International Journal of Man-Machine Studies*, *31*, 557–572.

Pritchard, P. (1988). Knowledge-based computer decision aids for general practice. In B. Göranzon & I. Josefson (Eds.), *Knowledge, skill, and artificial intelligence* (pp. 113–126). London: Springer-Verlag.

Qin, Y., & Simon, H. A. (1990). Laboratory replication of scientific discovery processes. *Cognitive Science*, *14*, 281–312.

Rabbit, P. M. A. (1978). Detection of errors by skilled typists. *Ergonomics*, *21*, 945–958.

Reber, A. S. (1976). Implicit learning of synthetic languages: The role of instructional set. *Journal of Experimental Psychology*: Human Learning and Memory, *2*, 88–94.

Reber, A. S. (1989). Implicit learning and tacit knowledge. *Journal of Experimental Psychology: General*, *118*, 219–235.

Reber, A. S., Allen, R., & Regan, S. (1985). Syntactical learning and judgment: Still unconscious and still abstract. *Journal of Experimental Psychology: General*, *114*, 17–24.

Reber, A. S., Cassin, S. M., Lewis, S., & Cantor, G. (1980). On the relation between implicit and explicit models of deriving a complex rule structure. *Journal of Experimental Psychology: Human Learning and Memory*, *6*, 492–502.

Reber, A. S., & Lewis, S. (1977). Implicit learning: An analysis of the form and structure of a body of tacit knowledge. *Cognition*, *5*, 333–361.

Reder, L. M. (1987). Beyond associations: Strategic components in memory retrieval. In D. S. Gorfein & R. R. Hoffman (Eds.), *Memory and learning: The Ebbinghaus Cetennial Conference* (pp. 203–220). Hillsdale, NJ: Erlbaum.

Reder, L. M., & Anderson, J. R. (1980). A partial resolution of the paradox of interference: The role of integrating knowledge. *Cognitive Psychology*, *12*, 447–472.

Reif, F., & Heller, J. I. (1982). Knowledge structures and problem solving in physics. *Educational Psychologist*, *17*, 102–127.

Reilly, B. A., & Doherty, M. E. (1989). A note on the assessment of self-insight in judgment research. *Organizational Behavior and Human Decision Processes*, *44*, 122–131.

Reitman, J. S. (1976). Skilled perception in GO: Deducing memory structures from inter-response times. *Cognitive Psychology*, *8*, 336–356.

Reitman, J. S., & Reuter, H. H. (1980). Organization revealed by recall orders and confirmed by pauses. *Cognitive Psychology*, *12*, 554–581.

Rennels, G. D., Shortliffe, E. H., Stockdale, F. E., & Miller, P. L. (1989, Spring). A computational model of reasoning from the clinical literature. *The AI Magazine*, 49–56.

Restle, F., & Davis, J. H. (1962). Success and speed of problem solving by individuals and groups. *Psychological Review*, *69*, 520–536.

Rhodes, G., Tan, S., Brake, S., & Taylor, K. (1989). Expertise and configural coding in face recognition. *British Journal of Psychology*, *80*, 313–331.

Roberts, G. H. (1968). The failure strategies of third-grade arithmetic pupils. *The Arithmetic Teacher*, *15*, 442–446.

Roe, A. (1951). A study of imagery in research scientists. *Journal of Personality*, *19*, 459–470.

Ross, B. H., & Kennedy, P. T. (1990). Generalizing from the use of earlier examples in problem solving. *Journal of Experimental Psychology: Learning, Memory, and Cognition*, *16*, 42–55.

Rumelhart, D., & Norman, D. A. (1981). Analogical processes in learning. In J. R. Anderson (Ed.), *Cognitive skills and their acquisition* (pp. 335–359). Hillsdale, NJ: Erlbaum.

Sanderson, P. M. (1989). Verbalizable knowledge and skilled task performance: Association, dissociation, and mental models. *Journal of Experimental Psychology: Learning, Memory, and Cognition*, *15*, 729–747.

Schartum, D. (1988). Delegation and decentralization: Computer systems as tools for instruction and improved service to clients. In B. Goranzon & I. Josefson (Eds.), *Knowledge, skill, and artificial intelligence* (pp. 59–168). London: Springer Verlag.

Schlatter, T. W. (1985). A day in the life of a mesoscale forecaster. *ESA Journal*, *9*, 235–256.

Schneider, W. (1985). Training high-performance skills: Fallacies and guidelines. *Human Factors*, *27*, 285–300.

Schneiderman, B., & Mayer, R. (1979). Syntactic/semantic interactions in programmer behavior: A model and experimental results. *International Journal of Computer and Information Sciences*, *8*, 219–239.

Schön, D. A. (1979). Generative metaphor: A perspective on problem-setting in social policy. In A. Ortony (Ed.), *Metaphor and thought* (pp. 254–283). Cambridge: Cambridge University Press.

Schön, D. A. (1983). *The reflective practitioner: How professionals think in action*. New York: Basic Books.

Schoenfeld, A., & Herrmann, D. (1982). Problem perception and knowledge structure in expert and novice mathematical problem solvers. *Journal of Experimental Psychology: Learning, Memory, Cognition*, *8*, 484–494.

Schvaneveldt, R. W., Durso, F. T., Goldsmith, T. E., Breen, T. J., Cooke, N. M., Tucker, R. G., & DeMaio, J. C. (1985). Measuring the structure of expertise. *International Journal of Man-Machine Studies*, *23*, 699–728.

Schvaneveldt, R. W., & Goldsmith, T. E. (1986). A model of air combat decisions. In E. Hollnagel, G. Mancini, & D. D. Woods (Eds.), *Intelligent decision support in process environments* (pp. 395–406). Berlin: Springer-Verlag.

Schwartz, S., & Griffin, T. (1986). *Medical thinking: The psychology of medical judgment and decision making*. New York: Springer-Verlag.

Scribner, S. (1984). Studying working intelligence. In B. Rogoff & S. Lave (Eds.), *Everyday cognition: Its development in social context* (pp. 9–40). Cambridge, MA: Harvard University Press.

Scribner, S. (1986). Thinking in action: Some characteristics of practical thought. In R. J. Sternberg & R. K. Wagner (Eds.), *Practical intelligence: Nature and origins of competence in the everyday world* (pp. 14–30). Cambridge: Cambridge University Press.

Seamon, J. G., Marsh, R. L., & Brody, N. (1984). Critical importance of exposure duration for affective discrimination of stimuli that are not recognized. *Journal of Experimental Psychology: Learning, Memory, and Cognition*, *10*, 465–469.

Sells, S. B. (1936). The atmosphere effect: An experimental study of reasoning. *Archives of Psychology* (Whole No. 200).

Shadbolt, N., & Burton, A. M. (1990). Knowledge elicitation techniques: Some experimental results. In K. L. McGraw & C. R. Westphal (Eds.), *Readings in knowledge acquisition* (pp. 21–33). New York: Ellis Horwood.

Shanteau, J. (1987). A psychologist looks at accounting research: Implications for doctoral programs. In American Accounting Association (Ed.), *A passion for excellence in accounting doctoral education* (pp. 1–12). Sarasota, FL: American Accounting Association.

Shanteau, J. (1989). Cognitive heuristics and biases in behavioral auditing: Review, comments, and observations. *Accounting Organizations and Society*, *14*, 165–177.

Shanteau, J., Grier, M., Johnson, J., & Berner, E. (1990). Teaching decision making skills to student nurses. In J. Barron & R. D. Brown (Eds.), *Teaching decision making to adolescents* (pp. 185–206). Hillsdale, NJ: Erlbaum.

Shanteau, J., & Phelps, R. H. (1977). Judgment and swine: Approaches in applied judgment analysis. In M. F. Kaplan & S. Schwartz (Eds.), *Human judgment and decision processes in applied settings* (pp. 255–272). New York: Academic Press.

Shaw, M. (1981). *On becoming a personal scientist.* London: Academic Press.

Shaw, M. (1982). A comparison of individuals and small groups in the rational solution of complex problems. *American Journal of Psychology, 44*, 491–504.

Shields, M. D., Solomon, I., & Waller, W. S. (1987). Effects of alternative sample space representations on the accuracy of auditors' uncertainty judgements. *Accounting Organizations and Society*, 375–385.

Siegler, R. S., & Shrager, J. (1984). Strategy choices in addition: How do children know what to do? In C. Sophian (Ed.), *Origins of cognitive skill.* Hillsdale, NJ: Erlbaum.

Silver, E. A. (1979). Student perceptions of relatedness among math verbal problems. *Journal for Research in Mathematics Education, 10*, 195–210.

Silver, E. A. (1981). Recall of math problems information: Solving related problems. *Journal for Research in Mathematics Education, 12*, 54–65.

Simon, D. P., & Simon, H. A. (1978). Individual differences in solving physics problems. In R. Siegler (Ed.), *Children's thinking: What develops?* (pp. 325–348). Hillsdale, NJ: Erlbaum.

Simon, H. A., & Gilmartin, K. (1973). A simulation of memory for chess positions. *Cognitive Psychology, 5*, 29–46.

Sisson, J. C., Schoomaker, E. B., & Ross, J. C. (1976). Clinical decision analysis: The hazard of using additional data. *Journal of the American Medical Association, 13*, 1259–1263.

Sleeman, D. H. (1984). An attempt to understand students' understanding of basic algebra. *Cognitive Science, 8*, 387–412.

Sleeman, D. H., & Smith, M. J. (1981). Modeling student problem solving. *Artificial Intelligence, 16*, 171–187.

Sloboda, J. A. (1976). Visual perception of musical notation: Registering pitch symbols in memory. *Quarterly Journal of Experimental Psychology, 8*, 1–16.

Slovic, P. (1969). Analyzing the expert judge: A description of stockbrokers' decision processes. *Journal of Applied Psychology, 53*, 255–263.

Slovic, P., Fischoff, B., & Lichtenstein, S. (1982). Response mode, framing, and information-processing effects in risk assessment. In R. Hogarth (Ed.), *New directions for methodology of social and behavioral science: Question framing and response consistency* (pp. 21–36). San Francisco: Jossey-Bass.

Slovic, P., Fischoff, B., & Lichtenstein, S. (1988). Response model, framing, and information-processing effects in risk assessment. In D. E. Bell, H. Raiffa, & A. Tversky (Eds.), *Decision making: Descriptive, normative, and prescriptive interactions* (pp. 152–166). New York: Cambridge University Press.

Slovic, P., & Lichtenstein, S. (1971). Comparison of Bayesian and regression approaches to the study of information processing in judgment. *Organizational Behavior and Human Performance, 6*, 649–744.

Smith, E. E., Adams, N., & Schor, D. (1978). Fact retrieval and the paradox of interference. *Cognitive Psychology, 10*, 438–464.

Spilich, G. J., Vesonder, G. T., Chiesi, H. L., & Voss, J. F. (1979). Text-processing of domain-related information for individuals with high and low domain knowledge. *Journal of Verbal Learning and Verbal Behavior, 18*, 275–290.

Spiro, R. J., Feltovich, P. J., Coulson, R. L., & Anderson, D. K. (1989). Multiple analogies for complex concepts: Antidotes for analogy-induced misconception in advanced knowledge acquisition. In S. Vosniadou & A. Ortony (Eds.), *Similarity and analogical reasoning* (pp. 498–531). Cambridge: Cambridge University Press.

Stadler, M. A. (1989). On learning complex procedural knowledge. *Journal of Experimental psychology: Learning, Memory, and Cognition, 15*, 1061–1069.

Staszewski, J. J. (1988). Skilled memory and expert mental calculation. In M. T. H. Chi, R. Glaser, & M. J. Farr (Eds.), *The nature of expertise* (pp. 71–128). Hillsdale, NJ: Erlbaum.

Ste-Marie, D. M., & Lee, T. D. (1991). Prior processing effects in gymnastic judging. *Journal of Experimental Psychology: Learning, Memory, and Cognition, 17*, 126–136.

Streitz, N. A. (1983). The importance of knowledge representation in problem solving: An example from text comprehension and problem solving. In G. Luer (Ed.), *Bericht über den 33: Kongress der Deutschen Gesellschaft für Psychologie* (pp. 403–407). Gottingen, West Germany: Hogrefe.

Strizenec, M. (1974). On research into the process operator's thinking and decision-making. In E. Edwards & F. P. Lees (Eds.), *The human operator in process control* (pp. 165–177).

Svenson, O. (1979). Process descriptions of decision making. *Organizational Behavior and Human Performance, 23*, 86–112.

Sweller, J., Mawer, R. F., & Ward, M. R. (1983). Development of expertise in mathematical problem solving. *Journal of Experimental Psychology: General, 112*, 638–661.

Thaler, R. (1980). Toward a positive theory of consumer choice. *Journal of Economic Behavior and Organization, 1*, 39–60.

Thomas, J. L. (1974). An analysis of behavior in the hobbits-orcs problem. *Cognitive Psychology, 6*, 257–269.

Tolcott, M. A., Marvin, F. F., & Lehner, P. E. (1989). Expert decision making in evolving situations. *IEEE Transactions on Systems, Man, and Cybernetics, 19*, 606–615.

Troutman, C. M., & Shanteau, J. (1977). Inferences based on nondiagnostic information. *Organizational Behavior and Human Performance, 19*, 43–55.

Turner, D. S. (1990). The role of judgment in macroeconomic forecasting. *Journal of Forecasting, 9*, 315–346.

Tversky, A., & Kahneman, D. (1973). Availability: A heuristic for judging frequency and probability. *Cognitive Psychology, 5*, 207–232.

Tversky, A., & Kahneman, D. (1974). Judgment and uncertainty: Heuristics and biases. *Science, 185*, 1124–1131.

Umbers, I. G., & King, P. J. (1981). An analysis of human decision-making in cement kiln control and the implications for automation. In E. H. Mamdani & B. R. Gaines (Eds.), *Fuzzy reasoning and its applications* (pp. 369–381). London: Academic Press.

VanLehn, K. (1983). Human skill acquisition: Theory, model, and psychological validation. In *Proceedings of AAAI–83* (pp. 420–423). Los Altos, CA: Kaufmann.

VanLehn, K. (1989). *Cognitive procedures: The acquisition and mental representation of basic Mathematical skills*. Cambridge, MA: MIT Press.

Veldehuyzen, W., & Stassen, H. G. (1977). The internal model concept: An application modeling human control of large ships. *Human Factors, 19*, 367–380.

Voss, J. F., Greene, J. R., Post, T. A., & Penner, B. C. (1983). Problem-solving skill in the social sciences. In G. H. Bower (Ed.), *The psychology of learning and motivation: Advances in research and theory* (Vol. 17, pp. 165–213). New York: Academic Press.

Voss, J. F., & Post, T. A. (1988). On the solving of ill structured problems. In M. T. H. Chi, R. Glaser, & M. J. Farr (Eds.), *The nature of expertise* (pp. 261–285). Hillsdale, NJ: Erlbaum.

Voss, J. F., Tyler, S. & Vengo, L. (1983). Individual differences in social science problem solving. In R. F. Dillon & R. R. Schmeck (Eds.), *Individual differences in cognitive processes* (Vol. 1, pp. 205–232). New York: Academic Press.

Wagenaar, W. A., & Keren, G. B. (1986). Does the expert know? The reliability of predictions and confidence ratings of experts. In E. Hollnagel, G. Mancini, & D. D. Woods (Eds.), *Intelligent decision support in process environments* (pp. 87–103). Berlin: Springer-Verlag.

Waldron, M. B., & Waldron, K. L. (1988). A time sequence study of a complex mechanical system design. *Design Studies, 9*, 95–106.

Wallace, H. A. (1923). What is in the corn judge's mind? *Journal of the American Society of Agronomy, 15*, 300–324.

Wallsten, T. S. (1981). Physician and medical student bias in evaluating diagnostic information. *Medical Decision Making, 1*, 145–164.

Wason, P. C. (1968). Reasoning about a rule. *Quarterly Journal of Experimental Psychology, 20*, 273–281.

Waterman, D., & Peterson, M. (1981). *Models of legal decision making* (Rand Report R–2717–ICJ). Santa Monica, CA: Rand Corporation.

Weiser, M., & Carey, S. (1983). When heat and temperature were one. In D. Gentner & A. L. Stevens (Eds.), *Mental models* (pp. 267–297). Hillsdale, NJ: Erlbaum.

Weitzenfeld, J. (1984). Valid reasoning by analogy: Technical reasoning. *Philosophy of Science, 51*, 137–149.

Wertheimer, M. (1945). *Productive thinking*. New York: Harper & Row.

West, M., & Harrison, J. (1989). Subjective intervention in formal models. *Journal of Forecasting, 8*, 33–54.

Whittlesea, B. W. A., & Brooks, L. R. (1988). Critical influence of particular experiences in the perception of letters, words, and phrases. *Memory & Cognition, 16*, 387–399.

Wickelgren, W. (1974). *How to solve problems*. San Francisco: Freeman.

Wickens, C. D. (1984). Processing resources in

attention. In R. Parasuraman & D. R. Davies (Eds). *Varieties of attention* (pp. 63–102). New York: Academic Press.

Wickens, C. D., & Kessel, C. (1980). Processing resource demands of failure detection in dynamic systems. *Journal of Experimental Psychology: Human Perception and Performance*, *6*, 564–577.

Wickens, C. D., & Kessel, C. (1981). Failure detection in dynamic systems. In J. Rasmussen & W. B. Rouse (Eds.), *Human detection and diagnosis of system failures* (pp. 155–169). New York: Plenum.

Wilder, L., & Harvey, D. (1971). Overt and covert verbalization in problem solving. *Speech Monographs*, *38*, 171–176.

Williams, M. D., & Hollan, J. D. (1981). The process or retrieval from very long-term memory. *Cognitive Science*, *5*, 87–119.

Williams, M. D., Hollan, J. D., & Stevens, A. L. (1983). Human reasoning about a simple physical system. In D. Gentner & A. L. Stevens (Eds.), *Mental models* (pp. 131–154). Hillsdale, NJ: Erlbaum.

Willingham, D. B., Nissen, M. J., & Bullemer, P. (1989). On the development of procedural knowledge. *Journal of Experimental Psychology: Learning, Memory, and Cognition*, *15*, 1047–1060.

Wiser, M., & Carey, S. (1983). When heat and temperature were one. In D. Gentner & A. L. Stevens (Eds.), *Mental models* (pp. 267–298). Hillsdale, NJ: Erlbaum.

Wolfe, C., & Flores, B. (1990). Judgmental adjustment of earnings forecasts. *Journal of Forecasting*, *9*, 389–406.

Woods, D. D., & Hollnagel, E. (1987). Mapping cognitive demands in complex problem solving worlds. *International Journal of Man-Machine Studies*, *26*, 257–275.

Zakay, D., & Wooler, S. (1984). Time pressure, training, and decision effectiveness. *Ergonomics*, *27*, 273–284.

Zhu, X., & Simon, H. A. (1987). Learning mathematics from examples and by doing. *Cognition and Instruction*, *4*, 137–166.

Appendix C
Bibliography: Knowledge Elicitation

Adams-Webber, J. R. (1984). Repertory grid technique. In R. Corsini (Ed.), *Encyclopedia of psychology* (p. 225). New York: Wiley Interscience.

Adelman, L. (1989). Measurement issues in knowledge engineering. *IEEE Transactions on Systems, Man, and Cybernetics*, *19*, 448–461.

Agarwal, R., & Tanniru, M. R. (1990). Knowledge acquisition using structured interviewing: An empirical investigation. *Journal of Management Information Systems*, *7*, 123–140.

Alexander, J. H., Freiling, M. J., Shulman, S. J., Rehfuss, S., & Messick, S. L. (1987). Ontological analysis: An ongoing experiment. *International Journal of Man-Machine Studies*, *26*, 473–485.

Arinze, B. (1989). A natural language front-end for knowledge acquisition. In C. R. Westphal & K. L. McGraw (Eds.), *Special Issue of ACM SIGART Newsletter on Knowledge Acquisition*, No. 108 (pp. 106–114). New York: Special Interest Group on Artificial Intelligence, Association for Computing Machinery.

Atwater, E. (1981). *I hear you: How to use listening skills for profit*. Englewood Cliffs, NJ: Prentice Hall.

Bailey, W. A., & Kay, D. J. (1987). Structural analysis of verbal data. In J. M. Carroll & P. P. Tanner (Eds.), *Human factors in computing systems and graphics interface* (pp. 297–301). New York: Association for Computing Machinery.

Bainbridge, L. (1974). Analysis of verbal protocols from a process control task. In E. Edwards & F. P. Lees (Eds.), *The human operator in process control* (pp. 146–158). New York: Halsted Press.

Bainbridge, L. (1979). Verbal reports as evidence of the process operator's knowledge. *International Journal of Man-Machine Studies*, *11*, 411–436.

Bainbridge, L. (1986). Asking questions and accessing knowledge. *Future Computing Systems*, *1*, 143–149.

Bannister, D. (1962). Personal construct theory: A summary and experimental paradigm. *Acta Psychologica*, *20*, 104–120.

Belkin, N. J., Brooks, H. M., & Daniels, P. J. (1987). Knowledge elicitation using discourse analysis. *International Journal of Man-Machine Studies*, *27*, 127–144.

Benjafield, J. (1969). Evidence that "thinking aloud" constitutes an externalization of inner speech. *Psychonomic Science*, *15*, 83–84.

Bereiter, C., & Bird, M. (1985). Use of thinking aloud in identification and teaching of reading comprehension strategies. *Cognition and Instruction*, *2*, 131–156.

Berry, D. C. (1987). The problem of implicit knowledge. *Expert Systems*, *4*, 144–151.

Berry, D. C., & Broadbent, D. E. (1984). On the relationship between task performance and associated verbalizable knowledge. *Quarterly Journal of Experimental Psychology*, *36A*, 209–231.

Blosser, P. (1973). *Handbook of effective questioning techniques*. Worthington, OH: Educational Association, Inc.

Bradley, J. V., & Harbison-Briggs, K. (1989, April). The symptom-component approach to knowledge acquisition. In C. R. Westphal & K. L. McGraw (Eds.), *Special Issue of ACM SIGART Newsletter on Knowledge Acquisition* (pp. 70–76). New York: Special Interest Group on Artificial Intelligence, Association for Computing Machinery.

Breuker, B., & Weilenga, B. (1987). Use of models in the interpretation of verbal data. In A. Kidd (Ed.), *Knowledge acquisition for expert systems* (pp. 17–44). New York: Plenum.

Brown, B. (1989, April). The taming of an expert: An anecdotal report. In C. R. Westphal & K. L. McGraw (Eds.), *Special Issue of ACM SIGART Newsletter on Knowledge Acquisition* (No. 108, pp. 133–135). New York: Special Interest Group on Artificial Intelligence, Association for Computing Machinery.

Brulé, J. F., & Blount, A. (1989). *Knowledge acquisition*. New York: McGraw-Hill.

Burton, A. M., Shadbolt, N. R., Hedgecock, A. P., & Rugg, G. (1987). A formal evaluation of knowledge elicitation techniques for expert systems: Domain 1. *Proceedings of the First European Workshop on Knowledge Acquisition for Knowledge-Based Systems* (D3, pp. 1–21). Reading, UK: Reading University.

Burton, A. M., Shadbolt, N. R., Hedgecock, A. P., & Rugg, G. (1987). A formal evaluation of knowledge elicitation techniques for expert systems. In D. S. Morabee (Ed.), *Research and developments in expert systems* (Vol. 4, pp. 35–46). Cambridge: Cambridge University Press.

Burton, A. M., Shadbolt, N. R., Rugg, G., & Hedgecock, A. P. (1988). Knowledge elicitation techniques in classification domains. In Y. Kodratoff (Ed.), *ECAI–88: Proceedings of the Eighth European Conference on AI* (pp. 85–90). London: Pitman.

Burton, A. M., Shadbolt, N. R., Rugg, G., & Hedgecock, A. P. (1990). The efficacy of knowledge elicitation techniques: A comparison across domains and levels of expertise. *Journal of Knowledge Acquisition*, *2*, 167–178.

Bylander, T., & Chandrasekaran, B. (1987). Generic tasks for knowledge-based reasoning: The "right" level of abstraction for knowledge acquisition. *International Journal of Man-Machine Studies*, *26*, 231–243.

Carney, T. F. (1972). *Content analysis*. Manitoba: University of Manitoba Press.

Chung, H. M. (1989). Empirical analysis of inductive knowledge acquisition methods. In C. R. Westphal & K. L. McGraw (Eds.), *Special Issue of ACM SIGART Newsletter on Knowledge Acquisition* (No. 108, pp. 156–159). New York: Special Interest Group on Artificial Intelligence, Association for Computing Machinery.

Clancey, W. J. (1988). The knowledge engineer a student: Metacognitive bases for asking good questions. In H. Mandl & A. Lesgold (Eds.), *Learning issues for intelligent tutoring systems* (pp. 80–113). New York: Springer-Verlag.

Claparède, E. (1902). *L'association des idées*. Paris: Doin.

Cleaves, D. A. (1987). Cognitive biases and corrective techniques: Proposals for improving knowledge elicitation Procedures for knowledge-based systems. *International Journal of Man-Machine Systems*, *27*, 155–166.

Cochran, E. L., Bloom, C. P., & Bullemer, P. T. (1990). Increasing the end-user acceptance of expert systems by using multiple experts: Case studies in knowledge acquisition. In K. L. McGraw & C. R. Westphal (Eds.), *Readings in knowledge acquisition* (pp. 73–89). New York: Ellis Horwood.

Conrad, C. (1985). *Strategic organizational communication: Cultures, situations, and adaptation*. New York: Holt, Rinehart, and Winston.

Cooke, N. M. (1989). The elicitation of domain-related ideas: Stage one of the knowledge acquisition process. In C. Ellis (Ed.), *Expertise and explanation: The knowledge language interface* (pp. 58–75). Chichester, England: Ellis Horwood.

Cooke, N. M. (1990). Link interpretation: Using Pathfinder as a knowledge elicitation tool. In R. Schvaneveldt (Ed.), *Pathfinder associative networks* (pp. 227–240). Norwood, NJ: Ablex.

Cooke, N. M., & McDonald, J. E. (1986). A formal methodology for acquiring and representing expert knowledge. *Proceedings of the IEEE*, *74*(10), 1422–1430.

Cooke, N. M., & McDonald, J. E. (1987). The application of psychological scaling techniques to knowledge elicitation for knowledge-based systems. *International Journal of Man-Machine Studies*, *26*, 533–550.

Cragan, J., & Wright, D. (1980). *Communication in small group discussions: A case study approach*. New York: West Publishing Company.

Crandall, B. W. (1989). A comparative study of think out loud and critical decision knowledge elicitation methods. In C. R. Westphal & K. L. McGraw (Eds.), *Special Issue of ACM SIGART Newsletter on Knowledge Acquisition* (No. 108, pp. 144–146). New York: Special Interest Group on Artificial Intelligence, Association for Computing Machinery.

Cullen, J., & Bryman, A. (1988). The knowledge acquisition bottleneck: Time for a reassessment? *Expert Systems*, *5*, 216–255.

Cunningham, J. (1985). Comprehension by model-building as a basis for an expert system. In M. Merry (Ed.), *Expert systems-85* (pp. 259–272). Cambridge: Cambridge University Press.

Davies, M., & Hakiel, S. (1988). Knowledge harvesting: A practical guide to interviewing. *Expert Systems*, *5*, 42–50.

Diaper, D. (Ed.). (1989). *Knowledge elicitation: Principles, techniques, and applications*. Chichester, England: Ellis Horwood.

Downs, C. G. (1988). Representing the structure of jobs in job analysis. *International Journal of Man-Machine Studies*, *28*, 363–390.

DuPreez, P. D., & Ward, D. G. (1970). Personal constructs of modern and traditional Zhosa. *Journal of Social Psychology*, *82*, 149–160.

Easterby-Smith, M. (1981). The design, analysis, and interpretation of repertory grids. In M. L. G. Shaw (Ed.), *Recent advances in personal construct psychology* (pp. 9–30). London: Academic Press.

Eden, C., & Jones, S. (1984). Using repertory grids for problem construction. *Journal of the Operational Research Society*, *35*(9), 779–790.

Eliot, L. B. (1987). Investigating the nature of expertise: Analogical thinking, expert systems, and thinkback. *Expert Systems*, *4*(3), 190–195.

Ericsson, K. A., & Simon, H. A. (1980). Verbal reports as data. *Psychological Review*, *87*, 215–251.

Ericsson, K. A., & Simon, H. A. (1984). *Protocol analysis: Verbal reports as data*. Cambridge, MA: The MIT Press.

Evans, J. St. B. T. (1988). The knowledge elicitation problem: A psychological perspective. *Behavior and Information Technology*, *7*, 111–130.

Fischer, P. M., & Mandl, H. (1988). Improvement of the acquisition of knowledge by informing feedback. In H. Mandl & A. Lesgold (Eds.), *Learning issues for intelligent tutoring systems*. New York: Springer-Verlag.

Fischoff, B. (1989). Eliciting knowledge for analytical representation. *IEEE Transactions on Systems, Man, and Cybernetics*, *19*, 448–461.

Flanagan, J. C. (1954). The critical incident technique. *Psychological Bulletin*, *51*, 327–358.

Forsythe, D. E., & Buchanan, B. C. (1989). Knowledge acquisition for expert systems: Some pitfalls and suggestions. *IEEE Transactions on Systems, Man, & Cybernetics*, *19*, 435–442.

Fox, J., Myers, C. D., Greaves, M. F., & Pegram, S. (1985). Knowledge acquisition for expert systems: Experience in leukemia diagnosis. *Methods of Information in Medicine*, *24*, 65–72.

Fox, J., Myers, C. D., Greaves, M. F., & Pegram, S. (1987). A systematic study of knowledge base refinement in the diagnosis of leukemia. In A. L. Kidd (Ed.), *Knowledge acquisition for expert systems: A practical handbook* (pp. 73–90). New York: Plenum.

Frank, A. (1982). *Communicating on the job*. Glenview, IL: Scott, Foresman.

Fransella, F., & Bannister, D. (1977). *A manual for repertory grid technique*. London: Academic Press.

Frieling, M., Alexander, J., Messick, S., Rehfuss, S., & Schulman, S. (1985, Fall). Starting a knowledge engineering project: Step-by-step approach. *The AI Magazine*, 150–164.

Gaines, B. R., & Sharp, M. (1987, September). A knowledge acquisition extension to notecards. *Proceedings of the First European Workshop on Knowledge Acquisition for Knowledge-Based Systems*. Reading, UK: Reading University.

Gale, W. A. (1987). Knowledge based knowledge: Acquisition for a statistical consulting system. *International Journal of Man-Machine Studies*, *26*, 55–64.

Gammack, J. G. (1987). Different techniques and different aspects on declarative knowledge. In A. Kidd (Ed.), *Knowledge acquisition for expert systems* (pp. 137–164). New York: Plenum.

Gammack, J. G., & Young, R. (1985). Psychological techniques for eliciting expert knowledge. In M. Bramer (Ed.), *Research and development in expert systems* (pp. 105–112). London: Cambridge University Press.

Garg-Janardan, C., & Salvendy, G. (1987). A conceptual framework for knowledge elicitation. *International Journal of Man-Machine Studies*, *26*, 521–532.

Gordon, S. E., & Gill, R. T. (1989). Question probes: A structured method for eliciting declarative knowledge. *AI Applications in Natural Resource Management*, *3*, 13–20.

Graesser, A., & Black, J. (Eds.). (1985). *The psychology of questions*. Hillsdale, NJ: Erlbaum.

Graesser, A., & Clark, L. F. (1985). *Structures and procedures for implicit knowledge*. Norwood, NJ: Ablex.

Graesser, A., Lang, K., & Horgan, D. (1988). A taxonomy for question generation. *Questioning Exchange*, *2*, 3–15.

Gregg, L. W. (1976). Methods and models for task analysis in instructional design. In D. Klahr (Ed.), *Cognition and instruction* (pp. 109–115). Hillsdale, NJ: Erlbaum.

Gruber, T. R. (1988). Acquiring strategic knowledge from experts. *International Journal of Man-Machine Studies*, *26*(2), 579–597.

Gruber, T. R. (1989). *The acquisition of strategic knowledge*. New York: Academic Press.

Hart, A. (1985). Knowledge elicitation: Issues and

methods. *Computer-Aided Design, 17*(9), 455–462.

Hart, A. (1986). *Knowledge acquisition for expert systems*. New York: McGraw-Hill.

Hart, A. (1987). Role of induction in knowledge elicitation. In A. Kidd (Ed.), *Knowledge acquisition for expert systems* (pp. 165–189). New York: Plenum.

Hart, A. (1988). Knowledge acquisition for expert systems. In B. Göranzon & I. Josefson (Eds.), *Knowledge, skill, and artificial intelligence* (pp. 103–112). London: Springer-Verlag.

Hart, A. (1989). *Knowledge acquisition for expert systems*. New York: McGraw-Hill.

Hawkins, D. (1983). An analysis of expert thinking. *International Journal of Man-Machine Studies, 18*(1), 1–47.

Hays, T. (1976). An empirical method for the identification of covert categories in ethnobiology. *American Ethnologist, 3*, 489–507.

Hilton, J. (1988). Skill, education, and social value: Some thoughts on the metonymy of skill and skill transfer. In B. Göranzon & I. Josefson (Eds.), *Knowledge, skill, and artificial intelligence* (pp. 93–102). London: Springer-Verlag.

Hoffman, R. R. (1987). The problem of extracting the knowledge of experts from the perspective of experimental psychology. *AI Magazine, 8*(2), 53–64.

Hoffman, R. R. (1989, April). A review of methods for extracting and characterizing the knowledge of experts. In C. R. Westphal & K. L. McGraw (Eds.), *Special Issue of ACM SIGART Newsletter on Knowledge Acquisition* (pp. 19–27). New York: Special Interest Group on Artificial Intelligence, Association for Computing Machinery.

Hoffman, R. R. (1990). A survey of methods for eliciting the knowledge of experts. In K. L. McGraw & C. R. Westphal (Eds.), *Readings in knowledge acquisition* (pp. 7–20). New York: Ellis Horwood.

Hyman, R. T. (1979). *Strategic questioning*. Englewood Cliffs, NJ: Prentice-Hall.

Jacoby, J. (1980). *The handbook of questionnaire construction*. Boston, MA: Ballinger.

Jagacinski, R. J., & Miller, R. A. (1978). Describing the human operator's internal model of a dynamic system. *Human Factors, 20*, 425–433.

Jardine, N., & Sibson, R. (1968). The construction of hierarchic and non-hierarchic classifications. *Computing Journal, 11*, 117–184.

Johnson, L., & Johnson, N. E. (1987). Knowledge elicitation involving teachback interviewing. In A. Kidd (Ed.), *Knowledge acquisition for expert systems* (pp. 91–108). New York: Plenum.

Johnson, N. E. (1987, September). Mediating representations in knowledge elicitation. *Proceedings of the First European Workshop on Knowledge Acquisition for Knowledge-Based Systems*. Reading, UK: Reading University.

Johnson, N. E. (1989). Mediating representations in knowledge elicitation. In D. Diaper (Ed.), *Knowledge elicitation: Principles, techniques, and applications* (pp. 179–194). Chichester, England: Ellis Horwood.

Johnson, P., Zualkerman, I., & Garber, S. (1987). Specification of expertise: Knowledge acquisition for expert systems. *International Journal of Man-Machine Studies, 26*(2), 161–181.

Kahn, G., & Cannell, C. (1982). *The dynamics of interviewing*. New York: Wiley.

Kato, T. (1986). What question asking protocols can say about the user interface. *International Journal of Man-Machine Studies, 25*, 659–674.

Kellogg, C., Gargan, R. A., Jr., Mark, W., McGuire, J. G., Ponecorvo, M., Schlossberg, J. L., Sullivan, J. W., Genesereth, M. R., & Singh, N. (1989). The acquisition, verification, and explanation of design knowledge. In C. R. Westphal & K. L. McGraw (Eds.), *Special Issue of ACM SIGART Newsletter on Knowledge Acquisition* (No. 108, pp. 163–165). New York: Special Interest Group on Artificial Intelligence, Association for Computing Machinery.

Kelly, G. A. (1955). *The psychology of personal constructs*. New York: Norton.

Kidd, A. L. (1985). What do users ask? Some thoughts on diagnostic advice. In M. Merry (Ed.), *Expert systems-85* (pp. 9–19). Cambridge: Cambridge University Press.

Kidd, A. L. (1987). Knowledge acquisition: An introductory framework. In A. L. Kidd (Ed.), *Knowledge acquisition for expert systems* (pp. 1–16). New York: Plenum.

Kidd, A. L. (1987). *Knowledge acquisition for expert systems: A practical handbook*. New York: Plenum.

Kidd, A. L., & Welbank, M. (1984). Knowledge acquisition. In J. Fox (Ed.), *Infotech state of the art report on expert systems*. London: Pergamon.

Killin, J. L., & Hickman, F. R. (1986). The role of phenomenological techniques of knowledge elicitation in complex domains. In M. Merry (Ed.), *Expert systems–86*. Cambridge: Cambridge University Press.

Klein, G. A., Calderwood, R., & Clinton-Cirocco, A. (1986). Rapid decision making on the fire ground. *Proceedings of the Human Factors Society 30th Annual Meeting* (pp. 576–580). Santa Monica, CA: Human Factors Society.

Klein, G. A., Calderwood, R., & MacGregor, D. (1989). Critical decision method for eliciting knowledge. *IEEE Transactions on Systems, Man, and Cybernetics*, *19*, 462–472.

Klein, G. A., & Weitzenfeld, J. (1982). The use of analogues in comparability analysis. *Applied Ergonomics*, *13*, 99–104.

Knapp, M. (1978). *Nonverbal communication in human interaction*. New York: Holt, Rinehart, and Winston.

Kornecki, A. J. (1989). Operational knowledge acquisition problems for air traffic expert controller. In C. R. Westphal & K. L. McGraw (Eds.), *Special Issue of ACM SIGART Newsletter on Knowledge Acquisition* (No. 108, pp. 84–92). New York: Special Interest Group on Artificial intelligence, Association for Computing Machinery.

Kupiers, B. (1987). New reasoning methods for artificial intelligence in medicine. *International Journal of Man-Machine Studies*, *26*, 707–718.

Kupiers, B., & Kassirer, J. P. (1984). Causal reasoning in medicine: Analysis of a protocol. *Cognitive Science*, *8*, 363–385.

Kupiers, B., & Kassirer, J. P. (1987). Knowledge acquisition by analysis of verbatim protocols. In A. L. Kidd (Ed.), *Knowledge acquisition for expert systems* (pp. 45–71). New York: Plenum.

LaFrance, M. (1987). The knowledge acquisition grid: A method for training knowledge engineers. *International Journal of Man-Machine Studies*, *26*, 245–255.

LaFrance, M. (1989, April). The quality of expertise: Implications of expert-novice differences for knowledge acquisition. In C. R. Westphal & K. L. McGraw (Eds.), *Special Issue of ACM SIGART Newsletter on Knowledge Acquisition* (pp. 6–14). New York: Special Interest Group on Artificial Intelligence: Association for Computing Machinery.

Lancaster, J. S. (1990). Cognitively-based knowledge acquisition: Capturing categorical, temporal, and causal knowledge. In K. L. McGraw & C. R. Westphal (Eds.), *Readings in knowledge acquisition* (pp. 184–199). New York: Ellis Horwood.

Laskey, K. B., Cohen, M. S., Martin, A. W. (1989). Representing and eliciting knowledge for uncertain evidence and its implications. *IEEE Transactions on Systems, Man, and Cybernetics*, *19*, 536–545.

Lazarus, S. (1975). *Loud & clear: A guide to effective communication*. New York: AMACOM.

LeClair, S. R. (1990). Interactive learning: A multiexpert paradigm for acquiring new knowledge. In K. L. McGraw & C. R. Westphal (Eds.), *Readings in knowledge acquisition* (pp. 104–121). New York: Ellis Horwood.

Lefkowitz, L. S., & Lesser, V. R. (1988). Knowledge acquisition as knowledge assimilation. *International Journal of Man-Machine Studies*, *29*, 215–226.

Lehner, P. E. (1989). Toward an empirical approach to evaluating the knowledge base of an expert system. *IEEE Transactions on Systems, Man, and Cybernetics*, *19*, 658–662.

Leiter, K. (1980). *A primer on ethnomethodology*. London: Oxford University Press.

Leplat, J. (1986). The elicitation of expert knowledge. In E. Hollnagel, G. Mancini, & D. D. Woods (Eds.), *Intelligent decision support in process environments* (pp. 107–122). Berlin: Springer-Verlag.

Lerner, D. (1956). Interviewing Frenchmen. *American Journal of Sociology*, *61*, 187–194.

Likert, R. (1932). The method of constructing an attitude scale. In M. Fishbein (Ed.), *Readings in attitude theory and measurement*. New York: Wiley.

Likert, R. (1932). A technique for the measurement of attitudes. *Archives of Psychology*, *140*, 44–53.

Mancaster-Ramer, A., & Lindsay, R. K. (1988). Linguistic knowledge as expertise. *International Journal of Expert Systems*, *1*, 329–343.

Martin, J. D. (1985). Knowledge acquisition through natural language dialogue. *Proceedings of the Second Conference on Artificial Intelligence Applications* (pp. 582–586). New York: IEEE Computer Society.

Martin, J. D., & Redmond, M. (1989). Acquiring knowledge by explaining observed problem solving. In C. R. Westphal & K. L. McGraw (Eds.), *Special Issue of ACM SIGART Newsletter on Knowledge Acquisition* (No. 108, pp. 77–83). New York: Special Interest Group on Artificial Intelligence, Association for Computing Machinery.

McGraw, K. L., & Harbison-Briggs, K. (1989). *Knowledge acquisition: Principals and guidelines*. Englewood Cliffs, NJ: Prentice Hall.

McGraw, K. L., & Riner, A. (1987). Task analysis: Structuring the knowledge acquisition process. *Texas Instruments Technical Journal*, *4*(6), 16–21.

McGraw, K. L., & Seale, M. R. (1988). Knowledge

elicitation with multiple experts: Considerations and techniques. *Artificial Intelligence Review*, *2*, 29–42.

Merton, R. K., Fiske, M., & Kendall, P. L. (1956). *The focused interview*. Glencoe, IL: The Free Press.

Merton, R. K., & Kendall, P. L. (1946). The focused interview. *American Journal of Sociology*, *51*, 541–557.

Micciche, P. F., & Lancaster, J. S. (1989, April). Application of neurolinguistic techniques to knowledge acquisition. In C. R. Westphal & K. L. McGraw (Eds.), *Special Issue of ACM SIGART Newsletter on Knowledge Acquisition* (pp. 28–33). New York: Special Interest Group on Artificial Intelligence, Association for Computing Machinery.

Mitchell, A. A. (1987). The use of alternative knowledge-acquisition procedures in the development of a knowledge-based media planning system. *International Journal of Man-Machine Studies*, *26*, 399–411.

Mittal, S., & Dym, C. L. (1985). Knowledge acquisition from multiple experts. *The AI Magazine*, *6*(2), 32–36.

Monk, A. (1985). How and when to collect behavioral data. In A. Monk (Ed.), *Fundamentals of human-computer interaction* (pp. 69–79). New York: Academic Press.

Moray, N., & Reeves, T. (1987). Hunting the homomorph: A theory of mental models and a method by which they may be identified. *Proceedings of the International Conference on Systems, Man, and Cybernetics* (pp. 594–597). New York: IEEE.

Morik, K. (1989). Integration issues in knowledge acquisition systems. In C. R. Westphal & K. L. McGraw (Eds.), *Special Issue of ACM SIGART Newsletter on Knowledge Acquisition* (No. 108, pp. 124–131). New York: Special Interest Group on Artificial Intelligence, Association for Computing Machinery.

Morrison, D. G. (1967). Measurement problems in cluster analysis. *Management Science*, *13*, 775–780.

Nisbett, R. E., & Wilson, T. D. (1977). Telling more than we can know: Verbal reports on mental processes. *Psychological Review*, *84*, 231–259.

Noble, D. (1989). Schema-based knowledge elicitation for planning and situation assessment aids. *IEEE Transactions on Systems, Man, & Cybernetics*, *19*, 473–482.

Norros, L. (1986). Expert knowledge, its acquisition and elicitation in developing intelligent tools for process control. In E. Hollnagel, G. Mancini, & D. D. Woods (Eds.), *Intelligent decision support in process environments* (pp. 137–144). Berlin: Springer-Verlag.

Nott, H. R., Peterson, S., & Nott, F. (1983). *Communication processes in the organization*. Dubuque, IA: Kendall/Hunt.

O'Leary, D. E. (1990). Soliciting weights or probabilities from experts for rule-based expert systems. *International Journal of Man-Machine Studies*, *32*, 293–302.

Olson, J. R., & Rueter, H. H. (1987). Extracting expertise from experts: Methods for knowledge acquisition. *Expert Systems*, *4*, 152–168.

Oppenheim, A. (1966). *Questionnaire design and attitude measurement*. New York: Basic Books.

Pau, L. F., & Nielsen, S. S. (1989). Conceptual graphs as a visual language for knowledge acquisition in architectural expert systems. In C. R. Westphal & K. L. McGraw (Eds.), *Special Issue of ACM SIGART Newsletter on Knowledge Acquisition* (No. 108, p. 151). New York: Special Interest Group on Artificial Intelligence, Association for Computing Machinery.

Prerau, D. S. (1987). Knowledge acquisition in the development of a large expert system. *AI Magazine*, *8*(2), 43–51.

Rasmussen, J. (1986). A framework for cognitive task analysis in systems design. In E. Hollnagel, G. Mancini, & D. D. Woods (Eds.), *Intelligent decision support in process environments* (pp. 175–196). Berlin: Springer-Verlag.

Regoczei, S., & Plantinga, E. P. O. (1987). Creating the domain of discourse: Ontology and inventory. *International Journal of Man-Machine Studies*, *27*, 235–250.

Reitman, W. (1989). Integrated design teams: Knowledge engineering for large-scale commercial expert system development. *IEEE Transactions on Systems, Man, & Cybernetics*, *19*, 443–447.

Robertson, S., & Swartz, M. (1988). Why do we ask ourselves questions? Question generation during skill acquisition. *Questioning Exchange*, *2*, 47–51.

Rodi, L. L., Pierce, J. A., & Dalton, R. E. (1989, April). Putting the expert in charge: Graphical knowledge acquisition for fault diagnosis and repair. In C. R. Westphal & K. L. McGraw (Eds.), *Special Issue of ACM SIGART Newsletter on Knowledge Acquisition* (pp. 56–62). New York: Special Interest Group on Artificial Intelligence, Association for Computing Machinery.

Rolandi, W. G. (1986). Knowledge engineering in practice. *AI Expert*, *1*, 58–62.

Ruberg, K., Cornick, S. M., & James, K. A. (1989). House call: Building and maintaining a rule base. *Knowledge Acquisition: An International Journal*, *1*, 379–401.

Rusk, R. A., & Krone, R. M. (1984). The Crawford Slip Method (CSM) as a tool for extraction of expert knowledge. In G. Salvendy (Ed.), *Human-computer interaction* (pp. 279–282). Amsterdam: Elsevier.

Saaty, T. L. (1981). *The analytic hierarchy process*. New York: McGraw-Hill.

Schvaneveldt, R. (Ed.). (1990). *Pathfinder associative networks*. Norwood, NJ: Ablex.

Schvaneveldt, R., Durso, F., Goldsmith, T., Breen, T., Cooke, N., Tucker, R., & DeMaio, J. (1985). Measuring the structure of expertise. *International Journal of Man-Machine Studies*, *23*, 699–728.

Schweickert, R., Burton, A. M., Taylor, N. K., Corlett, E. N., Shadbolt, N. R., & Hedgecock, A. P. (1987). Comparing knowledge elicitation techniques: A case study. *Artificial Intelligence Review*, *1*, 245–253.

Senjen, R. (1988). Knowledge acquisition by experiment: Developing test cases for an expert system. *AI Applications in Environmental Science*, *2*, 52–55.

Shachter, R. D., & Heckerman, D. E. (1987). Thinking backward for knowledge acquisition. *AI Magazine*, *8*(3), 231–259.

Shadbolt, N. R., & Burton, A. M. (1989). The empirical study of knowledge elicitation techniques. In C. R. Westphal & K. L. McGraw (Eds.), *Special Issue of ACM SIGART Newsletter on Knowledge Acquisition* (No. 108, pp. 15–18). New York: Special Interest Group on Artificial Intelligence, Association for Computing Machinery.

Shadbolt, N. R., & Burton, A. M. (1990). Knowledge acquisition. In E. N. Wilson & J. R. Corlett (Eds.), *Evaluation of human work: Practical ergonomics methodology* (pp. 321–345). London: Taylor and Francis.

Shadbolt, N. R., & Burton, A. M. (1990). Knowledge elicitation techniques: Some experimental results. In K. L. McGraw & C. R. Westphal (Eds.), *Readings in knowledge acquisition* (pp. 21–33). New York: Ellis Horwood.

Shaw, M. L. G. (1976). *Group dynamics: The psychology of small group behavior*. New York: McGraw-Hill.

Shaw, M. L. G. (Ed.). (1981). *Recent advances in personal construct technology*. New York: Academic Press.

Shaw, M. L. G., & McKnight, C. (1981). *Think again*. Englewood Cliffs, NJ: Prentice-Hall.

Shaw, M. L. G., & Woodward, J. B. (1988). Validation in a knowledge support system: Construing and consistency with multiple experts. *International Journal Man-Machine Studies*, *29*, 329–350.

Shulman, L. S., Loupe, M. J., & Piper, R. M. (1968). *Studies of the inquiry process: Inquiry patterns of students in teacher-training programs*. East Lansing, MI: Michigan State University Educational Publications Services.

Silverman, B. G., Wenig, R. G., & Wu, T. (1989). Coping with ongoing knowledge acquisition from collaborating hierarchies of experts. In C. R. Westphal & K. L. McGraw (Eds.), *Special Issue of ACM SIGART Newsletter on Knowledge Acquisition* (No. 108, pp. 170–171). New York: Special Interest Group on Artificial Intelligence, Association for Computing Machinery.

Silvestro, K. (1988). Using explanations for knowledge-base acquisition. *International Journal of Man-Machine Studies*, *29*, 159–170.

Sjoberg, G., & Nett, R. (1968). *A methodology for social research*. New York: Harper & Row.

Slater, B. M. (1989). Extracting lexical knowledge from dictionary text. In C. R. Westphal & K. L. McGraw (Eds.), *Special Issue of ACM SIGART Newsletter on Knowledge Acquisition* (No. 108, pp. 168–169). New York: Special Interest Group on Artificial Intelligence, Association for Computing Machinery.

Spradley, J. P. (1979). *The ethnographic interview*. New York: Holt, Rinehart, and Winston.

Stefik, M., Atkins, J., Balzer, R., Benoit, J., Birnbaum, L., Hayes-Roth, F., & Sacerdoti, E. (1983). Basic concepts for building expert systems. In F. Hayes-Roth, D. A. Waterman, & D. B. Lenat (Eds.), *Building expert systems* (pp. 59–86). Reading, MA: Addison-Wesley.

Steiner, I. (1972). *Group process and productivity*. New York: Academic Press.

Stewart, C., & Cash, W. (1985). *Interviewing: Principles and practices* (4th ed.). Dubuque, IA: Wm. C. Brown.

Velardi, P. (1989). Acquisition of semantic patterns from a natural corpus of texts. In C. R. Westphal & K. L. McGraw (Eds.), *Special Issue of ACM SIGART Newsletter on Knowledge Acquisition* (No. 108, pp. 115–123). New York: Special Interest Group on Artificial Intelligence, Association for Computing Machinery.

Waldron, V. R. (1985). Process tracing as a means of collecting knowledge for expert systems. *Texas Instruments Engineering Journal*, *2*(6), 90–93.

Waldron, V. R. (1986). Interviewing for knowledge. *IEEE Transactions on Professional Communications* PC-29(2), 31–35.

Waldron, V. R. (1989). Investigating the communication problems encountered in knowledge acquisition. In C. R. Westphal & K. L. McGraw (Eds.), *Special Issue of ACM SIGART Newsletter on Knowledge Acquisition* (No. 108, pp. 143–144). New York: Special Interest Group on Artificial Intelligence, Association for Computing Machinery.

West, W. D. (1984). Extraction of expert knowledge from operators of dynamic computer assisted systems. In G. Salvendy (Ed.), *Human-computer interaction* (pp. 287–292). Amsterdam: Elsevier.

Westphal, C. R., & McGraw, K. L. (Eds.). (1989, April). *Special Issue of ACM SIGART Newsletter on Knowledge Acquisition* (No. 108 pp. 1–197). New York: Special Interest Research Group on Artificial Intelligence, Association for Computing Machinery.

Westphal, C. R., & Reeker, L. H. (1990). Reasoning and representation mechanisms for multiple-expert knowledge acquisition. In K. L. McGraw & C. R. Westphal (Eds.), *Readings in knowledge acquisition* (pp. 90–103). New York: Ellis Horwood.

Wielinga, B. J., & Breuker, J. A. (1985). Interpretation of verbal data for knowledge acquisition. In T. O'Shea (Ed.), *Advances in artificial intelligence* (pp. 3–12). Amsterdam: North-Holland.

Witten, I. H., & McDonald, B. A. (1988). Using concept learning for knowledge acquisition. *International Journal of Man-Machine Studies*, *29*, 171–196.

Wolf, W. A. (1989). Knowledge acquisition from multiple experts. In C. R. Westphal & K. L. McGraw (Eds.), *Special Issue of ACM SIGART Newsletter on Knowledge Acquisition* (No. 108, pp. 138–140). New York: Special Interest Group on Artificial Intelligence, Association for Computing Machinery.

Wolvin, A., & Coakley, C. (1982). *Listening*. Dubuque, IA: Wm. C. Brown.

Woodward, B. (1990). Knowledge acquisition at the front end: Defining the domain. *Knowledge Acquisition: An International Journal*, *2*, 73–94.

Wright, G., & Ayton, P. (1987). Eliciting and modeling expert knowledge. *Decision Support Systems*, *3*, 13–26.

Yamada, S., & Tsuji, S. (1989). Acquisition of macro-operators from worked examples in problem solving. In C. R. Westphal & K. L. McGraw (Eds.), *Special Issue of ACM SIGART Newsletter on Knowledge Acquisition* (No. 108, pp. 171–172). New York: Special Interest Group on Artificial Intelligence, Association for Computing Machinery.

Young, R. M. (1988). Role of intermediate representations in knowledge elicitation. In D. S. Moralee (Ed.), *Research and development in experts systems IV* (pp. 287–288). Cambridge: Cambridge University Press.

Zaragoza, M. S., & Koshmider, J. W. (1989). Misled subjects may know more than their performance implies. *Journal of Experimental Psychology: Learning, Memory, and Cognition*, *15*, 246–255.

Zeisel, J. (1981). *Inquiry by design: Tools for environment-behavior research*. Monterey, CA: Brooks-Cole.

Zunin, L., & Zunin, N. (1975). *Contact: The first four minutes*. Los Angeles: Nash.

Zwick, R. (1988). The evaluation of verbal models. *International Journal of Man-Machine Studies*, *29*, 149–158.

Appendix D
Bibliography: Automated Knowledge Elicitation, Representation, and Instantiation ("Knowledge Acquisition")

Abrett, G., & Burstein, M. H. (1986, April). The BBN Laboratories knowledge acquisition project: KREME knowledge editing environment. *Proceedings from the DARPA-Sponsored Expert Systems Workshop* (pp. 1–21). Pacific Grove, CA.

Abrett, G., & Burstein, M. H. (1987). The KREME knowledge editing environment. *International Journal of Man-Machine Studies*, *27*, 103–126.

Adams-Webber, J. R. (1984). Repertory grid technique. In R. Corsini (Ed.), *Encyclopedia of psychology* (p. 225). New York: Wiley Interscience.

Adelman, L. (1987, October). Toward considering psychological measurement issues in developing expert systems. *Proceedings of the 1987 IEEE International Conference on Systems, Man, and Cybernetics* (Cat. No. 87CH2503-1). NY: IEEE.

Adelman, L. (1989). Measurement issues in knowledge engineering. *IEEE Transactions on Systems, Man, and Cybernetics*, *19*, 483–488.

Aikins, J. S. (1981). Representation of control knowledge in expert systems. *Proceedings of the First Meeting of the AAAI* (pp. 121–123). Stanford, CA: The American Association for Artificial Intelligence.

Alexander, J. H., Freiling, M. J., Shulman, S. J., Rehfuss, S., & Messick, S. L. (1987). Ontological analysis: An ongoing experiment. *International Journal of Man-Machine Studies*, *26*, 473–486.

Arbab, B., & Michie, D. (1988). Generating expert rules from examples in PROLOG. In J. E. Hayes, D. Michie, & J. Richards (Eds.), *Machine intelligence 11* (pp. 289–304). Oxford, England: Clarendon Press.

Arciszewski, T., Mustafa, M., & Ziarko, W. (1987). A methodology of design knowledge acquisition for use in learning expert systems. *International Journal of Man-Machine Studies*, *27*, 23–32.

Arinze, B. (1989, April). A natural language front-end for knowledge acquisition. In C. R. Westphal & K. L. McGraw (Eds.), *Special Issue of ACM SIGART Newsletter on Knowledge Acquisition* (No. 108, pp. 106–114). New York: Special Interest Group on Artificial Intelligence, Association for Computing Machinery.

Bailey, W. A., & Kay, E. J. (1986, October). Toward the structural analysis of verbal data. *Proceedings in the 1986 IEEE International Conference on Systems, Man, and Cybernetics* (Cat. No. 86CH2364-8). New York: IEEE.

Banning, R. W. (1984). Knowledge acquisition and system validation in expert systems for management. *Human systems managements* (Vol. 4, pp. 280–285). New York: Elsevier.

Banthay, B. H. (1988). Matching design methods to system type. *Systems Research*, *5*, 27–34.

Bareiss, E. R. (1989). *Exemplar-based knowledge acquisition: A unified approach to concept representation, classification, and learning*. New York: Academic Press.

Bareiss, E. R., Porter, B. W., & Wier, C. C. (1988). PROTOS: An exemplar-based learning apprentice. *International Journal of Man-Machine Studies*, *29*, 549–562.

Barnard, P., Wilson, M., & Maclean, A. (1986, April). The elicitation of system knowledge by picture probes. In M. Mantei & P. Orbeton (Eds.), *Proceedings of CHI '86 Conference* (pp. 235–240). New York: Association for Computing Machinery.

Becker, L. A., Bartlett, R., & Soroushian, F. (1989, April). Using simulation to compile diagnostic rules from a manufacturing process representation. In C. R. Westphal & K. L. McGraw (Eds.), *Special Issue of ACM SIGART Newsletter on Knowledge Acquisition* (No. 108, pp. 172–173).

New York: Special Interest Group on Artificial Intelligence, Association for Computing Machinery.

Bennett, J. S. (1985). ROGET: A knowledge-based system for acquiring the conceptual structure of a diagnostic expert system. *Journal of Automated Reasoning*, *1*, 49–74.

Berg-Cross, G., & Price, M. E. (1989). Acquiring and managing knowledge using a conceptual structures approach: Introduction and framework. *IEEE Transactions on Systems, Man, and Cybernetics*, *19*, 513–527.

Berwick, R. (1985). *The acquisition of syntactic knowledge*. Cambridge, MA: The MIT Press.

Birmingham, W. P., Sieworek, D. P. (1989). Automated knowledge acquisition for a computer hardware synthesis system. *Knowledge Acquisition: An International Journal*, *1*, 321–340.

Bloomfield, B. P. (1986). Capturing expertise by rule induction. *Knowledge Engineering Review*, *1*, 30–36.

Bonner, R. (1964). On some clustering techniques. *IBM Journal of Research and Development*, *9*, 22–32.

Boose, J. H. (1984). Personal construct theory and the transfer of human expertise. *Proceedings of AAAI–84* (pp. 27–33). Los Altos: CA: William Kaufmann.

Boose, J. H. (1984). A framework for transferring human expertise. In G. Salvendy (Ed.), *Human-computer interaction* (pp. 247–254). Amsterdam: Elsevier.

Boose, J. H. (1985). A knowledge acquisition program for expert systems based on personal construct psychology. *International Journal of Man-Machine Studies*, *23*, 495–525.

Boose, J. H. (1986). *Expertise transfer for expert system design*. New York: Elsevier.

Boose, J. H. (1986). Rapid acquisition and combination of knowledge from multiple experts in the same domain. *Future Computing Systems Journal*, *1*, 191–216.

Boose, J. H. (1988). Uses of repertory grid-centered knowledge acquisition tools for knowledge-based systems. *International Journal of Man-Machine Studies*, *29*, 287–310.

Boose, J. H., & Bradshaw, J. M. (1987). Expertise transfer and complex problems: Using AQUINAS as a knowledge-acquisition workbench for knowledge-based systems. *International Journal of Man-Machine Studies*, *26*, 3–28.

Boose, J. H., & Gaines, B. R. (1988). *Knowledge acquisition tools for expert systems* (Vol. 2). Cambridge, MA: Academic Press.

Boyle, C. D. B. (1985). Acquisition of control and domain knowledge by watching in a blackboard environment. In M. Merry (Ed.), *Expert systems-85* (pp. 273–286). Cambridge: Cambridge University Press.

Bradley, J. V., & Harbison-Briggs, K. (1989, April). The symptom-component approach to knowledge acquisition. In C. R. Westphal & K. L. McGraw (Eds.), *Special Issue of ACM SIGART Newsletter on Knowledge Acquisition* (No. 108, pp. 70–76). New York: Special interest Group on Artificial Intelligence, Association for Computing Machinery.

Bradshaw, J. M., & Boose, J. H. (1990). Decision analysis techniques for knowledge acquisition: Combining information and preferences using Aquinas and Axotl. *International Journal of Man-Machine Studies*, *32*, 121–186.

Branaghan, R. J. (1990). Pathfinder networks and multidimensional spaces: Relative strengths in representing strong associates. In R. W. Schvaneveldt (Ed.), *Pathfinder associative networks: Studies in knowledge organization* (pp. 11–120). Norwood, NJ: Ablex.

Bratko, I., Mozetič, I., & Lavarač, N. (1988). Automatic synthesis and compression of cardiological knowledge. In J. Hayes, D. Michie, & J. Richards (Eds.), *Machine intelligence* (pp. 435–454). Oxford: Oxford University Press.

Bratko, I., Mozetič, I., & Lavarač, N. (1989). *KARDIO: A study in deep and qualitative knowledge for expert systems*. Boston: The MIT Press.

Breuker, K., & Wielinga, B. (1987). Use of models in the interpretation of verbal data. In A. Kidd (Ed.), *Knowledge elicitation for expert systems: A practical handbook* (pp. 17–44). New York: Plenum.

Brown, D. (1989). A knowledge acquisition tool for decision support systems. In C. R. Westphal & K. L. McGraw (Eds.), *Special Issue of ACM SIGART Newsletter on Knowledge Acquisition* (No. 108, pp. 92–97). New York: Special Interest Group on Artificial Intelligence, Association for Computing Machinery.

Buchanan, B. G. (1983). Constructing an expert system. In F. Hayes-Roth & D. A. Waterman (Eds.), *Building expert systems* (pp. 127–167). Reading, MA: Addison-Wesley.

Butler, K., Carter H. (1986). The use of psychometric tools for knowledge acquisition: A case study. In W. Gale (Ed.), *Artificial intelligence and statistics*. Reading, MA: Addison-Wesley.

Bylander, T., & Chandrasekaran, B. (1987). Generic tasks in knowledge-based reasoning: The "right" level of abstraction for knowledge acquisition. *International Journal of Man-Machine Studies*, *26*, 231–244.

Caviedes, J. E., & Reed, M. K. (1989). Viewpoint: A troubleshooting-specific knowledge acquisition tool. In C. R. Westphal & K. L. McGraw (Eds.), *Special Issue of ACM SIGART Newsletter on Knowledge Acquisition* (No. 108, pp. 155). New York: Special Interest Group on Artificial Intelligence, Association for Computing Machinery.

Cendrowska, J. (1987). PRISM: An algorithm for inducing modular rules. *International Journal of Man-Machine Studies*, *27*, 349–370.

Chandrasekaran, B. (1990, Winter). Design problem solving: A task analysis. *The AI Magazine*, 59–71.

Chang, P. J., Ford, K. M., & Petry, F. E. (1987). The production of expert system rules from repertory grid data based on a logic of confirmation. *Seventh International Congress on Personal Construct Psychology*, Memphis, TN.

Cheng, Y., & Fu, K. (1985). Conceptual clustering in knowledge organization. *IEEE Transactions on Pattern Analysis and Machine Intelligence*, *7*, 592–598.

Chung, H. M. (1989). Empirical analysis of inductive knowledge acquisition methods. In C. R. Westphal & K. L. McGraw (Eds.), *Special Issue of ACM SIGART Newsletter on Knowledge Acquisition* (No. 108, pp. 156–159). New York: Special Interest Group on Artificial Intelligence, Association for Computing Machinery.

Clancey, W. J. (1983). The advantages of abstract control knowledge in expert system design. *Proceedings of the National Conference on Artificial Intelligence* (pp. 74–78), Washington, DC. Los Altos, CA: Morgan-Kaufmann.

Clancey, W. J. (1984). Knowledge acquisition for classification expert systems. *Proceedings of ACM Annual Conference* (pp. 11–14). New York: Association for Computing Machinery.

Clancey, W. J. (1986). Heuristic classification. In J. Kowalik (Ed.), *Knowledge-based problem solving*. Englewood Cliffs, NJ: Prentice-Hall.

Clancey, W. J. (1988). Acquiring, representing, and evaluating a competence model of diagnosis. In M. Chi, R. Glaser, & M. Farr (Eds.), *The nature of expertise*. Hillsdale, NJ: Erlbaum.

Cleary, J. G. (1987). Acquisition of uncertain rules in probablistic logic. *International Journal of Man-Machine Studies*, *27*, 145–154.

Cohen, P. R., DeLisio, J., & Hart, D. (1989). A declarative representation of control knowledge. *IEEE Transactions on Systems, Man, and Cybernetics*, *19*, 546–557.

Cooke, N. J. (1990). Empirically defined semantic relatedness and category judgment time. In R. W. Schvaneveldt (Ed.), *Pathfinder associative networks: Studies in knowledge organization* (pp. 101–110). Norwood, NJ: Ablex.

Cooke, N. J. (1990). Using pathfinder as a knowledge elicitation tool: Link interpretation. In R. W. Schvaneveldt (Ed.), *Pathfinder associative networks: Studies in knowledge organization* (pp. 227–240). Norwood, NJ: Ablex.

Cooke, N. M., & McDonald, J. E. (1986). A formal methodology for acquiring and representing expert knowledge. *Proceedings of the IEEE: Special Issue on Knowledge Representation*, *74*, 1422–1430.

Cooke, N. M., & McDonald, J. E. (1987). The application of psychological scaling techniques to knowledge elicitation for knowledge-based systems. *International Journal of Man-Machine Studies*, *26*, 533–550.

Cookson, M., Holman, J., & Thompson, D. (1985). Knowledge acquisition for medical expert systems: A system for eliciting diagnostic decision-making histories. In M. A. Bramer (Ed.), *Research and development in expert systems* (pp. 113–116). Cambridge: Cambridge University Press.

Cross, G. B., & Price, M. E. (1989). Acquiring and managing knowledge using a conceptual structures approach: Introduction and framework. *IEEE Transactions on Systems, Man, and Cybernetics*, *19*, 513–527.

Cvagren, B. J., & Stendel, H. J. (1987). A decision-table-used processor for checking completeness and consistency in rule-based expert systems. *International Journal of Man-Machine Studies*, *26*, 633–648.

Davis, R. (1979). Interactive transfer of expertise: Acquisition of new inference rules. *Artificial Intelligence*, *12*, 121–157.

Dayton, T., Durso, F. T., & Shepard, J. D. (1990). A measure of the knowledge reorganization underlying insight. In R. W. Schvaneveldt (Ed.), *Pathfinder associative networks: Studies in knowledge organization* (pp. 207–278). Norwood, NJ: Ablex.

Dean, T., & Kanazawa, K. (1989). Persistence and probabilistic projection. *IEEE Transactions on Systems, Man, and Cybernetics*, *19*, 574–585.

Dearholt, D. W., & Schvaneveldt, R. W. (1990). Properties of pathfinder networks. In R. W.

Schvaneveldt (Ed.), *Pathfinder associative networks: Studies in knowledge organization* (pp. 1–30). Norwood, NJ: Ablex.

Delgrande, J. P. (1987). A formal approach to learning from examples. *International Journal of Man-Machine Studies*, *26*, 123–141.

Diedrich, J. (1987). Knowledge-based elicitation. *Proceedings of the Tenth International Joint Conference on Artificial Intelligence* (pp. 201–204). Los Altos, CA: Morgan Kaufmann.

Diedrich, J., Ruhmann, L., & May, M. (1987). KRITON: A knowledge acquisition tool for expert systems. *International Journal of Man-Machine Studies*, *26*, 29–40.

Dieterich, T. (1981). Learning and inductive inference. In P. Cohen & E. Feigenbaum (Eds.), *The handbook of artificial intelligence*. Los Altos, CA: William Kaufmann.

Dong, W.-M., & Wong, F. S. (1987). Propagation of evidence in rule-based systems. *International Journal of Man-Machine Studies*, *26*, 551–566.

Dougherty, J., & Keller, C. (1982). Taskonomy: A practical approach to knowledge structures. *American Ethnologist*, *9*, 763–774.

Durso, F. T., & Coggins, K. A. (1990). Graphs in the social and psychological sciences: Empirical contributions of pathfinder. In R. W. Schvaneveldt (Ed.), *Pathfinder associative networks: Studies in knowledge organization* (pp. 31–52). Norwood, NJ: Ablex.

Eden, C., & Jones, S. (1984). Using repertory grids for problem construction. *Journal of the Operational Research Society*, *35*, 779–790.

Emde, W. (1988). An inference engine for representing multiple theories. In K. Morik (Ed.), *Knowledge representation and organization in machine learning*. New York: Springer-Verlag.

Eshelman, L. (1988). MOLE: A knowledge acquisition tool that buries certainty factors. *International Journal of Man-Machine Studies*, *29*, 563–577.

Eshelman, L., Ehret, D., McDermott, J., & Tan, M. (1987). MOLE: A tenacious knowledge acquisition tool. *International Journal of Man-Machine Studies*, *26*, 41–54.

Esposito, C. (1990). Fuzzy PFNETs: Coping with variability in proximity data. In R. W. Schvaneveldt (Ed.), *Pathfinder associative networks: Studies in knowledge organization* (pp. 53–60). Norwood, NJ: Ablex.

Esposito, C. (1990). A graph-theoretic approach to concept clustering. In R. W. Schvaneveldt (Ed.), *Pathfinder associative networks: Studies in knowledge organization* (pp. 89–100). Norwood, NJ: Ablex.

Everitt, B. (1980). *Cluster analysis*. New York: Halsted Press.

Fass, L. F. (1989). Learnability of CFLS: Inferring syntactic models from constituent structure. In C. R. Westphal & K. L. McGraw (Eds.), *Special Issue of ACM SIGART Newsletter on Knowledge Acquisition* (No. 108, pp. 175–176). New York: Special Interest Group on Artificial Intelligence, Association for Computing Machinery.

Finley, M. R., & Hausen-Trooper, E. B. (1989). A system for the representation of theorems and proofs. In C. R. Wesphal & K. L. McGraw (Eds.), *Special Issue of ACM SIGART Newsletter on Knowledge Acquisition* (No. 108, pp. 178–179). New York: Special Interest Group On Artificial Intelligence, Association for Computing Machinery.

Fischoff, B., MacGregor, D., & Blackshaw, L. (1987). Creating categories for databases. *International Journal of Man-Machine Studies*, *27*, 33–64.

Fisher, D. (1987). Knowledge acquisition via incremental conceptual clustering. *Machine Learning*, *2*, 139–172.

Fisher, D. (1987). Improving inference through conceptual clustering. *Proceedings of AAAI '87* (pp. 461–465). Los Altos, CA: Morgan Kaufmann.

Ford, K. M., Adams-Webber, J. R., Stahl, H. A., & Bringmann, M. W. (1990). Constructivist approaches to automated knowledge acquisition. In K. L. McGraw & C. R. Westphal (Eds.), *Readings in knowledge acquisition* (pp. 34–54). New York: Ellis Horwood.

Ford, K. M., & Chang, P. J. (1989). An approach to automated knowledge acquisition founded on personal construct theory. In M. Fishman (Ed.), *Advances in artificial intelligence research*. (Vol. 1, pp. 83–132). Greenwich, CT: JAI Press.

Ford, K. M., & Petry, F. E. (1989). Knowledge acquisition from repertory grids using a logic of confirmation. In C. R. Westphal & K. L. McGraw (Eds.), *Special Issue of ACM SIGART Newsletter on Knowledge Acquisition* (No. 108, pp. 146–147). New York: Special Interest Group on Artificial Intelligence, Association for Computing Machinery.

Ford, K. M., Petry, F. E., Adams-Webber, J. R., & Chang, P. J. (1991). An approach to knowledge acquisition based on the structure of personal construct systems. *IEEE Transactions on Knowledge and Data Engineering*, *3*, 78–88.

Ford, K. M., Stahl, H., Adams-Webber, J. R., Cañas, A. J., Novak, J., & Jones, J. C. (1991). ICONKAT: An integrated constructivist knowledge acquisition tool. *Knowledge Acquisition*, *3*, 215–236.

Forsyth, R. (1986). Software review: Personal consultant plus. *Expert Systems, 3*(4), 244–247.

Forsyth, R. (1987). Software review: Expertech Xi Plus, *Expert Systems*, *4*(1) 48–51.

Forsythe, D. E., & Buchanan, B. G. (1989). Knowledge acquisition for expert systems: Some pitfalls and suggestions. *IEEE Transactions on Systems, Man, and Cybernetics*, *19*, 435–442.

Fowler, R. H., & Dearholt, D. W. (1990). Information retrieval using pathfinder networks. In R. W. Schvaneveldt (Ed.), *Pathfinder associative networks: Studies in knowledge organization* (pp. 165–178). Norwood, NJ: Ablex.

Freedman, R. (1987). Evaluating shells. *AI Expert*, *9*, 70–74.

Fu, L. M., & Buchanan, B. G. (1985). *Inductive knowledge acquisition for rule-based expert systems* (Report No. KSL–85–42). Stanford, CA: Stanford University.

Gaines, B. R. (1987). An overview of knowledge acquisition and transfer. *International Journal of Man-Machine Studies*, *26*, 453–472.

Gaines, B. R., & Boose, J. H. (Eds.). (1988). *Foundations of knowledge acquisition*. London: Academic Press.

Gaines, B. R., & Boose, J. H. (Eds.). (1988). *Knowledge acquisition tools for expert systems*. London: Academic Press.

Gaines, B. R., & Shaw, M. (1986). Induction of inference rules for expert systems. *Fuzzy Sets and Systems*, *18*, 315–328.

Gale, W. A. (1987). Knowledge-based knowledge acquisition for a statistical consulting system. *International Journal of Man-Machine Studies*, *26*, 55–64.

Galloway, T. (1987). TAXI: A taxonomic assistant. *Proceedings of AAAI '87* (pp. 426–420). Los Altos, CA: Morgan Kaufmann.

Gammack, J. G. (1987). Different techniques and different aspects on declarative knowledge. In A. L. Kidd (Ed.), *Knowledge acquisition for expert systems: A practical handbook* (pp. 137–164). New York: Plenum.

Gammack, J. G. (1990). Expert conceptual structure: The stability of pathfinder representations. In R. W. Schvaneveldt (Ed.), *Pathfinder associative networks: Studies in knowledge organization* (pp. 213–226). Norwood, NJ: Ablex.

Garg-Janardan, C., & Salvendy, G. (1987). A conceptual framework for knowledge elicitation. *International Journal of Man-Machine Studies*, *26*, 521–526.

Garg-Janardan, C., & Salvendy, G. (1988). A structured knowledge elicitation methodology for building expert systems. *International Journal of Man-Machine Studies*, *29*, 377–406.

Gilbert, N. (1989). Explanation and dialog. *The Knowledge Engineering Review*, *4*, 235–248.

Gold, J. (1988). Do it yourself expert systems. *Computer Decisions*, *18*, 76–81.

Golden, B., Wasil, E., & Harker, P. (Eds.). (1989). *The analytic hierarchy process: Applications and studies*. New York: Springer-Verlag.

Goldsmith, J. E., & Johnson, P. J. (1990). A structural assessment of classroom learning. In R. W. Schvaneveldt (Ed.), *Pathfinder associative networks: Studies in knowledge organization* (pp. 241–254). Norwood, NJ: Ablex.

Goldsmith, T. F., & Davenport, D. M. (1990). Assessing structural similarity of graphs. In R. W. Schvaneveldt (Ed.), *Pathfinder associative networks: Studies in knowledge organization* (pp. 75–88). Norwood, NJ: Ablex.

Goodman, R. M., & Smyth, P. (1990). Decisition tree design using information theory. *Knowledge Acquisition: An International Journal*, *2*, 1–20.

Gruber, T. R. (1988). Acquiring strategic knowledge from experts. *International Journal of Man-Machine Studies*, *29*, 579–597.

Gruber, T. R., & Cohen, P. R. (1987). Design for acquisition: Principles of knowledge system design to facilitate knowledge acquisition. *International Journal of Man-Machine Studies*, *26*, 143–160.

Gruber, T. R., & Cohen, P. R. (1989, April). The design of an automated assistant for acquiring strategic knowledge. In C. R. Westphal & K. L. McGraw (Eds.), *Special Issue of ACM SIGART Newsletter on Knowledge Acquisition* (No. 108, pp. 147–151). New York: Special Interest Group on Artificial Intelligence, Association for Computing Machinery.

Grzymala-Busse, J. W. (1988). Knowledge acquisition under uncertainty: A rough set approach. *Journal of Intelligent and Robotic Systems*, *1*, 3–16.

Hall, L. O., & Bandler, W. (1985). Relational knowledge acquisition. *Proceedings of Second*

Conference on Artificial Intelligence Applications (pp. 509–513). New York: IEEE.

Handa, K.-I., Ishizaki, S. (1989). Learning importance of concepts: Construction of representative network. *Knowledge Acquisition: An International Journal, 1*, 365–378.

Hannan, J. J., & Politakis, P. (1985). ESSA: An approach to acquiring decision rules for diagnostic expert systems. *Proceedings of Second Conference on Artificial Intelligence Applications* (pp. 520–525). New York: IEEE.

Harbison-Briggs, K., & Briggs, A. B. (1990). The role of knowledge acquisition in the verification and validation of knowledge based systems. In K. L. McGraw & C. R. Westphal (Eds.), *Readings in knowledge acquisition* (pp. 233–250). New York: Ellis Horwood.

Hart, A. (1985). The role of induction in knowledge elicitation. *Expert Systems, 2*, 24–28.

Hart, A. (1985). Experience in the use of an inductive system in knowledge engineering. In M. A. Bramer (Ed.), *Research and development in expert systems* (pp. 117–126). Cambridge: Cambridge University Press.

Hart, A. (1986). *Knowledge acquisition for expert systems*. London: Kogan Page.

Hart, A. (1987). Role of induction in knowledge elicitation. In A. L. Kidd (Ed.), *Knowledge acquisition for expert systems: A practical handbook* (pp. 165–190). New York: Plenum.

Hayes-Roth, F. (1976). Patterns of induction and related knowledge acquisition algorithms. In C. Chen (Ed.), *Pattern recognition and artificial intelligence*. New York: Academic Press.

Hayes-Roth, F. (1977). Uniform representations of structured patterns and an algorithm for the induction of contingency-response rules. *Information and Control, 33*, 87–116.

Hayes-Roth, F. (1985). Codifying human knowledge for machine reading. *Next generation computers* (pp. 51–53). New York: IEEE.

Hayward, S. A., Wielinga, B. J., & Breuker, J. A. (1987). Structured analysis of knowledge. *International Journal of Man-Machine Studies, 26*, 487–498.

Hollan, J. D., Huchins, E. L., & Weitzman, L. (1984). STEAMER: An interactive inspectable simulation-based training system. *AI Magazine*, 15–27.

Huang, D. (1989). A framework for the credit-apportionment process in rule-based systems. *IEEE Transactions on Systems, Man, and Cybernetics, 19*, 489–498.

Jacobson, C., & Freiling, M. J. (1988). ASTEK: A multi-paradigm knowledge acquisition tool for complex structured knowledge. *International Journal of Man-Machine Systems, 29*, 311–328.

Jagannathan, V., & Elmaghraby, A. S. (1985, October). MEDKAT: Multiple expert delphi-based knowledge acquisition tool. *Proceedings of the Second Annual ACM Northeast Regional Conference* (pp. 30–34). New York: Association for Computing Machinery.

Jerrams-Smith, J. (1987). An expert system within a supportive interface for UNIX. *Behavior and Information Technology, 6*, 37–41.

Johnson, N. E. (1985). Varieties of representation in eliciting and representing knowledge for IKBS. *International Journal of Systems Research and Information Science, 1*, 69–90.

Johnson, S. (1967). Hierarchical clustering schemes. *Psychometrika, 32*, 241–254.

Jones, K., & Jackson, D. (1967). Current approaches to classification and clump finding at the Cambridge Language Research Unit. *Computing Journal*, 29–37.

Kahn, G. S., Breaux, E., Deklerk, P., & Joseph, R. (1987). A mixed-initiative workbench for knowledge acquisition. *International Journal of Man-Machine Studies, 27*, 167–179.

Kahn, G. S., Nowlan, S., & McDermott, J. (1985). Strategies for knowledge acquisition. *IEEE Transactions of Pattern Analysis and Machine Intelligence, PAMI–7*(5), 511–522.

Kellogg, W. A., & Breen, T. J. (1990). Using pathfinder to evaluate user and system models. In R. W. Schvaneveldt (Ed.), *Pathfinder associative networks: Studies in knowledge organization* (pp. 179–196). Norwood, NJ: Ablex.

Kidd, A. L. (1987). Knowledge acquisition: An introductory framework. In A. L. Kidd (Ed.), *Knowledge acquisition for expert systems: A practical handbook* (pp. 1–16). New York: Plenum.

Kidd, A. L. (1987). *Knowledge acquisition for expert systems: A practical handbook*. New York: Plenum.

Kieras, D., & Polson, P. (1985). An approach to the formal analysis of user complexity. *International Journal of Man-Machine Studies, 22*, 365–394.

Kinoshita, T. A. (1989, April). Knowledge acquisition model with applications for requirements specification and definition. In C. R. Westphal & K. L. McGraw (Eds.), *Special Issue of ACM SIGART Newsletter on Knowledge Acquisition* (pp. 166–168). New York: Special Interest Group on Artificial Intelligence, Association for Computing Machinery.

Kitto, C., & Boose, J. (1987). Heuristics for expertise transfer: The automatic management of complex knowledge acquisition dialogs. *International Journal of Man-Machine Studies*, *26*, 183–202.

Kitto, C., & Boose, J. (1987, June). Choosing knowledge acquisition strategies for application tasks. *WESTEX-87 Proceedings of the Western Conference on Expert Systems* (pp. 96–103). New York: IEEE.

Kitto, C. M., & Boose, J. H. (1989). Selecting knowledge acquisition tools and strategies based on application characteristics. *International Journal of Man-Machine Studies*, *31*, 149–160.

Klinker, G., Bentolila, J., Genetet, S., Grimes, M., & McDermott, J. (1987). KNACK: Report-driven knowledge acquisition. *International Journal of Man-Machine Studies*, *26*, 65–79.

Klinker, G., Genetet, S., & McDermott, J. (1988). Knowledge acquisition for evaluation systems. *International Journal of Man-Machine Studies*, *29*, 715–732.

Kolodner, J. L. (1984). *Retrieval and organizational strategies in conceptual memory: A computer model*. Hillsdale, NJ: Erlbaum.

Kopec, D., & Latour, L. (1989, April). Towards an expert/novice learning system with application to infectious disease. In C. R. Westphal & K. L. McGraw (Eds.), *Special Issue of ACM SIGART Newsletter on Knowledge Acquisition* (No. 108, pp. 165–166). New York: Special Interest Group on Artificial Intelligence, Association for Computing Machinery.

Kornecki, A. J. (1989, April). Operational knowledge acquisition problems for air traffic expert controller. In C. R. Westphal & K. L. McGraw (Eds.), *Special Issue of ACM SIGART Newsletter on Knowledge Acquisition* (No. 108, pp. 165–166). New York: Special Interest Group on Artificial Intelligence, Association for Computing Machinery.

Kornell, J. (1983). A VAX tuning expert built using automated knowledge acquisition. *Proceedings of First Conference on Artificial Intelligence Applications* (pp. 38–41). New York: IEEE.

Kornell, J. (1987). Formal thought and narrative thought in knowledge acquisition. *International Journal of Man-Machine Studies*, *26*, 203–212.

Kowalik, J. (1986). *Knowledge-based problem-solving*. Englewood Cliffs, NJ: Prentice-Hall.

Kraemer, K. L., & King, J. L. (1988). Computer-based systems for cooperative work and group decision making. *ACM Computing Surveys*, *20*, 115–146.

Kraiss, K.-F. (1986). Knowledge-based classification with interactive graphics. In E. Hollnagel, G. Mancini, & D. D. Woods (Eds.), *Intelligent decision support in process environments* (pp. 363–369). Berlin: Springer-Verlag.

Krishamurti, M., & Underbrink, A. J. (1989). Knowledge acquisition in a machine fault diagnosis shell. In C. R. Westphal & K. L. McGraw (Eds.), *Special Issue of ACM SIGART Newsletter on Knowledge Acquisition* (No. 108, pp. 84–92). New York: Special Interest Group on Artificial Intelligence, Association for Computing Machinery.

Krisnamurti, M., & Underbrink, A. J. (1990). Knowledge acquisition in a machine fault diagnosis shell. In K. L. McGraw & C. R. Westphal (Eds.), *Readings in knowledge acquisition* (pp. 124–136). New York: Ellis Horwood.

LaFrance, M. (1987). The knowledge acquisition grid: A method for training knowledge engineers. *International Journal of Man-Machine Studies*, *26*, 245–255.

Laird, J. E., Rosenbloom, P. S., & Newell, A. (1986). Chunking in SOAR: An architecture for general intelligence. *Machine Learning*, *1*, 11–46.

Laird, J. E., Rosenbloom, P. S., & Newell, A. (1987). SOAR: An architecture for general intelligence. *Artificial Intelligence*, *33*, 1–64.

Lance, G., & Williams, W. (1966). Computer programs for hierarchical polythetic classification. *Computing Journal*, *9*, 60–64.

Lancaster, J. S. (1990). Cognitively-based knowledge acquisition: Capturing categorical, temporal, and causal knowledge. In C. R. Westphal & K. L. McGraw (Eds.), *Readings in knowledge acquisition: Current practices and trends* (pp. 184–199). Chichester, UK: Horwood.

Lancaster, J., Westphal, C. R., & McGraw, K. L. (1989). A cognitively valid knowledge acquisition tool. In C. R. Westphal & K. L. McGraw (Eds.), *Special Issue of ACM SIGART Newsletter on Knowledge Acquisition* (No. 108, pp. 152–154). New York: Special Interest Group on Artificial Intelligence, Association for Computing Machinery.

Landau, J. A., Norwich, K. H., Evans, S. J., & Pich, B. (1987). An error correcting protocol for medical expert systems. *International Journal of Man-Machine Studies*, *26*, 617–626.

Laskey, K. B., Cohen, M. S., & Martin, A. W. (1989). Representing and eliciting knowledge for uncertain evidence and its implications. *IEEE Transactions on Systems, Man, and Cybernetics*, *19*, 558–573.

Lavrac, N. (1989). Methods for knowledge acquisition and refinement in second generation expert systems. In C. R. Westphal & K. L. McGraw (Eds.), *Special Issue of ACM SIGART Newsletter on Knowledge Acquisition* (No. 108, pp. 63–69). New York: Special Interest Group on Artificial Intelligence, Association for Computing Machinery.

LeClair, S. R. (1989). Interactive learning: A multi-expert paradigm for acquiring new knowledge. In C. R. Westphal & K. L. McGraw (Eds.), *Special Issue of ACM SIGART Newsletter on Knowledge Acquisition* (No. 108, pp. 34–44). New York: Special Interest Group on Artificial Intelligence, Association for Computing Machinery.

LeClair, S. R. (1990). Interactive learning: A multi-expert paradigm for acquiring new knowledge. In K. L. McGraw & C. R. Westphal (Eds.), *Readings in knowledge acquisition* (pp. 104–121). New York: Ellis Horwood.

Lee, J. K., Lee, I. K., Choi, R., & Ahn, S. M. (1990). Automatic rule generation by the transformation of an expert's diagram: LIFT. *International Journal of Man-Machine Studies*, *32*, 275–292.

Lefkowitz, L. S., & Lesser, V. R. (1988). Knowledge acquisition as knowledge assimilation. *International Journal of Man-Machine Studies*, *29*, 215–226.

Lehner, P. E., & Adelman, L. (1990). Behavioral decision theory and its implications for knowledge engineering. *The knowledge Engineering Review*, *5*, 5–14.

Lenat, D. B., Prakash, M., & Shepherd, M. (1986). CYC: Using common sense knowledge to overcome brittleness and knowledge acquisition bottlenecks. *The AI Magazine*, *7*, 65–85.

Littman, D. C. (1987). Modeling human expertise in knowledge engineering: Some preliminary observations. *International Journal of Man-Machine Studies*, *26*, 81–92.

Locke, C. (1989). Text-management tools for knowledge acquisition. *Expert Systems: Planning, implementation, integration*, *1*, 58–61.

Lucas, H. C., Jr., & Kaplan, R. B. (1974). A structured programming experiment. *Computer Journal*, *19*, 136–138.

MacDonald, B. A., & Witten, I. H. (1989). A framework for knowledge acquisition through techniques of concept learning. *IEEE Transactions on Systems, Man, and Cybernetics*, *19*, 499–512.

Malec, J. (1989). Knowledge elicitation during dynamic scene description. In C. R. Westphal & K. L. McGraw (Eds.), *Special Issue of ACM SIGART Newsletter on Knowledge Acquisition* (No. 108, pp. 162–163). New York: Special Interest Group on Artificial Intelligence, Association for Computing Machinery.

Manago, M., & Blythe, J. (1988). Learning disjunctive concepts. In K. Morik (Ed.), *Knowledge representation and organization in machine learning*. New York: Springer-Verlag.

Marcus, S. (1987). Taking backtracking with a grain of SALT. *International Journal of Man-Machine Studies*, *26*, 383–398.

Marcus, S. (Ed.). (1988). *Automating knowledge acquisition for expert systems*. Boston: Kluwer Academic.

Marcus, S., Stout, J., & McDermott, J. (1988). VT: An expert elevator designer that uses knowledge-based reasoning. *The AI Magazine*, *9*, 95–112.

Marshall, S. P. (1980). Procedural networks and production systems in adaptive diagnosis. *Instructional Science*, *9*, 129–143.

Martin, J. D., & Redmond, M. (1989, April). Acquiring knowledge by explaining observed problem solving. In C. R. Westphal & K. L. McGraw (Eds.), *Special Issue of ACM SIGART Newsletter on Knowledge Acquisition* (No. 108, pp. 77–83). New York: Special Interest Group on Artificial Intelligence, Association for Computing Machinery.

McDonald, J. E., Paap, K. R., & McDonald, D. R. (1990). Hypertext perspectives: Using pathfinder to build hypertext systems. In R. W. Schvaneveldt (Ed.), *Pathfinder associative networks: Studies in knowledge organization* (pp. 197–212). Norwood, NJ: Ablex.

McDonald, J. E., Plate, T. A., & Schvaneveldt, R. W. (1990). Using pathfinder to extract semantic information from text. In R. W. Schvaneveldt (Ed.), *Pathfinder associative networks: Studies in knowledge organization* (pp. 149–164). Norwood, NJ: Ablex.

McGraw, K. L. (1990). HyperKAT: A tool to manage and document knowledge acquisition. In K. L. McGraw C. R. Westphal & (Eds.), *Readings in knowledge acquisition* (pp. 164–181). New York: Ellis Horwood.

McGraw, K. L., & Westphal, C. R. (Eds.), (1990). *Readings in knowledge acquisition: Current prac-*

tices and trends. Chichester, England: Ellis Horwood.

Merrem, F. H. (1989). Automatic generation of knowledge structures: In C. R. Westphal & K. L. McGraw (Eds.), *Sepcial Issue of ACM SIGART Newsletter on Knowledge Acquisition* (No. 108, pp. 160–162). New York: Special Interest Group on Artificial Intelligence, Association for Computing Machinery.

Mettrey, W. (1987). An assessment of tools for building large knowledge-based systems. *The AI Magazine*, *8*(4), 81–89.

Micciche, P. F., & Lancaster, J. S. (1989). Applications of neurolinguistic techniques to knowledge acquisition. In C. R. Westphal & K. L. McGraw (Eds.), *Sepcial Issue of ACM SIGART Newsletter on Knowledge Acquisition* (No. 108, pp. 28–33). New York: Special Interest Group on Artificial Intelligence, Association for Computing Machinery.

Michalski, R. (1980). Knowledge acquisition through conceptual clustering: A theoretical framework and algorithm for partitioning data into conjunctive concepts. *International Journal of Policy Analysis and Information Systems*, *4*, 219–243.

Michalski, R. (1980). Pattern recognition as rule-guided inductive inference. *IEEE Transactions on Pattern Analysis and Machine Intelligence*, *2*, 349–361.

Michalski, R., Carbonell, J. G., & Mitchell, T. M. (Eds.). (1983). *Machine learning: An artificial intelligence approach*. Palo Alto, CA: Tioga Press.

Michalski, R., & Chilausky, R. L. (1980). Knowledge acquisition by encoding expert rules versus computer induction from examples: A case study involving soybean pathology. *International Journal of Man-Machine Studies*, *12*(1), 63–87.

Michalski, R., & Chilausky, R. L. (1981). Knowledge acquisition by encoding expert rules versus computer induction from examples: A case study involving soybean pathology. In E. H. Mamdani & B. R. Gaines (Eds.), *Fuzzy reasoning and its applications* (pp. 247–271). London: Academic Press.

Michalski, R., & Steep, R. (1983). Automated construction of classifications: conceptual clustering versus numerical taxonomy. *IEEE Transactions on Pattern Analysis and Machine Intelligence*, *5*, 396–409.

Michie, D. (1982). The state of the art in machine learning. In D. Michie (Ed.), *Introductory readings in expert systems* (pp. 208–228). London: Gordon and Breach.

Michie, D. (1983). Mind-like capabilities in computers: A note on computer induction. *Cognition*, *12*, 97–108.

Michie, D. (1983). Machine acquisition of expert rules: Do present methods face a "scaling up" issue? *Newsletter of the Specialist Group on Expert Systems*, *8*, 16–20. Cambridge: British Computer Society.

Mingers, J. (1987). Expert systems-rule induction with statistical data. *Journal of the Operational Research Society*, *38*, 39–47.

Mitchell, A. A. (1987). The use of alternative knowledge-acquisition procedures in the development of a knowledge-based media planning system. *International Journal of Man-Machine Studies*, *26*, 399–411.

Modesitt, K. L. (1990). Inductive knowledge acquisition: A case study of SCOTTY. In K. L. McGraw & C. R. Westphal (Eds.), *Readings in knowledge acquisition* (pp. 200–212). New York: Ellis Horwood.

Moore, E. A., & Agogino, A. M. (1987). INFORM: An architecture for expert-directed knowledge acquisition. *International Journal of Man-Machine Studies*, *26*, 213–230.

Morik, K. (1987). Acquiring domain models. *International Journal of Man-Machine Studies*, *26*, 93–104.

Morik, K. (1990). Integrating manual and automatic knowledge acquisition: BLIP. In K. L. McGraw & C. R. Westphal (Eds.), *Readings in knowledge acquisition* (pp. 213–232). New York: Ellis Horwood.

Motta, E., Rajan, T., & Eisenstadt, M. (1990). Knowledge acquisition as a process of model refinement. *Knowledge Acquisition: An International Journal*, *2*, 21–50.

Mozetic, I. (1986). Knowledge extraction through learning from examples. In T. M. Mitchell, J. G. Carbonell, & R. S. Michalski (Eds.), *Machine learning: A guide to current research* (pp. 227–231). Boston, MA: Kluwer Academic.

Mozetic, I. (1987). The role of abstractions in learning qualitative models. *Proceedings of the Fourth International Workshop on Machine Learning* (pp. 242–255). Irvine, CA: Morgan Kaufmann.

Mrozek, A. (1989). Rough sets and dependency analysis among attributes in computer implementations of expert's inference models. *International Journal of Man-Machine Studies*, *30*, 457–472.

Musen, M. A. (1989). *Automated generation of model-based knowledge-acquisition tools*. London: Pitman.

Musen, M. A. (1989). Knowledge acquisition at the metalevel: Creation of custom-tailored knowledge acquisition tools. In C. R. Westphal & K. L. McGraw (Eds.), *Special Issue of ACM SIGART Newsletter on Knowledge Acquisition* (No. 108, pp. 45–55). New York: Special Interest Group on Artificial Intelligence, Association for Computing Machinery.

Musen, M. A. (1990). Creating custom-tailored knowledge acquisition tools. In K. L. McGraw & C. R. Westphal (Eds.), *Readings in knowledge acquisition* (pp. 150–163). New York: Ellis Horwood.

Musen, M. A., Fagan, L. M., Combs, D. M., & Shortliffe, E. H. (1987). Use of a domain model to drive an interactive knowledge-editing tool. *International Journal of Man-Machine Studies*, *26*, 105–122.

Musen, M. A., Fagan, L. M., & Shortliffe, E. H. (1988). Graphical specification of procedural knowledge for an expert system. In J. Hendler (Ed.), *Expert systems: The user interface* (pp. 15–35). Norwood, NJ: Ablex.

Myler, H. R., & Gonzales, A. J. (1989). Automated design data capture using relaxation techniques. In C. R. Westphal & K. L. McGraw (Eds.), *Special Issue of ACM SIGART Newsletter on Knowledge Acquisition* (No. 108, pp. 169–170). New York: Special Interest Group on Artificial Intelligence, Association for Computing Machinery.

Narayanan, N. H., & Viswanadham, N. (1987). A methodology for knowledge acquisition and reasoning in failure analysis of systems. *IEEE Transactions on Systems, Man, and Cybernetics*, *SMC-17*, 274–288.

Neale, I. M. (1989). First generation expert systems: A review of knowledge acquisition methodologies. *The Knowledge Engineering Review*, *3*, 105–145.

Neves, D., & Anderson, J. R. (1981). Compilation: A mechanism for the automitization of cognitive skills. In J. R. Anderson (Ed.), *Cognitive skills and their acquisition*. Hillsdale, NJ: Erlbaum.

Niblett, T. B. (1988). Yapes: Yet another PROLOG expert system. In J. E. Hayes, D. Michie, & J. Richards (Eds.), *Machine intelligence 11* (pp. 167–192). Oxford, England: Clarendon Press.

Nobel, D. (1989). Schema-based knowledge elicitation for planning and situation assessment aids. *IEEE Transactions on Systems, Man, and Cybernetics*, *19*, 473–482.

Norros, L. (1986). Expert knowledge, its acquisition and elicitation in developing intelligent tools for process control. In E. Hollnagel, G. Mancini, & D. D. Woods (Eds.), *Intelligent decision support in process environments* (pp. 137–144). Berlin: Springer-Verlag.

Olson, J., & Reuter, H. (1987). Extracting expertise form experts: Methods for knowledge acquisition. *Expert Systems*, *4*, 152–168.

O'Neill, J. L. (1987). Knowledge acquisition for radar classification. In J. R. Quinlan (Ed.), *Application of expert systems* (pp. 184–199). Reading, MA: Addison-Wesley.

Onorato, L. A. (1990). Representation of problem schemata. In R. W. Schvaneveldt (Ed.), *Pathfinder associative networks: Studies in knowledge organization* (pp. 255–266). Norwood, NJ: Ablex.

Pau, L. F. (1986). Knowledge engineering techniques applied to fault detection test generation and maintenance. In J. K. Skwirzynski (Ed.), *Software system design methods*. Berlin: Springer-Verlag.

Pau, L. F., & Nielsen, S. S. (1989, April). Conceptual graphs as a visual language for knowledge acquisition in architectural expert systems. In C. R. Westphal & K. L. McGraw (Eds.), *Special Issue of ACM SIGART Newsletter on Knowledge Acquisition* (No. 108, p. 151). New York: Special Interest Group on Artificial Intelligence, Association for Computing Machinery.

Pau, L. F., Xie, X., & Westphal, C. R. (1989). A knowledge based editor for sensor fusion. In J. K. Aggarwal (Ed.), *Multisensor fusion for computer vision*. New York: Springer-Verlag.

Pazzani, M. J. (1987). Explanation-based learning for knowledge-based systems. *International Journal of Man-Machine Studies*, *26*, 413–434.

Pearce, D. A. (1988). The induction of fault diagnosis systems from qualitative models. *Proceedings of the National Conference on Artificial Intelligence, AAAI-88* (pp. 353–357). St. Paul, MN: Morgan Kaufmann.

Phillips, B., Messick, S. L., Freiling, M. J., & Alexander, J. H. (1985). INKA: The English knowledge acquisition interface for electronic instruments' troubleshooting systems. *Proceedings of the Second Conference on Artificial Intelligence Applications* (pp. 676–681). New York: IEEE.

Plaza, E., & de Mantaras, R. L. (1989). Model-based knowledge acquisition for heuristic classifications systems. In C. R. Westphal & K. L. McGraw (Eds.), *Special Issue of ACM SIGART Newsletter on Knowledge Acquisition* (No. 108, pp. 98–105). New York: Special Interest Group on Artificial Intelligence, Association for Computing Machinery.

Prerau, D. (1985). Selection of an appropriate domain for an expert system. *The AI Magazine*, *6*, 26–30.

Quinlan, J. R. (1982). Learning efficient classification procedures and their applications to chess end-games. In R. Michalski, J. G. Carbonell, & T. M. Mitchell (Eds.), *Machine learning: An artificial intelligence approach* (pp. 463–482). Palo Alto, CA: Tioga Press.

Quinlan, J. R. (1982). Semi-autonomous acquisition for pattern-based knowledge. In D. Michie (Ed.), *Introductory readings in expert systems* (pp. 192–207). New York: Gordon and Breach.

Quinlan, J. R. (1986). Induction of decision tress. *Machine Learning*, *1*, 81–106.

Quinlan, J. R. (1987). Inductive knowledge acquisition: A case study. In J. R. Quinlan (Ed.), *Applications of expert systems* (pp. 157–173). Reading, MA: Addison-Wesley.

Ralescu, A. L., & Baldwin, J. F. (1989). Concept learning from examples and counter examples. *International Journal of Man-Machine Studies*, *30*, 329–354.

Rappaport, A. T. (1987). Multiple-problem subspaces in the knowledge-design process. *International Journal of Man-Machine Studies*, *26*, 435–452.

Rappaport, A. T. (1988). Cognitive primitives. *International Journal of Man-Machine Studies*, *29*, 733–747.

Rappaport, A. T., & Gaines, B. R. (1990). Integrated knowledge base building environments. *Knowledge Acquisition: An International Journal*, *2*, 51–72.

Rasmus, D. W. (1988, January). Expert input. *MacUser*, pp. 136–150.

Rasmussen, J. (1986). A framework for cognitive task analysis in systems design. In E. Hollnagel, G. Mancini, & D. D. Woods (Eds.), *Intelligent decision support in process environments* (pp. 175–196). Berlin: Springer-Verlag.

Rendell, L. A. (1983, Winter). Toward a unified approach for conceptual knowledge acquisition. *The AI Magazine*, 19–27.

Richer, M. H., & Clancey, W. J. (1985). GUIDON-WATCH: A graphic interface for viewing a knowledge-based system. *IEEE Computer Graphics and Applications*, *5*, 51–64.

Robinson, V., Hardy, N. W., Barnes, D. P., Pace, C. J., & Lee, M. H. (1987). Experiences with a knowledge engineering toolkit: An assessment in industrial robotics. *The Knowledge Engineering Review*, *2*, 43–54.

Rodi, L. L., Pierce, J. A., & Dalton, R. E. (1989). Putting the expert in charge: Graphical knowledge acquisition for fault diagnosis and repair. In C. R. Westphal & K. L. McGraw (Eds.), *Special Issue of ACM SIGART Newsletter on Knowledge Acquisition* (No. 108, pp. 56–62). New York: Special Interest Group on Artificial Intelligence, Association for Computing Machinery.

Rodi, L. L., Pierce, J. A., & Dalton, R. E. (1990). Graphical knowledge acquisition for fault diagnosis and repair. In K. L. McGraw & C. R. Westphal (Eds.), *Readings in knowledge acquisition* (pp. 137–149). New York: Ellis Horwood.

Rook, F. W., & Crogan, J. W. (1989). The knowledge acquisition activity matrix: A systems engineering conceptual framework. *IEEE Transactions on Systems, Man, and Cybernetics*, *19*, 586–597.

Roske-Hofstrand, R. J., & Papp, K. R. (1990). Discriminating between degrees of low or high similarity: Implications for scaling techniques using semantic judgments. In R. W. Schvaneveldt (Ed.), *Pathfinder associative networks: Studies in knowledge organization* (pp. 61–74). Norwood, NJ: Ablex.

Rouse, W. B., Hammer, J. M., & Lewis, C. M. (1989). On capturing humans skills and knowledge algorithmic approaches to model identification. *IEEE Transactions on Systems, Man, and Cybernetics*, *19*, 558–573.

Rubin, D. C. (1990). Directed graphs as memory representations: The case of rhyme. In R. W. Schvaneveldt (Ed.), *Pathfinder associative networks: Studies in knowledge organization* (pp. 121–134). Norwood, NJ: Ablex.

Rusk, R. A., & Krone, R. M. (1984). The crawford slip method (CSM) as a tool for extraction of expert knowledge. In G. Salvendy (Ed.), *Human-computer interaction* (pp. 279–282). New York: Elsevier.

Saaty, T. L. (1981). *The analytic hierarchy process*. New York: McGraw-Hill.

Schvaneveldt, R. W. (1990). Proximities, networks, and schemata. In R. W. Schvaneveldt (Ed.), *Pathfinder associative networks: Studies in knowledge organization* (pp. 135–148). Norwood, NJ: Ablex.

Schweickert, R. (1987). Comparing knowledge elicitation techniques: A case study. *Artificial Intelligence Review*, *1*, 245–253.

Senjen, R. (1988). Knowledge acquisition by experiment: Developing test cases for an expert system. *AI Applications*, *2*, 52–55.

Shalin, V. L., Wisniewski, E. J., Levi, K. R., & Scott, P. D. (1988). A journal analysis of machine learning systems for knowledge elicitation. *International Journal of Man-Machine Studies*, *29*, 429–446.

Shapiro, A. D. (1987). *Structured induction in expert systems*. Reading, MA: Addison-Wesley.

Shaw, M. (1981). *Recent advances in personal construct technology*. New York: Academic Press.

Shaw, M. (1989). A grid-based tool for knowledge acquisition: Validation with multiple experts. In C. R. Westphal & K. L. McGraw (Eds.), *Special Issue of ACM SIGART Newsletter on Knowledge Acquisition* (No. 108, pp. 168–169). New York: Special Interest Group on Knowledge Acquisition, Association for Computing Machinery.

Shaw, M., & Gaines, B. R. (1987). KITTEN: Knowledge initiation and transfer tools for experts and novices. *International Journal of Man-Machine Studies*, *27*, 251–280.

Shaw, M., & Gaines, B. R. (1987). An interactive knowledge elicitation technique using personal construct technology. In A. L. Kidd (Ed.), *Knowledge acquisition for expert systems: A practical handbook*. New York: Plenum.

Shaw, M., & Gaines, B. R. (1989). Comparing conceptual structures: Consensus, conflict, correspondence, and contrast. *Knowledge Acquisition: An International Journal*, *1*, 341–346.

Shema, D. B., & Boose, J. H. (1988). Refining problem-solving knowledge in repertory grids using a consultation mechanism. *International Journal of Man-Machine Studies*, *29*, 447–460.

Siler, W., & Tucker, D. (1989). Patterns of inductive reasoning in a parallel expert system. *International Journal of Man-Machine Studies*, *30*, 113–127.

Silverman, B. G., Wenig, R. G., & Wu, T. (1989). COPEing with ongoing knowledge acquisition from collaborating hierarchies of experts. In C. R. Westphal & K. L. McGraw (Eds.), *Special Issue of ACM SIGART Newsletter on Knowledge Acquisition* (No. 108, pp. 170–171). New York: Special Interest Group on Artificial Intelligence, Association for Computing Machinery.

Silvestro, K. (1988). An explanation-based approach to knowledge-base acquisition. *International Journal of Man-Machine Studies*, *29*, 159–169.

Silvestro, K. (in press). An explanation-based approach to knowledge-based acquisition. *International Journal of Man-Machine Studies*.

Slator, B. M. (1989). Extracting lexical knowledge from dictionary text. In C. R. Westphal & K. L. McGraw (Eds.), *Special Issue of ACM SIGART Newsletter on Knowledge Acquisition* (No. 108, pp. 173–174). New York: Special Interest Group on Knowledge Acquisition, Association for Computing Machinery.

Stefik, M., Foster, G., Bobrow, D., Kahn, K., Lanning, S., & Suchman, L. (1987). Beyond the chalkboard: Computer support for collaboration and problem solving in meetings. *Communications of the ACM*, *30*, 32–47.

Stewart, V., & Stewart, A. (1981). *Business applications of repertory grids*. London: McGraw-Hill.

Subramanian, S., & Freuder, E. C. (1989). Compiling rules from constraint satisfaction problem solving. In C. R. Westphal & K. L. McGraw (Eds.), *Special Issue of ACM SIGART Newsletter on Knowledge Acquisition* (No. 108, pp. 177–178). New York: Special Interest Group on Artificial Intelligence, Association for Computing Machinery.

Swartout, W. R. (1983). XPLAIN: A system for creating and explaining expert consulting programs. *Artificial Intelligence*, *21*, 285–325.

Thieme, S. (1988). The acquisition of model knowledge for a model driven machine learning approach. In K. Morik (Ed.), *Knowledge representation and organization in machine learning*. New York: Springer-Verlag.

Thomas, M. (1988). Structured techniques for the development of expert systems. In D. S. Moralee (Ed.), *Research and development in expert systems IV* (pp. 258–266). Cambridge: Cambridge University Press.

Tiemann, P. W., & Markle, S. (1984). On getting expertise into an expert system. *Performance and Instruction Journal*, *23*, 25–29.

Tranowski, D. (1988). A knowledge acquisition environment for scene analysis. *International Journal of Man-Machine Studies*, *29*, 197–214.

VanLehn, K. (1987). Learning one subprocedure per lesson. *Artificial Intelligence*, *31*, 1–40.

Velardi, P. (1989). Acquisition of semantic patterns from a natural corpus of texts. In C. R. Westphal & K. L. McGraw (Eds.), *Special Issue of ACM SIGART Newsletter on Knowledge Acquisition* (No. 108, pp. 115–123). New York: Special Interest Group on Artificial Intelligence, Association for Computing Machinery.

Virkar, R. S., & Reach, J. W. (1988). Direct assimilation of expert level knowledge by automatically parsing research paper abstracts. *International Journal of Expert Systems*, *1*, 281–306.

Waldron, V. (1985). Process tracing as a means of collecting knowledge for expert systems. *Texas Instruments Engineering Journal*, *2*, 90–93.

Weilinga, B. J., & Breuker, J. A. (1984). Interpretation models for knowledge acquisition. In T. O'Shea (Ed.), *Advances in artificial intelligence* (pp. 3–12). Amsterdam: North-Holland.

Weitzel, J. R., & Kerschberg, L. (1989). A system development methodology for knowledge-based systems. *IEEE Transactions on Systems, Man, and Cybernetics*, *19*, 598–605.

Welbank, M. (1987). Perspectives on knowledge acquisition. In C. J. Pavelin & M. D. Wilson (Eds.), *Knowledge acquisition for engineering applications* (pp. 13–19).

Wells, T. L. (1989). Hypertext as a means for knowledge acquisition. In C. R. Westphal & K. L. McGraw (Eds.), *Special Issue of ACM SIGART Newsletter on Knowledge Acquisition* (No. 108, pp. 136–138). New York: Special Interest Group on Artificial Intelligence, Association for Computing Machinery.

Westphal, C. R., & Recker, L. H. (1990). Reasoning and representation mechanisms for multiple-expert knowledge acquisition. In K. L. McGraw & C. R. Westphal (Eds.), *Readings in knowledge acquisition* (pp. 90–103). New York: Eills Horwood.

Wick, M. R., & Slagle, J. R. (1989). The partitioned support network for expert system justification. *IEEE Transactions on Systems, Man, and Cybernetics*, *19*, 528–536.

Wilkins, D. C., Clancey, W. J., & Buchanan, B. G. (1987). Knowledge base retirement by monitoring abstract control language. *International Journal of Man-Machine Studies*, *27*, 281–294.

Williams, C. (1986). Expert systems, knowledge engineering, and AI tools: An overview: *IEEE Expert*, 66–70.

Winston, P. H. (1982). Learning new principles from precedents and exercises. *Artificial Intelligence*, *19*, 321–350.

Wisniewski, E., Winston, H., Smith, R., & Kleyn, M. (1987). A conceptual clustering program for rule generation. *International Journal of Man-Machine Studies*, *27*, 281–294.

Witten, I. H., & MacDonald, B. (1988). Using concept learning for knowledge acquisition. *International Journal of Man-Machine Studies*, *29*, 171–196.

Wright, G., & Ayton, P. (1987). Eliciting and modelling expert knowledge. *Decision Support Systems*, *3*(1), 13–26.

Woods, D. D., & Hollnagel, E. (1987). Mapping cognitive demands in complex problem-solving worlds. *International Journal of Man-Machine Studies*, *26*, 257–276.

Woods, W. A. (1983). What's important about knowledge representation? *Computer*, *16*, 22–27.

Yamada, S., & Tsuji, S. (1989). Acquisition of macro-operators from worked examples in problem solving. In C. R. Westphal & K. L. McGraw (Eds.), *Special Issue of ACM SIGART Newsletter on Knowledge Acquisition* (No. 108, pp. 171–172). New York: Special Interest Group on Artificial Intelligence, Association for Computing Machinery.

Yaski, R., & Ziarbo, W. (1988). An expert system for conceptual schema design: A machine learning approach. *International Journal of Man-Machine Studies*, *29*, 351–376.

Appendix E
Bibliography: Expertise in Programming

Adelson, B. (1981). Problem solving and the development of abstract categories in programming languages. *Memory & Cognition*, *9*, 422–433.

Adelson, B. (1984). When novices surpass experts: The difficulty of a task may increase with expertise. *Journal of Experimental Psychology: Learning, Memory and Cognition*, *10*, 483–495.

Adelson, B., & Soloway, E. (1988). A model of software design. In M. T. H. Chi, R. Glaser, & M. J. Farr (Eds.), *The nature of expertise* (pp. 185–208). Hillsdale, NJ: Erlbaum.

Allwood, C. M. (1986). Novices on the computer: A review of the literature. *International Journal of Man-Machine Studies*, *25*, 633–658.

Allwood, C. M., & Eliasson, M. (1987). Analogy and other sources of difficulty in novices' very first text-editing. *International Journal of Man-Machine Studies*, *27*, 1–22.

Allwood, C. M., & Eliasson, M. (1988). Question-asking when learning a text-editing system. *International Journal of Man-Machine Studies*, *29*, 63–80.

Alty, J., & Coombs, M. (1981). Communicating with university computer users: A case study. In M. Coombs & J. Alty (Eds.), *Computing skills and the user interface* (pp. 7–71). London: Academic Press.

Anderson, J. R. (1989). Analogical origins of errors in problem-solving. In D. Klahr & K. Kotovsky (Eds.), *Complex infromation processing: The impact of Herbert A. Simon* (pp. 343–372). Hillsdale, NJ: Erlbaum.

Anderson, J. R., Pirolli, P., & Farrell, R. (1988). Learning to program recursive functions. In M. T. H. Chi, R. Glaser, & M. J. Farr (Eds.), *The nature of expertise* (pp. 153–184). Hillsdale, NJ: Erlbaum.

Atwood, M., & Ramsey, H. R. (1978). *Cognitive structures in the comprehension and memory of computer programs: An investigation of computer program debugging* (Report No. SAI–77–194–DEN). Englewood, NJ: Science Applications, Inc.

Atwood, M., Turner, A., Ramsey, H. R., & Hooper, J. (1979). *An exploratory study of the congitive structures underlying the comprehension of software design problems* (Report No. SAI–79–100–DEN). Englewood, NJ: Science Applications, Inc.

Barfield, W. (1986). Expert-novice differences for software: Implications for problem solving and knowledge acquisition. *Behaviour and Information Technology*, *5*, 15–29.

Bateson, A. G., Alexander, R. A., & Murphy, M. D. (1987). Cognitive processing differences between novice and expert computer programmers. *International Journal of Man-Machine Studies*, *26*, 649–661.

Bonar, J., & Soloway, E. (1985). Pre-programming knowledge: A major source of misconceptions in novice programmers. *Human-Computer Interaction*, *1*, 133–161.

Briggs, P. (1990). Do they know what they're doing? An evaluation of word-processor users' implicit and explicit task-relevant knowledge, and its role in self-directed learning. *International Journal of Man-Machine Studies*, *32*, 385–398.

Campbell, R. L., & Brickhard, M. H. (1986). *Knowing levels and developmental stages*. Basel: Karger.

Card, S. K., Moran, J. P., & Newell, A. (1983). *The psychology of human-computer interaction*. Hillsdale, NJ: Erlbaum.

Carroll, J. M. (Ed.). (1987). *Interfacing thought: Cognitive aspects of human-computer interaction*. Cambridge, MA: Bradford Books/MIT Press.

Carroll, J. M., & Mack, R. L. (1983). Actively learning to use a word processor. In W. Cooper (Ed.), *Cognitive aspects of skilled typewriting* (pp. 259–282). New York: Springer-Verlag.

Carroll, J. M., & Mack, R. L. (1984). Learning to use a word processor: By doing, thinking, and knowing. In M. Schneider & J. Thomas (Eds.), *Human factors in computer systems*. Hillsdale, NJ: Ablex.

Carroll, J. M., & Mack, R. L. (1985). Metaphor, computing systems, and active learning. *International Journal of Man-Machine Studies, 91*, 39–57.

Carroll, J. M., Mack, R. L., & Kellogg, W. (1988). In M. Helander (Ed.), *Handbook of human-computer interaction* (pp. 67–85). New York: Springer-Verlag.

Carroll, J. M., Mack, R. L., Lewis, C., Grischkowsky, N., & Robertson, S. (1985). Exploring a word processor. *Human-Computer Interaction, 1*, 283–307.

Carroll, J. M., & Mazur, S. (1986, November). Lisa learning. *Computer Magazine*, pp. 35–49.

Carver, D. L. (1989). Programmer variations in software debugging approaches. *International Journal of Man-Machine Studies, 31*, 315–322.

Charney, D. H., & Reder, L. M. (1986). Designing interface tutorials for computer users. *Human-Computer Interaction, 2*, 297–317.

Cooke, N. M., & Schvaneveldt, R. (1988). Effects of computer programming experience on network representations of abstract programming concepts. *International Journal of Man-Machine Studies, 29*, 407–427.

Cramer, M. (1990). Structure and mnemonics in computer and command languages. *International Journal of Man-Machine Studies, 32*, 707–722.

Cross, S. E. (1983). A qualitative reasoning approach to mathematical and heuristic knowledge integration. *Proceedings of CHI'83 Conference on Human Factors in Computing Systems* (pp. 186–189). New York: Association for Computing Machinery.

Coventry, L. (1989). Some effects of cognitive style on learning UNIX. *International Journal of Man-Machine Studies, 31*, 349–366.

Curry, R. E. (1981). A model of human fault detection for computer dynamic processes. In J. Rasmussen & W. B. Rouse (Eds.), *Human detection and diagnosis of system failures* (pp. 171–183). New York: Plenum.

Curtis, B. (1986). By the way, did anyone study any real programmers? In E. Soloway & S. Iyengar (Eds.), *Empirical studies of programmers* (pp. 256–262). Norwood, NJ: Ablex.

Curtis, B. (1988). The impact of individual differences in programmers. In G. C. van der Veer, T. R. G. Green, J. M. Hoc, & D. M. Murray (Eds.), *Working with computers: Theory versus outcome* (pp. 279–294). London: Academic Press.

Curtis, B., Krasner, H., & Iscoe, N. (1988). A field study of the software design process for large systems. *Communications of the ACM, 31*, 1268–1287.

Davies, S. P. (1990). The nature and development of programming plans. *International Journal of Man-Machine Studies, 32*, 461–482.

Douglas, S. & Moran, T. (1983). Learning text-editing semantics by analogy. *Proceedings of CHI'83 Conference on Human Factors in Computer Systems* (pp. 207–211). New York: Association for Computing Machinery.

Ehrlich, K., & Soloway, E. (1984). An empirical investigation or the tacit plan knowledge in programming. In J. C. Thomas & M. L. Schneider (Eds.), *Human factors in computer systems* (pp. 113–133). Norwood, NJ: Ablex.

Gomez, L. M., & Dumais, S. T. (1986). Putting cognition to work: Examples from computer system design. In T. J. Knapp & L. C. Robertson (Eds.), *Approaches to cognition: Contrasts and controversies* (pp. 267–290). Hillsdale, NJ: Erlbaum.

Göranzon, B. (1988). The practice and use of computers: A paradoxical encounter between different traditions of knowledge. In B. Göranzon & I. Josefson (Eds.), *Knowledge, skill, and artificial intelligence* (pp. 9–18). London: Springer-Verlag.

Gould, J. D. (1975). Some psychological evidence on how people debug computer programs. *International Journal of Man-Machine Studies, 7*, 151–182.

Green, A. J. K., & Gilhooly, K. J. (1990). Individual differences and effective learning procedures: The case of statistical computing. *International Journal of Man-Machine Studies, 32*, 97–105.

Greene, S. L., Devlin, S. J., Cannata, P. E., & Gomez, L. M. (1990). No IFs, ANDs, or ORs: A study of database querying. *International Journal of Man-Machine Studies, 32*, 303–326.

Guindon, R., Krasner, H., & Curtis, B. (1987). Breakdowns and processes during the early activities of software design by professionals. In G. M. Olson, S. Sheppard, & E. Soloway (Eds.), *Empirical studies of programmers: Second workshop* (pp. 65–82). Norwood, NJ: Ablex.

Hanisch, K. A., Kramer, A. F., Hulin, C. L., & Schumacher, R. (1988). Novice-expert differences in the cognitive representation of computing systems: Mental models and verbalizable knowledge. *Procedings of the Human Factors Society 3rd Annual Meeting* (pp. 219–223). Santa Monica, CA: Human Factors Society.

Jeffries, R., Turner, A., Polson, P., & Atwood, M. (1981). The processes involved in designing software. In J. R. Anderson (Ed.), *Cognitive skills and their acquisition* (pp. 255–283). Hillsdale, NJ: Erlbaum.

Kagan, D. M., & Pietron, L. R. (1986). Aptitude for computer literacy. *International Journal of Man-Machine Studies*, *26*, 649–661.

Kato, T. (1986). What "question asking protocols" can say about the user interface. *International Journal of Man-Machine Studies*, *25*, 659–673.

Khalil, O. E. M., & Clark, J. D. (1989). The influence of programmers' cognitive complexity on program comprehension and modification. *International Journal of Man-Machine Studies*, *31*, 219–236.

Kieras, D. (1988). Towards a practical GOMS model methodology for user interface design. In M. Helander (Ed.), *Handbook of human-computer interaction* (pp. 135–158). Amsterdam: Elsevier Science.

Kieras, D., & Polson, P. (1985). An approach to the formal analysis of user complexity. *International Journal of Man-Machine Studies*, *22*, 365–394.

Knox, S. T., Bailey, W. A., & Lynch, E. F. (1989, April). Directed dialogue protocols: Verbal data for user interface design. *Proceedings of CHI, 1989* (pp. 283–288). New York: Association for Computing Machinery.

Lammers, S. (1986). *Programmers at work: Interviews*. Redmond, WA: Microsoft Press.

Larsen, S. F. (1986). Procedural thinking, programming, and computer use. In E. Hollnagel, G. Mancini, & D. D. Woods (Eds.), *Intelligent decision support in process environments* (pp. 145–150). Berlin: Springer-Verlag.

Leventhal, L. M. (1988). Experience of programming beauty: Some patterns of programming aesthetics. *International Journal of Man-Machine Studies*, *28*, 525–550.

Littman, D. C. (1987). Modeling human expertise in knowledge engineering: Some preliminary observations. *International Journal of Man-Machine Studies*, *26*, 81–92.

Mack, R. (1989). Understanding and learning text-editing skills: Observations on the role of new user expectations. In S. Robertson, W. Zachery, & J. Black (Eds.), *Cognition, computers and interaction* (pp. 304–357). Hillsdale, NJ: Erlbaum.

Mack, R., Lewis, C., & Carroll, J. (1983). Learning to use a word processor: Problems and prospects. *ACM Transactions on Office Information Systems*, *1*, 254–271.

Mayer, R. E. (1976). Some conditions of meaningful learning for computer programming: Advance organizers and subject control of frame order. *Journal of Educational Psychology*, *67*, 725–734.

Mayer, R. E. (1988). From novice to expert. In M. Helander (Ed.), *Handbook of human computer interaction* (pp. 569–580). Amsterdam: North-Holland.

McKeithen, K. B., Reitman, J. S., Reuter, H. H., & Hirtle, S. C. (1981). Knowledge organization and skill differences in computer programmers. *Cognitive Psychology*, *13*, 307–325.

Mynatt, B. T., Smith, K. H., Kamouri, A. L., & Tykodi, T. A. (1986). Which way to computer literacy, programming or applications experience? *International Journal of Man-Machine Studies*, *25*, 557–572.

Pennington, N. (1987). Stimulus structures and mental representations in explicit comprehension of computer programs. *Cognitive Psychology*, *19*, 295–341.

Pilgrim, J. (1990). On the training of EDP novices on the personal computer. *International Journal of Man-Machine Studies*, *32*, 399–422.

Redmond, R. T., & Gasen, J. B. (1989). Measuring change in the programming process. *International Journal of Man-Machine Studies*, *30*, 697–715.

Reitman-Olson, J. (1988). Cognitive analysis of people's use of software. In J. Carroll (Ed.), *Interfacing thought: Cognitive aspects of human-computer interaction* (pp. 260–293). Cambridge, MA: The MIT Press.

Robertson, S. W., Zachery, W., & Black, J. (Eds.). (1989). *Cognition, computers, and interaction*. Hillsdale, NJ: Erlbaum.

Rosson, M. B., & Gold, E. (1989). Problem-solution mapping in object-oriented design. In N. Meyrowitz (Ed.), *OOPSLA'89 Conference Proceedings* (pp. 7–10). New York: Association for Computing Machinery.

Schmidt, A. L. (1986). Effects of experience and comprehension on reading time and memory for computer programs. *International Journal of Man-Machine Studies*, *25*, 399–410.

Schneiderman, B. (1976). Exploratory experiments in programmer behavior. *International Journal of Computer and Information Sciences*, *5*, 123–143.

Schneiderman, B., & Mayer, R. (1979). Syntactic/semantic interactions in programmer behavior: A model and experimental results. *International Journal of Computer and Information Sciences*, *8*, 219–239.

Sheil, B. A. (1981). The psychological study of programming. *ACM Computing Surveys*, *13*, 101–120.

Sheppard, S. B., Bailey, J. W., & Bailey, E. K. (1984). An empirical evaluation of software documentation formats. In J. C. Thomas & M. L. Schneider (Eds.), *Human factors in computer systems* (pp. 135–164). Norwood, NJ: Ablex.

Siddiqi, I. A., & Ratcliff, B. (1989). Specification influences in program design. *International Journal of Man-Machine Studies*, *31*, 393–404.

Soloway, E., Adelson, B., & Ehrlich, K. (1988). Knowledge and processes in the comprehension of computer programs. In M. T. H. Chi, R. Glaser, & M. J. Farr (Eds.), *The nature of expertise* (pp. 129–152). Hillsdale, NJ: Erlbaum.

Soloway, E., Ehrlich, K., Bonar, J., & Greenspan, J. (1982). Tapping into TACIT programming knowledge. *Proceedings of CHI'82 Conference on Human Factors in Computing Systems* (pp. 52–57). New York: Association for Computing Machinery.

Soloway, E., & Iyengar, S. (Eds.), *Empirical studies of programmers*. Norwood, NJ: Ablex.

Soloway, E., & Spohrer, J. C. (Eds.). (1988). *Studying the novice programmer*. Hillsdale, NJ: Erlbaum.

Spavold, J. (1990). The child as naive user: A study of database use with young children. *International Journal of Man-Machine Studies*, *32*, 603–626.

Stone, D. N., Jordan, E. W., & Wright, M. K. (1990). The impact of Pascal education on debugging skill. *International Journal of Man-Machine Studies*, *32*, 81–96.

Thomas, J. C., & Schneider, M. L. (Eds.). (1984). *Human factors in computer systems*. Norwood, NJ: Ablex.

Van der Veer, G. C., Green, T. R., Hoc, J. M., & Murray, D. M. (Eds.). (1988). *Working with computers: Theory versus outcome*. London: Academic Press.

Vessey, I. (1985). Expertise in debugging computer programs: A process analysis. *International Journal of Man-Machine Studies*, *23*, 459–494.

Vessey, I. (1987). On matching programmers' chunks with program structures: An empirical investigation. *International Journal of Man-Machine Studies*, *27*, 65–90.

Vessey, I. (1988). Expert-novice organization: An empirical investigation using computer program recall. *Behaviour and Information Technology*, *7*, 153–171.

Vessey, I. (1989). Toward a theory of computer program bugs: An empirical test. *International Journal of Man-Machine Studies*, *30*, 23–46.

Weiderback, S. (1986). Beacons in computer program comprehension. *International Journal of Man-Machine Studies*, *25*, 697–710.

Weiser, M., & Shertz, J. (1983). Programming problem representation in novice and expert programmers. *International Journal of Man-Machine Studies*, *14*, 391–396.

Wickens, C. D., & Kessel, C. (1979). The effects of participatory mode and task workload on the detection of dynamic system failures. *IEEE Transactions on Systems, Man, and Cybernetics*, *13*, 24–31.

Young, R. M. (1983). Surrogates and mappings: Two kinds of conceptual models for interactive devices. In D. Gentner & A. Stevens (Eds.), *Mental models* (pp. 35–52). Hillsdale, NJ: Erlbaum.

Youngs, E. A. (1974). Human errors in programming. *International Journal of Man-Machine Studies*, *6*, 361–374.

Zissos, A., & Witten, I. (1985). User modelling for a computer coach: A case study. *International Journal of Man-Machine Studies*, *23*, 729–750.

Appendix F
Bibliography: AI Theory, Philosophy, and Reviews of Expert Systems

Adelman, L. (1989). Measurement issues in knowledge engineering. *IEEE Transactions on Systems, Man, and Cybernetics, 19*, 483–488.

Adelson, B., & Soloway, E. (1988). A model of software design. In M. T. H. Chi, R. Glaser, & M. J. Farr (Eds.), *The nature of expertise* (pp. 185–208). Hillsdale, NJ: Erlbaum.

Aikins, J. (1983). Prototypical knowledge for expert systems. *Artificial Intelligence, 20*, 163–210.

Aleksander, I. (1984). *Designing intelligent systems: An introduction*. London: Kogan Page.

Alexander, T. (1984, August). Why computers can't outthink the experts. *Fortune*, pp. 105–118.

Allen, R. B. (1990). User models: Theory, method, and practice. *International Journal of Man-Machine Studies, 32*, 511–544.

Anderson, J. R., Boyle, C. I., Corbett, A. T., & Lewis, M. W. (1990). Cognitive modeling and intelligent tutoring. *Artificial Intelligence, 42*, 7–49.

Banning, R. W. (1984). Knowledge acquisition and system validation in expert systems for management. *Human Systems, 4*, 280–285. New York: Elsevier (North-Holland).

Barr, A. (Ed.). (1983). A1: Cognition as computation. In F. Machlup & U. Mansfield (Eds.), *The study of information: Interdisciplinary messages* (pp. 237–262). New York: Wiley.

Barr, A., & Feigenbaum, E. (1981). *The handbook of artificial intelligence*. Los Altos, CA: Kaufman.

Berry, D. C., & Broadbent, D. E. (1986). Expert systems and the man-machine interface. *Expert Systems, 3*, 228–231.

Bierre, P. (1985, Winter). The professor's challenge. *The AI Magazine*, 60–70.

Bloomfield, B. P. (1986). Epistemology for knowledge engineers. *Communication and Cognition, 3*, 305–320.

Bloomfield, B. P. (Ed.). (1987). *The question of artificial intelligence: Philosophical and sociological perspectives*. London: Croon Helm.

Bloomfield, B. P. (1988). Expert systems and human knowledge: A view from the philosophy of science. *AI and Society, 2*, 17–29.

Bobrow, D. G., Mittal, S., & Stefik, M. J. (1986). Expert systems: Perils and promises. *Communications of the ACM, 29*(9), 880–894.

Brachman, R. J., Amarel, S., Engelman, C., Engelmore, R. S., Feigenbaum, E. A., & Wilkins, D. E. (1983). What are expert systems? In F. Hayes-Roth, D. A. Waterman, & D. B. Lenat (Eds.), *Building expert systems* (pp. 31–57). Reading, MA: Addison-Wesley.

Bramer, M. A. (Ed.). (1985). Research and development in expert systems. *Proceedings of the Fourth Technical Conference of the British Computer Society Specialist Group on Expert Systems*. Cambridge: Cambridge University Press.

Bramer, M. A. (Ed.). (1987). *Research and development in expert systems*. Cambridge: Cambridge University Press.

Bramer, M. A. (Ed.). (1989). *Research and development in expert systems* (Vol. 2). Cambridge: Cambridge University Press.

Brownston, L., & Farrell, R., Kant, E., & Martin, N. (1985). *Programming Expert Systems in OPS5*. Reading, MA: Addison-Wesley.

Buchanan, B. G. (1986). Expert systems: Working systems and the research literature. *Expert systems, 3*(1), 32–51.

Buchanan, B. G., Barstow, D., Bechtal, R, Bennett, J., Clancey, W. Kulikowski, C., Mitchell, T., & Waterman, D. (1983). Constructing an expert system. In F. Roth, D. Waterman, & D. Lenat (Eds.), *Building expert systems*. Reading, MA: Addison-Wesley.

Buchanan, B. G., & Duda, R. O. (1983). Principles of rule-based expert systems. *Advances in Computers*, *22*, 163–216.

Bylander, T., & Chandrasekaran, B. (1987). Generic tasks in knowledge-based reasoning: The "right" level of abstraction for knowledge acquisition. *International Journal of Man-Machine Studies*, *26*, 231–244.

Cebrzynski, G. (1987). Expert systems are seen as replacements for humans. *Marketing News*, *21*, 1.

Chandrasekaran, B. (1983). Towards a taxonomy of problem-solving types. *The AI Magazine*, *4*(1), 9–17.

Chandrasekaran, B. (1984). Expert systems: Matching techniques to tasks. In W. Reitman (Ed.), *AI applications for business* (pp. 116–132). Norwood, NJ: Ablex.

Chandrasekaran, B. (1986). Generic tasks in knowledge-based reasoning: High-level building blocks for expert system design. *IEEE Expert*, *1*, 22–30.

Chandrasekaran, B. (1988). Generic tasks as building blocks for knowledge-based systems: The diagnosis and routine design examples. *The Knowledge Engineering Review*, *3*, 183–210.

Chandrasekaran, B. (1988). An answer to the commentators on "Generic tasks as building blocks for knowledge-based systems: The diagnosis and routine design examples." *The Knowledge Engineering Review*, *3*, 271–220.

Chandrasekaran, B., & Goel, A. (1988). From numbers to symbols to knowledge structures: Artificial intelligence perspectives on the classification task. *IEEE Transactions on Systems, Man and Cybernetics*, *18*, 415–424.

Chandrasekaran, B., & Mittal, S. (1983). Deep versus compiled knowledge approaches to diagnostic problem solving. *International Journal of Man-Machine Studies*, *19*, 425–436.

Chandrasekaran, B., & Mittal, S. (1984). Deep versus compiled knowledge approaches to diagnostic problem solving. In J. Coombs (Ed.), *Developments in expert systems* (pp. 23–34). New York: Academic Press.

Chignell, M. H., & Peterson, J. G. (1988). Strategic issues in knowledge engineering. *Human Factors*, *30*, 381–394.

Clancey, W. J. (1983). The epistemology of a rule-based expert system: A framework for explanation. *Artificial Intelligence*, *20*, 215–251.

Clancey, W. J. (1985). Heuristic classification. *Artificial Intelligence*, *27*, 289–350.

Clancey, W. J. (1988). Acquiring, representing, and evaluating a competence model of diagnosis. In M. T. H. Chi, R. Glaser, & M. Farr (Eds.). *The nature of expertise*. Hillsdale, NJ: Erlbaum.

Clancey, W. J. (1988). The knowledge engineer as student: Metacognitive bases for asking good questions. In H. Mandl, & A. Lesgold (Eds.), *Learning issues for intelligent tutoring systems*, (pp. 80–112). New York: Springer-Verlag.

Clancey, W. J. (1989). Representing control knowledge as abstract tasks and metarules. In M. J. Coombs & L. Bolc (Eds.), *Computer expert systems*, New York: Springer-Verlag.

Cohen, P. R. (1987). The control of reasoning under uncertainty: A discussion of some programs. *The Knowledge Engineering Review*, *2*, 5–26.

Cohen, P. R., & Howe, A. E. (1988, Winter). How evaluation guides AI research. *The AI Magazine*, 35–43.

Cohen, P. R., & Howe, A. E. (1989). Toward AI research methodology: Three case studies in evaluation. *IEEE Transactions on Systems, Man, and Cybernetics*, *19*, 634–646.

Collins, H. M., Green, R. H., & Draper, R. C. (1985). Where's the expertise? Expert systems as a medium of knowledge transfer. In M. Merry (Ed.), *Expert Systems '85* (pp. 323–334). Cambridge: Cambridge University Press.

Cooley, M. (1988). Creativity, skill, and human-centered systems. In B. Göranzon & I. Josefson (Eds.), *Knowledge, skill, and artificial intelligence* (pp. 127–138). London: Springer-Verlag.

Coombs, M. J. (1986). Artificial intelligence and cognitive technology: foundations and perspectives. In E. Hollnagel, G. Mancini, & D. D. Woods (Eds.), *Intelligent decision support in process environments* (pp. 407–420). Berlin: Springer-Verlag.

Coombs, M., & Alty, J. (1984). Expert Systems: An alternative paradigm. In M. J. Coombs (Ed.), *Developments in expert systems* (pp. 135–158). London: Academic Press.

Davis, R. D. (1989, Spring). Expert systems: How far can they go? (Part 1). *The AI Magazine*, pp. 61–67.

Davis, R. D. (1989, Summer). Expert systems: How far can they go? (Part 2). *The AI Magazine*, pp. 65–76.

Davis, R. D., & Lenat, B. (1982). *Knowledge based systems in AI*. New York: McGraw-Hill.

Denning, P. J. (1986, Summer). Towards a science of expert systems. *IEEE Expert*, 80–83.

Dougherty, J., & Keller, C. (1982). Taskonomy: A practical approach to knowledge structures. *American Ethnologist*, *9*, 763–774.

Dreyfus, H. L., & Dreyfus, S. E. (1986). *Mind over machine: The power of human intuition and exper-*

tise in the era of the computer. New York: The Free Press.

Dreyfus, H., & Dreyfus, S. (1986, Summer). Why expert systems do not exhibit expertise. *IEEE Expert*, pp. 86–90.

Dreyfus, W. E. (1989, Spring). Presentation. *The AI Magazine*, pp. 64–67.

Drummond, M., Macintosh, A., & Tate, A. (1987). A framework for technology transfer within AI. *The Knowledge Engineering Review*, *2*, 159–168.

Duda, R. O., & Gashnig, J. G. (1981, September). Knowledge-based expert systems come of age. *Byte*, pp. 238–281.

Duda, R. O., & Shoftliffe, E. H. (1983). Expert systems research. *Science*, *220*, 261–276.

Eberts, R. E., Nof, S. Y., Zimolong, B., & Salvendy, G. (1984). Dynamic process control: Cognitive requirements and expert systems. In G. Salvendy (Ed.) *Human-computer interaction* (pp. 215–228). Amsterdam: Elsevier.

Eliot, L. B. (1986, Summer). Analogical problem solving and expert systems. *IEEE Expert*, 17–26.

Ennals, R. (1988). Can skills be transferable? In B. Göranzon & I. Josefson (Eds.), *Knowledge, skill, and artificial intelligence* (pp. 67–76). London: Springer-Verlag.

Feigenbaum, E. A. (1977). The art of artificial intelligence. *Proceedings of the 5th International Joint Conference on Artificial Intelligence* (pp. 1013–1029). Cambridge, MA: International Joint Conferences on Artificial Intelligence.

Feigenbaum, E. A. (1984). Knowledge engineering: The applied side of artificial intelligence. *Annals of the New York Academy of Sciences*, *426*, 91–107.

Feigenbaum, E. A., & McCorduck, p. (1983). *The fifth generation: Artificial intelligence and Japan's computer challenge to the world*. Reading, MA: Addison-Wesley.

Finley, M. R., & Hausen-Tropper, E. B. (1989). A system for the representation of theorems and proofs. In C. R. Westphal & K. L. McGraw (Eds.), *Special Issue of ACM SIGART NEWSLETTER on Knowledge Acquisition* (No. 108, pp. 178–179). New York: Special Interest Group on Artificial Intelligence, Association for Computing Machinery.

Fischer, G. (1981). Computational models of skill acquisition processes. *Computers in Education: 3rd World Conference on Computers and Education* (pp. 477–481). Lausanne, Switzerland.

Fox, J. (1987). The evolution of knowledge systems: Ideological confusion or healthy pragmatism? *The Knowledge Engineering Review*, *2*, 73–74.

Fox, J. (1990). Safe expert systems: Simulating experts or building formal theories? *The Knowledge Engineering Review*, *5*, 1–4.

Freedman, R. (1987, September). *Evaluating shells. AI Expert*, 70–74.

Freiling, M. J., Alexander, S., Messick, S., Rehfuss, S. & Shulman, S. (1985) Starting a knowledge engineering project: A step-by-step approach. *AI Magazine*, *6*, 150–163.

Frost, R. (1986). *Introduction to knowledge-based systems*. New York: Macmillan.

Gaines, B. R. (1987). Foundations of knowledge engineering. In M. A. Bramer (Ed.), *Research and development in expert systems III* (pp. 13–24). Cambridge: Cambridge University Press.

Garg-Janardan, C., & Salvendy, G. (1988). The contributions of cognitive engineering to the design and use of expert systems. *Behavior and Information Technology*, *7*, 323–342.

Geisel, L. K. (1989, Summer). Knowledge engineering is the key to successful technology transfer. *Expert Systems: Planning, Implementation, Integration*, *1*, 28–32.

Giuse, D. (1990). Efficient knowledge representation systems. *The Knowledge Engineering Review*, *5*, 35–50.

Glorioso, R. M., & Osorio, F. C. C. (1980). *Engineering intelligent systems: Concepts, theory, and applications*. Bedford, MA: Digital Press.

Glymor, C. (1989). When less is more. In D. A. Evans & V. L. Patel (Eds.), *Cognitive science in medicine: Biomedical modeling* (pp. 349–371). Cambridge, MA: Bradford Books/MIT Press.

Gold, J. (1988). Do it yourself expert systems. *Computer Decisions*, *18*, 76–81.

Goodall, A. *The guide to expert systems*. Oxford: Learned Information.

Göranzon, B., & Josefson, I. (Eds.). (1988). *Knowledge, skill, and artificial intelligence*. London: Springer-Verlag.

Grant, J., & Minker, J. (1989). Deductive database theories. *The Knowledge Engineering Review*, *4*, 267–304.

Graubard, S. R. (Ed.). (1988). *The artificial intelligence debate: False starts, real foundations*. Cambridge, MA: The MIT Press.

Gray, P. M. D. (1989). Interfacing a knowledge based system to a large data base. *The Knowledge Engineering Review*, *4*, 31–52.

Gregory, D. (1986). Delimiting expert systems. *IEEE Transactions on Systems, Man and Cybernetics*, *SMC-16*, 834–843.

Gullers, P. (1988). Automation-skill-apprenticeship. In B. Göranzon & I. Josefson (Eds.),

Knowledge, skill, and artificial intelligence (pp. 31–38). London: Springer-Verlag.

Haddawy, P., & Rendell, L. (1990). Planning and decision theory. *The Knowledge Engineering Review*, *5*, 15–43.

Hagglund, S. (1987). The Linkoping approach to technology transfer in knowledge engineering. *The Knowledge Engineering Review*, *2*, 153–158.

Hall, R. P., & Kibler, D. F. (1985, Fall). Differing methodological perspectives in artificial intelligence research. *AI Magazine*, 166–178.

Hamburger, H. (1988). Natural language and expert systems. *International Journal of Expert Systems*, *1*, v–vii.

Harbison-Briggs, K., & Briggs, A. B. (1990). The role of knowledge acquisition in the verification and validation of knowledge-based systems. In K. L. McGraw & C. R. Westphal (Eds.), *Readings in knowledge acquisition* (pp. 233–250). New York: Ellis Horwood.

Harmon, P., & King, D. (1985). *Expert Systems*. New York: Wiley.

Harris, S. D., & Helander, M. G. (1984). Machine intelligence in real systems: Some ergonomics issues. In G. Salvendy (Ed.), *Human-computer interaction* (pp. 267–278). Amsterdam: Elsevier.

Hart, A. (1988). *Expert systems for managers*. London: Kogan Page.

Hartley, R. T. (1981, August). How expert should an expert system be? *Proceedings of the Seventh International Joint Conference on Artificial Intelligence* (pp. 862–867), Vancouver, BC: International Joint Conferences in Artificial Intelligence.

Hartley, R. T. (1985). Representation of procedural knowledge for expert systems. *Proceedings of Second Conference on Artificial Intelligence Applications*, (pp. 526–531). IEEE Computer Society.

Hasling, D. W., Clancey, W. J., & Rennels, G. (1984). Strategic explanations for a diagnostic consultation system. In M. J. Coombs (Ed.), *Developments in expert systems* (pp. 117–134). New York: Academic Press.

Haugeland, J. (1985). *Artificial intelligence: The very idea*. Cambridge, MA: The MIT Press.

Hayes-Roth, F. (1985, October). Engineering system of knowledge: The great adventure ahead. *Expert Systems in Government Symposium* (pp. 678–692). New York: IEEE.

Hayes-Roth, F., Klahr, P., & Mostow, D. J. (1986). Knowledge acquisition, knowledge programming, and knowledge refinement. In P. Klahr, & D. A. Waterman (Eds.), *Expert systems: Techniques, tools, and applications* (pp. 310–349). Reading, MA: Addison Wesley.

Hayes-Roth, F., Waterman, D. A., & Lenat, D. (1983). *Building expert systems*. Reading, MA: Addison-Wesley.

Hayward, S. A. (1985). Is a decision tree an expert system? In M. A. Bramer (Ed.), *Research and development in expert systems* (pp. 185–192). Cambridge: Cambridge University Press.

Hendler, J., Chandrasekaran, B., Adelson, B., Alterman, R., Bylander, T., & Dyer, M. (1988, Winter). Theoretical issues in conceptual information processing. *The AI Magazine*, pp. 71–76.

Hern, L. E. C. (1988). On distributed artificial intelligence. *The Knowledge Engineering Review*, *3*, 21–58.

Hertvik, J. (1989). Bad beginnings make bad endings. *Expert Systems: Planning, Implementation, Integration*, *1*, 11–15.

Hertvik, J. (1989). The only difference. *Expert Systems: Planning, Implementation, Integration*, *1*, 50–53.

Hollnagel, E. (1987). Information and reasoning in intelligent decision support systems. *International Journal of Man-Machine Studies*, *27*, 665–678.

Hollnagel, E. (1987). Commentary: Issues in knowledge-based decision support. *International Journal of Man-Machine Studies*, *27*, 743–751.

Jackson, A. (1985, October). Teaching an American to play cricket. *Datalink*, p. 14.

Jackson, A. (1986, January). The cost of coconuts. *Datalink*, p. 9.

Jackson, A. (1986, February). A little learning is a dangerous thing. *Datalink*, p. 10.

Jackson, A. (1986, April). Inadmissible evidence. *Datalink*, p. 10.

Jackson, P. C. (1974). *Introduction to artificial intelligence*. New York: Petrocelli Books.

Jackson, P. C. (1986). *Introduction to expert systems*. Wokingham, England: Addison-Wesley.

Jackson, P. C., Reichgelt, H., & van Harmelen, F. (1989). *Logic-based knowledge representation*. Cambridge: MIT Press.

Johnson, P. E. (1983). What kind of expert should a system be? *The Journal of Medicine and Philosophy*, *8*, 77–97.

Johnson, P. E., Zualkerman, I. A., David, J. M., Iwasaki, Y., Keller, R., & Feigenbaum, E. (1988). Responses to: "Generic tasks as building blocks for knowledge-based systems: The diagnosis and routine design examples" by B. Chandrasekaran. *The Knowledge Engineering Review*, *3*, 211–216.

Johnson, P. E., Zualkerman, I. A., & Garber, S. (1987). Specification of expertise: Knowledge acquisition for expert systems. *International Journal of Man-Machine Studies*, *26*, 161–182.

Johnson, T., Hewett, J., Guilfoyle, C., & Jeffcoate, J. (1988). Expert systems: The second wave. *The Knowledge Engineering Review*, *3*, 177–182.

Jones, S. (1988). Graphical interfaces for knowledge engineering: An overview of the relevant literature. *The Knowledge Engineering Review*, *3*, 221–248.

Josefson, I. (1987). Knowledge and experience. *Applied Artificial Intelligence*, *1*(2), 173–180.

Kanal, L., & Lemmer, J. (1986). *Uncertainty in artificial intelligence*. Amsterdam: North-Holland.

Keravnou, E. T., & Washbrook, J. (1989). What is a deep expert system? An analysis of the architectural requirements of second-generation expert systems. *The Knowledge Engineering Review*, *4*, 205–234.

Keyes, J. (1989, November). Why expert systems fail. *AI Expert*, 50–53.

Kidd, A. L. (1987). Knowledge acquisition: An introductory framework. In A. L. Kidd (Ed.), *Knowledge acquisition for expert systems: A practical handbook* (pp. 1–16). New York: Plenum.

Kidd, A. L., & Cooper, M. B. (1985). Man-machine interface issues in the construction and use of an expert system. *International Journal of Man-Machine Studies*, *22*, 91–102.

Klahr, D., Langley, P., & Neches, R. (Eds.). (1987). *Production system models of learning and development*. Cambridge, MA: MIT Press.

Kline, P., & Dollin, S. (1986). Problem features that influence the design of expert systems. *Proceedings of the American Association for Artificial Intelligence* (pp. 956–962). Stanford, CA: AAAI.

Kolodner, J. L. (1984). Towards an understanding of the role of experience in the evolution from novice to expert. In M. J. Coombs (Ed.), *Developments in expert systems* (pp. 77–94). New York: Academic Press.

Kornell, J. (1987). Formal thought and narrative thought in knowledge acquisition. *International Journal of Man-Machine Studies*, *26*, 203–212.

Kurzweil, R. (1985). What is artificial intelligence anyway? *American Scientist*, *73*, 258–264.

Lamberti, M., & Newsome, S. L. (1989). Presenting abstract versus concrete information in expert systems: What is the impact on user performance? *International Journal of Man-Machine Studies*, *31*, 27–45.

Laufmann, S. C. (1987). A strategy for near-term success using knowledge-based systems. *The Knowledge Engineering Review*, *2*, 179–184.

Laurent, J.-P. (1987). Types of control structure in expert systems. *The Knowledge Engineering Review*, *2*, 123–136.

Lefkowitz, L. S., & Lesser, V. R. (1988). Knowledge acquisition as knowledge assimilation. *International Journal of Man-Machine Studies*, *29*, 215–226.

Lehner, P. E. (1989). Toward an empirical approach to evaluating the knowledge base of an expert system. *IEEE Transactions on Systems, Man, and Cybernetics*, *19*, 658–662.

Lehner, P. E., & Adelman, L. (1987, October). Biases in knowledge engineering. In *Proceedings of the 1987 IEEE International Conference on Systems, Man, and Cybernetics* (pp. 376–377). New York: IEEE.

Lehner, P. E., & Adelman, L. (1989). Perspectives in knowledge engineering. *IEEE Transactions on Systems, Man, and Cybernetics*, *19*, 423–434.

Levi, K. (1989). Expert systems should be more accurate than human experts: Evaluation procedures from human judgment and decision making. *IEEE Transactions on Systems, Man, and Cybernetics*, *19*, 645–647.

Littman, D. C. (1987). Modeling human expertise in knowledge engineering: Some preliminary observations. *International Journal of Man-Machine Studies*, *26*, 81–92.

Long, D. (1989). A review of temporal logics. *The Knowledge Engineering Review*, *4*, 141–162.

Madni, A. M. (1988). The role of human factors in expert systems design and acceptance. *Human Factors*, *30*, 395–414.

Maes, P. (1988). Computational reflection. *The Knowledge Engineering Review*, *3*, 1–20.

Mayer, R. F. (1988). From novice to expert. In M. Holander (Ed.), *Handbook of human-computer interaction* (pp. 569–580). Amsterdam: North-Holland.

McCarthy, J. (1983). Some expert systems need common sense. *Annals of the New York Academy of Sciences*, *426*, 129–137.

McCarthy, J. (1984, Fall). President's message. *The AI Magazine*, 7–8.

McDermott, J. (1986). Making expert systems explicit. In *Proceedings of the Tenth Congress of the International Federation of Information Processing Societies* (pp. 539–544). Amsterdam: Elsevier.

McGraw, K. L. (1986). Artificial intelligence: The competitive edge in integrated systems development. *Texas Instruments Engineering Journal, 3*, 12–16.

McGraw, K. L., & Briggs, K. H. (1989). *Knowledge acquisition: Principles and guidelines*. Englewood Cliffs, NJ: Prentice Hall.

Merry, M. (Ed.). (1985). *Expert systems '85*. Cambridge: Cambridge University Press.

Merry, M. (1985). Expert systems—some problems and opportunities. In M. Merry (Ed.) *Expert systems-85* (pp. 1–8). Cambridge: Cambridge University Press.

Merry, M. (Ed.) (1986). *Expert systems-86*. Cambridge: Cambridge University Press.

Minsky, M. (1975). A framework for representing knowledge. In P. H. Winston (Ed.), *The psychology of computer vision* (pp. 211–277). New York: McGraw-Hill.

Minsky, M. (1986). *The society of mind*. New York: Simon & Schuster.

Minsky, M., & Papert, S. (1974). *Artificial intelligence*, Eugene, OR: Oregon State System of Higher Education.

Mishkoff, H. (1985). *Understanding artificial intelligence*. Dallas, TX: Texas Instruments Inc.

Moralee, D. S. (Ed.). (1988). *Research and development in expert systems*. Cambridge: Cambridge University Press.

Moray, N. (1986). Intelligent decision aids, mental models, and the theory of machines. In E. Hollnagel, G. Mancini, & D. D. Woods (Eds.), *Intelligent decision support in process environments* (pp. 273–291). New York: Springer-Verlag.

Moray, N. (1987). Intelligent aids, mental models, and the theory of machines. *International Journal of Man-Machine Studies, 27*, 619–630.

Morgan, R. (1989, Summer). Expert systems development strategies. *Expert Systems: Planning, Implementation, Integration, 1*, 5–10.

Morik, K. (1989, April). Integration issues in knowledge acquisition systems. In C. R. Westphal & K. L. McGraw (Eds.), *Special Issue of ACM SIGART Newsletter on Knowledge Acquisition* (No. 108, pp. 124–131). New York: Special Interest Group on Knowledge Acquisition, Association for Computing Machinery.

Muir, B. M. (1987). Trust between humans and machines and the design of decision aids. *International Journal of Man-Machine Studies, 27*, 527–540.

Nazareth, D. L. (1989). Issues in the verification of knowledge rule-based systems. *International Journal of Man-Machine Studies, 30*, 255–272.

Newell, A. (1981). The knowledge level. *Artificial Intelligence, 18*, 87–127.

Newell, A. (1983). Some intellectual issues in the history of artificial intelligence. In F. Machlup & U. Mansfield (Eds.), *The study of information: Interdisciplinary messages* (pp. 187–227). New York: Wiley.

Newell, A. (1987). Discussion: Comments on expert systems and their use. In T. B. Sheridan, D. S. Kruser, & S. Deutsch (Eds.), *Human factors in automated and robotic space systems: Proceedings of a symposium* (pp. 142–146). Washington, DC: National Research Council.

Norman, D. (1986). Cognitive engineering. In D. Norman & S. Draper (Eds.), *User-centered system design* (pp. 31–61). Hillsdale, NJ: Erlbaum.

O'Keefe, R. M., Balci, O., Smith, E. P. (1987, Winter). Validating expert system performance. *IEEE Expert*, 81–89.

O'Leary, D. E. (1988). On the representation and the impact of reliability on expert system weights. *International Journal of Man-Machine Studies, 29*, 637–646.

Patton, C. (1985, May). Knowledge engineering: Tapping the experts. *Electronic Design*, 93–100.

Pau, L. F. (1987). Prototyping, validation, and maintenance of knowledge based system. *Third Expert Systems in Government Conference* (pp. 248–253). Washington, DC.

Prerau, D. S. (1985, Summer). Selection of an appropriate domain for an expert system. *The AI Magazine*, 26–30.

Prerau, D. S. (1987, Summer). Knowledge acquisition in the development of a large expert system. *AI Magazine*, 26–30.

Prerau, D. S. (1990). *Developing and managing expert systems: Proven techniques for business and industry*. Reading, MA: Addison-Wesley.

Rasmus, D. W. (1988). Expert input. *MacUser*, 136–150.

Rasmussen, J. (1983). Skills, rules, knowledge: Signals, signs, and symbols and other distinctions in human performance models. *IEEE Transactions on Systems, Man, and Cybernetics, 13*, 257–267.

Rasmussen, J. (1986). *Information processing and human-machine interaction: An approach to cognitive engineering*. New York: Elsevier.

Reddy, R. (1988, Winter). Foundations and grand challenges of artificial intelligence. *AI Magazine*, 9–12.

Reeker, L. H., Blaxton, T. A., & Westphal, C. R. (1988). Applying software engineering to knowledge engineering (and vice versa), *Proceedings of the 27th Annual Technical Symposium of the Washington, D.C. ACM Chapter* (pp. 107–114).

New York: Association for Computing Machinery.

Refenes, A. N. (1989). Parallelism in knowledge-based machines. *The Knowledge Engineering Review*, *4*, 53–72.

Regoczei, S., & Plantinga, E. (1987). Creating the domain of discourse: Ontology and inventory. *International Journal of Man-Machine Studies*, *27*, 235–250.

Reichgelt, H. (1989). Logics for reasoning about knowledge and beliefs. *The Knowledge Engineering Review*, *4*, 119–140.

Reichgelt, H., & van Harmelen, F. (1986). Criteria for choosing representation languages and control regimes for expert systems. *Knowledge Engineering Review*, *1*, 2–17.

Rich, E. (1983). *Artificial intelligence*. New York: McGraw-Hill.

Riesbeck, C. K. (1984). Knowledge reorganization and reasoning style. In M. J. Coombs (Ed.), *Developments in expert systems* (pp. 159–176). New York: Academic Press.

Ringwood, G. A. (1988). Metalogic machines: A retrospective rationale for the Japanese Fifth Generation. *The Knowledge Engineering Review*, *3*, 303–320.

Ringwood, G. A. (1989). A comparative exploration of concurrent logic languages. *The Knowledge Engineering Review*, *4*, 305–332.

Rolandi, W. G. (1986, December). Knowledge engineering in practice. *AI Expert*, 58–62.

Rolandi, W. G. (1988, April). A practical approach to knowledge engineering. *AI Expert*, 60–65.

Rouse, W. B. (1984). Design and evaluation of computer-based decision support systems. In G. Salvendy (Ed.), *Human-computer interaction* (pp. 229–246). Amsterdam: Elsevier.

Rouse, W. B., Hammer, J. M., & Lewis, C. M. (1989). On capturing human skills and knowledge: Algorithmic approaches to model identification. *IEEE Transactions on Systems, Man, and Cybernetics*, *19*, 538–573.

Saffiotti, A. (1987). An AI view of the treatment of uncertainty. *The Knowledge Engineering Review*, *2*, 75–98.

Saffiotti, A. (1988). The treatment of uncertainty in AI: Is there a better vantage point? *The Knowledge Engineering Review*, *3*, 87–92.

Schneiderman, B. (1980). *Software psychology: Human factors in computers and information systems*. Boston, MA: Little, Brown & Co.

Sell, P. S. (1985). *Expert systems: A practical introduction*. New York: Wiley.

Shaw, M. L. G., & Woodward, J. B. (1988). Validation in a knowledge support system: Construing and consistency with multiple experts. *International Journal of Man-Machine Studies*, *29*, 329–350.

Shin, D., & Bera, P. B. (1989). Computer architectures for logic-oriented data/knowledge bases. *The Knowledge Engineering Review*, *4*, 1–30.

Slatter, P. E. (1987). *Building expert systems: Cognitive emulation*. Chichester, England: Ellis Horwood.

Slatter, P. E. (1987). Cognitive emulation in expert system design. *The Knowledge Engineering Review*, *2*, 27–42.

Smart, G. (1987, June). How to make expert systems . . . and stay in business. *Third International Expert Systems Conference* (pp. 215–222). Oxford: Learned Information.

Sowa, J. (1984). *Conceptual structures: Information processing in mind and machine*. Reading, MA: Addison-Wesley.

Sparck-Jones, K. (1985). Natural language interfaces for expert systems: An introductory note. In M. A. Bramer (Ed.), *Research and development in expert systems* (pp. 85–94). Cambridge: Cambridge University Press.

Sridharan, N. S. (1985, Fall). Evolving systems of knowledge. *AI Magazine*, pp. 108–120.

Steels, L. (1985). Second generation expert systems. *Future Generation Expert Systems*, *1*, 213–221.

Steels, L. (1987). Second generation expert systems. In M. A. Bramer (Ed.), *Research and development in expert systems* (pp. 175–183). Cambridge: Cambridge University Press.

Stefik, M., Aikins, A., Balzer, R., Benoit, J., Birnbaum, L., Hayes-Roth, F., & Sacerdoti, E. (1982). The organization of expert systems: A tutorial. *Artificial Intelligence*, *18*, 135–173.

Stefik, M., Aikins, J., Balzer, R., Benoit, J., Birnbaum, L., Hayes-Roth, F., & Sacerdoti, E. (1983). Basic concepts for building expert systems. In F. Hayes-Roth, D. A. Waterman, & D. B. Lenat (Eds.), *Building expert systems* (pp. 59–86). Reading, MA: Addison-Wesley.

Stefik, M., & Conway, L. (1982, Summer). Towards the principled engineering of knowledge. *AI Magazine*, 4–16.

Stenton, S. P. (1987). Dialog management for cooperative knowledge-based systems. *The Knowledge Engineering Review*, *2*, 99–122.

Stephanou, H. (1987). Perspectives on imperfect information processing. *IEEE Transactions on Systems, Man, and Cybernetics*, *17*, 780–798.

Tiemann, P. W., & Markle, S. (1984, November). On getting expertise into an expert system. *Performance and Instruction Journal*, 25–29.

Voelker, J. A., & Ratica, G. B. (1986). The genesis of a knowledge-based expert system. *IBM Systems Journal*, *25*, 181–189.

Walker, E., & Stevens, A. (1986). Human and machine knowledge in intelligent systems. In E. Hollnagel, G. Mancini, & D. D. Woods (Eds.), *Intelligent decision support in process environments* (pp. 421–433). Berlin: Springer-Verlag.

Walters, J. R., & Nielsen, N. R. (1988). *Crafting knowledge-based systems*. New York: Wiley.

Waterman, D. A. (1986). *A guide to expert systems*. Reading, MA: Addison-Wesley.

Weiss, S., & Kulikowski, C. (1984). *A practical guide to designing expert systems*. Totowa, NJ: Rowman & Allanheld.

Wexelblat, R. L. (1989, Fall). On interface requirements for expert systems. *The AI Magazine*, pp. 66–78.

Williams, C. (1986, Winter). Expert systems, knowledge engineering, and AI tools: An overview. *IEEE Expert*, pp. 66–70.

Wood, S. (1986). Expert systems for theoretically ill-formulted domains. In M. A. Bramer (Ed.), *Research and development in expert systems* (Vol. 3, pp. 132–139). Cambridge: Cambridge University Press.

Woods, D. D. (1987). Commentary: Cognitive engineering in complex and dynamic worlds. *International Journal of Man-Machine Studies*, *27*, 571–585.

Woods, W. A. (1983). What's important about knowledge representation? *Computer*, *16*, 22–27.

Woods, W. A. (1985, Winter). Cognitive technologies: The design of joint human-machine cognitive systems. *The AI Magazine*, 86–92.

Wyatt, J. C., & Emerson, P. A. (1988, May). Pragmatic knowledge engineering: A new approach for difficult domains. *Conference on Human and Organizational Issues of Expert Systems*. Stratford-on-Avon, UK.

Young, R. M. (1979). Production systems for modeling human cognition. In D. Michie (Ed.), *Expert systems in a microelectronic age* (pp. 35–45). Edinburgh: Edinburgh University Press.

Appendix G
Bibliography: Applications of Expert Systems

Alden, B. E. P., & Bramer , M. A. (1988). An expert system for solving retrograde-analysis chess problems. *International Journal of Man-Machine Studies*, *29*, 97–112.

Ali, M. (1990). Intelligent systems in aerospace. *The Knowledge Engineering Review*, *5*, 147–166.

Allgayer, J., Harbusch, K., Kobsa, A., Reddig, C., Reithinger, N., & Schmauks, D. (1989). XTRA: A natural-language access system to expert systems. *International Journal of Man-Machine Studies*, *31*, 161–196.

Amendola, A., Bersini, V., Cacciabve, P. C., & Mancini, G. (1987). Modeling operators in accident conditions: Advances and perspectives on a cognitive model. *International Journal of Man-Machine Studies*, *27*, 599–612.

Apte, C. W., & Weiss, S. M. (1987). An expert systems methodology for control and interpretation of applications software. *International Journal of Expert Systems*, *1*, 17–38.

Banning, R. W. (1984). Knowledge acquisition and system validation in expert systems for management. *Human Systems Management*, *4*, 280–285. New York: Elsevier (North-Holland).

Barrett, M. L., & Beerel, A. C. (1988). Expert systems in business: *A practical approach*. Chichester, England: Ellis Horwood.

Basden, A. (1984). On the application of expert systems. In M. J. Coombs (Ed.), *Developments in expert systems* (pp. 59–76). New York: Academic Press.

Becker, L. A., Bartlett, R., & Soroushian, F. (1989, April). Using simulation to compile diagnostic rules from a manufacturing process representation. In C. R. Westphal & K. L. McGraw (Eds.), *Special Issue of ACM SIGART Newsletter on Knowledge Acquisition* (No. 108, pp. 172–173). New York: Special Interest Group on Artificial intelligence, Association for Computing Machinery.

Birmingham, W. P., & Sieworek, D. P. (1989). Automated knowledge acquisition for a computer hardware synthesis system. *Knowledge Acquisition: An Integrated Journal*, *1*, 321–340.

Biswas, G., Oliff, M., & Abramczyk, R. (1988). OASES (Operations Analysis Expert System): An application in fiberglass manufacturing. *International Journal of Expert Systems*, *1*, 193–216.

Boy, G. A. (1987). Operator assistant systems. *International Journal of Man-Machine Studies*, *27*, 541–554.

Bratko, I., Mozetič, I., & Lavarač, N. (1988). Automatic synthesis and compression of cardiological knowledge. In J. E. Hayes, D. Michie, & J. Richards (Eds.), *Machine intelligence 11* (pp. 435–454). Oxford, England: Clarendon Press.

Buchanan, B. G. (1987). Expert systems: Applications in space. In T. B. Sheridan, D. S. Kruser, & S. Deutsch (Eds.), *Human factors in automated and robotic space systems: Proceedings of a symposium* (pp. 113–146). Washington, DC: National Research Council.

Buchanan, B. G., & Feigenbaum, E. A. (1978). DENDRAL and Meta-DENDRAL: Their applications dimension. *Artificial Intelligence*, *11*, 5–24.

Buchanan, B. G., & Shortliffe, E. H. (1984). *Rule-based expert systems: The MYCIN experiments of the Stanford Heuristic Programming Project.* Reading, MA: Addison-Wesley.

Buchanan, B. G., Sutherland, G., & Feigenbaum, E. A. (1969). Heuristic DENDRAL: A program for generating explanatory hypotheses in organic chemistry. In B. Meltzer & D. Michie (Eds.),

Machine intelligence (Vol. 4, pp. 209–254). Edinburgh: Edinburgh University Press.

Butler, A. R., & Chamberlin, G. (1988). The ARIES Club-experience of expert systems in insurance and investment. In D. S. Moralee (Ed.), *Research and development in expert systems IV* (pp. 246–257). Cambridge: Cambridge University Press.

Carter, I. M. (1990). Applications and prospects for AI in mechanical engineering design. *The Knowledge Engineering Review*, *5*, 167–180.

Chandrasekaran, B. (1990, Winter). Design problem solving: A task analysis. *The AI Magazine*, 59–71.

Clancey, W. J. (1984). Methodology for building an intelligent tutoring system. In W. Kintsch, J. R. Miller, & P. G. Polson (Eds.), *Method and tactics in cognitive science* (pp. 51–83). Hillsdale, NJ: Erlbaum.

Cochran, E. L., Bloom, C. P., & Bullemer, P. T. (1990). Increasing the end-user acceptance of expert systems by using multiple experts. In K. L. McGraw & C. R. Westphal (Eds.), *Readings in knowledge acquisition* (pp. 73–89). New York: Ellis Horwood.

Cochran, E. L., & Hutchins, B. L. (1987). Testing, verifying, and releasing an expert system: The case history of Mentor. *Proceedings of Third Conference on Artificial Intelligence Applications* (pp. 163–167). Kissimmee, FL: IEEE Computer Society.

Collier, J. T., Evans, M., Hier, D., & Li, P.-Y. (1988). Generating case reports for a medical expert systems. *International Journal of Expert Systems*, *1*, 307–328.

Connell, N. A. D. (1987). The current impact of expert systems on the accounting profession and some reasons for hesitancy in the adoption of such systems. *The Knowledge Engineering Review*, *2*, 207–212.

Cookson, M., Holman, J., & Thompson, D. (1985). Knowledge acquisition for medical expert systems: a system for eliciting diagnostic decision making histories. In M. A. Bramer (Ed.), *Research and development in expert systems* (pp. 113–116). Cambridge: Cambridge University Press.

Coombs, M. J. (1986). Artificial intelligence and cognitive technology: foundations and perspectives. In E. Hollnagel, G. Mancini, & D. D. Woods (Eds.), *Intelligent decision support in process environments* (pp. 407–420). Berlin: Springer-Verlag.

Coombs, M. J., & Hartley, R. T. (1988). Explaining novel events in process control through model generative reasoning. *International Journal of Expert Systems*, *1*, 87–110.

Crawford, S. L., Fung, R. M., & Tse, E. (1989). Data-driven assessment and decision making. In L. F. Pau, Y. H. Pao, J. Motiwalla, & H. H. Teh (Eds.), *Expert systems in economics, banking, and management*. Amsterdam: North-Holland.

Cross, T. B. (1988). *Knowledge engineering: The use of artificial intelligence in business*. New York: Brady Books.

D'Agapeyeff, A. (1988). The nature of expertise and its elicitation for business expert systems: A commentary. *The Knowledge Engineering Review*, *3*, 147–158.

D'Agapeyeff, A., & Hawkins, C. J. B. (1987). Expert systems in UK business: A critical assessment. *The Knowledge Engineering Review*, *2*, 185–202.

Dallman, B., Pieper, W., & Richardson, J. (1983). A graphics simulation system: Task emulation, not equipment modeling. *Journal of Computer-Based Instruction*, *10*(3–4), 70–72.

Davis, R. D. (1984). Reasoning from first principles in electronic trouble-shooting. In M. J. Coombs (Ed.), *Developments in expert systems* (pp. 1–22). New York: Academic Press.

Dillard, J. F., & Mutchler, J. F. (1987). Expertise in assessing solvency problems. *Expert Systems*, *4*(3), 170–179.

Duda, R. O., Gashnig, J., & Hart, P. (1979). Model design in the PROSPECTOR expert system for mineral exploration. In D. Michie (Ed.), *Expert systems in the micro-electronic age* (pp. 153–167). Edinburgh: Edinburgh University press.

Elio, R., & deHaan, J. (1986). Representing quantitative and qualitative knowledge in a knowledge-based storm forecasting system. *International Journal of Man-Machine Studies*, *25*, 523–548.

Feigenbaum, E. A., Buchanan, B. G., & Lederberg, J. (1971). On generality and problem solving: A case study using the DENDRAL program. In B. Metzler & D. Michie (Eds.), *Machine intelligence 6* (pp. 165–190). Edinburgh: Edinburgh University Press.

Fenn, J. A., Worden, R. P., Foote, M. H., & Willison, R. G. (1986). An expert assistant for electromyography. *Biomedical Measurement Information and Control*, *1*(4), 210–214.

Finley, M. R., & Hausen-Tropper, E. B. (1989). A system for the representation of theorems and proofs. In C. R. Westphal & K. L. McGraw (Eds.), *Special Issue of ACM SIGART Newsletter on Knowledge Acquisition* (No. 108, pp. 178–179). New York: Special Interest Group on Artificial Intelligence, Association for Computing Machinery.

Fischoff, B. (1986). Decision making in complex systems. In E. Hollnagel, G. Mancini, & D. D. Woods (Eds.), *Intelligent decision support in process environments* (pp. 61–85). Berlin: Springer-Verlag.

Fox, J., Fieschi, M., & Engelbrecht, R. (Eds.) (1987). Lecture notes in medical informatics: Proceedings of the AIME 1987 European Conference on artificial intelligence in medicine. Berlin: Springer-Verlag.

Fox, J., Myers, C. D., Greaves, M. F., & Pegram, S. (1987). A systematic study of knowledge base refinement in the diagnosis of leukemia. In A. L. Kidd (Ed.), *Knowledge acquisition for expert systems: A practical handbook* (pp. 73–90).

Freeman, P. (1985, October). Knowledge elicitation: A commercial perspective. *Proceedings of the 1st International Expert System Conference* (pp. 339–349), London: Oxford Learned Information.

Gale, W. A. (1986). *Artificial intelligence and statistics*. Reading, MA: Addison-Wesley.

Gale, W. A. (1987). Statistical applications of artificial intelligence and knowledge engineering. *The Knowledge Engineering Review, 2*, 227–248.

Gallanti, M., & Guda, G. (1986). Intelligent decision aids for process environments: An expert system approach. In E. Hollnagel, G. Mancini, & D. D. Woods (Eds.), *Intelligent decision support in process environments* (pp. 373–394). Berlin: Springer-Verlag.

Galloway, T. (1987). TAXI: A taxonomic assistant. *Proceedings of AAAI–87* (pp. 416–420). Los Altos, CA: Morgan Kaufmann.

Gevarter, W. B. (1987). The nature and evaluation of commercial expert system building tools. *Computer, 20*, 24–41.

Gill, K. S. (1988). Artificial intelligence and social action: Education and training. In B. Goranzon & I. Josefson (Eds.), *Knowledge, skill, and artificial intelligence* (pp. 77–92). London: Springer-Verlag.

Gleeson, J. F. J., & West, M. L. J. (1988). CLINTE: Coopers & Lybrand International Tax Expert System. In D. S. Moralee (Ed.), *Research and development in expert systems IV* (pp. 18–31). Cambridge: Cambridge University Press.

Goyals, S., Prerau, D., Lemmon, A., Gunderson, A., & Reinke, R. (1985). COMPASS: An expert system for telephone switch maintenance, *Expert Systems: The International Journal of Knowledge Engineering, 2*, 112–126.

Hall, L. O., & Kandel, A. (1988). Toward a methodology for building expert systems for imprecise domains. *International Journal of Expert Systems, 1*, 237–252.

Hart, A. (1988). *Expert systems for managers*. London: Kogan Page.

Hasling, D. W., Clancey, W. J., & Rennels, G. (1984). Strategic explanations for a diagnostic consultation system. In M. J. Coombs (Ed.), *Developments in expert systems* (pp. 117–134). New York: Academic Press.

Herrod, R., & Smith, M. (1986). The Campbell Soup story: An application of AI technology in the food industry. *Texas Instruments Engineering Journal, 3*, 16–19.

Hollan, J. D., Hutchins, E. L., & Weitzman, L. (1984). STEAMER: An interactive inspectable simulation-based training system, *AI Magazine*, pp. 15–27.

Hollnagel, E. (1987). Information and reasoning in intelligent decision support systems. *International Journal of Man-Machine Studies, 27*, 665–678.

Hollnagel, E. (1987). Commentary: Issues in knowledge-based decision support. *International Journal of Man-Machine Studies, 27*, 743–751.

Huang, D. (1989). A framework for the credit-apportionment process in rule-based systems. *IEEE Transactions on Systems, Man, and Cybernetics, 19*, 489–498.

Hutton, D., Ponton, J. W., & Waters, A. (1990). AI applications in process design, operation, and safety. *The Knowledge Engineering Review, 5*, 69–96.

Jackson, P., & Lefrere, P. (1984). On the application of rule-based techniques to the design of advice-giving systems. In M. J. Coombs (Ed.), *Developments in expert systems* (pp. 177–200). New York: Academic Press.

Jenkins, J. P. (1984). An application of an expert system to problem solving in process control displays. In G. Salvendy (Ed.), *Human-computer interaction* (pp. 255–260). Amsterdam: Elsevier.

Jerrams-Smith, J. (1987). An expert system within a supportive interface for UNIX. *Behaviour and Information Technology, 6*(1), 37–41.

Jerrams-Smith, J. (1989). An attempt to incorporate expertise about users into an intelligent interface for Unix. *International Journal of Man-Machine Studies*, *31*, 269–292.

Johnson, L., & Johnson, N. E. (1987). Research and methods in building knowledge systems at Brunel University. In C. J. Pavelin & M. D. Wilson (Eds.), *Knowledge acquisition for engineering applications* (pp. 84–95).

Johnson, P., Nachstheim, C., & Zualkerman, I. (1987). Consultant expertise. *Expert Systems*, *4*, 180–188.

Johnson, T. (1984). *The commercial application of expert systems technology*. London: Ovum.

Karimi, J. (1988). Software engineering for knowledge-based systems software. *Proceedings of the Twenty-First Annual Hawaii International Conference on System Sciences* (Vol. 3, pp. 168–173). New York: IEEE.

Karlsen, T. K., & Oppen, M. (1988). Professional knowledge and the limits of automation in administrations. In B. Goranzon & I. Josefson (Eds.), *Knowledge, skill, and artificial intelligence* (pp. 139–150). London: Springer-Verlag.

Kellogg, C., Gargan, Jr., R. A., Mark, W., McGuire, J. G., Pontecorvo, M., Schlossberg, J. L., Sullivan, J. W., Genesereth, M. R., & Singh, N. (1989). The acquisition, verification, and explanation of design knowledge. In C. R. Westphal & K. L. McGraw (Eds.), *Special Issue of ACM SIGART Newsletter on Knowledge Acquisition* (No. 108, pp. 163–165). New York: Special Interest Group on Artificial Intelligence, Association for Computing Machinery.

Keravnou, E. T., & Johnson, L. (1986). *Competent expert systems: A case study in fault diagnosis*. London: Kogan Page.

Kodratoff, Y., & Tecuci, G. (1987). Techniques of design and DISCIPLE learning apprentice. *International Journal of Expert Systems*, *1*, 39–66.

Kopec, D., & Latour, L. (1989). Towards and expert/novice learning systems with application to infectious disease. In C. R. Westphal & K. L. McGraw (Eds.), *Special Issue of ACM SIGART Newsletter on Knowledge Acquisition* (No. 108, pp. 120–132). New York: Special Interest Group on Artificial Intelligence, Association for Computing Machinery.

Kraemer, K. L., & King, J. L. (1988). Computer-based systems for cooperative work and group decision making. *ACM Computing Surveys*, *20*, 115–146.

Krisnamurthi, M., & Underbrink, A. J. (1990). Knowledge acquisition in a machine fault diagnosis shell. In K. L. McGraw & C. R. Westphal (Eds.), *Readings in knowledge acquisition* (pp. 124–138). New York: Ellis Horwood.

Kuipers, B. (1987). New reasoning methods for artificial intelligence in medicine. *International Journal of Man-Machine Studies*, *26*, 707–718.

Kunz, J. C., Fallat, R. J., McClung, D. H., Osborn, J. J., Votteri, B. A., Nii, H. P., Aikins, J. S., Fagan, L. M., & Feigenbaum, E. A. (1979). A physiological rule-based system for interpreting pulmonary function test results. *Proceedings of the Conference on Computers in Critical Care and Pulmonary Medicine* (pp. 375–379). New York: IEEE.

Lamberti, D. M., & Newsome, S. L. (1989). Presenting abstract versus concrete information in expert systems: What is the impact on user performance? *International Journal of Man-Machine Studies*, *31*, 27–46.

Lambird, B. A., Lavine, D., & Kanal, L. N. (1984). Distributed architecture and parallel non-directed search for knowledge-based cartographic feature extraction systems. In M. J. Coombs (Ed.), *Development in expert systems* (pp. 221–234). New York: Academic Press.

Lan, M. S., Panos, R. M., & Balban, M. S. (1987). Experience using S.1: An expert system for newspaper printing press configuration. *The Knowledge Engineering Review*, *2*, 277–286.

Langlotz, C. P., & Shortliffe, E. H. (1984). Adapting a consultation system to critique user plans. In M. J. Coombs (Ed.), *Developments in expert systems* (pp. 77–94). New York: Academic Press.

LeClair, S. R. (1990). Interactive learning: A multi-expert paradigm for acquiring new knowledge. In K. L. McGraw & C. R. Westphal (Eds.), *Readings in knowledge acquisition* (pp. 104–121). New York: Ellis Horwood.

Lederberg, J., & Feigenbaum, E. A. (1968). Mechanization of inductive inference in organic chemistry. In B. Kleinmuntz (Ed.), *Formal representation of human judgment* (pp. 187–267). New York: Wiley.

Lenat, D., Prakash, M., & Shepherd, M. (1986). CYC: Using common sense knowledge to overcome brittleness and knowledge acquisition bottlenecks. *The AI Magazine*, *7*, 65–85.

Levitt, R. E., & Kartam, N. A. (1990). Expert systems in construction engineering and management: State of the art. *The Knowledge Engineering Review*, *5*, 97–126.

Linkens, D. A. (1990). AI in control systems engineering. *The Knowledge Engineering Review*, *5*, 181–214.

Looney, C. G. (1988). Expert control design with fuzzy rule matrices. *International Journal of Expert Systems*, *1*, 159–168.

Marcus, S., Stout, J., & McDermott, J. (1988). VT: An expert elevator designer that uses knowledge-based reasoning. *AI Magazine*, *9*(1), 95–112.

Marshall, S. P. (1980). Procedural networks and production systems in adaptive diagnosis, *Instructional Science*, *9*, 129–143.

McDermott, J. (1981). R1: The formative years, *The AI Magazine*, pp. 21–28.

McDermott, J. (1982). R1: A rule-based configurer of computer systems, *Artificial Intelligence*, *19*, 39–88.

McGraw, K. L. (1986). Artificial intelligence: The competitive edge in integrated systems development. *Texas Instruments Engineering Journal*, *3*, 12–16.

McGraw, K. L. (1990). HyperKAT: A tool to manage and document knowledge acquisition. In K. L. McGraw & C. R. Westphal (Eds.), *Readings in knowledge acquisition* (pp. 164–181). New York: Ellis Horwood.

Miller, P. L., & Rennels, G. D. (1988, Fall). Prose generation from expert systems. *The AI Magazine*, pp. 37–44.

Miller, R. H., Pople, H., & Myers, J. (1982, August). INTERNIST-I, an experimental computer-based diagnostic consultant for general internal medicine. *New England Journal of Medicine*, pp. 468–476.

Mitchell, A. A. (1987). The use of alternative knowledge-acquisition procedures in the development of a knowledge-based media planning system. *International Journal of Man-Machine Studies*, *26*, 399–412.

Mittal, S., Bobrow, D. G., & de Kleer, J. (1984). *DARN: A community memory for a diagnosis and repair task*. Palo Alto, CA: Xerox PARC.

Mori, F. (1987). Expert systems in business: A Japanese experience. *The Knowledge Engineering Review*, *2*, 203–206.

Murray, S. B. (1989). Preserving corporate knowledge. *Expert Systems: Planning, Implementation, Integration*, *1*, 62–64.

Musen, M. A. (1990). Creating custom-tailored knowledge acquisition tools. In K. L. McGraw & C. R. Westphal (Eds.), *Readings in knowledge acquisition* (pp. 150–163). New York: Ellis Horwood.

Niwa, K. (1987). Knowledge transfer: A key to successful application of knowledge-based systems. *The Knowledge Engineering Review*, *2*, 147–152.

Norros, L. (1986). Expert knowledge, its acquisition and elicitation in developing intelligent tools for process control. In E. Hollnagel, G. Mancini, & D. D. Woods (Eds.), *Intelligent decision support in process environments* (pp. 137–144). Berlin: Springer-Verlag.

O'Neill, J. L. (1987). Knowledge acquisition for radar classification. In J. R. Quinlan (Ed.), *Applications of expert systems* (pp. 184–199). Reading, MA: Addison-Wesley.

Ostberg, O. (1988). Applying expert systems techology: Division of labor and division of knowledge. In B. Goranzon & I. Josefson (Eds.), *Knowledge, skill, and artificial intelligence* (pp. 169–184). London: Springer-Verlag.

Partridge, D. (1986). *Artificial intelligence: Applications in the future of software engineering*. Chichester, England: Ellis Horwood.

Patel, V. L., & Groen, G. J. (1986). Knowledge based solution strategies in medical reasoning. *Cognitive Science*, *10*, 91–116.

Pau, L. F. (1986). Knowledge engineering techniques applied to fault detection test generation and maintenance (NATO ASI Series, Vol. F22). In J. K. Skwirzynski (Ed.), *Software system design methods* (pp. 53–79). Berlin, Heidelberg: Springer-Verlag.

Pau, L. F., Pai, Y. H., Motiwalla, J., & Teh, H. H. (Eds.). (1989). *Expert systems in economics*, banking, and management. Amsterdam: North-Holland.

Prerau, D. S., Gunderson, A. S., Reinke, R. E., & Goyal, S. K. (1985). The COMPASS expert system: Verification, technology transfer, and expansion, *Proceeding of Second Conference on Artificial Intelligence Applications* (pp. 597–602). New York: IEEE Computer Society.

Rada, R., & Barlow, J. (1988). Expert systems and hypertext. *The Knowledge Engineering Review*, *3*, 285–302.

Ragheb, M., & Tsoukalas, L. (1988). Monitoring performance of devices using a coupled probability-possibilities method. *International Journal of Expert Systems*, *1*, 111–130.

Randhawa, S. U., & McDowell, E. D. (1990). An investigation of the applicability of expert systems to job shop scheduling. *International Journal of Man-Machine Studies*, *36*, 203–214.

Rauch-Hindin, W. (1985). *Artificial intelligence in business, science, and industry* (Vol. 2). Englewood Cliffs, NJ: Prentice-Hall.

Rault, J.-C. (1987). A survey of french developments in knowledge-based systems. *The Knowledge Engineering Review*, *2*, 287–296.

Reiner, J. (1985). Applications of expert systems to sensor fusion. *IEEE National Aerospace and Electronics Conference* (pp. 1444–1450). Dayton, OH.

Reitman, W. (Ed.). (1984). *AI applications for business*. Norwood, NJ: Ablex.

Reitman, W. (1989). Integrated Design Teams: Knowledge Engineering for Large Scale Commercial Expert System Development. *IEEE Transactions on Systems, Man, and Cybernetics, 19*, 443–447.

Rodi, L. L., Pierce, J. A., & Dalton, R. E. (1990). Graphical knowledge for fault diagnosis and repair. In K. L. McGraw & C. R. Westphal (Eds.), *Readings in knowledge acquisition* (pp. 137–149). New York: Ellis Horwood.

Rodolitz, N. S., & Clancey, W. J. (1989). GUIDON-MANAGE: Teaching the process of medical diagnosis. In D. A. Evans & V. L. Patel (Eds.), *Cognitive science in medicine: Biomedical modeling* (pp. 313–348). Cambridge, MA: Bradford Books/MIT Press.

Rolandi, W. G. (1986, December). Knowledge engineering in practice. *AI Expert*, pp. 58–62.

Scarl, E. A., Jamieson, J. R., & Delaune, C. I. (1987, May/June). Diagnosis and sensor validation through knowledge of structure and function. *IEEE transactions on Systems, Man, and Cybernetics, SMC-17*, 360–368.

Schmneider, M., & Kandel, A. (1988). Applications of fuzzy expected intervals to fuzzy expert systems. *International Journal of Expert Systems, 1*, 169–186.

Schvaneveldt, R. W., & Goldsmith, T. E. (1986). A model of air combat decisions. In E. Hollnagel, G. Mancini, & D. D. Woods (Eds.), *Intelligent decision support in process environments* (pp. 395–406). Berlin: Springer-Verlag.

Shortliffe, E. (1976). *Computer-Based Medical Consultations: MYCIN*. New York: Elsevier.

Shortliffe, E., Buchanan, B., & Feigenbaum, E. (1979). Knowledge engineering for medical decision making: A review of computer-based clinical decision aids. *Proceedings of the IEEE, 67*, 1204–1224.

Siler, W., & Tucker, D. (1989). Patterns of inductive reasoning in a parallel expert system. *International Journal of Man-Machine Studies, 30*, 113–127.

Skuce, D., Stanley, R., & Tauzovich, B. (1988). An expert advisor that answers coding questions about commercial fourth-generation software. *International Journal of Expert Systems, 1*, 217–236.

Slagle, J. R., & Wick, M. R. (1988, Winter). A method for evaluating candidate expert system applications. *The AI Magazine*, pp. 44–53.

Smart, G. (1987, June). How to make expert systems . . . and stay in business. *Third International Expert Systems Conference* (pp. 215–222). Oxford: Learned Information.

Stassen, H. G. (1987). Human supervision modeling: Some new developments. *International Journal of Man-Machine Studies, 27*, 613–618.

Stauffer, L. (1989). Eliciting and analyzing data about the engineering-design process. *Proceedings of the NSF Design Theory and Methodology Conference*. Amherst, MA: University of Massachusetts.

Stewart, V., & Stewart, A. (1981). *Business applications of repertory grids*. London: McGraw-Hill.

Susskind, R. (1987). Critical notes: Artificial intelligence and legal reasoning. *The Knowledge Engineering Review, 2*, 213–218.

Swigger, K. M., & Brazile, R. P. (1989). Experimental comparison of design/documentation formats for expert systems. *International Journal of Man-Machine Studies, 31*, 47–60.

Teschler, L. (1987). Trends in the marketing of expert system technology. *The Knowledge Engineering Review, 2*, 175–178.

Voelker, J. A., & Ratica, G. B. (1986). The genesis of a knowledge-based expert system. *IBM Systems Journal, 25*, 181–189.

Walker, N., & Fox, J. (1987). Knowledge-based interpretation of images: A biomedical perspective. *The Knowledge Engineering Review, 2*, 249–264.

Ward, R. D., & Sleeman, D. (1987). Learning to use the S.1 knowledge engineering tool. *The Knowledge Engineering Review, 2*, 265–276.

Weiner, J. L. (1980). BLAH: A system which explains its own reasoning. *Artificial Intelligence, 15*, 19–48.

Weintraub, J. (1989, Spring). Expert systems in government administration. *The AI Magazine*, pp. 69–70.

Weiss, S., & Kulikowski, C. (1984). *A practical guide to designing expert systems*. Towata, NJ: Rowman and Allanheld.

Wess, B. P. (1987). Commentary on the commercialization of knowledge engineering enterprise and product development. *The Knowledge Engineering Review, 2*, 169–174.

Wilson, M., Duce, D., & Simpson, D. (1989). Lifecycles in software and knowledge engineering: A comparative review. *The Knowledge Engineering Review, 4*, 189–204.

Wong, K. P. (1990). Applications of artificial intelligence and expert systems in power engineering. *The Knowledge Engineering Review*, *5*, 127–140.

Wood, S. (1986). Expert systems for theoretically ill-formulated domains. In M. A. Bramer (Ed.), *Research and development in expert systems* (Vol. 3, pp. 132–139). Cambridge: Cambridge University Press.

Woods, D. D. (1987). Commentary: Cognitive engineering in complex dynamic worlds. *International Journal of Man-Machine Studies*, *27*, 571–586.

Zarri, G. P. (1984). Expert systems and information retrieval: An experiment in the domain of biographical data management. In M. J. Coombs (Ed.), *Developments in expert systems* (pp. 201–220). New York: Academic Press.

Zimolong, B., & Nof, S. Y., Ebertsand, R. E., & Salvendy, O. (1987). On the limits of expert systems and engineering models in process control. *Behaviour and Information Technology*, *6*, 15–36.

Zubrick, S. (1988). Validation of a weather forecasting expert system. In J. E. Hayes, D. Michie, & J. Richards (Eds.), *Machine Intelligence 11* (pp. 391–422). Oxford, England: Clarendon Press.

Appendix H
Bibliography: Programming, Building, and Verifying Expert Systems

Adelson, B., & Soloway, E. (1988). A model of software design. In M. T. Chi, R. Glaser, & M. J. Farr (Eds.), *The nature of expertise* (pp. 185–208). Hillsdale, NJ: Erlbaum.

Aikins, J. (1983). Prototypical knowledge for expert systems. *Artificial Intelligence, 20*, 163–210.

Aleksander, I. (1984). *Designing intelligent systems: An introduction*. London: Kogan Page.

Bramer, M. A. (Ed.). (1985). *Research and development in expert systems: Proceedings of the Fourth Technical Conference of the British Computer Society Annual Technical Conference of the British Computer Society Specialist Group on Expert Systems*. Cambridge: Cambridge University Press.

Bramer, M. A. (Ed.), (1987). *Research and development in expert systems: Proceedings of Expert Systems '86, the Sixth Annual Technical Conference of the British Computer Society Specialist Group on Expert Systems*. Cambridge: Cambridge University Press.

Bramer, M. A. (Ed.). (1989). *Research and development in expert systems*. Cambridge: Cambridge University Press.

Brownston, L., Farrell, R., Kant, E., & Martin, N. (1985). *Programming expert systems in OPS5*. Reading, MA: Addison-Wesley.

Buchanan, B. G., Barstow, D., Bechtal, R., Bennett, J., Clancey, W., Kulikowski, C., Mitchell, T., & Waterman, D. (1983). Constructing an expert system. In F. Roth, D. Waterman, & D. Lenat (Eds.), *Building expert systems* (pp. 127–168). Reading, MA: Addison-Wesley.

Buckley, J. J. (1988). Managing uncertainty in a fuzzy expert system. *International Journal of Man-Machine Studies, 29*, 129–148.

Buxton, R. (1989). Modeling uncertainty in expert systems. *International Journal of Man-Machine Studies, 31*, 415–476.

Chandrasekaran, B. (1988). Generic tasks as building blocks for knowledge-based systems: The diagnosis and routine design examples. *The Knowledge Engineering Review, 3*, 183–210.

Chandrasekaran, B. (1988). An answer to the commentators on "Generic tasks as building blocks for knowledge-based systems: The diagnosis and routine design examples." *The Knowledge Engineering Review, 3*, 211–220.

Chignell, M. H., & Peterson, J. G. (1988). Strategic issues in knowledge engineering. *Human Factors, 30*, 381–394.

Clancy, W. J. (1983). The advantages of abstract control knowledge in expert system design. *Proceedings of the National Conference on Artificial Intelligence* (pp. 74–78). Los Altos, CA: Morgan-Kaufmann.

Clark, D., Baldwin, J., Berenji, H., Cohen, P., Dubois, D., Fox, J., Lemmer, J., Prade, H., Spigelhalter, D., Smets, P., & Zadeh, L. (1988). Responses to "An AI view of the treatment of uncertainty" by A. Saffiotti. *The Knowledge Engineering Review, 3*, 59–86.

Cochran, E. L., Bloom, C. P., & Bullemer, P. T. (1990). Increasing the end-user acceptance of expert systems by using multiple experts. In K. L. McGraw & C. R. Westphal (Eds.), *Readings in knowledge acquisition* (pp. 73–89). New York: Ellis Horwood.

Cochran, E. L., & Hutchins, B. L. (1987). Testing, verifying, and releasing an expert system: The case history of Mentor. *Proceedings of the Third Conference on Artificial Intelligence Applications* (pp. 162–167). New York: IEEE.

Cohen, P. R. (1987). The control of reasoning under uncertainty: A discussion of some programs. *The Knowledge Engineering Review*, *2*, 5–26.

Cohen, P. R., DeLisio, J., & Hart, D. (1989). A declarative representation of control knowledge. *IEEE Transactions on Systems, Man, and Cybernetics*, *19*, 546–557.

Cohn, A. G. (1985). Deep knowledge representation techniques. In M. Merry (Ed.), *Expert systems-85* (pp. 299–306). Cambridge: Cambridge University Press.

Coombs, M. J. (Ed.). (1984). *Developments in expert systems*. London: Academic Press.

Dahlstrom, D. (1989). Worlds of knowing and nonmonotonic reasoning. *IEEE Transactions on Systems, Man, and Cybernetics*, *19*, 626–633.

Davis, J. S. (1990). Effect of modularity on maintainability of rule-based systems. *International Journal of Man-Machine Studies*, *32*, 439–448.

Davis, R. D., & Lenat, B. (1982). *Knowledge based systems in AI*. New York: McGraw-Hill.

Dean, T., & Kanazawa, K. (1989). Persistence and probabilistic projection. *IEEE Transactions on Systems, Man, and Cybernetics*, *19*, 574–585.

Diaper, D. (1986, December). Identifying the knowledge requirements of an expert system's natural language processing interface. *Proceedings of the Second Conference of the British Computer Society Human Computer Interaction Specialist Group* (pp. 263–280). Cambridge: Cambridge University Press.

Diaper, D. (1989). Designing expert systems: From Dan to Beersheba. In D. Diaper (Ed.), *Knowledge elicitation: Principles, techniques, and applications* (pp. 15–46). Chichester, England: Ellis Horwood.

Drummond, M., MacIntosh, A., & Tate, A. (1987). A framework for technology transfer within AI. *The Knowledge Engineering Review*, *3*, 159–168.

Dym, C. L. (1987). Issues in the design and implementation of expert systems. *Artifical Intelligence for Engineering Design, Analysis and Manufacturing*, *1*, 37–46.

Ellis, C. (1989). Explanation in intelligent systems. In C. Ellis (Ed.), *Expert knowledge and explanation: The knowledge-language interface* (pp. 108–126). Chichester, England: Ellis Horwood.

Feigenbaum, E. A. (1977). The art of artificial intelligence. *Proceedings of the fifth International Joint Conference on Artificial Intelligence* (pp. 1013–1029). Cambridge, MA: International Joint Conferences on Artificial Intelligence.

Feigenbaum, E. A. (1984). Knowledge engineering: The application of artificial intelligence. *Annals of the New York Academy of Sciences*, *426*, 91–107.

Finley, M. R., & Hausen-Tropper, E. B. (1989). A system for the representation of theorems and proofs. In C. R. Westphal & K. L. McGraw (Eds.), *Special Issue of ACM SIGART NEWSLETTER on Knowledge Acquisition* (No. 108, pp. 178–179). New York: Special Interest Group on Artificial Intelligence, Association for Computing Machinery.

Freiling, M. J., Alexander, S., Messick, S., Rehfuss, S., & Shulman, S. (1985). Starting a knowledge engineering project: A step-by-step approach. *AI Magazine*, *6*, 150–163.

Frost, R. (1986). *Introduction to knowledge based systems*. New York: Macmillan.

Gaines, B. R. (1987). Foundations of knowledge engineering. In M. A. Bramer (Ed.), *Research and development in expert systems* (pp. 13–24). Cambridge: Cambridge University Press.

Gaines, B. R. (1989). Integration issues in knowledge support systems. *International Journal of Man-Machine Studies*, *31*, 495–516.

Garg-Janardan, C., & Salvendy, G. (1988). The contributions of cognitive engineering to the design and use of expert systems. *Behavior and Information Technology*, *7*, 323–342.

Gaschnig, J., Klahr, P., Pople, H., Shortliffe, E., & Terry, A. (1983). Evaluation of expert systems: Issues and case studies. In F. Hayes-Roth, D. A. Waterman, & D. B. Lenat (Eds.), *Building expert systems* (pp. 241–280). Reading, MA: Addison-Wesley.

Glorioso, R. M., & Osorio, F. C. C. (1980). *Engineering intelligent systems: Concepts, theory, and applications*. Bedford, MA: Digital Press.

Goodall, A. *The guide to expert systems*. Oxford: Learned Information.

Guise, D. (1990). Efficient knowledge representation systems. *The Knowledge Engineering Review*, *5*, 35–50.

Haddawy, P., & Rendell, L. (1990). Planning and decision theory. *The Knowledge Engineering Review*, *5*, 15–43.

Hagglund, S. (1987). The Linkoping approach to technology transfer in knowledge engineering. *The Knowledge Engineering Review*, *2*, 153–158.

Harbison-Briggs, K., & Briggs, A. B. (1990). The role of knowledge acquisition in the verification and validation of knowledge-based systems. In K. L. McGraw & C. R. Westphal (Eds.), *Readings in knowledge acquisition* (pp. 233–250). New York: Ellis Horwood.

Harmon, P., & King, D. (1985). *Expert systems*. New York: Wiley.

Hayes-Roth, F. (1985, October). Engineering systems of knowledge: The great adventure ahead. *Expert Systems in Government Symposium* (pp. 678–692). New York: IEEE.

Hayes-Roth, F., Klahr, P., & Mostow, D. J. (1986). Knowledge acquisition, knowledge programming and knowledge refinement. In P. Klahr, & D. A. Waterman, (Eds.), *Expert systems: Techniques, tools and applications* (pp. 310–349). Reading, MA: Addison-Wesley.

Hayes-Roth, F., Waterman, D., & Lenat, D. (1983). *Building expert systems*. Reading, MA: Addison-Wesley.

Hern, L. E. C. (1988). On distributed artificial intelligence. *The Knowledge Engineering Review*, *3*, 21–58.

Hollnagel, E. (1987). Information and reasoning in intelligent decision support systems. *International Journal of Man-Machine Studies*, *27*, 665–678.

Hollnagel, E. (1987). Commentary: Issues in knowledge-based decision support. *International Journal of Man-Machine Studies*, *27*, 743–751.

Jackson, P. C. (1974). *Introduction to artificial intelligence*. New York: Petrocelli Books.

Jackson, P. C. (1986). *Introduction to expert systems*. Wokingham, England: Addison-Wesley.

Jackson, P. C. (1989). Applications of non-monotonic logic to diagnosis. *The Knowledge Engineering Review*, *4*, 97–118.

Jackson, P. C., & Lefrere, P. (1984). On the application of rule-based techniques to the design of advice-giving systems. In M. J. Coombs (Ed.), *Developments in expert systems* (pp. 177–200). New York: Academic Press.

Jackson, P. C., Reichgelt, H., & van Harmelen, F. (1989). *Logic-based knowledge representation*. Cambridge: MIT Press.

Johnson, P. E., Zualkerman, I. A., David, J. M., Iwasaki, Y., Keller, R., & Feigenbaum, E. (1988). Responses to: "Generic tasks as building blocks for knowledge-based systems: The diagnosis of routine design examples" by B. Chandrasekaran. *The Knowledge Engineering Review*, *3*, 211–216.

Jones, S. (1988). Graphical interfaces for knowledge engineering: An overview of the relevant literature. *The Knowledge Engineering Review*, *3*, 221–248.

Kanal, L., & Lemmer, J. (1986). *Uncertainty in artificial intelligence*. Amsterdam: North Holland.

Keravnou, E. T., & Washbrook, J. (1989). What is a deep expert system? An analysis of the architectural requirements of second-generation expert systems. *The Knowledge Engineering Review*, *4*, 205–234.

Kline, P., & Dollin, S. (1986). Problem features that influence the design of expert systems. *Proceedings of the American Association for Artificial Intelligence* (pp. 956–962). Stanford, CA: AAAI.

Laufmann, S. C. (1987). A strategy for near-term success using knowledge-based systems. *The Knowledge Engineering Review*, *2*, 179–184.

Laurent, J.-P. (1987). Types of control structure in expert systems. *The Knowledge Engineering Review*, *2*, 123–136.

Lehner, P. E. (1989). Toward an empirical approach to evaluating the knowledge base of an expert system. *IEEE Transactions on Systems, Man, and Cybernetics*, *19*, 658–662.

Lehner, P. E., & Adelman, L. (1987, October). Biases in knowledge engineering. *Proceedings of the 1987 IEEE International Conference on Systems, Man, and Cybernetics* (pp. 376–377). New York: IEEE.

Lehner, P. E., & Adelman, L. (1989). Perspectives in knowledge engineering. *IEEE Transactions on Systems, Man, and Cybernetics*, *19*, 423–434.

Leibowitz, J., & DeSalvo, D. (Eds.). (1989). *Structured methodologies for expert systems development*. Englewood Cliffs, NJ: Yourdan Press.

Lind, M. (1986). Decision models and the design of knowledge-based systems. In E. Hollnagel, G. Mancini, & D. D. Woods (Eds.), *Intelligent decision support in process environments* (pp. 197–210). Berlin: Springer-Verlag.

Littman, D. C. (1987). Modeling human expertise in knowledge engineering: Some preliminary observations. *International Journal of Man-Machine Studies*, *26*, 81–92.

Long, D. (1989). A review of temporal logics. *The Knowledge Engineering Review*, *4*, 141–162.

Madni, A. M. (1988). The role of human factors in expert systems design and acceptance. *Human Factors*, *30*, 395–414.

McDonald, J. E., & Schvaneveldt, R. W. (1988). The application of user knowledge to interface design. In R. Guindon (Ed.), *Cognitive science and its applications for human-computer interaction* (pp. 289–338). Hillsdale, NJ: Erlbaum.

McGraw, K. L. (1986). Producing user documentation for expert systems. *IEEE Transactions on Professional Communications*, *29*, 42–47.

McGraw, K. L. (1989). User documentation for expert systems. In J. Liebowitz & D. DeSalvo (Eds.), *Structured methodologies for expert sys-*

tems development. Englewood Cliffs, NJ: Yourdon Press.

McGraw, K. L., & Briggs, K. H. (1989). *Knowledge acquisition: Principles and guidelines*. Englewood Cliffs, NJ: Prentice Hall.

Merry, M. (Ed.). (1985). *Expert systems '85*. Cambridge: Cambridge University Press.

Minsky, M. (1975). A framework for representing knowledge. In P. H. Winston (Ed.), *The psychology of computer vision* (pp. 211–277). New York: McGraw-Hill.

Minsky, M., & Papert, S. (1974). *Artificial intelligence*. Eugene, OR: Oregon State System of Higher Education.

Mishkoff, H. (1985). *Understanding artificial intelligence*. Dallas, TX: Texas Instruments Inc.

Moralee, D. S. (Ed.). (1988). Research and development in expert systems. Cambridge: Cambridge University Press.

Morgan, R. (1989, Summer). Expert systems development strategies. *Expert Systems: Planning, Implementation, Integration*, *1*, 5–10.

Morik, K. (1989, April). Integration issues in knowledge acquisition systems. In C. R. Westphal & K. L. McGraw (Eds.), *Special Issue of ACM SIGART Newsletter onn Knowledge Acquisition* (No. 108, pp. 124–131). New York: Special Interest Group on Knowledge Acquisition, Association for Computing Machinery.

Mott, P. (1988). Default non-monotonic logic. *The Knowledge Engineering Review*, *3*, 265–284.

Niwa, K. (1987). Knowledge transfer: A key to successful application of knowledge-based systems. *The Knowledge Engineering Review*, *2*, 147–152.

O'Keefe, R. M., Balci, O., & Smith, E. P. (1987, Winter). Validating expert system performance. *IEEE Expert*, pp. 81–89.

O'Leary, D. E. (1988). On the representation and the impact of reliability on expert system weights. *International Journal of Man-Machine Studies*, *29*, 637–646.

Prerau, D. (1990). *Developing and managing expert systems: Proven techniques for business and industry*. Reading, MA: Addison-Wesley.

Rada, R., & Barlow, J. (1988). Expert systems and Hypertext. *The Knowledge Engineering Review*, *3*, 285–302.

Rasmussen, J. (1986). *Information processing and human-machine interaction: An approach to cognitive engineering*. New York: Elsevier.

Rasmussen, J. (1986). A framework for cognitive task analysis in systems design. In E. Hollnagel, G. Mancini, & D. D. Woods (Eds.), *Intelligent decision support in process environments* (pp. 175–196). Berlin: Springer-Verlag.

Refenes, A. N. (1989). Parallelism in knowledge-based machines. *The Knowledge Engineering Review*, *4*, 53–72.

Reggia, J. A., Nau, D. S., & Wang, P. Y. (1984). Diagnostic expert systems based on a set covering model. In M. J. Coombs (Ed.), *Developments in expert systems* (pp. 35–58). New York: Academic Press.

Reichgelt, H. (1985). Relevant criteria for choosing an inference engine in expert systems. In M. Merry (Ed.), *Expert systems-85* (pp. 21–30). Cambridge: Cambridge University Press.

Reichgelt, H. (1989). Logics for reasoning about knowledge and beliefs. *The Knowledge Engineering Review*, *4*, 119–140.

Reichgelt, H., & van Harmelen, F. (1986). Criteria for choosing representation languages and control regimes for expert systems. *Knowledge Engineering Review*, *1*, 2–17.

Rich, E. (1983). *Artificial intelligence*. New York: McGraw-Hill.

Ringwood, G. A. (1989). A comparative exploration of concurrent logic languages. *The Knowledge Engineering Review*, *4*, 305–332.

Rosove, P. E. (1967). *Developing computer-based information systems*. New York: Wiley.

Rouse, W. B., Hammer, J. M., & Lewis, C. M. (1989). On capturing human skills and knowledge: Algorithmic approaches to model identification. *IEEE Transactions on Systems, Man, and Cybernetics*, *19*, 538–573.

Roycroft, A. J. (1988). Expert against Oracle. In J. E. Hayes, D. Michie, & J. Richards (Eds.), *Machine intelligence 11* (pp. 347–374). Oxford, England: Clarendon Press.

Saffiotti, A. (1987). An AI view of the treatment of uncertainty. *The Knowledge Engineering Review*, *2*, 75–98.

Saffiotti, A. (1988). The treatment of uncertainty in AI: Is there a better vantage point? *The Knowledge Engineering Review*, *3*, 87–92.

Schneiderman, B. (1980). *Software psychology: Human factors in computers and information systems*. Boston, MA: Little, Brown & Co.

Sell, P. S. (1985). *Expert systems: A practical introduction*. New York: Wiley.

Shin, D., & Bera, P. B. (1989). Computer architectures for logic-oriented data/knowledge bases. *The Knowledge Engineering Review*, *4*, 1–30.

Slatter, P. E. (1987). *Building expert systems: Cognitive emulation*. Chichester, England: Ellis Horwood.

Sparck-Jones, K. (1985). Natural language interfaces for expert systems: An introductory note. In M. A. Bramer (Ed.), *Research and development in expert systems* (pp. 85–94). Cambridge: Cambridge University Press.

Stefik, M., Aikins, J., Balzer, R., Benoit, J., Birnbaum, L., Hayes-Roth, F., & Sacerdoti, E. (1982). The organization of expert systems: A tutorial. *Artificial Intelligence*, *18*, 135–173.

Stefik, M., Aikins, J., Balzer, R., Benoit, J., Birnbaum, L., Hayes-Roth, F., & Sacerdoti, E. (1983). Basic concepts for building expert systems. In F. Hayes-Roth, D. A. Waterman, & D. B. Lenat (Eds.), *Building expert systems* (pp. 59–86). Reading, MA: Addison-Wesley.

Stefik, M., & Conway, L. (1982, Summer). Towards the principled engineering of knowledge. *The AI Magazine*, pp. 4–16.

Stenton, S. P. (1987). Dialog management for cooperative knowledge-based systems. *The Knowledge Engineering Review*, *2*, 99–122.

Tiemann, P. W., & Markle, S. (1984, November). On getting expertise into an expert system. *Performance and Instruction Journal*, pp. 25–29.

Voelker, J. A., & Ratica, G. B. (1986). The genesis of a knowledge-based expert system. *IBM Systems Journal*, *25*, 181–189.

Walters, J. R., & Nielsen, N. R. (1988). *Crafting knowledge-based systems*. New York: Wiley.

Waterman, D. A. (1986). *A guide to expert systems*. Reading, MA: Addison-Wesley.

Weiss, S., & Kulikowski, C. (1984). *A practical guide to designing expert systems*. Totowa, NJ: Rowman & Allanheld.

Wexelblat, R. L. (1989, Fall). On interface requirements for expert systems. *The AI Magazine*, pp. 66–78.

Wilson, M., Duce, D., & Simpson, D. (1989). Lifecycles in software and knowledge engineering: A comparative review. *The Knowledge Engineering Review*, *4*, 189–204.

Wood, S. (1986). Expert systems for theoretically ill-formulated domains. In M. A. Bramer (Ed.), *Research and development in expert systems* (Vol. 3, pp. 132–139). Cambridge: Cambridge University Press.

Woods, D. D. (1987). Commentary: Cognitive engineering in complex and dynamic worlds. *International Journal of Man-Machine Studies*, *27*, 571–585.

Woods, W. A. (1985, Winter). Cognitive technologies: The design of joint human-machine cognitive systems. *The AI Magazine*, pp. 86–92.

Young, R. M. (1979). Production systems for modeling human cognition. In D. Michie (Ed.), *Expert systems in a microelectronic age* (pp. 35–45). Edinburgh: Edinburgh University Press.

Zubrick, S. (1988). Validation of a weather forecasting expert system. In J. E. Hayes, D. Michie, & J. Richards (Eds.), *Machine intelligence 11* (pp. 391–422). Oxford, England: Clarendon Press.

Subject Index

Author Index